SCIENCE OF EVERYDAY THINGS

SCIENCE OF EVERYDAY THINGS

VOLUME 4: REAL-LIFE EARTH SCIENCE

EDITED BY NEIL SCHLAGER
WRITTEN BY JUDSON KNIGHT

A SCHLAGER INFORMATION GROUP BOOK

GALE®

THOMSON
GALE™

Detroit • New York • San Diego • San Francisco • Cleveland • New Haven, Conn. • Waterville, Maine • London • Munich

THOMSON
GALE

Science of Everyday Things
Volume 4: Real-Life Earth Science
A Schlager Information Group Book
Neil Schlager, Editor
Written by Judson Knight

Project Editor
Kimberley A. McGrath

Editorial
Mark Springer

Permissions
Kim Davis

Imaging and Multimedia
Robert Duncan, Leitha Etheridge-Sims, Mary K. Grimes, Lezlie Light, Dan Newell, David G. Oblender, Robyn V. Young

Product Design
Michelle DiMercurio, Michael Logusz

Manufacturing
Evi Seoud, Rhonda Williams

For more information contact
The Gale Group, Inc.
27500 Drake Rd.
Farmington Hills, MI 48331-3535
Or you can visit our Internet site at
http://www.gale.com

LIBRARY OF CONGRESS CATALOG-IN-PUBLICATION DATA

Knight, Judson.
Science of everyday things / written by Judson Knight, Neil Schlager, editor.
p. cm.
Includes bibliographical references and indexes.
Contents: v. 1. Real-life chemistry – v. 2 Real-life physics.
SBN 0-7876-5631-3 (set : hardcover) – ISBN 0-7876-5632-1 (v. 1) – ISBN 0-7876-5633-X (v. 2)
1. Science–Popular works. I. Schlager, Neil, 1966-II. Title.

Q162.K678 2001
500–dc21 2001050121

ISBN 0-7876-5631-3 (set), 0-7876-5632-1 (vol. 1), 0-7876-5633-X (vol. 2), 0-7876-5634-8 (vol. 3), 0-7876-5635-6 (vol. 4)

Printed in the United States of America
10 9 8 7 6 5 4 3 2 1

CONTENTS

40.00

Contents

INTRODUCTION

Overview of the Series

Welcome to *Science of Everyday Things*. Our aim is to explain how scientific phenomena can be understood by observing common, real-world events. From luminescence to echolocation to buoyancy, the series will illustrate the chief principles that underlay these phenomena and explore their application in everyday life. To encourage cross-disciplinary study, the entries will draw on applications from a wide variety of fields and endeavors.

Science of Everyday Things initially comprises four volumes:

Volume 1: *Real-Life Chemistry*
Volume 2: *Real-Life Physics*
Volume 3: *Real-Life Biology*
Volume 4: *Real-Life Earth Science*

Future supplements to the series will expand coverage of these four areas and explore new areas, such as mathematics.

Arrangement of Real-Life Earth Science

This volume contains 40 entries, each covering a different scientific phenomenon or principle. The entries are grouped together under common categories, with the categories arranged, in general, from the most basic to the most complex. Readers searching for a specific topic should consult the table of contents or the general subject index.

Within each entry, readers will find the following rubrics:

- **Concept:** Defines the scientific principle or theory around which the entry is focused.
- **How It Works:** Explains the principle or theory in straightforward, step-by-step language.
- **Real-Life Applications:** Describes how the phenomenon can be seen in everyday life.
- **Where to Learn More:** Includes books, articles, and Internet sites that contain further information about the topic.

In addition, each entry includes a "Key Terms" section that defines important concepts discussed in the text. Finally, each volume includes many illustrations and photographs throughout.

Included in this volume, readers will find (in addition to the volume-specific general subject index), a cumulative general index, as well as a cumulative index of "everyday things." This latter index allows users to search the text of the series for specific everyday applications of the concepts.

About the Editor, Author, and Advisory Board

Neil Schlager and Judson Knight would like to thank the members of the advisory board for their assistance with this volume. The advisors were instrumental in defining the list of topics, and reviewed each entry in the volume for scientific accuracy and reading level. The advisors include university-level academics as well as high school teachers; their names and affiliations are listed elsewhere in the volume.

Neil Schlager is the president of Schlager Information Group Inc., an editorial services company. Among his publications are *When*

Technology Fails (Gale, 1994); *How Products Are Made* (Gale, 1994); the *St. James Press Gay and Lesbian Almanac* (St. James Press, 1998); *Best Literature By and About Blacks* (Gale, 2000); *Contemporary Novelists*, 7th ed. (St. James Press, 2000); *Science and Its Times* (7 vols., Gale, 2000-2001); and *Science in Dispute* (Gale, 2002). His publications have won numerous awards, including three RUSA awards from the American Library Association, two Reference Books Bulletin/Booklist Editors' Choice awards, two New York Public Library Outstanding Reference awards, and a *CHOICE* award for best academic book.

Judson Knight is a freelance writer, and author of numerous books on subjects ranging from science to history to music. His work on science includes *Science, Technology, and Society, 2000* B.C.-A.D. 1799 (U•X•L, 2002), as well as extensive contributions to Gale's seven-volume *Science and Its Times* (2000-2001). As a writer on history, Knight has published *Middle Ages Reference Library* (2000), *Ancient Civilizations* (1999), and a volume in U•X•L's *African American Biography* series (1998). Knight's publications in the realm of music include *Parents Aren't Supposed to Like It* (2001), an overview of contemporary performers and genres, as well as *Abbey Road to Zapple Records: A Beatles Encyclopedia* (Taylor, 1999).

COMMENTS AND SUGGESTIONS

Your comments on this series and suggestions for future editions are welcome. Please write: The Editor, *Science of Everyday Things,* Gale Group, 27500 Drake Road, Farmington Hills, MI 48331-3535.

ADVISORY BOARD

SCIENCE OF EVERYDAY THINGS

REAL-LIFE EARTH SCIENCE

UNDERSTANDING THE EARTH SCIENCES

EARTH, SCIENCE, AND NONSCIENCE

GEOSCIENCE AND EVERYDAY LIFE

EARTH SYSTEMS

EARTH, SCIENCE, AND NONSCIENCE

CONCEPT

To understand the composition and structure of Earth, one must comprehend the forces that shaped it. Much the same is true of the earth sciences themselves, which originated from attempts to explain the origins of Earth and the materials of which it is composed. Before the modern era, such explanations had roots in religion, mythology, or philosophy and drew from preconceived ideas rather than from observed data. A turning point came with the development of the scientific method, a habit of thinking that spread from astronomy and physics to chemistry and the earth sciences.

HOW IT WORKS

ARISTOTLE'S FOUR CAUSES

Though the Greek philosopher Aristotle (384–322 B.C.) exerted a negative influence on numerous aspects of what became known as the physical sciences (astronomy, physics, chemistry, and the earth sciences), he is still rightly regarded as one of the greatest thinkers of the Western world. Among his contributions to thought was the identification of four causes, or four approaches to the question of how and why something exists as it does.

In Aristotle's system, which developed from ideas of causation put forward by his predecessors, the most basic of explanations is the *material cause,* or the substance of which a thing is made. In a house, for instance, the wood and other building materials would be the material cause. The builders themselves are the *efficient cause,* or the forces that shaped the house. More complex than these is a third variety of cause-effect relationship, the *formal cause*—that is, the design or blueprint on which something is modeled.

The first three Aristotelian causes provide a pathway for explaining *how;* the fourth and last cause approaches the much more challenging question of *why.* This is the *final cause,* or the reason why a thing exists at all—in other words, the purpose for which it was made. Even in the case of the house, this is a somewhat complicated matter. A house exists, of course, to provide a dwelling for its occupants, but general contractors would not initiate the building process if they did not expect to make a profit, nor would the subcontractors and laborers continue to work on it if they did not earn an income from the project.

RELIGION, SCIENCE, AND EARTH

The matter of final cause is almost unimaginably more complex when applied to Earth rather than to a house. The question "Why does Earth exist?" or "What is the ultimate reason for Earth's existence?" is not really a topic for science at all, but rather for theology and philosophy. Nor do the answers provided by religion and philosophical beliefs qualify as answers in the same sense that workable scientific theories do.

There has always been a degree of tension between religion and the sciences, and nowhere has this been more apparent than in the earth sciences. As will be discussed later in this essay, most early theories concerning Earth's structure and development were religious in origin, and even some modern explanations have theological roots. Certainly there is nothing wrong with a

ENGRAVING AFTER A MARBLE BUST OF ARISTOTLE. *(Library of Congress.)*

scientist having religious beliefs, as long as those beliefs do not provide a filter for all data. If they do, the theologically minded scientist becomes rather like a mathematician attempting to solve a problem on the basis of love rather than reason. Most people would agree that love is higher and greater than mathematics; nonetheless, it has absolutely no bearing on the subject.

SCIENTIFIC ANSWERS AND THE SEARCH FOR A DESIGNER. The third, or formal, cause is less fraught with problems than the final cause when applied to the study of Earth, yet it also illustrates the challenges inherent in keeping science and theology separate. Does Earth have a "design," or blueprint? The answer is yes, no, and maybe. Yes, Earth has a design in the sense that there is an order and a balance between its components, a subject discussed elsewhere with reference to the different spheres (geosphere, hydrosphere, biosphere, and atmosphere). The physical evidence, however, tends to suggest a concept of design quite different from the theistic notion of a deity who acts as creator.

Consider, for example, the ability of an animal to alter its appearance as a means of blending in with its environment, to ward off predators, to disguise itself while preying upon other animals, or for some other purpose. On the one hand, this seems like an example of conscious design by a loving creator, but as Charles Darwin (1809–1882) showed, it may simply be a matter of adaptability. According to Darwin, members of species unable to alter their appearance died out, leading to the dominance of those who could camouflage themselves.

In fact, science is not really capable of addressing the matter of a Designer (i.e., God), and thus, for scientists, the question of a deity's role in nature is simply irrelevant. This is not because scientists are necessarily atheists (many are and have been dedicated men and women of faith) but because the concept of a deity simply adds an unnecessary step to scientific analysis.

This is in line with Ockham's razor, a principle introduced by the medieval English philosopher William of Ockham (*ca.* 1285?–1349). According to Ockham, "entities must not be unnecessarily multiplied." In other words, in analyzing any phenomenon, one should seek the simplest and most straightforward explanation. Scientists are concerned with hard data, such as the evidence obtained from rock strata. The application of theological ideas in such situations would at best confuse and complicate the process of scientific analysis.

THE ARGUMENT FROM DESIGN. A few years before Ockham, the Italian philosopher Thomas Aquinas (1224 or 1225–1274) introduced a philosophical position known as the "argument from design." According to Aquinas, whose idea has been embraced by many up to the present day, the order and symmetry in nature indicate the existence of God. Some philosophers have conceded that this order does indeed indicate the existence of *a* god, though not necessarily the God of Christianity. Science, however, cannot afford to go even that far: where spiritual matters are concerned, science must be neutral.

Does any of this disprove the existence of God? Absolutely not. Note that science must be *neutral,* not in opposition, where spiritual matters are concerned. Indeed, one could not disprove God's existence scientifically if one wanted to do so; to return to the analogy given earlier, such an endeavor would be akin to using mathematics to disprove the existence of love. Religious matters are simply beyond the scope of science, and to use science against religion is as misinformed a position as its opposite.

SCIENCE AND THE FIRST TWO CAUSES. To return to Aristotle's causes, let us briefly consider the material and efficient cause as applied to the subject of Earth. These are much simpler matters than formal and final cause, and science is clearly able to address them. An understanding of Earth's material cause—that is, its physical substance—requires a brief examination of the chemical elements. The elements are primarily a subject for chemistry, though they are discussed at places throughout this book, inasmuch as they relate to the earth sciences and, particularly, geochemistry. Furthermore, the overall physical makeup of Earth, along with particular aspects of it, are subjects treated in much greater depth within numerous essays concerning specific topics, such as sedimentation or the biosphere.

Likewise the efficient cause, or the complex of forces that have shaped and continue to shape Earth, is treated in various places throughout this book. In particular, the specifics of Earth's origins and the study of these origins through the earth sciences are discussed in essays on aspects of historical geology, such as stratigraphy. Here the origins of Earth are considered primarily from the standpoint of the historical shift from mythological or religious explanations to scientific ones.

REAL-LIFE APPLICATIONS

MYTHOLOGY AND GEOLOGY

Most of what people believed about the origins and makeup of Earth before about 1700 bore the imprint of mythology or merely bad science. Predominant among these theories were the Creation account from the biblical Book of Genesis and the notion of the four elements inherited from the Greeks. These four elements—earth, air, fire, and water—were said to form the basis for the entire universe, and thus every object was thought to be composed of one or more of these elements. Thanks in large part to Aristotle, this belief permeated (and stunted) the physical sciences.

To call the biblical Creation story mythology is not, in this context at least, a value judgment. The Genesis account is not scientific, however, in the sense that it was not written on the basis of observed data but rather from religious principles. The concept of the four elements at least relates somewhat to observation, but specifically to untested observation; for this reason, it is hardly more scientific than the Genesis Creation story. The four elements were not, strictly speaking, a product of mythology, but they were mythological in the pejorative sense—that is, they had no real basis in fact.

GEOMYTHOLOGY. The biblical explanation of Earth's origins is but one of many creation myths, part of a larger oral and literary tradition that Dorothy B. Vitaliano, in her 1973 book *Legends of the Earth,* dubbed *geomythology.* Examples of geomythology are everywhere, and virtually every striking natural feature on Earth has its own geomythological backdrop. For instance, the rocky outcroppings that guard the western mouth of the Mediterranean, at Gibraltar in southern Spain and Ceuta in northern Morocco, are known collectively as the Pillars of Hercules because the legendary Greek hero is said to have built them.

Geomythological stories can be found in virtually all cultures. For instance, traditional Hawaiian culture explains the Halemaumau volcano, which erupted almost continuously from 1823 to 1924, as the result of anger on the part of the Tahitian goddess Pele. Native Americans in what is now Wyoming passed down legends concerning the grooves along the sides of Devils Tower, which they said had been made by bears trying to climb the sides to escape braves hunting them.

GREEK GEOMYTHOLOGY. In Western culture, among the most familiar examples of geomythology, apart from those in the Bible, are the ones that originated in ancient Greece and Rome. The Pillars of Hercules represents but one example. In particular, the culture of the Greeks was infused with geomythological elements. They believed, for instance, that the gods lived on Mount Olympus and spoke through the Delphic Oracle, a priestess who maintained a trancelike state by inhaling intoxicating vapors that rose through a fault in the earth.

Much of Greek mythology is actually geomythology. Most of the principal Greek deities ruled over specific aspects of the natural world that are today the province of the sciences, and many of them controlled realms now studied by the earth sciences and related disciplines. Certain branches of geology today are concerned with Earth's interior, which the Greeks believed was controlled by Hades, or the Roman god Pluto. Volcanoes and thunderbolts were the work of the blacksmith god Hephaestus (the Roman deity Vulcan), while Poseidon (known to the Romans as Neptune) oversaw the area studied today by oceanographers.

ATLANTIS. Among the most persistent geomyths with roots in Greek civilization is the story of Atlantis, a continent that allegedly sank into the sea. Over the years, the myth grew to greater and greater dimensions, and in a blurring between the Atlantis myth and the biblical story of Eden, Atlantis came to be seen as a lost utopia. Even today, some people believe in Atlantis, and for scholarly endorsement they cite a passage in the writings of Plato (427?–347 B.C.). The great Greek philosopher depicted Atlantis as somewhere beyond the Pillars of Hercules, and for this reason its putative location eventually shifted to the middle of the Atlantic—an ocean in fact named for the "lost continent."

Given the layers of mythology associated with Atlantis, it may come as a surprise that the story has a basis in fact and that accounts of it appear in the folklore of peoples from Egypt to Ireland. It is likely that the myth is based on a cataclysmic event, either a volcanic eruption or an earthquake, that took place on the island of Crete, as well as nearby Thíra, around 1500 B.C. This cataclysm, some eight centuries before the rise of classical Greek civilization, brought an end to the Minoan civilization centered around Knossos in Crete. Most likely it raised vast tidal waves, or tsunamis, that reached lands far away and may have caused other cities or settlements to disappear beneath the sea.

BIBLICAL GEOMYTHOLOGY. As important as such Greek stories are, no geomythological account has had anything like the impact on Western civilization exerted by the first nine chapters of the Bible. These chapters contain much more than geomythology, of course; in fact, they introduce the central themes of the Bible itself: righteousness, sin, redemption, and God's covenant with humankind. In these nine chapters (or, more properly, eight and a half chapters), which cover the period from Earth's creation until the Great Flood, events are depicted as an illustration of this covenant. Thus, in 9 Genesis, when God introduces the rainbow after the Flood, he does so with the statement that it is a sign of his promise never again to attempt to destroy humanity.

As with Atlantis, the story of the Great Flood appears in other sources as well. Its antecedents include the Sumerian Gilgamesh epic, which originated in about 2000 B.C., a millennium before the writing of the biblical account. Also as in the case of Atlantis, the biblical flood seems to have a basis in fact. Some modern scientists theorize that the Black Sea was once a freshwater lake, until floods covered the land barriers that separated it from saltwater.

The Flood occupies chapters 6 through 9 of Genesis, while chapters 3 through 5 are concerned primarily with human rather than geologic events. The story of Adam, Eve, the serpent, and the fruit of the Tree of Knowledge is a beautiful, complex, and richly symbolic explanation of how humans, born innocent, are prone to sin. It is the first conflict between God and human, just as Cain's murder of Abel is the first conflict between people. Both stories serve to illustrate the themes mentioned earlier: in both cases, God punishes the sins of the humans but also provides them with protection as a sign of his continued faithfulness.

THE BIBLE AND SCIENCE. In fact, the entire Creation story, source of centuries' worth of controversy, occupies only two chapters, and this illustrates just how little attention the writers of the Bible actually devoted to "scientific" subjects. Certainly, many passages in the Bible describe phenomena that conflict with accepted scientific knowledge, but most of these fall under the classification of miracles—or, if one does not believe them, alleged miracles. Was Jesus born of a virgin? Did he raise the dead? People's answers to those questions usually have much more to do with their religious beliefs than with their scientific knowledge.

Most of the biblical events related to the earth sciences appear early in the Old Testament, and most likewise fall under the heading of "miracles." Certain events, such as the parting of the Red Sea by Moses, even have possible scientific explanations: some historians believe that there was actually an area of dry land in the Red Sea region and that Moses led the children of Israel across it. The account of Joshua causing the Sun to stand still while his men marched around the city of Jericho is a bit more difficult to square with science, but a believer might say that the Sun (or rather, Earth) *seemed* to stand still.

In any case, the Bible does not present itself as a book of science, and certainly the Israelites of ancient times had little concept of science as we know it today. Some of the biblical passages mentioned here have elicited controversy, but few have inspired a great deal of discussion, precisely because they are generally regarded as accounts of miracles. The same is not true, however, of the first two chapters of Genesis, which even today remain a subject of dispute in some quarters.

SIX DAYS? Actually, 2 Genesis concerns Adam's life before the Fall as well as the creation of Eve from his rib, so the Creation story proper is confined to the first chapter. One of the most famous passages in Western literature, 1 Genesis describes God's creation of the universe in all its particulars, each of which he spoke into being, first by saying, "Let there be light." After six days of activity that culminated with the creation of the human being, he rested, thus setting an example for the idea of a Sabbath rest day.

As prose poetry, the biblical Creation story is among the great writings of all time. It is also a beautiful metaphoric description of creation by a loving deity; but it is not a guide to scientific study. Yet for many centuries, Western adherence to the Genesis account (combined with a number of other factors, including the general stagnation of European intellectual life throughout much of the medieval period) forced a virtual standstill of geologic study. The idea that Earth was created in 144 hours reached its extreme with the Irish bishop James Ussher (1581–1656), who, using the biblical genealogies from Adam to Christ, calculated that God finished making Earth at 9:00 A.M. on Sunday, October 23, 4004 B.C.

THE MYTH OF THE FOUR ELEMENTS

Religion alone is far from the only force that has slowed the progress of science over the years. Sometimes the ideas of scientists or philosophers themselves, when formed on the basis of something other than scientific investigation, can prove at least as detrimental to learning. Such is the case when thinkers become more dedicated to the theory than to the pursuit of facts, as many did in their adherence to the erroneous concept of the four elements.

Today scientists understand an element as a substance made up of only one type of atom, meaning that unlike a compound, it cannot be broken down chemically into a simpler substance. This definition developed over the period from about 1650 to 1800, thanks to the British chemist Robert Boyle (1627–1691), who originated the idea of elements as the simplest substances; the French chemist Antoine Lavoisier (1743–1794), who first distinguished between elements and compounds; and the British chemist John Dalton (1766–1844), who introduced the atomic theory of matter.

DEVILS TOWER, WITH THE BIG DIPPER VISIBLE IN THE NIGHT SKY. *(© Jerry Schad/Photo Researchers. Reproduced by permission.)*

During the twentieth century, with the discovery of the atomic nucleus and the protons within it, scientists further refined their definition of an element. Today elements are distinguished by atomic number, or the number of protons in the atomic nucleus. Carbon, for instance, has an atomic number 6, meaning that there are six protons in the carbon nucleus; therefore, any element with six protons in its atomic nucleus *must* be carbon.

ATOMIC THEORY VERSUS THE FOUR ELEMENTS. Atomic, or corpuscular, theory had been on the rise for some 150 years before Dalton, who built on ideas of predecessors that included Galileo Galilei (1564–1642)

and Sir Isaac Newton (1642–1727). In any case, the first thinker to conceive of atoms lived more than 2,000 years earlier. He was Democritus (*ca.* 460–*ca.* 370 B.C.), a Greek philosopher who described the world as being composed of indivisible particles—*atomos* in Greek. Democritus's idea was far from modern scientific atomic theory, but it came much closer than any other theory before the Scientific Revolution (*ca.* 1550–1700).

Why, then, did it take so long for Western science to come around to the atomic idea? The answer is that Aristotle, who exerted an almost incalculable impact on Muslim and Western thought during the Middle Ages, rejected Democritus' atomic theory in favor of the four elements theory. The latter had its roots in the very beginnings of Greek ideas concerning matter, but it was the philosopher Empedocles (*ca.* 490–430 B.C.) who brought the notion to some kind of maturity.

A NONSCIENTIFIC THEORY. According to the four elements theory, every object could be identified as a combination of elements: bone, for instance, was supposedly two parts earth, two parts water, and two parts fire. Of course, this is nonsense, and, in fact, none of the four elements are even really elements. Water comes the closest, being a compound of the elements hydrogen and oxygen. Earth and air are mixtures, while fire is the result of combustion, a form of oxidation-reduction chemical reaction.

Nonetheless, the theory had at least some basis in observation, since much of the physical world seems to include liquids, things that grow from the ground, and so on. Such observations alone, of course, are not enough to construct a theory, as would have become apparent if the Greeks had attempted to test their ideas. The ancients, however, tended to hold scientific experimentation in low esteem, and they were more interested in applying their intellects to the development of ideas than they were in getting their hands dirty by putting their concepts to the test.

THINKING IN FOURS. Aristotle explained the four elements as combinations of four qualities, or two pairs of opposites: hot/cold and wet/dry. Thus, fire was hot and dry, air was dry and cold, water was cold and wet, and earth was wet and hot. It is perhaps not accidental that there were four elements, four qualities, or even perhaps four Aristotelian causes.

Much earlier, the philosopher and mathematician Pythagoras (*ca.* 580–*ca.* 500 B.C.), who held that all of nature could be understood from the perspective of numbers, first suggested the idea of four basic elements because, he maintained, the number four represents perfection. This concept influenced Greek thinkers, including Empedocles and even Aristotle, and is also probably the reason for the expression *four corners of the world.*

That expression, which conveys a belief in a flat Earth, raises an important point that must be made in passing. Despite his many erroneous ideas, Aristotle was the first to prove that Earth is a sphere, which he showed by observing the circular shadow on the Moon during a lunar eclipse. This points up the fact that ancient thinkers may have been misguided in many regards, yet they still managed to make contributions of enormous value. In the same vein, Pythagoras, for all his strange and mystical ideas, greatly advanced scientific knowledge by introducing the concept that numbers can be applied to the study of nature.

In any case, the emphasis on fours trickled down through classical thought. Thus, the great doctors Hippocrates (*ca.* 460–*ca.* 377 B.C.) and Galen (129–*ca.* 199) maintained that the human body contains four "humors" (blood, black bile, green bile, and phlegm), which, when imbalanced, caused diseases. Humoral theory would exert an incalculable toll on human life throughout the Middle Ages, resulting in such barbaric medical practices as the use of leeches to remove "excess" blood from a patient's body. The idea of the four elements had a less clearly pernicious effect on human well-being, yet it held back progress in the sciences and greatly impeded thinkers' understanding of astronomy, physics, chemistry, and geology.

THE SHOWDOWN BETWEEN MYTH AND SCIENCE

Aristotle's teacher Plato had accepted the idea of the four elements, but proposed that space is made up of a fifth, unknown element. This meant that Earth and the rest of the universe are fundamentally different, a misconception that prevailed for two millennia. Aristotle adopted that idea, as well as Plato's concept of a Demiurge, or Prime Mover, as Aristotle called it. Cen-

turies later Aquinas equated Aristotle's Prime Mover with the Christian God.

Building on these and other ideas, Aristotle proceeded to develop a model of the cosmos in which there were two principal regions: a celestial, or heavenly, realm above the orbit of the Moon and a terrestrial, or earthly, one in what was known as the sublunary (below the Moon) region. Virtually everything about these two realms differed. The celestial region never changed, whereas change was possible on Earth. Earth itself consisted of the four elements, whereas the heavens were made up of a fifth substance, which he called *ether.*

If left undisturbed, Aristotle theorized, the four elements would completely segregate into four concentric layers, with earth at the center, surrounded by water, then air, and then fire, bounded at the outer perimeter by the ether. The motion of bodies above the Moon's sphere caused the elements to behave unnaturally, however, and thus they remained mixed and in a constant state of agitation.

The distinction between so-called natural and unnatural (or violent) motion became one of the central ideas in Aristotle's physics, a scientific discipline whose name he coined in a work by the same title. According to Aristotle, all elements seek their natural position. Thus, the element earth tends to fall toward the center of the universe, which was identical with the center of Earth itself.

THE SCIENTIFIC REVOLUTION. On these and other ideas, Aristotle built a complex, systematic, and almost entirely incorrect set of principles that dominated astronomy and physics as well as what later became the earth sciences and chemistry. The influence of Aristotelian ideas on astronomy, particularly through the work of the Alexandrian astronomer Ptolemy (*ca.* 100–170), was especially pronounced.

It was through astronomy, the oldest of the physical sciences, that the Aristotelian and Ptolemaic model of the physical world ultimately was overthrown. This revolution began with the proof, put forward by Nicolaus Copernicus (1473–1543), that Earth is not the center of the universe. The Catholic Church, which had controlled much of public life in Europe for the past thousand years, had long since accepted Ptolemy's geocentric model on the reasoning that if the human being is created in God's image, Earth must be at the center of the universe. Copernicus' heliocentric (Sun-centered) cosmology therefore constituted a challenge to religious authority—a very serious matter at a time when the Church held the power of life and death.

Copernicus died before he suffered the consequences of his ideas, but Galileo, who lived much later, found himself in the middle of a debate between the Church and science. This conflict usually is portrayed in simplistic terms, with Galileo as the noble scientific genius defending reason against the powers of reaction, but the facts are much more complex. For centuries, the Church had preserved and encouraged learning, and the reactionary response to Copernican ideas must be understood in light of the challenges to Catholic authority posed by the Protestant Reformation. Furthermore, Galileo was far from diplomatic in his dealings, for instance, deliberately provoking Pope Urban VIII (1568–1644), who had long been a friend and supporter.

In any case, Galileo made a number of discoveries that corroborated Copernicus' findings while pointing up flaws in the ideas of Aristotle and Ptolemy. He also conducted studies on falling objects that, along with the laws of planetary motion formulated by Johannes Kepler (1571–1630), provided the basis for Newton's epochal work in gravitation and the laws of motion. Perhaps most of all, however, Galileo introduced the use of the scientific method.

THE SCIENTIFIC METHOD. The scientific method is a set of principles and procedures for systematic study using evidence that can be clearly observed and tested. It consists of several steps, beginning with observation. This creates results that lead to the formation of a hypothesis, an unproven statement about the way things are. Up to this point, we have gone no further than ancient science: Aristotle, after all, was making a hypothesis when he said, for instance, that heavy objects fall faster than light ones, as indeed they seem to do.

Galileo, however, went beyond the obvious, conducting experiments that paved the way for modern understanding of the acceleration due to gravity. As it turns out, heavy objects fall faster than light ones only in the presence of resistance from air or another medium, but in a vacuum a stone and a feather would fall at the same rate. How Galileo arrived at this idea is not important

KEY TERMS

ATOM: The smallest particle of an element, consisting of protons, neutrons, and electrons. An atom can exist either alone or in combination with other atoms in a molecule.

ATOMIC NUMBER: The number of protons in the nucleus of an atom.

COMPOUND: A substance made up of atoms of more than one element, chemically bonded to one another.

COSMOLOGY: A branch of astronomy concerned with the origin, structure, and evolution of the universe.

COSMOS: The universe.

ELEMENT: A substance made up of only one kind of atom. Unlike compounds, elements cannot be broken chemically into other substances.

GEOCENTRIC: Earth-centered.

GEOMYTHOLOGY: Folklore inspired by geologic phenomena.

HELIOCENTRIC: Sun-centered.

HYPOTHESIS: An unproven statement regarding an observed phenomenon.

LAW: A scientific principle that is shown always to be the case and for which no exceptions are deemed possible.

PHYSICAL SCIENCES: Astronomy, physics, chemistry, and the earth sciences.

PROTON: A positively charged particle in an atom.

SCIENTIFIC METHOD: A set of principles and procedures for systematic study that includes observation; the formation of hypotheses, theories, and ultimately laws on the basis of such observation; and continual testing and reexamination.

SCIENTIFIC REVOLUTION: A period of accelerated scientific discovery that completely reshaped the world. Usually dated from about 1550 to 1700, the Scientific Revolution saw the origination of the scientific method and the introduction of such ideas as the heliocentric (Sun-centered) universe and gravity.

THEORY: A general statement derived from a hypothesis that has withstood sufficient testing.

VACUUM: An area devoid of matter, even air.

here; rather, his application of the scientific method, which requires testing of hypotheses, is the key point.

If a hypothesis passes enough tests, it becomes a theory, or a general statement. An example of a theory is uniformitarianism, an early scientific explanation of Earth's origins discussed elsewhere, in the context of historical geology. Many scientific ideas remain theories and are quite workable as such: in fact, much of modern physics is based on the quantum model of subatomic behavior, which remains a theory. But if something always has been observed to be the case and if, based on what scientists know, no exceptions appear possible, it becomes a law. An example is Newton's third law of motion: no one has ever observed or created a situation in which a physical action does not yield an equal and opposite reaction.

Even laws can be overturned, however, and every scientific principle therefore is subjected to continual testing and reexamination, making the application of the scientific method a cyclical process. Thus, to be scientific, a principle must be capable of being tested. It should also be said that one of the hallmarks of a truly scientific theory is the attitude of its adherents. True scientists are always attempting to *disprove* their own ideas by subjecting them to rigorous tests; the more such tests a theory survives, the stronger it becomes.

CREATIONISM: RELIGION UNDER A VEIL OF SCIENCE

During the twentieth century, a movement called creationism emerged at the fringes of science. Primarily American in origin, creationism is a fundamentalist Christian doctrine, meaning that it is rooted in a strict literal interpretation of the Genesis account of Creation. (For this reason, creationism has little influence among Christians and Christian denominations not prone to literalism.) From the 1960s onward, it has been called *creation science,* but even though creationism sometimes makes use of scientific facts, it is profoundly unscientific.

Again, the reference to creationism as unscientific does not necessarily carry a pejorative connotation. Many valuable things are unscientific; however, to call creationism unscientific is pejorative in the sense that its adherents *claim* that it is scientific. The key difference lies in the attitude of creationists toward their theory that God created the Earth if not in six literal days, then at least in a very short time.

If this were a genuine scientific theory, its adherents would be testing it constantly against evidence, and if the evidence contradicted the theory, they would reject the theory, not the evidence. Science begins with facts that lead to the development of theories, but the facts always remain paramount. The opposite is true of creationism and other nonscientific beliefs whose proponents simply look for facts to confirm what they have decided is truth. Conflicting evidence simply is dismissed or incorporated into the theory; thus, for instance, fossils are said to be the remains of animals who did not make it onto Noah's ark.

Creationism (for which *The Oxford Companion to the Earth* provides a cogent and balanced explanation) is far from the only unscientific theory that has pervaded the hard sciences, the social sciences, or society in general. Others, aside from the four elements, have included spontaneous generation and the phlogiston theory of fire as well as various bizarre modern notions, such as flat-Earth theory, Holocaust or Moon-landing denial, and Afrocentric views of civilization as a vast racial conspiracy. Compared with Holocaust denial, for instance, creationism is benign in the sense that its proponents seem to act in good faith, believing that any challenge to biblical literalism is a challenge to Christianity itself.

Still, there is no justification for the belief that Earth is very young; quite literally, mountains of evidence contradict this claim. Nor is the idea of an old Earth a recent development; rather, it has circulated for several hundred years—certainly long before Darwin's theory of evolution, the scientific idea with which creationists take the most exception. For more about early scientific ideas concerning Earth's age, see Historical Geology and essays on related subjects, including Paleontology and Geologic Time. These essays, of course, are concerned primarily with modern theories regarding Earth's history, as well as the observations and techniques that have formed the basis for such theories. They also examine pivotal early ideas, such as the Scottish geologist James Hutton's (1726–1797) principle of uniformitarianism.

WHERE TO LEARN MORE

Bender, Lionel. *Our Planet.* New York: Simon and Schuster Books for Young Readers, 1992.

Elsom, Derek M. *Planet Earth.* Detroit: Macmillan Reference USA, 2000.

Gamlin, Linda. *Life on Earth.* New York: Gloucester Press, 1988.

Hancock, Paul L., and Brian J. Skinner. *The Oxford Companion to the Earth.* New York: Oxford University Press, 2000.

Llamas Ruiz, Andrés. *The Origin of the Universe.* Illus. Luis Rizo. New York: Sterling Publishers, 1997.

Skinner, Brian J., Stephen C. Porter, and Daniel B. Botkin. *The Blue Planet: An Introduction to Earth System Science.* 2nd ed. New York: John Wiley and Sons, 1999.

The Talk. Origins Archive: Exploring the Creation/Evolution Controversy (Web site). <http://www.talkorigins.org/>.

Van der Pluijm, Ben A., and Stephen Marshak. *Earth Structure* (Web site). <http://www-personal.umich.edu/~vdpluijm/earthstructure.htm>.

Vitaliano, Dorothy B. *Legends of the Earth.* Bloomington: Indiana University Press, 1973.

Web Elements (Web site). <http://www.webelements.com/>.

Windows to the Universe (Web site). <http://www.windows.ucar.edu/win_entry.html>.

GEOSCIENCE AND EVERYDAY LIFE

CONCEPT

How can learning about rocks help us in our daily lives? The short answer is that geology and the related geologic sciences (sometimes referred to collectively as *geoscience*) give us a glimpse of the great complexity inherent in the natural world, helping us appreciate the beauty and order of things. This, in turn, makes us aware of our place in the scheme of things, so that we begin to see our own daily lives in their proper context. Beyond that, the study of geoscientific data can give us an enormous amount of information of practical value while revealing much about the world in which we dwell. The earth sciences are, quite literally, all around us, and by learning about the structures and processes of our planet, we may be surprised to discover just how prominent a place geoscience occupies in our daily lives and even our thought patterns.

After all, geoscience *is* the study of Earth, and therefore it relates to something of obvious and immediate practical value. We may think of a hundred things more important and pressing than studying Earth—romantic involvements, perhaps, or sports, or entertainment, or work (both inside and outside school)—yet without Earth, we would not even have those concerns. Without the solid ground beneath our feet, which provides a stage or platform on which these and other activities take place, life as we know it would be simply impossible. Our lives are bounded by the solid materials of Earth—rocks, minerals, and soil—while our language reflects the primacy of Earth in our consciousness. As we discuss later, everyday language is filled with geologic metaphors.

HOW IT WORKS

Why Study Geoscience?

One of the questions students almost always ask themselves or their teachers is "How will I use this?" or "What does all this have to do with everyday life?" It is easy enough to understand the application of classes involved in learning a trade or practical skill—for example, wood shop or a personal finance course. But the question of applicability sometimes becomes more challenging when it comes to many mathematical and scientific disciplines. Such is the case, for instance, with the earth sciences and particularly geoscience. Yet if we think about these concerns for just a moment, it should become readily apparent just why they are applicable to our daily lives.

Defining Geoscience

The geologic sciences—geology, geophysics, geochemistry, and related disciplines—are sometimes referred to together as geoscience. They are united in their focus on the solid earth and the mostly nonorganic components that compose it. In this realm of earth science, geology is the leading discipline, and it has given birth to many offshoots, including geophysics and geochemistry, which represent the union of geology with physics and chemistry, respectively.

Geology is the study of the solid earth, especially its rocks, minerals, fossils, and land formations. It is divided into historical geology, which is concerned with the processes whereby Earth was formed, and physical geology, or the study of the materials that make up the planet. Geophysics addresses Earth's physical processes as well as its gravitational, magnetic, and electric properties and the means by which energy is transmitted through its interior. Geochemistry is concerned with the chemical properties and processes of Earth—in particular, the abundance and interaction of chemical elements.

These subjects are of principal importance in this book. Though geology takes the lion's share of attention, geophysics and geochemistry each encompass areas of study essential to understanding our life on Earth: hence we look in separate essays at such geophysical subjects as Gravity and Geodesy or Geomagnetism as well as such geochemical topics as Biogeochemical Cycles, Carbon Cycle, and Nitrogen Cycle.

OTHER AREAS OF GEOSCIENCE. In addition to these principal areas of interest in geoscience, this book treats certain subdisciplines of geology as areas of interest in their own right. These include geomorphology and the studies of sediment and soil. Geomorphology is an area of physical geology concerned with the study of landforms, with the forces and processes that have shaped them, and with the description and classification of various physical features on Earth.

In contrast to geology, which normally is associated with rocks and minerals, geomorphology is concerned more with larger configurations, such as mountains, or with the erosive and weathering forces that shape such landforms. (See, for instance, essays on Mountains, Erosion, and Mass Wasting.) Erosion and weathering also play a major role in creating sediment and soil, areas that are of interest in the subdisciplines of sedimentology and soil science.

CONTRAST WITH OTHER DISCIPLINES AND SUBDISCIPLINES. Geoscience is distinguished sharply from the other branches of the earth sciences, namely, hydrologic sciences and atmospheric sciences. The first of these sciences, which is concerned with water, receives attention in essays on Hydrology and Hydrologic Cycle. The second, which includes meteorology (weather forecasting) and climatology, is the subject of the essays Weather and Climate.

In addition to the hydrologic and atmospheric sciences, there are areas of earth sciences study that touch on biology. Essays in this book that treat biosphere-related topics include Ecosystems and Ecology and Ecological Stress. There is one area or set of areas, however, in which geoscience and biology more or less overlap: sedimentology and soil science, since soil is a combination of rock fragments and organic material (see Soil).

THE TERRITORY OF GEOSCIENCE

The organic material in soil—dead plants and animals and parts thereof—has ceased to be part of the biosphere and is part of the geosphere. The geosphere encompasses the upper part of Earth's continental crust, or that portion of the solid earth on which human beings live and which provides them with most of their food and natural resources. (For more about the "spheres," see Earth Systems.)

Later in this essay, we discuss several areas of geoscientific study that take place close to the surface of Earth. Yet the territory of geoscience extends far deeper, going well below the geosphere into the interior of the planet. (For more on this subject, see Earth's Interior.) Geoscience even involves the study of "earths" other than our own; as discussed in such essays as Planetary Science and Sun, Moon, and Earth, there is considerable overlap between geoscience and astronomy.

REAL-LIFE APPLICATIONS

THE PRIMACY OF EARTH

We may not think about geoscience or earth science much, or at least we may not *think* that we think about these topics very much—and yet we spend our lives in direct contact with these areas. Certainly in a given day, every person experiences physics (the act of getting out of bed is an example of the third law of motion, discussed in Gravity and Geodesy) and chemistry (eating and digesting food, for instance), but the experience of geoscience is more direct: we can actually *touch* the earth.

Before the late nineteenth century and the introduction of processed foods, everything a person ate clearly either was grown in the soil or was part of an animal that had fed on plants grown in the soil. Even today, the most grotesquely processed products, such as the synthetic cream puffs sold at a convenience store, still hold a connection to the earth, inasmuch as they contain sugar—a natural product. In any case, most of what we eat (especially in a health-conscious diet) has a close connection to the earth.

GEOSCIENCE AND LANGUAGE. No wonder, then, that a number of creation stories, including the one in Genesis, depict humankind as coming from the soil—an account of origins reflected in the well-known graveside benediction "Ashes to ashes, dust to dust." Our language is filled with geoscientific metaphors, including such proverbs as "A rolling stone gathers no moss" or "Still waters run deep." (The latter aphorism, despite its hydrologic imagery, actually refers to the fact that in deeper waters, rock formations are, by definition, not likely to be near the surface. By contrast, in order for a "babbling brook" to make as much noise as it does, it must be flowing over prominent rocks.)

Then there are the countless geologic figures of speech: "rock solid," "making mountains out of molehills," "cold as a stone," and so on. When the rock musician Bob Seger sang, in a 1987 hit, about being "Like a Rock" as a younger man, listeners knew exactly what he meant: solid, strong, dependable. So established was the metaphor that a few years later, Chevrolet used the song in advertising their trucks and sport-utility vehicles (including, ironically, a vehicle whose name uses a somewhat less reassuring geologic image: the Chevy Avalanche).

THE GEOMORPHOLOGY OF RELIGIOUS FAITH. Rocks and other geologic features have long captured the imagination of humans; hence, we have the many uses of mountains in, for instance, religious imagery. There was the mystic mountain paradise of Valhalla in Norse mythology as well as Mount Olympus in Greek myths and legends. Unlike Valhalla, Olympus is a real place; so, too, is Kailas in southwestern Tibet, which ancient adherents of the Jain religion called Mount Meru, the center of the cosmos, and which Sanskrit literature identifies as the paradise of Siva, one of the principal Hindu deities.

There is also Sri Pada, or Adam's Peak, in Sri Lanka, a spot sacred to four religions. Buddhists believe the mountain is the footprint of the Buddha, while Hindus call it the footprint of Siva. Muslims and Christians believe it to be the footprint of Adam. Then there are the countless mountain locales of the Old Testament, including Ararat (in modern Turkey), where Noah's ark ran aground, and Sinai (in the Sinai Desert between Egypt and Israel), where Moses was called by God and later received the Ten Commandments.

The New Testament account of the life of Jesus Christ is punctuated throughout with geologic and geomorphologic details: the temptations in the desert, the Sermon on the Mount, and the Transfiguration, which probably took place atop Mount Tabor in Israel. He was crucified on a hill, buried in a cave, rolled a stone away at his Resurrection, and finally ascended to heaven from the Mount of Olives.

Arts, Media, and the Geosciences

From ancient times rocks and minerals have intrigued humans, not only by virtue of their usefulness but also because of their beauty. On one level there is the purely functional use of rock as a building material, and on another level there is the aesthetic appreciation for the beauty imparted by certain types of rock, such as marble.

Rock is an excellent building material when it comes to compression, as exerted by a great weight atop the rock; in the case of tension or stretching, however, rock is very weak. This shortcoming of stone, which was otherwise an ideal building material for the ancients (given its cheapness and relative abundance in some areas of the world), led to one of history's great innovations in architecture and engineering: the arch. A design feature as important for its aesthetic value as for its strength, the arch owed its physical power to the principle of weight redistribution. Arched Roman structures two thousand or more years old still stand in Europe, a tribute to the interaction of art, functionality, and geoscience.

THE VISUAL ARTS. *The Oxford Companion to the Earth* contains a number of excellent entries on the relationship between geoscience and the arts. In the essay "Art and the Earth Sciences," for instance, Andrew C. Scott notes four ways in which the earth sciences and the visual arts (including painting, sculpture, and photography) interact: through the depiction of such earth sciences phenomena as mountains or storms, through the use of actual geologic illustrations or even maps as forms of artwork, through the application of geologic materials in art (most notably, marble in sculpture), and

While stone is a strong building material in terms of compression, it is weak in terms of tension. The arch owes its strength to the principle of weight distribution, which overcomes this shortcoming of stone. Indeed, the Roman Coliseum has stood for more than two thousand years. *(© John Moss/Photo Researchers. Reproduced by permission.)*

through the employment of geology to investigate aspects of art objects (for instance, determining the origins of materials in ancient sculpture).

In the first category, visual depictions of geologic phenomena, Scott mentions works by unknown artists of various premodern civilizations (in particular, China and Japan) as well as by more recent artists whose names are hardly household words. On the other hand, some extremely well known figures produced notable works related to geoscience and the earth sciences. For example, the Italian artist and scientist Leonardo da Vinci (1452–1519), who happened to be one of the fathers of geology (see Studying Earth), painted many canvases in which he portrayed landscapes with a scientist's eye.

Another noteworthy example of earth sciences artwork and illustration is *The Great Piece of Turf* (1503), by Leonardo's distinguished contemporary the German painter and engraver Albrecht Dürer (1471–1528). A life-size depiction of grasses and dandelions, *Turf* belongs within the realm of earth sciences or even biological sciences rather than geoscience, yet it is significant as a historical milestone for all natural sciences.

In creating this work, Dürer consciously departed from the tradition, still strong even in

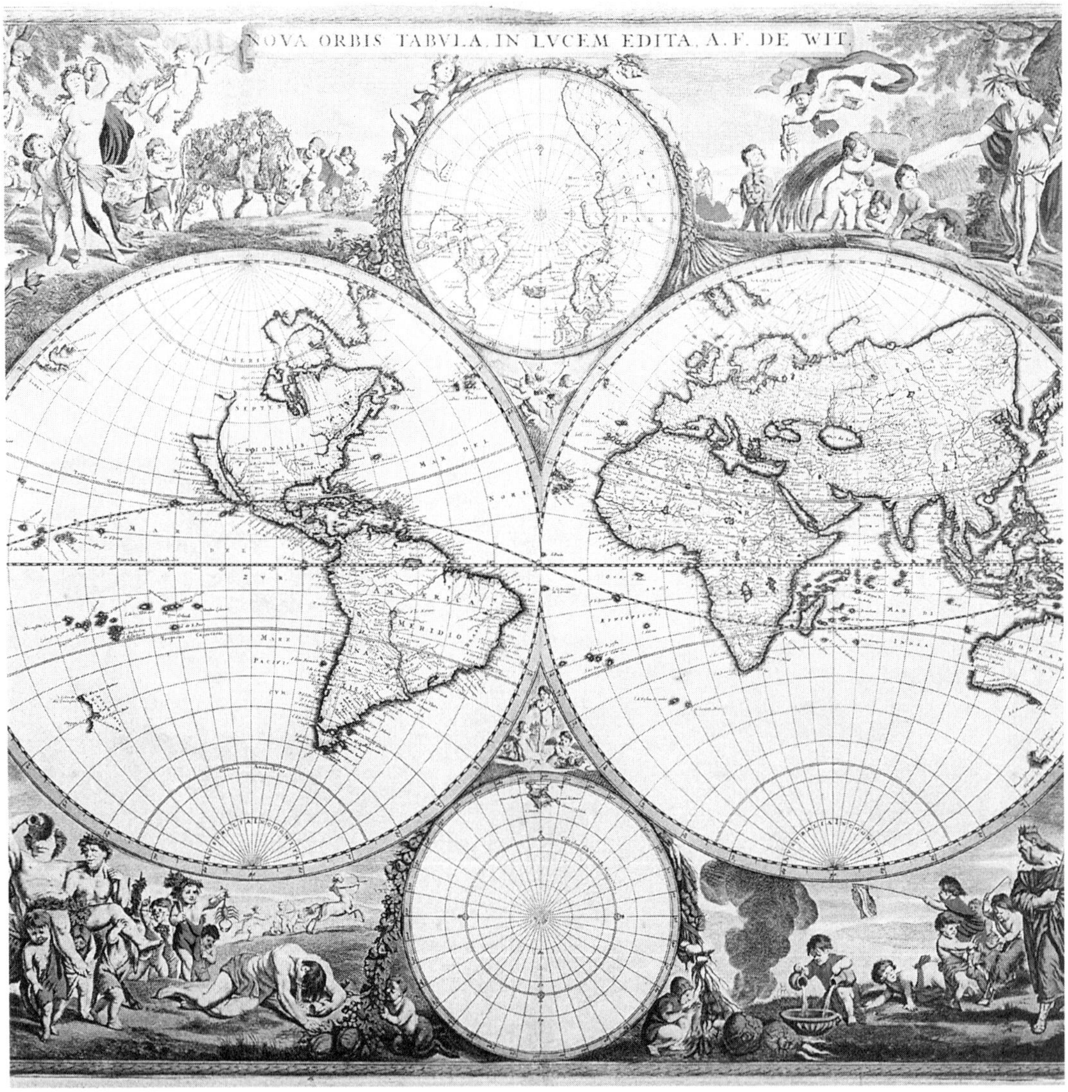

Scientific illustration became popular between 1500 and 1700, bridging the boundary between earth science and art. This map of the world, surrounded by allegorical scenes depicting the rewards and pitfalls of exploration, dates to 1689. *(© G. Bernard/Photo Researchers. Reproduced by permission.)*

the Renaissance, of representing "important" subjects, such as those of the Bible and classical mythology or history. By contrast, Dürer chose a simple scene such as one might find at the edge of any pond, yet his painting had a tremendous artistic and scientific impact. He set a new tone of naturalism in the arts and established a standard for representing nature as it is rather than in the idealized version of the artist's imagination.

As a result of Dürer's efforts, the period between about 1500 and 1700 saw the appearance of botanical illustrations whose quality far exceeded that of all previous offerings. Thus, he started a movement that spread throughout the world of scientific illustrations in general. Later, such geologists as England's William Smith (1769–1839) would produce maps that are rightly regarded as works of art in their own right (see Measuring and Mapping Earth).

Sometimes geologic phenomena have themselves become the basis for works of art, as Scott points out, observing that the modern American artist James Turrell once "set out to modify an extinct volcano, the Roden Crater [in northern Arizona], by excavating chambers and a tunnel to provide a visual experience of varying spatial relationships, the effects of light, and the perception of the sky." Elsewhere in the *Oxford Companion,* other writers show how evidence of a geoscientific influence has appeared in other arts and media, including music.

MUSIC. In "Music and the Earth Sciences," D. L. Dineley and B. Wilcock offer a fascinating overview of natural formations or materials that have their own musical qualities: for example, the "singing sands" of the Arabian peninsula and other regions, which produce musical tones when millions of grains are rubbed together by winds. The authors also discuss the effect of geologic phenomena on the sound and production of music—for instance, the acoustic qualities of music played in an auditorium built of stone.

Then there is the subject of musical compositions inspired by geoscientific or earth sciences phenomena. Among them are *The Hebrides; or, Fingal's Cave* by the German composer Felix Mendelssohn (1809–1847) as well as one the authors do not mention: *The Planets,* presented in 1918 by the German composer Gustav Holst (1874–1934). One also might list popular songs that refer to such phenomena, including "The White Cliffs of Dover." Written by Walter Kent and Nat Burton in 1941, the song epitomizes the longing for peace in a world torn by war. The cliffs themselves, which guard the eastern approaches of Britain, sometimes are referred to incorrectly as "chalk," though they are made of gypsum.

Ironically, rock music has few significant songs that refer to rocks. Usually the language is metaphoric, as was the case with the Bob Seger song discussed earlier. Hence, we have the name of the rock group Rolling Stones (with its implicit reference to the proverbial saying mentioned earlier) as well as the title to one of their earliest hits, "Heart of Stone." Jim Morrison's lyrics for the Doors include several references to the ground and things underneath it, including a gold mine in "The End." Coal mines have appeared in more than one song: "Working in the Coal Mine" was a hit for Lee Dorsey in the 1960s and was performed anew by the group Devo in 1981—not long after the Police song "Canary in a Coal Mine" appeared.

FILM. More significantly, the year 1981 marked the release of *Raiders of the Lost Ark,* a film cited as a major turning point by Ted Nield in the *Oxford Companion*'s "Geoscience in the Media" entry. The film is not about a geoscientist but an archaeologist, Indiana Jones (played by the actor Harrison Ford); however, the character of Jones is based on an American paleontologist, Roy Chapman Andrews (1884–1960). Earlier movies, Nield observes, had portrayed the typical scientist as an "egghead . . . an arrogant, unworldly, megalomaniac obsessive . . . But with Indiana Jones we saw the beginning of a reaction. Increasing audience sophistication is part of the reason."

Nield goes on to discuss the movie *Jurassic Park* (1993), which features three scientists, all of whom receive positive treatment. The actor Sam Neill, as a paleontologist, is described as "dedicated—perhaps a bit too educated—but also intuitive, a superb communicator, and above all, knowledgeable about dinosaurs." Laura Dern, playing a paleobiologist, is "strong-willed, independent, feminist, and sexy," while Jeff Goldblum's mathematician is "weird, roguish, and cool." Sparking a widespread interest in dinosaurs and paleontology, the film (a major box-office hit directed by Steven Spielberg) helped advance the cause of the geosciences.

The positive trend in movie portrayals of geoscientists, Nield states, continued in *Dante's Peak* (1997), in which even the casting of the ultra-handsome actor Pierce Brosnan as a geologist says a great deal about changing perceptions of scientists. Noting that audiences had come to differentiate between science and the misapplication thereof, Nield observes that "The heat seems to have come off those who are merely curious about Nature's workings." Additionally, "by being associated with the open air and fieldwork, [geoscientists] can take on some of the clichéd but healthy characteristics usually associated on film with oilmen and lumberjacks."

In an entirely different category is another fascinating example of geoscience in film, Australian director Peter Weir's *Picnic at Hanging Rock* (1975). Weir, who went on to make such well-known films as *The Year of Living Dangerously* (1982), *Witness* (1985), and *Dead Poets' Society* (1989), established his reputation—and that of Australian cinema in general—with *Picnic,* which concerns the disappearance of a group of schoolgirls and their teacher on Valentine's Day, 1900. The story itself is fictional, though it seems otherwise (*Picnic* later inspired *The Blair Witch Project,* which also presents fiction as fact); however, the rock in the title is very much a real place. In the film, Hanging Rock is by far the most striking character, a brooding presence whose foreboding features serve as a reminder of Earth's vastness and great age in the face of human insignificance.

The Work of Geoscientists

The work of the geoscientist indeed is associated with the open air to a much greater degree than that of the physicist or chemist; on the other hand, a geoscientist might very well work indoors, for instance, as a teacher. Prospective geoscientists who subscribe to a worldview of environmental utopianism can get a job "saving the world"—perhaps even working for starvation wages, so as to heighten the nobility of the undertaking. On the other hand, a pragmatist can go to work for an "evil" oil company and make a good living. The point is that there is a little of something for everyone in the world of geoscience.

Geoscientists may work for educational institutions, governments, or private enterprise. They may be involved in the search for energy resources, such as coal or oil (or even uranium for nuclear power), or they may be put to work searching for valuable and precious metals ranging from iron to gold. They even may be employed in the mining of diamonds or other precious gems in South Africa, Russia, or other locales. Other, perhaps less glamorous but no less important resources for which geoscientists in various roles search are water as well as rocks, clay, and minerals for building.

The majority of employed geoscientists work for industry but not always in the capacity of resource extraction. Some are involved in environmental issues; indeed, environmental geology—the application of geologic techniques to analyze, monitor, and control the environmental impact of natural and human phenomena—is a growing field. Among the areas of concern for environmental geologists are water management, waste disposal, and land-use planning.

ENVIRONMENTAL AND URBAN GEOLOGY. Many environmental geologists, as one might expect, are employed by governments. They may be involved in soil studies before the commencement of a building project, in analyzing the necessary thickness and materials for a particular stretch of road, or in designing and establishing specifications for a landfill. Many such concerns come into play when large populations gather together. In fact, a growing area of specialization in environmental geology is urban geology.

Urban geology can be defined as the application of geologic techniques to the study of the built environment. (The latter term is architectural and engineering jargon for any physical or geographic area containing human construction.) At first, "urban geology" might almost seem like an oxymoron, since the term geology usually calls to mind vast, unpopulated mountain ranges and rock formations—perhaps in South Dakota or Wyoming. In fact, geology is a major factor in the development of cities. Most are defined by their geomorphology: the hills of Athens and Rome, the mountains above Los Angeles, or the harbors of New York and other major ports, for instance.

Most cities have natural barriers to growth, and this is precisely because geomorphology originally dictated the location at which the city was established. A rare exception is Atlanta, Georgia, which grew around the point where several rail lines met. (In the 1840s, when it was established, it bore the name Terminus, a reference to the fact that it lay at the end of the rail line.) Bounded by no ocean, significant rivers, mountains, or other natural barriers, such as deserts, Atlanta began a period of explosive growth in the latter part of the twentieth century and has never stopped growing. Today Atlanta is a textbook example of urban sprawl: lacking a vital city center, it is a settlement of some four million people spread over an area much larger than Rhode Island, with no end to growth in sight.

Los Angeles often is cited as a case of urban sprawl, but its problems are quite different: it is rife with geomorphologic barriers, including oceans, mountains, and desert. The result is increasing growth within a limited area, resulting in heightened stress on existing resources. These are some of the issues confronted by urban geologists. Another example is the problem of determining the strength of bedrock, which dictates the viability of tall buildings. Urban geologists also are concerned with such issues as underground facilities for transportation, infrastructure, and even usable workspace—one possible solution to the problem of urban sprawl.

GEOARCHAEOLOGY AND RELATED FIELDS. At the opposite extreme, in many ways, from urban geology is geoarchae-

A GOLD MINE IN ZIMBABWE. GEOSCIENTISTS WORK FOR EDUCATIONAL INSTITUTIONS, GOVERNMENTS, AND PRIVATE ENTERPRISE IN SUCH FIELDS AS RESOURCE EXTRACTION, ENVIRONMENTAL STUDIES AND MANAGEMENT, AND EVEN ARCHAEOLOGY AND CRIMINOLOGY. *(© Peter Bowater/Photo Researchers. Reproduced by permission.)*

ology, or the application of geologic analysis to archaeology and related fields. Whereas urban geology is concerned with the here and now, geoarchaeology—like the larger field of historical geology—addresses the past. And whereas urban geologists are most likely to be employed by governments, geoarchaeologists and those in similar areas are typically on the payroll of universities.

In a different sense, geoarchaeology also contrasts with archaeological geology, which is the study of archaeological sites for data relevant to the geosciences; thus, archaeological geology stands the approach of geoarchaeology on its head. An example of a study in archaeological geology can be found in the work conducted around the Roman ruins at Hierapolis in what is now Turkey. There, investigation of walls and gutters reveals the fact that the city was sitting astride an earthquake fault zone—a fact unknown to its residents, except when they experienced seismic tremors.

By contrast, an example of geoarchaeology in action would be establishing an explanation for how people came to the Americas from Siberia near the end of the last ice age—by crossing a land bridge that existed at that time. Another example of geoarchaeology would be the realm of ecclesiastical geology, which involves the study of old church masonry walls with the purpose of identifying areas from which rocks, bricks, and other materials were derived. Studies of medieval churches in England, for instance, show varieties of rock from sometimes unexpected locations, often placed alongside bricks taken from older Roman structures.

From the explanation and examples given here, it may be a bit hard to discern the difference between geoarchaeology and archaeological geology. Certainly there is a great deal of overlap, and in practice the difference comes down to a question of who is leading the fieldwork—a geologist or an archaeologist. In any case, both realms are concerned with the relatively recent human past, as opposed to the vast stretches of time that are the domain of historical geology (see Geologic Time).

FORENSIC GEOLOGY. On October 7, 2001, the United States launched air strikes against Afghanistan in retaliation for the refusal of that country's Taliban regime to surrender Osama bin Laden, the suspected mastermind of the World Trade Center bombing on September 11. On the same day, bin Laden's al-Qaeda ter-

KEY TERMS

ATMOSPHERIC SCIENCES: A major division of the earth sciences, distinguished from geoscience and the hydrologic sciences by its concentration on atmospheric phenomena. Among the atmospheric sciences are meteorology and climatology.

BIOSPHERE: A combination of all living things on Earth—plants, mammals, birds, reptiles, amphibians, aquatic life, insects, viruses, single-cell organisms, and so on—as well as all formerly living things that have not yet decomposed.

EARTH SCIENCES: The entire range of scientific disciplines focused on the study of Earth, including not only geoscience but also the atmospheric and hydrologic sciences.

ENVIRONMENTAL GEOLOGY: A field of geology involved in the application of geologic techniques to analyze, monitor, and control environmental impact of both natural and human phenomena.

GEOCHEMISTRY: A branch of the earth sciences, combining aspects of geology and chemistry, that is concerned with the chemical properties and processes of Earth—in particular, the abundance and interaction of chemical elements and their isotopes.

GEOLOGY: The study of the solid earth, in particular, its rocks, minerals, fossils, and land formations.

GEOMORPHOLOGY: An area of physical geology concerned with the study of landforms, with the forces and processes that have shaped them, and with the description and classification of various physical features on Earth.

GEOPHYSICS: A branch of the earth sciences that combines aspects of geology and physics. Geophysics addresses the planet's physical processes as well as its gravitational, magnetic, and electric properties and the means by which energy is transmitted through its interior.

rorist organization released a videotape of their leader delivering a diatribe against the United States. Naturally, military and law-enforcement agencies involved in the hunt for bin Laden took an interest in the tape, and some specialists sought clues in an unexpected place: the rocks behind bin Laden, featured prominently in the tape.

Although the efforts to trace bin Laden's location by the rock formations in the area were not successful, the underlying premise—that geographic regions have their own specific types and patterns of rock—was both a fascinating and a plausible one. This was just another example of a specialty known as forensic geology, or the use of geologic and other geoscientific data in solving crimes. Forensic geology has it origins around the beginning of the twentieth century, but some historians cite Sherlock Holmes, the master sleuth created by the English physician and writer Sir Arthur Conan Doyle (1859–1930), as an early practitioner.

In *The Sign of Four,* for instance, Holmes uses geologic data to ascertain that Watson has been to the Wigmore Street Post Office: "Observation tells me that you have a little reddish mould adhering to your instep," he explains. "Just opposite the Wigmore Street Office they have taken up the pavement and thrown up some earth, which lies in such a way that it is difficult to avoid treading in it in entering. The earth is of this peculiar reddish tint which is found, as far as I know, nowhere else in the neighbourhood."

The true founder of forensic geology was probably the Austrian jurist and pioneer in

KEY TERMS CONTINUED

GEOSCIENCE: The geologic sciences (geology, geochemistry, geophysics, and related disciplines), as opposed to other earth sciences—that is, atmospheric sciences, such as meteorology, and hydrologic sciences, such as oceanography.

GEOSPHERE: The upper part of Earth's continental crust, or that portion of the solid earth on which human beings live and which provides them with most of their food and natural resources.

HISTORICAL GEOLOGY: The study of Earth's physical history. Historical geology is one of two principal branches of geology, the other being physical geology.

HYDROLOGIC SCIENCES: Areas of the earth sciences concerned with the study of the hydrosphere. Among these areas are hydrology, glaciology, and oceanography.

HYDROSPHERE: The entirety of Earth's water, excluding water vapor in the atmosphere but including all oceans, lakes, streams, groundwater, snow, and ice.

ORGANIC: At one time chemists used the term *organic* only in reference to living things. Now the word is applied to most compounds containing carbon, with the exception of carbonates (which are minerals) and oxides, such as carbon dioxide.

PHYSICAL GEOLOGY: The study of the material components of Earth and of the forces that have shaped the planet. Physical geology is one of two principal branches of geology, the other being historical geology.

PHYSICAL SCIENCES: Astronomy, physics, chemistry, and the earth sciences.

SEDIMENT: Material deposited at or near Earth's surface from a number of sources, most notably preexisting rock. Soil is derived from sediment, particularly the mixture of rock fragments and organic material.

criminology Hans Gross (1847–1915), whose *Handbuch für Untersuchungsrichter* (Handbook for examining magistrates, 1898) was a pivotal work in the field. "Dirt on shoes," wrote Gross, "can often tell us more about where the wearer of those shoes has last been than toilsome inquiries." Near the turn of the nineteenth century, Germany's Georg Popp, who operated a forensic laboratory in Frankfurt, used the new science effectively in two cases.

The first of these cases involved the murder of a woman named Eva Disch in October 1904. Among the items found at the murder scene was a dirty handkerchief containing traces of coal, snuff, and hornblende, a mineral. Popp matched the handkerchief with a suspect who worked at two locations that used a great deal of hornblende. In addition, the suspect's pants cuffs bore soil both from the murder scene and the victim's house.

Four years later, in investigating the murder of Margaethe Filbert in Bavaria, Popp ascertained that the soil at the crime scene was characterized by red quartz and red clay rich in iron. By contrast, the chief suspect had a farm whose fields were notable for their porphyry, milky quartz, and mica content. As it turned out, the suspect's shoes bore traces of quartz and red clay rather than those other minerals, even though he claimed he had been working in his fields when the crime occurred.

WHERE TO LEARN MORE

Career Information for Geology Majors (Web site). <http://www.uakron.edu/ascareer/Geology.html>.

Careers in the Geosciences (Web site). <http://www.agiweb.org/agi/careers.html>.

Geoarchaeology (Web site). <http://www.geoarchaeology.com/geoarc.htm>.

Geoarchaeology and Related Subjects (Web site). <http://archaeology.about.com/cs/earthsciences/>.

Geology in the City (Web site). <http://www.fortunecity.com/greenfield/ecolodge/60/cityhome.html>.

Hancock, Paul L., and Brian J. Skinner, eds. *The Oxford Companion to the Earth.* New York: Oxford University Press, 2000.

Murray, Raymond C. *Devil in the Details: The Science of Forensic Geology* (Web site). <http://www.forensicgeology.net/science.htm>.

Sarah Andrews Forensic Geology Pages (Web site). <http://www.sonoma.edu/Geology/andrews.htm>.

Sevil Atasoy's Links of Forensic Geology (Web site). <http://abone.turk.net/atasoy/forensicgeology.html>.

Urban Geology of the National Capital (Web site). <http://sts.gsc.nrcan.gc.ca/page1/urban/urb.htm>.

What Can I Do with a Degree in Geology? (Web site). <http://www.hartwick.edu/geology/alumni/careers.html>.

EARTH SYSTEMS

CONCEPT

A system is any set of interactions set apart from the rest of the universe for the purposes of study, observation, and measurement. Theoretically, a system is isolated from its environment, but this is an artificial construct, since nothing is ever fully isolated. Earth is largely a closed system, meaning that it exchanges very little matter with its external environment in space, but the same is not true of the systems within the planet—geosphere, hydrosphere, biosphere, and atmosphere—which interact to such a degree that they are virtually inseparable. Together these systems constitute an intricate balance, a complex series of interrelations in which events in one sector exert a profound impact on conditions in another.

HOW IT WORKS

SYSTEMS

An isolated system is one so completely sealed off from its environment that neither matter nor energy passes through its boundaries. This is an imaginary construct, however, an idea rather than a reality, because it is impossible to create a situation in which no energy is exchanged between the system and the environment. Under the right conditions it is perhaps conceivable that matter could be sealed out so completely that not even an atom could pass through a barrier, but some transfer of energy is inevitable. The reason is that electromagnetic energy, such as that emitted by the Sun, requires no material medium in which to travel.

In contrast to an isolated system is a closed system, of which Earth is an approximation. Despite its name, a closed system permits the exchange of energy with the environment but does not allow matter to pass back and forth between the external environment and the system. Thus, Earth absorbs electromagnetic energy, radiated from the Sun, yet very little matter enters or departs Earth's system. Note that Earth is an *approximation* of a closed system: actually, some matter does pass from space into the atmosphere and vice versa. The planet loses traces of hydrogen in the extremities of its upper atmosphere, while meteorites and other forms of matter from space may reach Earth's surface.

Earth more closely resembles a closed system than it does an open one—that is, a system that allows the full and free exchange of both matter and energy with its environment. The human circulatory system is an example of an open system, as are the various "spheres" of Earth (geosphere, hydrosphere, biosphere, and atmosphere) discussed later. Whereas an isolated system is imaginary in the sense that it does not exist, sometimes a different feat of imagination is required to visualize an open system. It is intricately tied to its environment, and therefore the concept of an open system as a separate entity sometimes requires some imagination.

USING SYSTEMS IN SCIENCE

To gain perspective on the use of systems in science as well as the necessity of mentally separating an open system from its environment, consider how these ideas are used in formulating problems and illustrating scientific principles. For example, to illustrate the principle of potential and kinetic energy in physics, teachers often use the example of a baseball dropping from a great height (say, the top of a building) to the ground.

At the top of the building, the ball's potential energy, or the energy it possesses by virtue of its position, is at a maximum, while its kinetic energy (the energy it possesses by virtue of its motion) is equal to zero. Once it is dropped, its potential energy begins to decrease, and its kinetic energy to increase. Halfway through the ball's descent to the ground, its potential and kinetic energy will be equal. As it continues to fall, the potential energy keeps decreasing while the kinetic energy increases until, in the instant it strikes the ground, kinetic energy is at a maximum and potential energy equals zero.

KEEPING OUT IRRELEVANT DETAILS. What has been described here is a system. The ball itself has neither potential nor kinetic energy; rather, energy is in the system, which involves the ball, the height through which it is dropped, and the point at which it comes to a stop. Furthermore, because this system is concerned with potential and kinetic energy only in very simple terms, we have mentally separated it from its environment, treating it as though it were closed or even isolated, though in reality it would more likely be an open system.

In the real world, a baseball dropping off the top of a building and hitting the ground could be affected by such conditions as prevailing winds. These possibilities, however, are not important for the purposes of illustrating potential and kinetic energy, and even if they were, they could be incorporated into the larger energy system.

THE "MAGIC" OF A SYSTEM. Since kinetic energy and potential energy are inversely related, the potential energy at the top of the building will always equal the kinetic energy at the point of maximum speed, just before impact. This is true whether the ball is dropped from 10 ft. (3 m) or 1,000 ft. (305 m). It may seem almost magical that the sum of potential and kinetic energy is always the same or that the two values are perfectly inverse. In fact, there is nothing magical here: the system has a certain total energy, and this does not change, though the distribution of that energy can and does vary.

Suppose one had a money jar known to contain $20. If one reaches in and grasps a five-dollar bill, two one-dollar bills, three quarters, a dime, and two nickels ($7.95), there must be $12.05 left in the jar. There is nothing magical in this; rather, what has been illustrated is the physical principle of conservation. In physics and other sciences, "to conserve" something means "to result in no net loss of" that particular component. It is possible that within a given system, the component may change form or position, but as long as the net value of the component remains the same, it has been conserved. Thus, the total energy is conserved in the situation involving the baseball, and the total amount of money is conserved in the money-jar.

APPLYING THE SYSTEM PRINCIPLE TO EARTH

In the baseball illustration, the distribution between types of energy varies, but the total amount is always the same. Likewise in the money-jar illustration, the total amount of money remains fixed even though the distribution according to various denominations may vary. The same is true of Earth, though here it is the total amount of matter. This includes valuable resources, among them materials that can be mined to produce energy—for instance, fossil fuels such as coal or petroleum—as well as waste products. Because Earth is a closed system, there are no additional resources, nor is there any dumping ground other than the one beneath our feet. Thus, the situation calls for prudence both in the use of the planet's material wealth and in the processing of materials that will leave a byproduct of waste.

The fact that a closed system is by definition finite leads to the principle that the relationships between its constituent parts are likewise finite, and therefore changes in one part of the system are liable to produce effects in another part. Conditions in the baseball or money-jar illustrations are so simple that it is easy to predict the effect of a change. For instance, if we substitute a basketball for a baseball, this will change the total energy, because the latter is a function of the ball's mass. If the denominations making up the $20 in the money jar are replaced with a collection of two-dollar bills and dimes, this will make it impossible to reach in and pull out an odd-numbered value in dollars or cents.

What about the changes that result when one aspect of Earth's system is altered? In some cases, it is easy to guess; in others, the interactions are so complex that prediction requires sophisticated mathematical models. It is perhaps no accident that chaos theory was developed by a meteorologist, the American Edward Lorenz (1917–). Chaos theory, the study of complex systems that appear to follow no orderly laws, involves the analysis of phenomena that appear connected by something than an ordinary cause and effect relationship. The classic example of this is the "butterfly effect," the idea that a butterfly beating its wings in China can change the weather in New York City. This, of course, is a far-fetched scenario, but sometimes changes in one sector of Earth's system can yield amazing consequences in an entirely different part.

THE FOUR "SPHERES"

The systems approach is relatively new to the earth sciences, themselves a group of disciplines whose diversity reflects the breadth of possible approaches to studying Earth (see Studying Earth). At one time, earth scientists tended to investigate specific aspects of Earth without recognizing the ways in which these aspects connect with one another; today, by contrast, the paradigm of the earth sciences favors an approach that incorporates the larger background.

Given the complexities of Earth itself, as well as the earth sciences, it is helpful to apply a schema (that is, an organizational system) for dividing larger concepts and entities into smaller ones. For this reason, earth scientists tend to view Earth in terms of four interconnected "spheres." One of these terms, atmosphere, is a familiar one, while the other three (geosphere, hydrosphere, and biosphere) may sound at first like mere scientific jargon.

UNDERSTANDING THE SPHERES. In fact, each sphere represents a sector of existence on the planet that is at once clearly defined and virtually inseparable from the others. Each is an open system within the closed system of Earth, and overlap is inevitable. For example, the seeds of a plant (biosphere) are placed in the ground (geosphere), from which they receive nutrients for growth. In order to sustain life, they receive water (hydrosphere) and carbon dioxide (atmosphere). Nor are they merely receiving: they also give back oxygen to the atmosphere, and by providing nutrition to an animal, they contribute to the biosphere.

Each of the spheres, or Earth systems, is treated in various essays within this book. These essays examine these subsystems of the larger Earth system in much greater depth; what follows, by contrast, is the most cursory of introductions. It should be noted also that while these four subsystems constitute the entirety of Earth as humans know and experience it, they are only a small part of the planet's entire mass. The majority of that mass lies below the geosphere, in the region of the mantle and core.

A CURIOUS AND INSTRUCTIVE POINT. As a passing curiosity, it is interesting to note that modern scientists have identified four subsystems and given them the name spheres. As discussed in the essay Earth, Science, and Nonscience, the ancient Greeks were inclined to divide natural phenomena into fours, a practice that reached its fullest expression in the model of the universe developed by the Greek philosopher Aristotle (384–322 B.C.) He even depicted the physical world as a set of spheres and suggested that the heaviest material would sink to the interior of Earth while the lightest would rise to the highest points.

These points of continuity with ancient science are notable because almost everything about Aristotle's system was wrong, and, indeed, the differences between his model of the physical world and the modern one are instructive. There are four spheres in the modern earth sciences because these four happen to be useful ways of discussing the larger Earth system—not, as in the case of the Greeks, because the number four represents spiritual perfection. Furthermore, scientists understand these spheres to be artificial constructs, at least to some extent, rather than a key to some deeper objective reality about existence, as the ancients would have supposed.

Nor are the spheres of the modern earth sciences literally spheres, as Aristotle's concentric orbits of the planets around Earth were. If anything, the use of the term sphere represents a holdover from the Greek way of viewing the material world. Finally, unlike such ancient notions as the concept of the four elements, four qualities, or four humors, the idea of the four spheres is not simply the result of pure conjecture. Instead, the concept of these four interrelated systems came about by application of the scientific method and entered the vocabulary of earth scientists because the ideas involved clearly reflected and illustrated the realities of Earth processes.

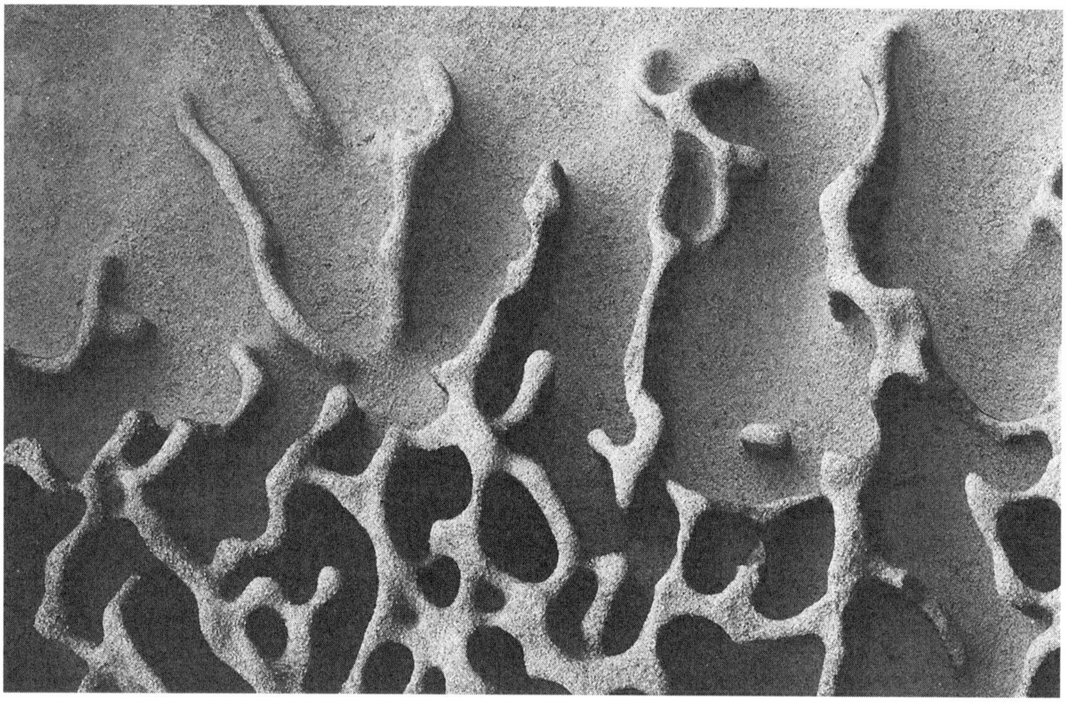

SANDSTONE ERODED BY WAVES. *(© Stephen Parker/Photo Researchers. Reproduced by permission.)*

THE SPHERES IN BRIEF

The geosphere itself may be defined as the upper part of the planet's continental crust, the portion of the solid earth on which human beings live, which provides them with most of their food and natural resources. Even with the exclusion of the mantle and core, the solid earth portion of Earth's system is still by far the most massive. It is estimated that the continental and oceanic crust to a depth of about 1.24 mi. (2 km) weighs 6×10^{21} kg—about 13,300 billion billion pounds. The mass of the biosphere, by contrast, is about one millionth that figure. If the mass of all four spheres were combined, the geosphere would account for 81.57%, the hydrosphere 18.35%, the atmosphere 0.08%, and the biosphere a measly 0.00008%. (Of that last figure, incidentally, animal life—of which humans are, of course, a very small part—accounts for less than 2%.)

Not only is the geosphere the largest, it is also by far the oldest of the spheres. Its formation dates back about four billion years, or within about 0.5 billion years of the planet's formation. As Earth cooled after being formed from the gases surrounding the newborn Sun, its components began to separate according to density. The heaviest elements, such as iron and nickel, drifted toward the core, while silicon rose to the surface to form the geosphere.

ATMOSPHERE, HYDROSPHERE, AND BIOSPHERE. In that distant time Earth had an atmosphere in the sense that there was a blanket of gases surrounding the planet, but the atmospheric composition was quite different from today's mixture of nitrogen (78%), oxygen (21%), argon (0.93%), and other substances that include water vapor, carbon dioxide, ozone, and noble gases such as neon, which together comprise 0.07%. The atmosphere then consisted largely of carbon dioxide from Earth's interior as well as gases brought to Earth by comets. Elemental hydrogen and helium escaped the planet, and much of the carbon was deposited in what became known as carbonate rocks. What remained was a combination of hydrogen compounds, including methane, ammonia, nitrogen- and sulfur-rich compounds expelled by volcanoes, and (most important of all) H_2O, or water.

Simultaneous with these developments, the gases of Earth's atmosphere cooled and condensed, taking the form of rains that, over millions of years, collected in deep depressions on the planet's surface. This was the beginning of the oceans, the largest but far from the only component of Earth's hydrosphere, which consists of all the planet's water except for water vapor in the atmosphere. Thus, the hydrosphere includes not only saltwater but also lakes, streams, groundwater, snow, and ice.

Water, of course, is necessary to life, and it was only after its widespread appearance that the first life-forms appeared. This was the beginning of the biosphere, which consists of all living organisms as well as any formerly living material that has not yet decomposed. (Typically, following decomposition an organism becomes part of the geosphere.) Over millions of years, plants formed, and these plants gradually began producing oxygen, helping to create the atmosphere as it is known today—an example of interaction between the open systems that make up the larger Earth system.

REAL-LIFE APPLICATIONS

EARTH AS AN ORGANISM

Clearly, a great deal of interaction occurs between spheres and has continued to take place for a long time. Earth often is described as a living organism, a concept formalized in the 1970s by the English meteorologist James Lovelock (1919–) and the American biologist Lynn Margulis (1938–), who developed the Gaia hypothesis. Sometimes called the Gaian hypothesis, this principle is named after the Greek earth goddess, a prototype for "Mother Earth," and is based on the idea that Earth possesses homeostatic or self-regulating mechanisms that preserve life. (Lovelock's neighbor William Golding [1911–1993], author of *Lord of the Flies,* suggested the name to him.)

Though the Gaia hypothesis seems very modern and even a bit "New Age" (that is, relating to a late twentieth-century movement that incorporates such themes as concern for nature and spirituality), it has roots in the ideas of the great Scottish geologist James Hutton (1726–1797), who described Earth as a "superorganism." A forward-thinking person, Hutton maintained that physiology provides the model for the study of Earth systems. Out of Hutton's and, later, Lovelock's ideas ultimately grew the earth science specialty of geophysiology, an interdisciplinary approach incorporating aspects of geochemistry, biology, and other areas.

The Gaia hypothesis is far from universally accepted, however, and remains controversial. One reason is that it seems to contain a teleologic, or goal-oriented, explanation of physical behaviors that does not fully comport with the findings of science. An animal responds to external conditions in such a way as to preserve life, but this is because it has instinctive responses "hardwired" into its brain. Clearly, if the Earth is an "organism, " it is an organism in quite a different sense than an animal, since it does not make sense to describe Earth as having a "brain."

HOMEOSTASIS AND CYCLES

Nonetheless, Lovelock, Margulis, and other supporters of the Gaia hypothesis have pointed to a number of anomalies that have yet to be explained fully and for which the Gaia hypothesis offers one possible solution. For example, it would have taken only about 80 million years for the present levels of salt in Earth's oceans to have been deposited there from the geosphere; why, then, is the sea not many, many times more salty than it is? Could it be that Earth has somehow regulated the salinity levels in its own seas?

Earth's systems unquestionably display a homeostatic and cyclical behavior typical of living organisms. Just as the human body tends to correct any stresses imposed on it, Earth likewise seeks equilibrium. And just as blood, for instance, cycles through the body's circulatory system, so matter and energy move between various spheres in the course of completing certain cycles of the Earth system. These include the energy and hydrologic cycles; a number of biogeochemical cycles, such as the carbon and nitrogen cycles; and a rock cycle of erosion, weathering, and buildup. (Each of these systems is discussed in a separate essay, or as part of a separate essay, in this book.)

FEEDBACK. Though particulars of the Gaia hypothesis remain a matter of question, it is clear that Earth regulates these cycles and does so through a process of feedback and corrections. To appreciate the idea of feedback, consider a financial example. In the early 1990s, the U.S. Congress placed a steep tax on luxury boats, presumably with the aim of getting more money from wealthy taxpayers. The result, however, was exactly the opposite: boat owners sold their crafts, and many of those considering purchases cancelled their plans. Rather than redistributing

AN OIL-COVERED BIRD, VICTIM OF THE 1989 EXXON VALDEZ'S OIL SPILL IN PRINCE WILLIAM SOUND, ALASKA. *(AP/Wide World Photos. Reproduced by permission.)*

wealth from the rich to those less fortunate, the tax resulted in the government's actually getting *less* money from rich yacht owners.

Whereas Congress expected the rich to provide positive feedback by giving up more tax money, instead the yacht owners responded by acting against the tax—a phenomenon known as negative feedback. Feedback itself is the return of output to a system, such that it becomes input which then produces further output. Feedback that causes the system to move in a direction opposite that of the input is negative feedback, whereas positive feedback is that which causes the system to move in the same direction as the input. The luxury tax would have made perfect sense if the purpose had been to halt the production and purchase of expensive boats, in which case the output would have been deemed positive.

In the luxury-tax illustration, negative feedback is truly "negative" in the more common sense of the word, but this is not typically the case where nature in general or Earth systems in particular are concerned. In natural systems negative feedback serves as a healthy corrective and tends to stabilize a system. To use an example from physiology, if a person goes into a cold environment, the body responds by raising the internal temperature. Likewise, in chemical reactions the system tends to respond to any stress placed on it by reducing the impact of the stress, a concept known as Le Châtelier's principle after the French chemist Henry Le Châtelier (1850–36).

Positive feedback, on the other hand, is often far from "positive" and is sometimes described as a "vicious cycle." Suppose rainwater erodes a portion of a hillside, creating a gully. Assuming the rains continue, the opening of this channel for the water facilitates the introduction of more water and therefore further erosion of the hillside. Given enough time, the rain can wash a deep gash into the hill or even wash away the hill entirely.

FAR-REACHING CONSEQUENCES

Given the interconnectedness of systems on Earth, it is easy to see how changes in one part of the larger Earth system can have far-reaching impacts on another sector. For example, the devastating Alaska earthquake of March 1964 produced tsunamis felt as far away as Hawaii, while the Exxon *Valdez* oil spill that afflicted Alaska exactly 25 years later had an effect on the biosphere and hydrosphere over an enormous area.

El Niño is a familiar example of far-reaching consequences produced by changes in Earth sys-

KEY TERMS

ATMOSPHERE: A blanket of gases surrounding Earth and consisting of nitrogen (78%), oxygen (21%), argon (0.93%), and other substances that include water vapor, carbon dioxide, ozone, and noble gases such as neon, which together comprise 0.07%.

BIOSPHERE: A combination of all living things on Earth—plants, mammals, birds, reptiles, amphibians, aquatic life, insects, viruses, single-cell organisms, and so on—as well as all formerly living things that have not yet decomposed. Typically, after decomposing, a formerly living organism becomes part of the geosphere.

CLOSED SYSTEM: A system that permits the exchange of energy with its external environment but does not allow matter to pass between the environment and the system. Compare with isolated system, on the one hand, and open system, on the other.

CONSERVATION: In physics and other sciences, "to conserve" something means "to result in no net loss of" that particular component. It is possible that within a given system, the component may change form or position, but as long as the net value of the component remains the same, it has been conserved.

ELECTROMAGNETIC ENERGY: A form of energy with electric and magnetic components, which travels in waves and, depending on the frequency and energy level, can take the form of long-wave and short-wave radio; microwaves; infrared, visible, and ultraviolet light; x rays; and gamma rays.

ENVIRONMENT: In discussing systems, the term *environment* refers to the surroundings—everything external to and separate from the system.

FEEDBACK: The return of output to a system, such that the output becomes input that produces further output. Feedback that causes the system to move in a direction opposite to that of the input is negative feedback, whereas positive feedback is that which causes the system to move in the same direction as the input.

GAIA HYPOTHESIS: The concept, introduced in the 1970s, that Earth behaves much like a living organism, possessing self-regulating mechanisms that preserve life. Sometimes called the Gaian hypothesis, it is named after Gaia, the Greek goddess of the earth.

GEOSPHERE: The upper part of Earth's continental crust, or that portion of the solid earth on which human beings live and which provides them with most of their food and natural resources.

HOMEOSTASIS: A tendency toward equilibrium.

HOMEOSTATIC: The quality of being self-regulating.

HYDROSPHERE: The entirety of Earth's water, excluding water vapor in the atmosphere but including all oceans, lakes, streams, groundwater, snow, and ice.

ISOLATED SYSTEM: A system that is so fully separated from the rest of the universe that it exchanges neither matter nor energy with its environment. This is an imaginary construct, since full isolation is impossible.

OPEN SYSTEM: A system that allows complete, or near-complete, exchange of matter and energy with its environment.

KEY TERMS CONTINUED

SCIENTIFIC METHOD: A set of principles and procedures for systematic study that includes observation; the formation of hypotheses, theories, and laws; and continual testing and reexamination.

SYSTEM: Any set of interactions that can be set apart mentally from the rest of the universe for the purposes of study, observation, and measurement.

TSUNAMI: A tidal wave produced by an earthquake or volcanic eruption. The term comes from the Japanese words for "harbor" and "wave."

tems. Spanish for "child" (because it typically occurs around Christmastime), El Niño begins on the western coast of South America. There, every few years, trade winds slacken, allowing the wind from the west to push warm surface water eastward. Lacking vital nutrients, this warm water brings about a decline in the local marine life. It also causes heavy rains and storms.

IMPACT OF EL NIÑO AROUND THE WORLD. To the extent described, El Niño is largely a local phenomenon. But it can affect the jet streams, or high-level winds, that push storms across the Western Hemisphere. This can result in milder weather for western Canada or the northern United States, as the winds push more severe storms into Alaska, but it also can bring about heavy rains in the Gulf of Mexico region. Nor are its effects limited to the Western Hemisphere. El Niño has been known to alter the pattern of monsoons, or rainy seasons, in India, Southeast Asia, and parts of Africa, thus producing crop failures that affect millions of people.

Aside from the indirect effects, such as the famines in the Eastern Hemisphere, the direct effects of the El Niño phenomenon can be devastating. The El Niño of 1982–83, which affected the United States, the Caribbean, western South America, Africa, and Australia, claimed some 2,000 lives and cost about $13 billion in property damage. It returned with a vengeance 15 years later, in 1997–98, killing more than 2,100 people and destroying $33 billion worth of property.

Years Without Summer

Whereas El Niño is an example of a disturbance in the hydrosphere that affects the atmosphere and ultimately the biosphere, an even more terrifying phenomenon can begin with an eruption in the geosphere, which spreads to the atmosphere and then the hydrosphere and biosphere. This phenomenon might be called "years without summer"; an example occurred in 1815–16.

In June of 1816 snow fell in New England, and throughout July and August temperatures hovered close to freezing. Frosts hit in September, and New Englanders braced themselves for an uncommonly cold winter, as that of 1816–17 turned out to be. It must have seemed as though the world were coming to an end, yet the summer of 1817 proved to be a normal one. The cause behind this year without summer in 1816 lay in what is now Indonesia, and it began a year earlier.

In 1815, Mount Tambora to the east of Java had erupted, pouring so much volcanic ash into the sky that it served as a curtain against the Sun's rays, causing a brutally cold summer in New England the following year. An eruption of Mount Katmai in Alaska in 1912 produced far-reaching effects, including some lowering of temperatures, but its impact was nothing like that of Tambora. Nor did the 1980 Mount Saint Helens eruption in Washington State prove nearly as potent in the long run as the eruption of Tambora did (though it produced a devastating immediate impact).

THE CATACLYSM OF A.D. 535. Even the eruption of Mount Tambora may have been overshadowed by another, similar event, known simply as the catastrophe, or cataclysm, of A.D. 535. In the late twentieth century, the British dendrochronologist Mike Baillie discovered a pattern of severely curtailed growth in tree rings dating to the period A.D. 535–541. More or less simultaneous with Baillie's work was that of the amateur archaeologist David Keys, who found a number of historical texts by Byzantine, Chinese, and Anglo-Saxon scholars of the era, all suggesting that something cataclysmic had happened in A.D. 535. For example, the Byzantine historian Procopius (d. 565) wrote, "The sun gave forth its light without brightness ... for the whole year."

Some geologists have maintained that the cataclysm resulted from the eruption of another Indonesian volcano, the infamous Krakatau, which had a devastating eruption in 1883 and which could have produced enough dust to cause an artificial winter. Whatever the cause, the cataclysm had an enormous impact that redounds from that time perhaps up to the present. The temperature drop may have sparked a chain of events, beginning in southern Africa, that ultimately brought a plague to the Byzantine Empire, forcing Justinian I (r. A.D. 483–565) to halt his attempted reconquest of western Europe. At the same time, the cataclysm may have been responsible for food shortages in central Asia, which spawned a new wave of European invasions, this time led by the Avars.

The result was that the fate of Europe was sealed. For a few years it had seemed that Justinian could reconquer Italy, thus reuniting the Roman Empire, whose western portion had ceased to exist in A.D. 476. Forced to give up their reconquest, with the Avars and others overrunning Europe while the plague swept through Greece, the Byzantines turned their attention to affairs at home and increasingly shut themselves off from western Europe. Thus the Dark Ages, the split between Catholicism and Eastern Orthodoxy, the Crusades—even the Cold War, which reflected the old east-west split in Europe—may have been the results of a volcano on the other side of the world.

WHERE TO LEARN MORE

Cox, Reg, and Neil Morris. *The Natural World.* Philadelphia: Chelsea House, 2000.

Earth's Energy Budget (Web site). <http://radar.metr.ou.edu/OK1/meteorology/EnergyBudget.html.>.

Farndon, John. *Dictionary of the Earth.* New York: Dorling Kindersley, 1994.

The Gaia Hypothesis (Web site). <http://www.magna.com/au/~prfbrown/gaia_jim.html>.

Geophysiology Online (Web site). <http://www.webtorre.com/geo/GeoPhysOnline.asp>.

Hancock, Paul L., and Brian J. Skinner. *The Oxford Companion to the Earth.* New York: Oxford University Press, 2000.

Knapp, Brian J. *Earth Science: Discovering the Secrets of the Earth.* Illus. David Woodroffe and Julian Baker. Danbury, CT: Grolier Educational, 2000.

Kump, Lee R., James F. Kasting, and Robert G. Crane. *The Earth System.* Upper Saddle River, NJ: Prentice Hall, 2000.

Lovelock, J. E. *Gaia: A New Look at Life on Earth.* New York: Oxford University Press, 2000.

Skinner, Brian J., Stephen C. Porter, and Daniel B. Botkin. *The Blue Planet: An Introduction to Earth System Science.* 2nd. ed. New York: John Wiley and Sons, 1999.

THE STUDY OF EARTH

STUDYING EARTH

CONCEPT

The physical sciences include astronomy, physics, chemistry, and the earth sciences, but the last of these sciences is quite unlike the other three. Whereas the objects of study in physics and chemistry often seem abstract to the uninitiated and astronomy is concerned with faraway planets and other bodies, the earth sciences are devoted to things that are both concrete and immediate. The focus of study for the earth sciences is literally underneath our feet, a planet at once vast and tiny, a world that (as far as we know) stands alone in the universe as the sole supporter of intelligent life. The earth sciences also differ from other disciplines in that their boundaries are not always defined clearly. The study of Earth is a multifarious array of specialties that includes a range of geologic, hydrologic, and atmospheric sciences that overlap with the other physical sciences, biology, and even the social sciences.

HOW IT WORKS

INTRODUCTION TO THE EARTH SCIENCES

At the simplest level, the earth sciences can be divided into three broad areas: the geologic, hydrologic, and atmospheric sciences. These specialties fit neatly with three of the "spheres," or subsystems within the larger Earth system: geosphere, hydrosphere, and atmosphere. (See Earth Systems for more about the spheres.) The fourth of these subsystems is the biosphere, and this illustrates the difficulty of stating exactly what is and what is not a part of the earth sciences.

Usually the earth sciences are considered part of the physical sciences, as opposed to the biological sciences, such as biology, botany, and zoology. Yet the earth sciences clearly overlap with biological sciences in a variety of areas, such as oceanography and various studies of complex biological environments. There is even a new (and as yet not fully formalized) discipline called geophysiology, built on the premise that Earth has characteristics of a living organism.

OVERLAP WITH OTHER PHYSICAL SCIENCES. The earth sciences also overlap with other physical sciences in several realms. There is geophysics, which addresses the planet's physical processes, including its magnetic and electric properties and the means by which energy is transmitted through its interior. There is also geochemistry, which is concerned with the chemical properties and processes of Earth. And there are numerous areas of confluence between the earth sciences and astronomy (among them, planetary geology), which fall under the heading of planetary science (sometimes called planetology or planetary studies).

These terms all refer to the same discipline, a branch of the earth sciences concerned with the study of other planetary bodies. This discipline, or set of disciplines, is concerned with the geologic, geophysical, and geochemical properties of other planets but also draws on aspects of astronomy, such as cosmology. Regardless of the name by which it is called, planetology is an example of the fact that the study of Earth is still very much an evolving set of disciplines. In many cases, the earth sciences are still in process of being defined.

SCIENTIFIC PARADIGMS

This last point is an important one to consider because of what it implies about the nature of scientific study. In the past, scientists tended to think that they were in the business of discovering some sort of objective truth that was waiting for them to discover it; in reality, however, the quest of the scientific thinker is much less guided. The natural world does not in any way speak to the scientist, telling him or her how to categorize data. In fact, the divisions of scientific knowledge with which we are familiar have come about not because they necessarily reflect an underlying truth, but because they have proved useful in separating certain aspects of the physical world from certain others.

When science had its beginnings in ancient times, scientists were simply collecting observations (including a lot of incorrect ones) and sometimes forming theories of a sort, but they did not think in terms of developing models for viewing their objects of study. Today, however, scientific thinkers are acutely conscious of the model, or paradigm, that governs a particular discipline, school of thought, or theory.

A paradigm may be likened to a lens. The lens does not change the actual object that is viewed through it; it can alter only the way in which it is viewed. As thinkers within a particular discipline or theory begin to define the governing paradigm, they are much like an eye doctor testing various lenses on a patient. In such a situation, there is no one lens that is right for all circumstances. Rather, it is a question of finding the lens that best suits the patient's vision needs.

All sciences are gradually changing, evolving models that better suit the data under their consideration. Chemistry, for instance, was once primarily a matter simply of mixing chemicals and observing their external processes. In fact, the definition of chemistry has expanded greatly since about 1800, and today it is more like what people tend to think of as physics; that is, it is concerned with atomic and subatomic structures and types of behavior. The earth sciences are in an even more transitional state, and the problem of defining the disciplines it comprises is a still more fundamental one.

THE EVOLVING EARTH SCIENCES. In the discussion that follows, we outline the broad parameters of the earth sciences, considering basic areas of study and specialties within them. This does not represent a definitive organizational scheme, nor does this brief review refer to every possible area of study in the wide-ranging earth sciences. To do so would require an entire book; rather, the purpose is to consider the most significant disciplines and subdisciplines.

To appreciate the way that these disciplines fit within the larger perspective, however, it is necessary to examine a few historical details. Most of these details concern the early history of the earth sciences, since much of the more modern history (for example, the development of plate tectonics theory during the 1960s) is treated in the relevant essays within this book. The purpose of this brief historical review, instead, is to impart an understanding about how the study of Earth emerged as a real science, as opposed to a merely descriptive undertaking concerned with recording observations. Important themes are the development of the scientific method as well as the search for proper ways of classifying the various studies under the heading of what became known as the earth sciences.

REAL-LIFE APPLICATIONS

THE SCIENTIFIC METHOD

As discussed later, the scientific method emerged during the seventeenth century and has remained in use ever since. It is a way of looking at facts and data, and its application is what truly separates science from nonscience. Nonscientific "theories" postulate answers based not on evidence but on pure conjecture, a habit of mind that was widespread before the development of the scientific method and is still all too common. A contemporary example would be the claim that intelligent extraterrestrial life-forms built the great pyramids of Egypt.

The only real basis for this belief is the fact that the pyramids are extremely sophisticated architectural achievements for a civilization that had no metal tools, no wheel, and virtually no understanding of geometry, as was the case in Egypt in about 2500 B.C. But is a huge conceptual leap to go from the observation of these anomalies to the claim that visitors from outer space built the pyramids. The same is true of other strange artifacts from ancient or prehistoric times, such as the great structures of Stonehenge in England or the Nazca lines in South America. They are curious, puzzling, and intriguing, and it may be fun to speculate about engineers from another world—but such speculation is *not* science. (For more about the methodological distinctions between science and its opposite, see Earth, Science, and Nonscience.)

It is no mistake that here we are talking about the application of the scientific method to something outside the "hard sciences," the study of the pyramids being the province of social sciences, such as archaeology and history. In fact, the method has had just as much impact in those areas as it has in the physical and biological sciences. The scientific method (along with the closely related philosophical principles of basic logic, handed down from the ancient Greeks) can be applied in many aspects of daily life, enabling a person to make sense of a complex world. Many of the controversies of the modern world, including those involving race, sex, religion, and politics, could be treated more constructively if people approached the topics with a genuine interest in understanding the facts rather than simply finding confirmation for their emotionally based preconceived notions.

APPLYING THE SCIENTIFIC METHOD. Scientists rigorously applying the scientific method begin their quest for understanding by looking solely at the facts that can be garnered by observation. On the basis of these data, they form hypotheses, or unproven statements regarding observed phenomena. This is usually as far as many people go in their thinking, and it is not far enough: up to this point, for instance, the theory that claims that visitors from another planet built the pyramids is in accordance with the scientific method. But such fanciful notions, as well as kindred ideas, such as conspiracy theories, never go to the next step, which marks the dividing line between science and pure opinion.

Having formed a hypothesis, the scientist subjects it to testing, the most critical component of the scientific method. By contrast, advocates of nonscientific ideas (which, of course, usually pose as scientific ideas) typically focus on searching for evidence that will confirm the hypothesis. Anything that supports the hypothesis is reported; anything that does not is simply ignored. By

NICOLAUS COPERNICUS *(Library of Congress.)*

contrast, a true scientist is constantly trying to *disprove* his or her own hypotheses.

If a hypothesis withstands enough repeated testing, it acquires the status of a theory, or a more general statement about nature. If the idea embodied in a theory is shown to be the case in every situation for which it is tested, it then becomes a law. The process is a bit like that involved in making metal stronger: the more abuse it endures, assuming it is able to recover, the more impervious it will be to further abuse. But every type of metal has its limits, a threshold of compression or tension beyond which it cannot retain its original shape, and likewise it is possible that a scientific law can be overturned. For this reason, all laws are subject to continual testing, and if a test disproves a scientific law, this opens the way for the development of a new paradigm.

THE HISTORICAL ROOTS OF THE EARTH SCIENCES

The earth sciences are both old and new. On the one hand, they address matters of fundamental importance to human beings, and for this reason, the rudiments of the earth sciences probably made their appearance before any of the other fields of scientific study except perhaps astronomy. From prehistoric times, societies have been concerned with obtaining metals from the ground to make tools and weapons, finding water to support human life and crops, and discerning the future of weather patterns that could greatly affect conditions for human populations. Thus were born, respectively, the geologic, hydrologic, and atmospheric sciences.

Much of what took place in the earth sciences before about 1800, however, was a matter of superstition, legend, guesswork, and a smattering of real science. Much of it was dominated by religious belief, which relied on a strict interpretation of the Bible. Based on the amount of time that elapsed between Adam and Jesus, combined with the fact that Genesis 1 states that Adam was created at the end of the first week of Earth's existence, the Catholic Church maintained that the planet could not possibly be more than a few thousand years old. (For more on this subject, see Earth, Science, and Nonscience.)

THE ANCIENT EARTH SCIENCES. Despite the many impediments to scientific study in ancient times, a few thinkers contributed significantly to our knowledge. For instance, the Greek philosopher Aristotle (384–322 B.C.) discovered that Earth is a sphere by noting the rounded shadow on the Moon during a lunar eclipse. His pupil, Theophrastus (372?–287? B.C.), wrote a highly competent work, *Concerning Stones,* that remained a guide to mineralogy for two millennia. A few centuries later, the Greek mathematician Eratosthenes of Cyrene (*ca.* 276–*ca.* 194 B.C.) made an astoundingly accurate measurement of Earth's circumference.

Much of what passed for science, however, was little more than entertaining, anecdotal misinformation. Such is the case, for instance, in the *Historia Naturalis* (Natural history) of the Roman scholar Pliny the Elder (A.D. 23–79), a work that, despite its many flaws, remained widely respected through the Renaissance. As for Eratosthenes' measurement of Earth, the Alexandrian astronomer Ptolemy (*ca.* A.D. 100–170) rejected it in favor of a much smaller, much less correct figure. Thus, Ptolemy may deserve some of the credit for discovering the New World: if Christopher Columbus (1451–1506) had known just how far it was around Earth, he might not have been so confident about sailing off into the seas to the west of Europe.

Ptolemy's rejection of Eratosthenes's measurement was far from his only negative contribution to the history of science. Influenced by highly misguided concepts handed down from Aristotle himself (see Earth, Science, and Nonscience), he developed a complex cosmology that depicted Earth as the center of the universe. By this he meant that Earth was the center of the solar system, because up until a few centuries ago, astronomers believed that space consisted only of Earth, Sun, Moon, the five planets visible to the naked eye, and the "fixed stars" in the night sky.

DAWN OF THE SCIENTIFIC METHOD. What made Ptolemy's cosmology so complex, of course, was the fact that Earth is not anywhere near the center of the universe, and therefore his system required intricate mathematical acrobatics to remain workable. This posed little problem during the early Middle Ages, when learning in Europe all but ceased, and even in the much more scientifically progressive Muslim world of that time, Ptolemy's word remained virtual holy writ. By the late Middle Ages, however, as scientific learning returned to Europe, thinkers began to notice increasing difficulties in using his system.

The watershed event in what became known as the Scientific Revolution was the proof, by the Polish astronomer Nicolaus Copernicus (1473–1543), that Earth and the other planets of the solar system revolve around the Sun. By that point, the Catholic Church had given its official approval to Ptolemy's geocentric model, because it comported well with the idea that God had created humankind in his own image to fulfill a specific destiny. Therefore, Copernicus's challenge to established teachings proved highly controversial, and the Italian astronomer Galileo Galilei (1564–1642) would be forced to recant his support of it or face punishment by death.

Yet Galileo paved the way for the full acceptance of Copernicus's work and for the Scientific Revolution that followed in its wake. His theories and experiments concerning gravitational acceleration greatly influenced the English natural philosopher Isaac Newton (1642–1727), leading to the latter's epochal work on gravitation and motion. But at least as important as Galileo's work was his methodology: Galileo virtually introduced the scientific method, providing a set of principles for systematic study.

THE FOUNDATIONS OF MODERN GEOLOGY

The scientific method had an enormous impact on all the sciences. Unlike earlier "scientific" principles, which were built on the teachings of religious prophets or the uninformed conjecture of philosophers, this one was established on a foundation of observation, and it opened the way for unprecedented progress in the sciences.

Until late in the eighteenth century, however, the relatively young field of geology centered primarily on mere observation rather than the development of theories. Thus, the discipline was not all that different from what it had been in ancient times, or when the Anglo-Saxon historian known as the Venerable Bede (673–735) coined the term geology. The latter term, a combination of the Greek *geo* and *logia,* means "study of Earth," and was intended to distinguish such pursuits from theology, or the study of heavenly things.

HISTORICAL AND PHYSICAL GEOLOGY. In modern times, geology is defined as the study of the solid earth, in particular, its rocks, minerals, fossils, and land formations. As for Bede's putative opposition between geology and theology, it would become more pronounced in the period from about 1500 to 1800, as the findings of geologists began increasingly to contradict the teachings in the biblical book of Genesis. Among the first to consider the age of Earth in scientific terms, rather than by recourse to the Scriptures, was one of the world's greatest thinkers: the Italian scientist and artist Leonardo da Vinci (1452–1519), who speculated that fossils might have been made by the remains of long-dead animals.

Less famous was Leonardo's German contemporary Georgius Agricola (1494–1555), the "father of mineralogy," who wrote extensively on mining, metallurgy, and minerals. Together, these two men represent the two principal strains of geology: historical geology, or the study of Earth's history, and physical geology. The latter discipline, of which Agricola was a key representative, is concerned with the material components of Earth and with the forces that have shaped the planet. All the areas of geology discussed here fall under one of those two headings.

Most of the important developments in geology during the period from 1500 to 1800 fall under the heading of historical geology, beginning with a key observation on strata, or layers of rock, made by the Danish geologist Nicolaus Steno (1638–1687). As Steno correctly hypothesized, the lower a layer of rock lies, the earlier the historical period it represents. These observations, later developed into a theory by the German geologist Johann Gottlob Lehmann (1719–1767), had several implications.

First of all, the ideas of Steno and Lehmann provided geologists with a method for dating the age of rock formations not unlike the rings observed by dendrochronologists studying the biography of a tree. As a result of study based on these findings, scientists were confronted with the growing realization that Earth is much, much older than a strict interpretation of the Bible would suggest. This finding, in turn, led to the first theories concerning the shaping of Earth and thus to the foundation of geology as a modern scientific discipline.

THREE IMPORTANT SCHOOLS OF THOUGHT. In the wake of this breakthrough, at least three schools of thought developed. One of them, catastrophism, centered around the foregone conclusion that Earth had been created in six literal days or, at the very least, in an extremely short time, through a series of catastrophes. Opposed to this view was the Neptunist stratigraphy of the German geologist Abraham Gottlob Werner (1750–1817), who maintained that Earth had been shaped by a vast ocean (hence the name *Neptune*) that once covered its entire surface. Finally, there was the Plutonist school of the Scottish geologist James Hutton (1726–1797). Named after the Greek god of the underworld, this theory held that volcanoes and other disturbances beneath Earth's surface had been the principal forces in shaping the planet.

Hutton's theory would prevail, and today he is regarded as the father of modern geology. In *Theory of the Earth* (1795), he introduced one of the key concepts underlying the study of the planet's history, the principle of uniformitarianism—the idea that the forces at work on Earth today have always been in operation and are the same ones that shaped it. Nonetheless, Neptunism and even catastrophism had their merits. Although Werner and his followers were incorrect, Neptunism was the first well-developed theory concerning Earth's origins and helped pave the way for others.

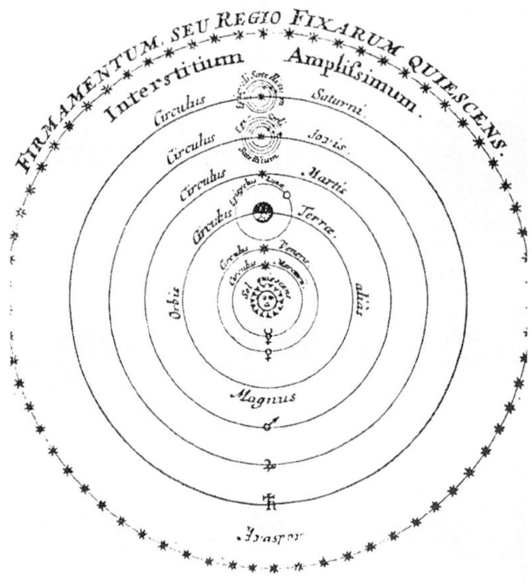

DIAGRAM SHOWING THE SOLAR SYSTEM AS COPERNICUS ENVISIONED IT, WITH THE SUN AT THE CENTER. THE COPERNICAN SYSTEM HERALDED THE START OF THE SCIENTIFIC REVOLUTION. *(© Dr. Jeremy Burgess/Photo Researchers. Reproduced by permission.)*

As for the advocates of catastrophism, they were correct inasmuch as they noted the role of sudden catastrophes in shaping the planet. These catastrophes (for instance, a comet about 66 million years ago, which may have destroyed the dinosaurs), however, can be explained within the framework of a very old Earth. Nor does the fact of the catastrophes themselves in any way suggest a planet that is only a few thousand years old.

LATER DEVELOPMENTS. As noted earlier, the later history of the earth sciences is discussed more properly within the context of specific subjects. Instead, our focus here is on the array of disciplines that proliferated alongside geology and on the need for a disciplinary paradigm larger than that of geology alone. By the mid–twentieth century, the range of disciplines involved in the study of Earth had become so complex and varied that it was a major achievement when the English geologist Arthur Holmes (1890–1965) developed a model that incorporated most of them. Holmes's system was not simply a "model" in the way that the term typically is used; Holmes also constructed a literal diagram that enabled students to visualize the relationships between subdisciplines.

Holmes's model was concerned primarily with the solid earth sciences, or the geologic sciences, meaning that it did not include the hydrologic sciences. Within its purview, however, it used a method of classification so broad (yet still targeted) that it has been adapted in recent years to include subdisciplines developed since Holmes's time. These changes serve to emphasize further the evolving nature of what came to be known as the earth sciences. The latter term came into use only during the 1960s and 1970s, when it became apparent that neither geology alone nor even a combination of geology, geophysics, and geochemistry could encompass all the areas of study devoted to Earth.

OVERVIEW OF THE EARTH SCIENCES

Throughout most of what remains of this essay, we very briefly sketch the outlines of the earth sciences. It should be reiterated that the organizational system used here is not necessarily definitive and is intended only to provide the reader with a general idea as to how the various earth sciences fit together.

GEOLOGY. At the core of the earth sciences, of course, is geology itself, which focuses on the study of the solid earth. As noted earlier, geology can be subdivided into historical and physical geology. The principle subdisciplines of historical geology are as follows.

- Stratigraphy: the study of rock layers, or strata, beneath Earth's surface
- Geochronology: the study of Earth's age and the dating of specific formations in terms of geologic time
- Sedimentology: the study and interpretation of sediments, including sedimentary processes and formations
- Paleontology: the study of fossilized plants and animals, or flora and fauna
- Paleoecology: the study of the relationship between prehistoric plants and animals and their environments.

Note that there are several other disciplines referred to by the prefix *paleo-* (or *palaeo-*), Greek for "very old." Two of the more well-known ones are paleobiology and paleobotany, but the subdisciplines can become very specialized, as evidenced by the existence of a field

KEY TERMS

BIOSPHERE: A combination of all living things on Earth—plants, mammals, birds, reptiles, amphibians, aquatic life, insects, viruses, single-cell organisms, and so on—as well as all formerly living things that have not yet decomposed. Typically, after decomposing, a formerly living organism becomes part of the geosphere.

COSMOLOGY: The study of the origin, structure, and evolution of the universe.

ECONOMIC GEOLOGY: The study of fuels, metals, and other materials from Earth that are of interest to industry or the economy in general.

GEOCENTRIC: Earth-centered.

GEOCHEMISTRY: A branch of the earth sciences, combining aspects of geology and chemistry, that is concerned with the chemical properties and processes of Earth.

GEOCHRONOLOGY: The study of Earth's age and the dating of specific formations in terms of geologic time.

GEOLOGY: The study of the solid earth, in particular, its rocks, minerals, fossils, and land formations.

GEOMORPHOLOGY: The study of landforms and of the forces and processes that have shaped them.

GEOPHYSICS: A branch of the earth sciences that combines aspects of geology and physics. Geophysics addresses the planet's physical processes as well as its magnetic and electric properties and the means by which energy is transmitted through its interior.

GEOSPHERE: The upper part of Earth's continental crust or that portion of the solid earth on which human beings live and which provides them with most of their food and natural resources.

HISTORICAL GEOLOGY: The study of Earth's physical history. Historical geology is one of two principal branches of geology, the other being physical geology.

HYDROSPHERE: The entirety of Earth's water, excluding water vapor in the atmosphere but including all oceans, lakes, streams, groundwater, snow, and ice.

HYPOTHESIS: An unproven statement regarding an observed phenomenon.

LAW: A scientific principle that is shown to always be the case and for which no exceptions are deemed possible.

MINERALOGY: The study of minerals (crystalline structures that make up rocks), which includes several smaller subdisciplines, such as crystallography.

PALEONTOLOGY: The study of fossilized plants and animals, or flora and fauna.

PHYSICAL GEOLOGY: The study of the material components of Earth and of the forces that have shaped the planet. Physical geology is one of two principal branches of geology, the other being historical geology.

PHYSICAL SCIENCES: Astronomy, physics, chemistry, and the earth sciences.

PLANETARY SCIENCE: The branch of the earth sciences, sometimes called planetology or planetary studies, that focuses on the study of other planetary bodies. This discipline, or set of disciplines, is concerned with the geologic, geophysical, and geochemical properties of other

KEY TERMS CONTINUED

planets but also draws on aspects of astronomy, such as cosmology.

SCIENTIFIC METHOD: A set of principles and procedures for systematic study that includes observation; the formation of hypotheses, theories, and laws; and continual testing and reexamination.

SCIENTIFIC REVOLUTION: A period of accelerated scientific discovery that completely reshaped the world. Usually dated from about 1550 to 1700, the Scientific Revolution saw the origination of the scientific method and the introduction of ideas such as the heliocentric (Sun-centered) universe and gravity.

SEDIMENTOLOGY: The study and interpretation of sediments, including sedimentary processes and formations.

STRATIGRAPHY: The study of rock layers, or strata, beneath Earth's surface.

STRUCTURAL GEOLOGY: The study of rock structures, shapes, and positions in Earth's interior.

THEORY: A general statement derived from a hypothesis that has withstood sufficient testing.

known as paleobiogeography, or the study of fossils' geographic distribution. The principle subdisciplines of physical geology are:

- Geomorphology: the study of landforms and of the forces and processes that have shaped them
- Structural geology: the study of rock structures, shapes, and positions in Earth's interior
- Mineralogy: the study of minerals (crystalline structures that make up rocks), which includes several smaller subdisciplines, such as crystallography
- Petrology: the study of rocks, which is divided into several smaller subdisciplines, most notably igneous, metamorphic, and sedimentary petrology
- Economic geology: the study of fuels, metals, and other materials from Earth that are of interest to industry or the economy in general
- Environmental geology: the study of the geologic impact of both natural and human activity on the environment.

It should be noted that there is some overlap between historical and physical geology. For instance, sedimentology often is placed under the heading of physical geology, while some sources include a third category of subdisciplines that overlap both historical and physical geology.

OTHER GEOLOGIC SCIENCES.

Geology occupies a central place among the geologic sciences or geosciences, but also important are those disciplines and subdisciplines formed, as Holmes pointed out, at the intersections between geology and astronomy, physics, and chemistry, respectively. (Some sources, on the other hand, consider these disciplines to be a part of geology itself. In the present context, the term geologic sciences is used to encompass not only geology but also these related areas of study.)

Planetary science applies the earth sciences paradigm to other planets. Among its important subdisciplines is astrogeology or planetary geology, or the study of the rock record on the Moon, the planets, and other bodies. Also significant is cosmology, the study of the origin, structure, and evolution of the universe, which often is treated as part of astronomy.

Geophysics, or an application of physics to the study of Earth, occupies a position of prominence within the earth sciences. Among the areas it addresses are the production, expenditure, and transmission of energy within Earth as well as the planet's magnetic, electric, and gravitational properties. Geophysics encompasses such areas as geodesy, the science of measuring Earth's shape and gravitational field. Seismology, or the study of the waves produced by earthquakes and volcanoes, is another important part of geophysics. (On the other hand, volcanology, or the study of volcanoes themselves, would fall more properly under physical geology.)

Geochemistry, which is concerned with the chemical properties and processes of Earth, covers a wide array of natural phenomena—from radioactive isotopes in the ground to life-forms in the biosphere. Under the heading of geochemistry fall several biogeochemical processes, such as the carbon cycle, whose study brings together aspects of the physical sciences geology and chemistry as well as various life sciences.

OTHER EARTH SCIENCES. The hydrological sciences are concerned with the hydrosphere and its principal component, water. These disciplines include hydrology, the study of the water cycle; glaciology, the study of ice in general and glaciers in particular; and oceanography. Clearly, oceanography overlaps with the life sciences; likewise, hydrogeology (the study of groundwater), as its name implies, overlaps with geology.

The atmospheric sciences, obviously, are devoted to the atmosphere. Most notable among these sciences is meteorology, the study of weather patterns, and climatology, the study of temperature and climate. (Paleoclimatology is an important subdiscipline of historical geology.) The atmospheric sciences also are concerned with phenomena ranging from pollution to the optical effects created by the interaction of the Sun's rays with the atmosphere.

Finally, there are miscellaneous areas of study that either are interdisciplinary or cross boundaries between the earth sciences and the social sciences. In the former category, for instance, would be environmental studies that involve aspects of the biosphere, atmosphere, geosphere, and hydrosphere. Examples of the second category are paleoarchaeology, the study of the earliest humans and humanoid forms, and, of course, geography. Also included in this group are such intriguing areas as urban geology, a branch of environmental geology concerned with human settlements.

WHERE TO LEARN MORE

Allaby, Ailsa, and Michael Allaby. *A Dictionary of Earth Sciences.* 2d ed. New York: Oxford University Press, 1999.

Athro: Your Source for High School and College Level Biology, Earth Science, and Geology on the Web (Web site). <http://www.athro.com/>.

Cox, Reg, and Neil Morris. *The Natural World.* Philadelphia: Chelsea House Publishers, 2000.

Dasch, E. Julius. *Earth Sciences for Students.* New York: Macmillan Reference USA, 1999.

Geology Entrance (Web site). <http://www.ucmp.berkeley.edu/exhibit/geology.html>.

Hancock, Paul L., and Brian J. Skinner. *The Oxford Companion to the Earth.* New York: Oxford University Press, 2000.

Illustrated Glossary of Geologic Terms (Web site). <http://www.geology.iastate.edu/new_100/glossary.html>.

Jobs in Earth Sciences (Web site). <http://geology.com/jobs.htm>.

Knapp, Brian J., David Woodroffe, and Julian Baker. *Earth Science: Discovering the Secrets of the Earth.* Danbury, CT: Grolier Educational, 2000.

Skinner, Brian J., Stephen C. Porter, and Daniel B. Botkin. *The Blue Planet: An Introduction to Earth System Science.* 2d ed. New York: John Wiley and Sons, 1999.

Stace, Alexa. *Atlas of Earth.* Milwaukee, WI: Gareth Stevens Publishing, 2000.

MEASURING AND MAPPING EARTH

CONCEPT

Today a sharp distinction exists between the earth sciences and geography, but this has not always been the case. In ancient times, when scientists lacked the theoretical or technological means to study Earth's interior, the two disciplines were linked much more closely. Even in the centuries since these disciplines parted ways, the earth sciences have continued to benefit from a foundation established in part by early geographers, whose work informed the geophysical subdiscipline of geodesy. Like geographers, earth scientists are interested in measuring and mapping Earth, though their interests are quite different. Among the areas of concern to earth scientists are the location of underground resources and the obtaining of data on the planet's gravitational and magnetic fields. In these and other pursuits, earth scientists use a number of techniques and technologies, ranging from the ancient discipline of surveying to the most modern forms of satellite-based remote sensing.

HOW IT WORKS

Geography and Geology

Whereas geology is the study of the solid earth, including its history, structure, and composition, geography is the study of Earth's surface. Geologists are concerned with the grand sweep of the planet's history over more than four billion years, whereas the work of geographers addresses the here and now—or at most, in the case of historical geography, a span of just a few thousand years.

Today the distinctions between these two disciplines are sharp, so much so that most books on the earth sciences barely even mention geography. In a modern university, chances are that the geography and geology/earth sciences departments will not even be located in the same building. Geography, after all, usually is classified among social sciences such as anthropology or archaeology, whereas geology is a "hard science," along with physics or biology.

It is true that the divisions between geography and geology are clear, symbolized by the features that appear or do not appear on the maps used by either discipline. Geography is somewhat concerned with natural features, but its interests include man-made boundaries, points, and such formations as population centers, roads, and so on. Certainly, an atlas may include physical maps, which are dominated by natural features and contain little or no evidence of man-made demarcations or points of interest. Nonetheless, the purpose of a geographical atlas is to identify locations of interest to humans, among them, cities, roads from one place to another, and borders that must be crossed.

By contrast, geologic maps contain detailed information about rock formations and other natural features, with virtually nothing to indicate the presence of humans except as it relates to natural features under study. An exception might be a map designed to be used by paleoarchaeologists, who study the earliest humans and humanoid forms. Their discipline, which combines aspects of the earth sciences and archaeology, is concerned with human settlements, but mostly only prehistoric human settlements.

Despite the depth and breadth of distinctions between them, it is significant that studies in both geography and geology make use of maps. Mapmaking, or cartography, is considered a subdiscipline of geography, yet it is actually an interdisciplinary pursuit (much like many of the earth sciences—see Studying Earth) and combines aspects of science, mathematics, technology, and even art. Although their interests are in most cases quite different from those of geographers, geologists rely heavily on the work of cartographers.

EARLY GEOGRAPHIC STUDIES

The history of the sciences has been characterized by the continual specialization and separation of disciplines. Thus, it should not be surprising to discover that to the ancients, the lines were blurred between geography, mathematics, astronomy, and what people today would call earth sciences. Most of the early advances in the study of Earth involved all of those disciplines, an example being the remarkable estimate of Earth's size made by Eratosthenes of Cyrene (*ca.* 276–*ca.* 194 B.C.).

A mathematician and librarian at Alexandria, Egypt, Eratosthenes discovered that at Syene, several hundred miles south along the Nile near what is now Aswan, the Sun shone directly into a deep well and upright pillars cast no shadow at noon on the summer solstice (June 21). By using the difference in angles between the Sun's rays in both locations as well as the distance between the two towns, he calculated Earth's circumference at about 24,662 mi. (39,459 km). This figure is amazingly close to the one used today: 24,901.55 mi. (39,842.48 km) at the equator. Eratosthenes published his results in a book whose Greek name, *Geographica* (Geography), means "writing about Earth." This was the first known use of the term.

PTOLEMY'S GEOGRAPHY. Unfortunately, the Alexandrian astronomer Ptolemy (*ca.* A.D. 100–170), one of the most influential figures of the ancient scientific world, rejected Eratosthenes' calculations and performed his own, based on faulty information. The result was a wildly inaccurate estimate of 16,000 mi. (25,600 km). More than thirteen centuries later, Christopher Columbus (1451–1506) relied on Ptolemy's figures rather than those of Eratosthenes, whose work was probably unknown to him. Thinking that the circumference of Earth was two-thirds what it actually is, Columbus set sail westward from Spain—something he might not have done if he had known about the "extra" 8,000 mi. (12,874 km) that lay to the west of Europe.

Nonetheless, in his *Hyphegesis geographike* (Guide to geography), Ptolemy did make useful contributions to geographical study. He helped popularize the use of latitude and longitude lines, first conceived by the Greek astronomer and mathematician Hipparchus (*fl.* 146–127 B.C.), and rejected the widespread belief that a vast ocean, known to the ancients as "the Ocean Sea," surrounded the entire world. He also presented a set of workable mathematical principles for representing the spherical surface of Earth on a flat page, always a problem for cartographers.

In addition, Ptolemy established the practice of orienting maps with north at the top of the page. Today this is taken for granted, but in his time cartographers depicted the direction of the rising Sun, east, at the top of their maps. Ptolemy used a northward orientation because the Mediterranean region that he knew extended twice as far east to west as it did north to south. To represent the area on a scroll, the form in which books appeared during his time, it was easier to make maps with north at the top.

OTHER ANCIENT GEOGRAPHERS. Most other ancient geographers of note were primarily historians rather than scientists, preeminent examples being the Greek Herodotus (*ca.* 484–*ca.* 430–420 B.C.), the father of history, and Strabo (*ca.* 64 B.C.–*ca.* A.D. 24), whose work provides some of the earliest Western descriptions of India and Arabia. An exception was Pomponius Mela (*fl. ca.* A.D. 44), whose work would fit into the subdiscipline known today as physical geography, concerned with the exterior physical features and changes of Earth.

In his three-volume *De situ orbis,* (A description of the world), Mela introduced a system of five temperature zones: northern frigid, northern temperate, torrid (very hot), southern temperate, and southern frigid. Unlike many scientific works of antiquity, Mela's geography has remained influential well into modern times, and his idea of the five temperature zones remains in use. Mela and Ptolemy, who followed him by about a century, were among the last geographers

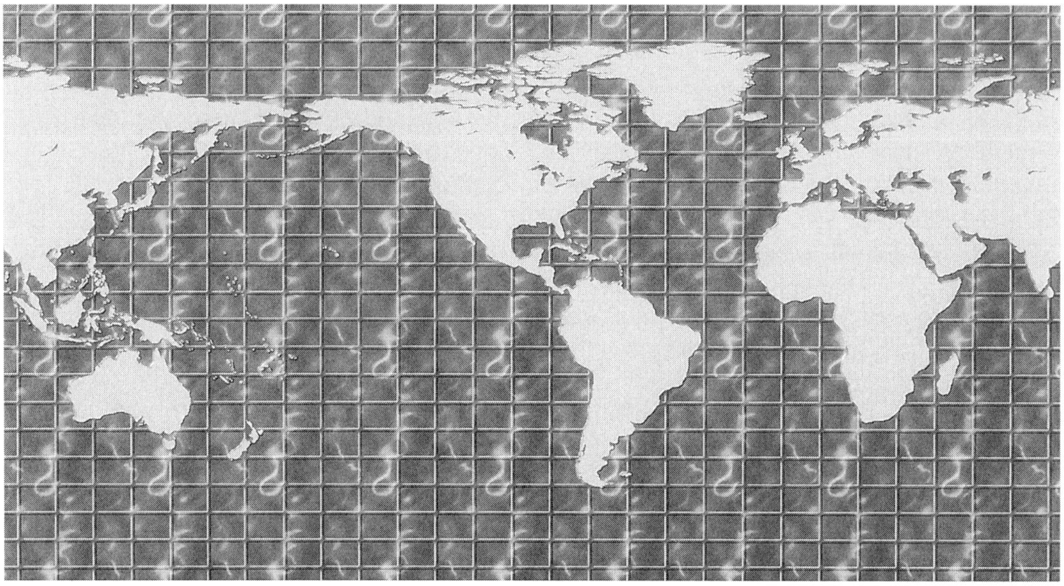

MERCATOR CYLINDRICAL PROJECTION OF THE CONTINENTS OF EARTH, SHOWING THE CHARACTERISTIC EXAGGERATION IN SCALE OF LAND MASSES NEAR THE POLES. *(© R. Winter/Photo Researchers. Reproduced by permission.)*

of note in the West for more than a thousand years.

THE SEPARATION OF GEOGRAPHY AND EARTH SCIENCES. During the first half of the Middle Ages significant work in cartography took place in the Muslim world and in the Far East, but not in Europe. Only in the course of the Crusades (1095–1291) did Europeans become interested in exploration again, and this interest grew as the Mongol invasions of the thirteenth century opened trade routes from Europe to China for the first time in a thousand years. The crusading and exploring spirits met in Henry the Navigator (1394–1460), the prince who, while he never actually took part in any voyages himself, quite literally launched the Age of Exploration from his navigation school at Sagres, Portugal.

In the two centuries that followed, European mariners and conquerors explored and mapped the continents of the world. Along the way, these nonscientists sometimes added knowledge to what would now be considered the earth sciences, as when the Italian explorer Christopher Columbus (1451–1506) became the first to notice magnetic declination. (See Geomagnetism for more on this subject.) Meanwhile, Columbus's contemporary, the Italian artist and scientist Leonardo da Vinci (1452–1519), as well as Georgius Agricola (1494–1555) of Germany, known as the father of mineralogy, conducted some of the first modern studies in the earth sciences. (See Studying Earth for more about Leonardo's and Agricola's contributions.)

In the sixteenth century, the Flemish cartographer Gerhardus Mercator (1512–1594) greatly advanced the science of mapmaking with his development of the Mercator projection. The latter method, still used in many maps today, provided an effective means of rendering the spherical surface of Earth on a two-dimensional map. By that time, cartography had emerged as a vital subdiscipline, and over the ensuing two centuries the separation between geography and the earth sciences became more and more distinct.

The full separation of disciplines took place in the eighteenth century, when the German geographer Anton Friedrich Büsching (1724–1793) pioneered modern scientific geography. Beginning in 1754, Büsching published the 11-volume *Neue Erdbeschreibung* (New description of the earth), which established a foundation for the study of geography in statistics rather than descriptive writing. During the same century geology was in the middle of its own paradigm shift. Until the time of the Scottish geologist James Hutton (1726–1797), a near exact contemporary of Büsching, geologists had been concerned primarily with explaining how the world began—a topic that almost inevitably led to conflict over religious questions (see Earth, Science, and Nonscience). Hutton, known as the father of geology, was the first to transcend this issue and instead offer a theory about the workings of Earth's internal mechanisms.

SURVEYING

The era of Büsching and Hutton coincided with that of the English astronomer Charles Mason (1730–1787) and the English surveyor Jeremiah Dixon (*d.* 1777), who in 1763 began their famous survey of the boundary between Pennsylvania and Maryland. It took them five years to survey the 233-mi. (373-km) Mason–Dixon Line, which eventually became known as the border between the free and slave states before the Civil War. Surveying, a profession practiced by such great Americans as the first president, George Washington (1732–1799), and the mathematician and astronomer Benjamin Banneker (1731–1806), eventually became associated with the United States, but it originated in Egypt as early as 2700 B.C.

Surveying is a realm of applied mathematics devoted to measuring and mapping areas of land. Though it is obviously of value to the earth sciences and geography, it has a great deal of importance economically and politically as well. For this reason, the Romans, who were not nearly as inclined toward theoretical study as the Greeks, became preeminent surveyors whose land parcels still can be seen from the air over parts of western Europe. Owing to its great practical importance, surveying—unlike virtually all other forms of learning—continued to thrive in Europe during the early Middle Ages. Nonetheless, surveying improved in the Renaissance and thereafter, as the result of the introduction of new tools and mathematical techniques.

Among these tools were the theodolite, used for measuring horizontal and vertical angles, and the transit, a type of theodolite that employs a hanging plumb bob to determine a level sight line. Mathematical techniques included triangulation, whereby the third side of a triangle can be determined from measurements of the other two sides and angles. The German mathematician Karl Friedrich Gauss (1777–1855), regarded as the father of geodesy, introduced the heliotrope, a mechanism that aids in triangulation. Other tools included the compass, level, and measuring tapes; modern surveying benefits from remote sensing. (Both remote sensing and geodesy are discussed later in this essay.)

REAL-LIFE APPLICATIONS

GEOLOGIC MAPS AND SURVEYS

A geologic map shows the rocks beneath Earth's surface, including their distribution according to type as well as their ages, relationships, and structural features. The first geologic map, made in 1743, depicted subterranean East Kent, England. Its creator, the English physician and geologist Christopher Packe (1686–1749), introduced the technique of hachuring, that is, representing relief (elevation) on a map by shading in short lines in the direction of the slopes.

Three years later, the French geologist Jean-Etienne Guettard (1715–1786) made the first geologic map that crossed national lines, thus illustrating the distinction between geology and geography. As Guettard discovered, the geologic features of the French and English coasts along the English Channel are identical, indicating that the areas are connected. At a time when France and England were still bitter political and military rivals, Guettard's work showed that they literally shared some of the same land.

Geologic mapmaking received a further boost in 1815, when the English geologist William Smith (1769–1839) produced what has been called the first geologic map based on scientific principles. Entitled *A Delineation of the Strata of England and Wales with Part of Scotland,* the map used different colors to indicate layers of sedimentation. Roughly 6 ft. by 9 ft. (1.8 × 2.7 m), the map linked paleontology with stratigraphy (the study of fossils and the study of rock layers, respectively) and proved to be a milestone in geologic cartography.

MODERN GEOLOGIC MAPMAKING. Geologic mapmaking changed considerably in the period after World War II. Before that time, geologists did most of their work with the use of topographical maps, or maps that showed only surface features. Following the war, however, aerial photography became much more common, giving rise to the technique of photogeology, the use of aerial photographic data to make determinations regarding the geologic characteristics of an area.

Petroleum companies, which often have taken the lead in developing advanced methods of geologic study, introduced the practice of creating three-dimensional images from the air.

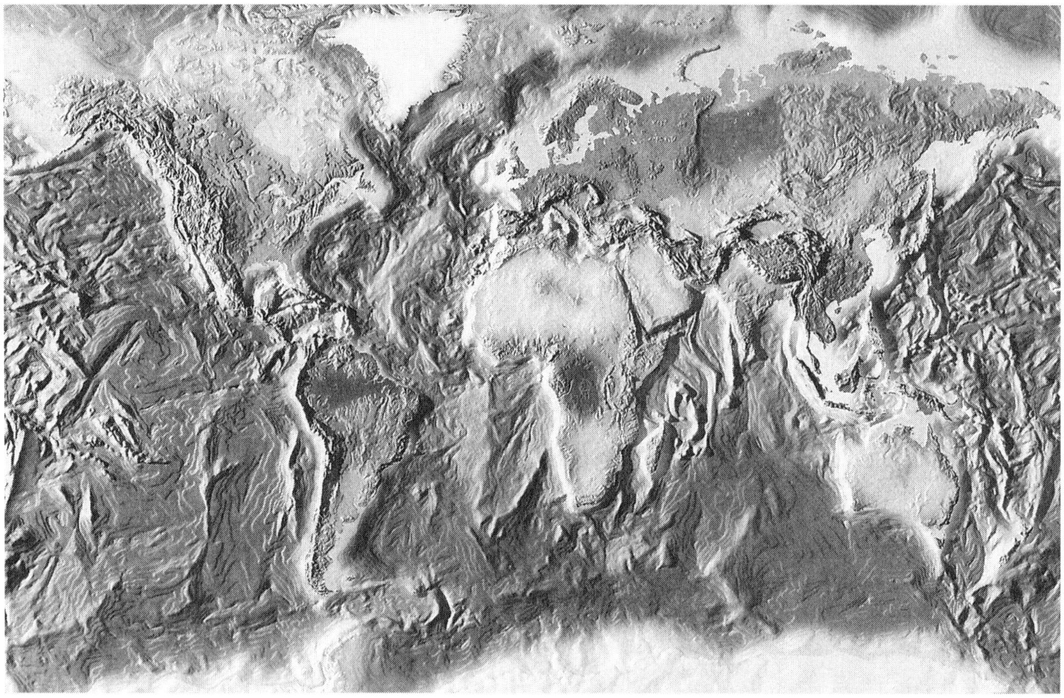

RELIEF-SHADED MAP OF EARTH, SHOWING THE CONTINENTS IN RAISED ELEVATION. *(© M. Agliolo/Photo Researchers. Reproduced by permission.)*

They did this by taking pairs of photographs which, when viewed through a stereoscope, provided images that could be studied in great detail for information about all manner of geologic features. Height proved a great advantage, revealing features that would not have been as clear to a geologist working on the ground.

Of course, a great deal of work on the ground was still necessary for confirming the data revealed by aerial surveillance and for other purposes, such as measurement and sample collection. Nonetheless, aerial photography provided an enormous boost to geologic studies, as did the use of satellite imaging from the 1970s onward. Also helpful were such new devices as the handheld magnetometer, which made it relatively easy to separate rocks containing magnetite from other, nonmagnetic samples.

WHY MAKE GEOLOGIC MAPS? One might wonder why any of this is important, aside from a purely academic interest in the structure of rocks under the ground. In fact, the need for precise geologic data goes far beyond "purely academic interests," as the reference to oil companies suggests. Geologic mapmaking is critical to the location of oil as well as minerals and other valuable natural resources—including the most useful one of all, water.

Likewise it is necessary to have accurate geologic information before undertaking a large engineering project, such as the building of a road or bridge. In such situations, geologic studies can quite literally be a matter of life and death, and eventually such studies may save more lives by aiding in the prediction of earthquakes or volcanoes. Geologic data is also a critical part of studies directed toward environmental protection, both for areas designed to remain natural habitats and for those designated for development. And, finally, there are studies whose purpose is purely, or mostly, academic but that reveal a great deal of useful information about the history of the planet, the forces that shaped it, and perhaps even future events.

GEOLOGIC SURVEYS. Geologic mapmaking is so vital, in fact, that national governments have undertaken large-scale and ongoing geologic studies since 1835. That was the year that Great Britain became the first country to establish a geologic survey, with the aim of preparing a geologic map of the entire British Isles. The project began with the mapping of Cornwall and southern Wales and has continued ever since, with the addition of new details as they have become available.

In the ensuing years several other countries established their own geologic surveys, including the United States in 1889. (The Web site of the U.S. Geological Survey is listed in the bibliography.) Other important national geologic surveys include those of France, Canada, China, and Russia. The last of these surveys is of particular interest, dating as it does from the "Stone Department," a mineralogical survey established in 1584. Today even much smaller nations, such as Uruguay, Slovakia, and Namibia, have their own national geologic surveys.

Over the years, techniques of information gathering have evolved, particularly with the development of satellite remote-sensing technology. So, too, have the areas under the purview of various national geologic surveys, which since the 1950s have undertaken the mapping of the continental shelves adjacent to their own shorelines. (In addition, the nations claiming territories in Antarctica, including Britain and the United States, have mapped the geologic features of that continent extensively, though mining or other economic development there is forbidden.) The U.S. Geological Survey has also seen its scope extended to include studies on such issues as radioactive waste disposal and prediction of natural hazards, among them, earthquakes in urban areas.

Geophysical Measurements

Geophysics is a branch of the earth sciences that combines aspects of geology and physics. Among the areas it addresses are Earth's physical processes as well as its gravitational, magnetic, and electric properties and the means by which energy is transmitted through its interior. Areas of geophysics with a particular focus on measurement and mapping include the study of geomagnetism, or Earth's magnetic field, and geodesy, which is devoted to the measurement of Earth's shape and gravitational field.

The measurement of gravitational fields involves the use of either weights dropped in a vacuum or mechanical force-balance instruments. The first of these techniques is much older than the other and provides an absolute measure of the gravitational field in a given area. As for force-balance instruments, they are similar in principle to scales and furnish a relative measure of the gravitational field. To compare the gravitational field at different positions, however, it is necessary to establish a frame of reference. This is known as the *geoid,* a surface of uniform gravitational potential covering the entire earth at a height equal to sea level. (See Gravity and Geodesy for more on these topics.)

GEODETIC MEASUREMENTS OF EARTH'S SURFACE. As noted, geodesy is concerned not only with Earth's gravitational field but also with its shape, and earth scientists working on this aspect of the subdiscipline employ many of the techniques and equipment described earlier with regard to geography and surveying. Eratosthenes's measurement of Earth's size is thought to be the first geodetic measurement, and in performing geodetic measurements today, earth scientists often employ concepts familiar to surveyors.

Among these concepts is triangulation, which was developed in the sixteenth century by the Dutch mathematician Gemma Frisius (1508–1555). Triangulation remained an important method of geodetic measurement until the development of satellite geodesy made possible simpler and more accurate measurements through remote sensing. Even today triangulation is still used by geologists without access to satellite data. In performing measurements using triangulation, geologists employ the theodolite, and typically at least one triangulation point is highly visible—for instance, the top of a mountain.

Until the 1950s scientists used a measuring tape of a material called Invar, a nickel-iron alloy noted for its tendency not to expand or contract with changes in temperature. From that time, however, electronic distance measurement (EDM) systems, which employ microwaves or visible light, came into use. EDM helped overcome some of the possibilities for error inherent in using any kind of tape, for example, the likelihood that it would sag and thus render incorrect measurements. Furthermore, EDM tended to reduce errors caused by atmospheric refraction.

With the advent of the United States program in the 1960s, increasingly more sophisticated forms of geodesic remote-sensing technology came into use. Among these techniques is satellite laser ranging, which relies on measurements of the amount of time required for a laser pulse to travel from a ground station to a satellite and back. Before the development of the global posi-

KEY TERMS

CARTOGRAPHY: The creation, production, and study of maps. Cartography is a subdiscipline of geography and involves not only science but also mathematics, technology, and even art.

DOPPLER EFFECT: The change in the observed frequency of a wave when the source of the wave is moving with respect to the observer.

FIELD: A region of space in which it is possible to define the physical properties of each point in the region at any given moment in time.

GEODESY: An area of geophysics devoted to the measurement of Earth's shape and gravitational field.

GEOGRAPHY: A social science concerned with the description of physical, biological, and cultural aspects of Earth's surface and with the distribution and interaction of these features. Compare with geology.

GEOID: A surface of uniform gravitational potential covering the entire earth at a height equal to sea level.

GEOLOGIC MAP: A map showing the rocks beneath Earth's surface, including their distribution according to type as well as their ages, relationships, and structural features.

GEOLOGY: The study of the solid earth, in particular, its rocks, minerals, fossils, and land formations.

GEOPHYSICS: A branch of the earth sciences that combines aspects of geology and physics. Geophysics addresses the planet's physical processes as well as its gravitational, magnetic, and electric properties and the means by which energy is transmitted through its interior.

GEOMAGNETISM: A term referring to the magnetic properties of Earth as a whole, rather than those possessed by a single object or place on Earth.

tioning system (GPS), discussed later, the use of satellite systems necessitated tracking through the Doppler effect, or the change in the observed frequency of a wave when the source of the wave is moving with respect to the observer. Thanks to GPS, put into operation by the U.S. Department of Defense, satellite tracking is much simpler and more accurate today.

GEOMAGNETIC MEASUREMENTS. Ever since the ancient Chinese discovered that pieces of lodestone (magnetite) tend to point north, mariners have used the compass for navigation. The compass was augmented by other navigational devices until it was supplanted by the gyroscope in modern times and still later by more sophisticated devices and methods, such as GPS. Yet a compass still works fine for many a hiker, and its use serves to emphasize the importance of Earth's magnetic field.

Earth has an overall geomagnetic field, and specific areas on the planet have their own local magnetic fields. Thanks in large part to the contributions of Gauss, who developed a standardized local magnetic coordinate system in the early nineteenth century, it became possible to perform reasonably accurate measurements of local magnetic data while correcting for the influence of Earth's geomagnetic field. Indeed, one of the challenges in measuring magnetic fields is the fact that the Earth system possesses magnetic force from so many sources: the molten core, from whence originates the preponderance of Earth's magnetic field; external fields, such as the magnetosphere and ionosphere; local materials, such as magnetite, hematite, or pyrrhotite;

KEY TERMS CONTINUED

HACHURING: A method of representing relief (elevation) on a map by shading in short lines in the direction of the slopes.

MAGNETOSPHERE: An area surrounding Earth, reaching far beyond the atmosphere, in which ionized particles (i.e., ones that have lost or gained electrons so as to acquire a net electric charge) are affected by Earth's magnetic field.

PALEOMAGNETISM: An area of historical geology devoted to studying the direction and intensity of magnetic fields in the past, as discerned from the residual magnetization of rocks.

PALEONTOLOGY: The study of fossilized plants and animals, or flora and fauna.

PHOTOGEOLOGY: The use of aerial photographic data to make determinations regarding the geologic characteristics of an area.

PHYSICAL GEOGRAPHY: A subdiscipline of geography concerned with the exterior physical features and changes of Earth.

POTENTIAL: Position in a field, such as a gravitational force field.

REFRACTION: The bending of light as it passes at an angle from one transparent material into a second transparent material.

REMOTE SENSING: The gathering of data without actual contact with the materials or objects being studied.

STRATIGRAPHY: The study of rock layers, or strata, beneath Earth's surface.

SURVEYING: An area of applied mathematics devoted to measuring and mapping areas of land.

TRIANGULATION: A technique in surveying whereby the third side of a triangle can be determined from measurements of the other two sides and angles.

and even man-made sources of magnetic or electric force.

After making calculations that correct for interfering sources of magnetism, geophysicists study the remaining magnetic anomalies, which can impart extremely valuable information. Classic examples include the discovery that Earth's magnetic polarity has reversed many times, a finding that led to the development of the geophysical subdiscipline known as paleomagnetism. Paleomagnetic studies, in turn, served as a highly significant confirmation of plate tectonics, which originated in the middle of the twentieth century and remains the dominant theory regarding geologic processes. (See Plate Tectonics. For more about geomagnetism, including some of the topics mentioned here, see Geomagnetism.)

Knowing One's Location

In these and other types of studies that involve mapping and measurement, it is important for scientists conducting surveys to be aware of the frame of reference from which they are operating—that is, the perspective from which they view data. Simply put, one must know first where one is before one can measure and map geophysical or other data for surrounding areas. This requires knowledge of latitude and longitude, or east–west and north–south positions, respectively.

From earliest times, mariners and scientists have been able to ascertain latitude with relative ease, simply by observing the angle of the Sun and other stars. Determination of longitude, however, proved much more difficult, because it required highly accurate timepieces. Only in the late eighteenth century, with the breakthroughs achieved by the British horologist John Harrison (1693–1776) did such calculations become possible. As a result, many a ship's crew was saved from the misfortunes that could result from inaccurate estimates of location.

GLOBAL POSITIONING SYSTEM. By the latter part of the twentieth century, navigational technology had become vastly more sophisticated than it was in Harrison's day. From the 1957 launch of the Soviet satellite *Sputnik 1,* the skies over Earth became increasingly populated with satellites, such that within half a century dozens of countries had payloads in space. Aside from governments and scientific research establishments, even cable television companies used satellites to beam programming to homes all over the industrialized world, a fact that in itself says much about the spread of satellite technology.

Among the most impressive uses of satellites is the GPS, developed by the U.S. Department of Defense to assist in surveillance. GPS consists of 24 satellites orbiting at an altitude of 12,500 mi. (20,000 km). They move in orbital paths such that an earthbound receiver can obtain signals from four or more satellites at any given moment. On board are atomic clocks, which provide exact time data with each signal and eliminate the necessity of the receiver's having such an accurate clock. By receiving data from these satellites, persons on the ground can compute their own positions in terms of latitude and longitude as well as altitude. Not all receivers have access to the most accurate data possible: in line with the strategic mission for which it initiated GPS in the first place, the Defense Department ensures that only authorized personnel receive the most precise information.

Thus, GPS has built-in errors so that civilian users can calculate locations with an accuracy of "only" 328-492 ft. (100–150 m). This, of course, is amazingly accurate, but not as accurate as the data available to those authorized to receive normally encrypted information on the P (Precise) code. The latter provides an accuracy of 3.28-16.4 ft. (1–5 m) instantaneously, and more detailed measurements based on GPS data can be used to achieve accuracy of up to 0.2 in. (5 mm).

Remote Sensing

Many of the methods used by geologists and geophysicists to map and measure Earth make use of remote sensing, the gathering of data without actual contact with the materials or objects being studied. Without remote sensing, it would be impossible to discuss many physical phenomena intelligently, because it is unlikely that any technology ever will make it possible to explore many areas underneath the planet's surface directly.

An example of remote sensing is photogeology, described briefly earlier; so, too, is satellite imaging for data collection. The earth scientist of the twenty-first century likewise has other highly sophisticated forms of technology, such as radar systems or infrared imaging, at his or her disposal. As with many other aspects of geologic mapping and measurement, this one has value far beyond the classroom: remote-sensing studies make it possible, for instance, to observe the environmental impact of deforestation in large geographical areas. (For much more about this subject, see Remote Sensing.)

WHERE TO LEARN MORE

Brooks, Susan. *The Geography of the Earth.* New York: Oxford University Press, 1996.

Erickson, John. *Exploring Earth from Space.* Blue Ridge Summit, PA: TAB Books, 1989.

Geologic Maps (Web site). <http://http://www.aqd.nps.gov/grd/usgsnps/gmap/gmap1.html>.

Hancock, Paul L., and Brian J. Skinner. *The Oxford Companion to the Earth.* New York: Oxford University Press, 2000.

Internet Resources for Geography and Geology (Web site). <http://www.uwsp.edu/geo/internet/geog_geol_resources.html>.

Moffitt, Francis H., and Harry Bouchard. *Surveying.* 8th ed. New York: Harper and Row, 1987.

Remote Sensing Data and Information (Web site). <http://rsd.gsfc.nasa.gov/rsd/RemoteSensing.html>.

U.S. Geological Survey (Web site). <http://www.usgs.gov/>.

Virtual Museum of Surveying (Web site). <http://www.surveyhistory.org/index.htm>.

Wilford, John Noble. *The Mapmakers.* New York: Knopf, 2000.

REMOTE SENSING

CONCEPT

Scientists of many disciplines are accustomed to studying data that cannot be observed through direct contact. Physicists and chemists, for instance, know a great deal about the structure of the atom, even though even the most high-powered microscope cannot make an atom visible to the human eye. The objects of study for earth scientists are often similarly remote, though not necessarily because they are small. In some cases, the problem is quite the opposite: an area selected for study is too large to provide understanding to geologists working only on the ground. Other areas are simply inaccessible to human beings or even their equipment. This has necessitated the development of remote sensing equipment and techniques, primarily involving views from the air or from space and utilizing electromagnetic radiation across a wide spectrum.

HOW IT WORKS

AN INTRODUCTION TO REMOTE SENSING

The work of geologists would be much easier if Earth were transparent and they could simply look down into the ground as they would into the sky. But the ground is not transparent; nor, for that matter, is the sky, to which meteorologists look for information regarding atmospheric and weather patterns. Some places are hard to see, and many are difficult or even impossible to visit physically. Some places, such as the Sun or the Earth's core, could not be approached physically even by unmanned technology.

Hence the need for remote sensing, or the gathering of data without actual contact with the materials or objects being studied. Some earth scientists define the term more narrowly, restricting "remote sensing" to the use of techniques involving radiation on the electromagnetic spectrum. The latter category includes visible, infrared, and ultraviolet light as well as lower-frequency signals in the microwave range of the spectrum. This definition excludes the study of force fields involving gravitational or electromagnetic force. In general, in this essay we abide by that more narrow definition, primarily because most forms of remote sensing in use today involve electromagnetic radiation.

Remote sensing is used for a variety of measuring and mapping applications. The reader therefore is encouraged to consult the essay Measuring and Mapping Earth for more on this subject. Applications of remote sensing go far beyond cartography (mapmaking) and measurement, however. As suggested already, remote sensing makes it possible for earth scientists to collect data from places they could not possibly go. In addition, it allows for data collection in places where a human being would be "unable to see the forest for the trees"—which in places such as the Amazon valley is quite literally the case.

THE MILITARY INFLUENCE

Scientists' understanding of the electromagnetic spectrum was still in its infancy in 1849, when the French army engineer Aimé Laussedat (1819–1907) introduced what was then called iconometry, from the Greek words *icon* ("image") and *-metry* ("measurement"). Laussedat, who experimented with aerial photography

A HAND-HELD GLOBAL POSITIONING DEVICE. *(© Ken M. Johns/Photo Researchers. Reproduced by permission.)*

by means of cameras mounted on balloons or kites, is regarded as a pioneer of photogrammetry, the use of aerial or satellite photography to provide measurements of or between objects on the ground.

A few years later, the United States armies of the Civil War adopted the use of aerial photography for surveillance purposes, mounting cameras on balloons to provide intelligence regarding federal or Confederate positions and troop strength. This fact, combined with Laussedat's status as an army engineer, hints at one of the underlying themes in the history of remote sensing, and indeed of many another technological advance: the influence of the military. It is a fact of human existence that nations from at least the time of the Assyrians, if not the Egyptians of the New Kingdom, have devoted far more attention and resources to military applications than they have to peacetime activities.

On the other hand, societies have benefited enormously from technological and organizational innovations with military origins, innovations whose application later spread to a variety of peacetime uses. Some examples include the adoption of the chariot by the Egyptian army after the Hyksos invasion (*ca.* 1670 B.C.); the Assyrian introduction of logistics in an effort to supply imperial troops (*ca.* 800 B.C.); the Persian development of the postal service (*ca.* 600 B.C.); numerous Roman innovations, particularly in road building (*ca.* 200 B.C.–*ca.* A.D. 200); and the Chinese invention of the wheelbarrow (*ca.* 100 B.C.). And so the list goes, right up to such latter-day American developments as the Internet and GPS, or global positioning system.

MILITARY CONTRIBUTIONS TO REMOTE SENSING. Forms of technology pioneered by military forces and now used in remote sensing include infrared photography, thermal imagery, radar scanning, and satellites. The first of these types of technology makes use of light in the infrared portion of the electromagnetic spectrum—a region that, as its name suggests, is adjacent to the red portion of visible light. Red has the longest wavelength and the lowest frequency of all colors, and infrared has an even longer wavelength and lower frequency. Military forces use infrared photography to distinguish between vegetation and camouflage designed to look like vegetation: live plants reflect infrared radiation, whereas dead ones and camouflaged material absorb it.

Whereas infrared photography measures reflection of infrared radiation, thermal imaging indicates the amount of such radiation that is emitted by the source. Its military origins lie in its use for reconnaissance during night bombing missions. Similarly, radar scanning makes it possible to view targets on the ground, regardless of lighting or cloud cover. Finally, there are satellites, which have extensive surveillance applications. Among the most important examples of military activity above Earth's atmosphere are the 24 satellites of GPS, which make allow U.S. forces to plot positions with amazing accuracy. Less accurate GPS intelligence is also available to civilians. (See Measuring and Mapping Earth for more on GPS.)

REAL-LIFE APPLICATIONS

Photogeology

All of these innovations introduced by the military, of course, have found application for civilian purposes. Thanks in part to improvements in aircraft during World War II, for instance, photogeologic data gathering has increased dramatically in the years since then. Efforts at gaining information by means of airborne sensing devices underwent enormous improvements throughout the middle and latter part of the twentieth century, with the development of technology that made it possible for earth scientists to gather information using techniques beyond ordinary photography, visible light, and airplanes.

Still, much of the remote-sensing activity that takes place today is performed aboard airplanes rather than satellites, using ordinary analogue photography within the visible spectrum. Stereoscopic techniques aid in the visualization of relief, or elevation and other inequalities on a land surface. Humans are used to seeing stereoscopically: the distance between the two eyes on our faces results in a difference between the two images each eye sees. The brain corrects for this difference, rendering a stereoscopic image that is more full and dimensional than anything a single eye could produce. The use of multiple cameras and stereoscopic technology replicates this activity of the human brain and thus provides earth scientists with much more information than they could gain simply by looking at "flat" photographs taken from an airplane.

The materials studied by a geologist, of course, are primarily underground, but Earth's surface furnishes many clues that a trained observer can interpret. Uplands and lowlands tend to suggest different types of rocks, while the direction of a dip in the land can supply volumes of information regarding the stratigraphic characteristics of the region. The presence of vegetation can make it harder to discern such clues, but a careful study of plant life can reveal much regarding minerals in the soil, local water resources, and so on.

DIGITAL PHOTOGRAPHY

Within both photogeology and the larger realm of remote sensing, several innovations from the 1960s onward have underpinned more effective methods of observation. One of these is digital photography, which is as much of an improvement over old-fashioned photography as compact discs are over phonograph records. In both cases, the contrast is between analog technology and digital technology. In analog photography, for instance, the image is recorded by a camera and stored on photosensitive materials in a film emulsion. In digital photography the image is recorded on a solid-state device called an image sensor and stored in the camera's memory for transfer to a computer.

A COMMUNICATIONS SATELLITE IN ORBIT AROUND EARTH. *(© ESA/Photo Researchers. Reproduced by permission.)*

An analogue (the preferred spelling for the word as a noun) is just that, a "close copy," whereas digital methods make possible a more exact reproduction of images by assigning to each shade of color a number between 0 and 255. Instead of storing the image in a medium that can be destroyed or lost easily, as is the case with ordinary film, digital images can be saved on a computer, backed up, and sent anywhere in the world via the Internet. Furthermore, these images can be adjusted with the use of a computer, so as to make it easier to see certain features.

Computers and digital photography aid in the creation of false-color imaging, a means of representing invisible electromagnetic data by assigning specific colors to certain wavelengths. An example would be the use of red to depict areas of high energy. This is certainly a false use of color, since red actually has the lowest energy

An aerial photograph shows the Colorado River delta in the Gulf of California. The river itself is the dark hemisphere at the bottom, with its waters branching out through sandbars like the boughs of a tree. *(© Photo Researchers. Reproduced by permission.)*

in the visible spectrum, with purple possessing the highest energy. (The reason we associate red, orange, and yellow with heat and green, blue, and purple with coldness is that in either case, these are the colors objects *reflect,* not the ones they absorb.)

RADAR. Most remote-sensing technology uses light, whether infrared or visible, that falls at the middle to high end of the electromagnetic spectrum. By contrast, at least one important means of remote detection uses microwaves, which are much lower in energy levels. Microwaves carry FM radio and television signals, as well as radar, or *RA*dio *D*etection *A*nd *R*anging.

Radar makes it possible for pilots to "see" through clouds, rain, fog, and all manner of natural phenomena—not least of which is darkness. It also can identify objects, both natural and man-made, on the ground. In addition to its application in remote sensing, radar using the Doppler effect (the change in the observed frequency of a wave when the source of the wave is moving with respect to the observer) helps meteorologists track storms.

In the simplest model of radar operation, a sensing unit sends out microwaves toward the target, and the waves bounce back off the target to the unit. In a monostatic unit—one in which the transmitter and receiver are in the same location—the radar unit has to be switched continually between sending and receiving modes. Clearly, a bistatic unit—one in which the transmitter and receiver antennas are at locations remote from one another—is generally preferable, but on an airplane, for instance, there is no choice but to use a monostatic unit.

SATELLITE DATA

The term satellite refers to any object orbiting a larger one; thus, Earth's Moon and all the other moons of the solar system are satellites, as are the many artificial satellites that orbit Earth. In practice, however, most people use the term to refer only to artificial satellites, of which there are many hundreds, launched by entities ranging from national governments to international associations to independent firms. Artificial satellites typically are intended for the purposes of gathering information (i.e., scientific research or military surveillance) or disseminating it (i.e., through satellite television broadcasting).

In launching a satellite, it is necessary to overcome the enormous pull of Earth's gravitational field. This is done by providing the satellite with power through rocket boosters that launch it far above Earth's atmosphere. At a height of 200 mi. (320 km) or more, the satellite is far above the dense gases of the atmosphere yet well within the gravitational field of the planet. The craft is then in a position to orbit Earth indefinitely without the need for additional power from man-made sources; instead, Earth's own gravitational energy keeps the satellite in orbit for as long as the satellite's structure remains intact. (See Gravity and Geodesy for more about the mechanics of orbit.)

The greater the altitude, the longer it takes a satellite to complete a single revolution. One of the most commonly used altitudes is at 22,500 mi. (36,000 km), at which height a satellite takes 24 hours to orbit Earth. Thus, it is said to be in geosynchronous orbit, meaning that it revolves at the same speed as the planet itself and therefore remains effectively stationary over a given area. Some satellites revolve at even higher altitudes—25,000 mi. (40,225 km), which, while it is far beyond the atmosphere, is well within Earth's gravitational field.

LANDSAT. One of the most impressive undertakings in the field of satellite research is Landsat, an Earth-monitoring satellite designed specifically for the use of earth scientists and resource managers. Conceived by the United States Department of the Interior in the mid-1960s, the Landsat project soon came to involve the National Aeronautics and Space Administration (NASA) and the U.S. Geological Survey (USGS; see Measuring and Mapping Earth for more about geologic surveys.) *Landsat 1* went into orbit on July 23, 1972.

Over the years, Landsat has gone into six subsequent generations. *Landsat 6*, launched in 1993, was unable to achieve orbit, but *Landsat 1* lasted more than five times as long as its projected life expectancy of one year. Since 1972 at least one Landsat satellite has been in orbit over Earth, and as of early 2001 both *Landsat 5* (launched in March 1984) and *Landsat 7* (launched in April 1999) were on line. (*Landsat 5* was decommissioned in June 2001.) Over the course of the years, the Landsat governing body has changed. In the 1980s, NOAA (National Oceanic and Atmospheric Administration) took over from NASA, and in October 1985 the Landsat system came under the direction of a commercial organization, the Earth Observation Satellite Company (EOSat).

In contrast to communication satellites, which tend to maintain geosynchronous orbits, Landsat moves at a much lower altitude and therefore orbits Earth much more quickly. *Landsat 7* takes approximately 99 minutes to orbit the planet, thus making 14 circuits in a 24-hour period. Though it never quite passes over the poles, it covers the rest of Earth in swaths 115 mi. (185 km) wide, meaning that eventually it passes over virtually all other spots on the planet.

SATELLITES AT WORK. Landsat and other satellites, such as France's SPOT (Satellite Positioning and Tracking), provide data for governments, businesses, scientific institutions, and even the general public. Following the September 11, 2001, terrorist bombing of the World Trade Center in New York City, for instance, the SPOT U.S. Web site (<http://www.spot.com>) provided viewers with "Images of Infamy": views of downtown Manhattan before and just a few hours after the bombing.

Data from Landsat has been used to study disasters and potential disasters with particular application to the earth sciences. An example is the area of the tropical rainforest in Brazil's Amazon River valley, a region of about 1.9 million sq.

KEY TERMS

CARTOGRAPHY: The creation, production, and study of maps. Cartography is a subdiscipline of geography and involves not only science but also mathematics, technology, and even art.

DOPPLER EFFECT: The change in the observed frequency of a wave when the source of the wave is moving with respect to the observer.

ELECTROMAGNETIC RADIATION: See Electromagnetic spectrum and Radiation.

ELECTROMAGNETIC SPECTRUM: The complete range of electromagnetic waves on a continuous distribution from a very low range of frequencies and energy levels, with a correspondingly long wavelength, to a very high range of frequencies and energy levels, with a correspondingly short wavelength. Included on the electromagnetic spectrum are long-wave and short-wave radio; microwaves; infrared, visible, and ultraviolet light; x rays; and gamma rays.

FALSE-COLOR IMAGING: A means of representing invisible electromagnetic data by assigning specific colors to certain wavelengths.

FREQUENCY: The number of waves, measured in Hertz, passing through a given point during the interval of one second. The higher the frequency, the shorter the wavelength.

GEODESY: An area of geophysics devoted to the measurement of Earth's shape and gravitational field.

HERTZ: A unit for measuring frequency equal to one cycle per second. High frequencies are expressed in terms of kilohertz (kHz; 10^3, or 1,000 cycles per second), megahertz (MHz; 10^6, or one million cycles per second), and gigahertz (GHz; 10^9, or one billion cycles per second).

PHOTOGEOLOGY: The use of aerial photographic data to make determinations regarding the geologic characteristics of an area.

PHOTOGRAMMETRY: The use of aerial or satellite photography to provide measurements of or between objects on the ground.

RADIATION: The transfer of energy by means of electromagnetic waves, which require no physical medium (for example, water or air) for the transfer. Earth receives the Sun's energy, via the electromagnetic spectrum by means of radiation.

RELIEF: Elevation and other inequalities on a land surface.

REMOTE SENSING: The gathering of data without actual contact with the materials or objects being studied.

STRATIGRAPHY: The study of rock layers, or strata, beneath Earth's surface.

WAVELENGTH: The distance between a crest and the adjacent crest or a trough and the adjacent trough of a wave. Wavelength is inversely related to frequency, meaning that the shorter the wavelength, the higher the frequency.

mi. (five million sq km), in which deforestation is claiming between 4,250 sq. mi. and 10,000 sq. mi. (11,000–26,000 sq km) a year. This is an extremely serious issue, because the Amazon basin represents approximately one-third of the total rainforest area on Earth. Earlier estimates, however, had suggested that deforestation was claiming up to three times as much as it actually is, and Landsat provided a more accurate figure.

Because of its acute spatial resolution (98 ft., or 30 m, compared with more than 0.6 mi., or 1 km), Landsat is much more effective for this purpose than other satellite systems operated by NOAA or other organizations. It is also cheaper to obtain images from it than from SPOT. Over the years, Landsat has provided data on urban sprawl in areas as widely separated as Las Vegas, Nevada, and Santiago, Chile. It has offered glimpses of disasters ranging from the eruption of Mount Saint Helens, Washington, in 1980 to some of the most potent recent examples of destruction caused by humans, including the nuclear disaster at Chernobyl, Ukraine, in 1986 and the fires and other effects of the Persian Gulf War of 1990–1991. (For more on this subject, see the Earthshots Web site, operated by USGS.)

Not all the news from Landsat is bad, as a visit to the *Landsat 7* Web site (<http://landsat.gsfc.nasa.gov/>) in late 2001 revealed. Certainly there were areas of concern, among them, flooding in Mozambique and runaway development in Denver, Colorado. But images taken over the Aldabra atoll in the Seychelles showed the world's largest refuge for giant tortoises. And shots taken from Landsat over Lake Nasser in southern Egypt during the latter part of 2000 showed four lakes created by excess water from Nasser. As a result, that region of the Sahara had new lakes for the first time in 6,000 years.

WHERE TO LEARN MORE

Burtch, Robert. *A Short History of Photogrammetry* (Web site). <http://users.netonecom.net/~rburtch/sure340/history.html>.

"Earthshots: Satellite Images of Environmental Change," U.S. Geological Survey (Web site). <http://edcwww.cr.usgs.gov/earthshots/slow/tableofcontent.

Hancock, Paul L., and Brian J. Skinner. *The Oxford Companion to the Earth.* New York: Oxford University Press, 2000.

NASA EOS IDS Volcanology (Web site). <http://eos.pgd.hawaii.edu/>.

Remote Sensing Data and Information (Web site). <http://rsd.gsfc.nasa.gov/rsd/RemoteSensing.html>.

Skinner, Brian J., Stephen C. Porter, and Daniel B. Botkin. *The Blue Planet: An Introduction to Earth System Science,* 2d ed. New York: John Wiley and Sons, 1999.

Smith, David G. *The Cambridge Encyclopedia of Earth Sciences.* New York: Cambridge University Press, 1981.

Strain, Priscilla, and Frederick Engle. *Looking at Earth.* Atlanta: Turner Publishing, 1992.

Visualization of Remote Sensing Data (Web site). <http://rsd.gsfc.nasa.gov/rsd/>.

The WWW Virtual Library: Remote Sensing (Web site). <http://www.vtt.fi/aut/rs/virtual/>.

SCIENCE OF EVERYDAY THINGS
REAL-LIFE EARTH SCIENCE

PLANETOLOGY

PLANETARY SCIENCE

SUN, MOON, AND EARTH

PLANETARY SCIENCE

CONCEPT

The term *planetary science* encompasses a whole range of studies involving a combination of earth sciences and astronomy. Sometimes known as planetology or planetary studies, these disciplines are concerned primarily with the geologic, geophysical, and geochemical properties of other planets. They also draw on areas of astronomy, such as cosmology, a fascinating discipline devoted to the study of the origin, structure, and evolution of the universe. As always when considering realms beyond our Earth, there are many surprises. Indeed, the more one learns about Earth's relationship to the rest of the cosmos, the harder it is to say which is more intriguing: the many factors that make Earth different or the myriad ways that our home planet is just like the rest of the known universe.

HOW IT WORKS

EARTH AND THE COSMOS

Most of us spend our daily lives without devoting a great deal of thought to what lies beyond Earth. People who live outside cities are perhaps more attuned to the cosmos than are their urban counterparts, simply because they see the vast oceans of stars that cover the sky on a clear night. But a person who lives in the city, where bright lights and smog conspire to cover all but the brightest heavenly bodies, rarely finds a reason to look up into the night sky.

One reason people spend little time thinking about the cosmos is that to do so ultimately fills one with a sense of awe bordering on dread. We know that Earth is but one planet of nine, revolving around an average-sized star, the Sun, somewhere between the center and the edge of a galaxy called the Milky Way—itself just one of many galaxies in the universe. This awareness naturally makes a person feel small and almost inevitably raises questions about the nature of the soul, divinity, and the afterlife.

RELIGION, PHILOSOPHY, MYSTICISM, AND SCIENCE. Such questions are a natural accompaniment to of our feeling that if one person is so truly insignificant in this vast cosmos, there must be something else that gives meaning to the structure of reality. These vast issues, of course, are properly addressed not by science but by theology and philosophy. Science, on the other hand, is concerned simply with the facts of how the universe emerged and how Earth fits into the larger picture.

Yet it is easy to see how ancient peoples would have perceived no distinction between religion and science where the study of the cosmos was concerned. The Babylonians, for instance, had no concept of any difference between scientific astronomy and astrology, which today is recognized as a superstitious and thoroughly unscientific pursuit. The Greeks modeled the cosmos on their philosophical systems, which provided a hierarchy of material forms and an ordered arrangement of causes and substances. And the Judeo-Christian tradition depicts a universe fashioned by a loving, all-powerful creator who designed the human being in his own image.

In the belief systems of Judaism and Christianity, handed down through the Bible, the cosmos is depicted as the setting of a vast spiritual drama centered around the themes of free will, sin, and redemption. The Bible never says that Earth is the center of the physical universe, but it clearly presents it as the center of the spiritual one. This is understandable enough, especially if human beings truly are the only intelligent lifeforms; unfortunately, these spiritual ideas eventually informed an erroneous cosmology that depicted Earth as the physical center of the cosmos.

Cosmology

In fairness to Christianity, it should be said that most religious, philosophical, and even scientific traditions before about 1500 depicted Earth as the center of the universe. Indeed, it required a great feat of insight to discern that Earth is not the center. The same is true of many other discoveries about the cosmos, where nothing is as it appears when simply gazing into the night sky.

In a scene from his great novel *The Adventures of Huckleberry Finn* (1884), Mark Twain aptly illustrated the impossibility of understanding the universe simply on the basis of unaided intellect. Huck and the runaway slave Jim have just finished supper and are lying on their backs and staring up at the stars, speculating as to their origins. One of them comes up with a theory that seems altogether plausible on the face of it: the Moon, because it looks larger than the stars, must have laid them like eggs. A similar scene occurs in the children's movie *The Lion King* (1994), in which one character postulates that the stars have become stuck to the sky like flies on flypaper. When another character, the warthog Pumbaa, correctly suggests that the stars are actually great balls of burning gas billions of miles away, his companions laugh this off as preposterous.

ARISTARCHUS AND HIPPARCHUS. Although they lacked telescopes, the Greeks developed rather sophisticated (though in many cases wrong) ideas concerning the arrangement of the cosmos. Most notable among these early thinkers was the astronomer Aristarchus of Samos (*ca.* 320–*ca.* 250 B.C.), who proposed that Earth rotates on its axis once every day and revolves with other planets around the Sun. He also correctly suggested that the Sun is larger than Earth.

Unfortunately, the astronomer Hipparchus (146–127 B.C.) rejected this heliocentric, or Sun-centered, cosmology in favor of a geocentric, or Earth-centered, model. Among Hipparchus's later followers was the Alexandrian Ptolemy (*ca.* A.D. 100–170), destined to become the most influential astronomer of ancient and medieval times, who established geocentric cosmology as a guiding principle of astronomy.

THE PTOLEMAIC SYSTEM. The influence of Ptolemy's erroneous ideas is partly an accident of history. He lived, as it turned out, in the last great era of civilization: ten years after his death came that of the Roman emperor Marcus Aurelius (A.D. 121–180), whose passing marked the beginning of Rome's decline over the next three centuries. Learning in western Europe virtually ceased until about 1200, and even though the Muslim world produced several thinkers of note during this period, most of them worked within the tradition established by Ptolemy. Muslim thinkers' respect for Ptolemy is reflected in the name that Arab translators gave to his most important writing: *al-majisti* or "majesty." When this work made its way to Europe, it became known as the *Almagest.*

The Ptolemaic system proves that it is possible to prove anything, if one creates a methodology elaborate enough. Of course, as we know now, Earth is not the center of the universe, but pure observation alone did not reveal this, and Ptolemy's cosmology worked because he developed mathematics and ideas of planetary motion that made it workable. For instance, not only did planets orbit around Earth in Ptolemy's cosmology, but they also moved in circles around the paths of their own orbits. Of course, they do revolve on their axes, but that was not part of Ptolemy's model. In fact, it is hard to find an analogy in the real world, with the exception of some bizarre amusement park ride, for the form of motion Ptolemy was describing.

He was trying to explain retrograde motion, or the fact that other planets seem to speed up and slow down. Retrograde motion makes perfect sense once one understands that Earth is moving even as the other planets are moving, thus creating the optical illusion that the others are changing speeds. Since the Ptolemaic system depicted a still Earth in the middle of a moving universe, however, the explanation of retrograde motion required mental acrobatics.

CHALLENGING PTOLEMY. Although it is incorrect, the Ptolemaic system was a creation of genius; otherwise, it could not have survived for as long as it did. Even with the recovery of learning in Europe during the late Middle Ages, scientists continued to uphold Ptolemy's ideas. Instead of discarding his system, or at least calling it into question, astronomers simply adjusted the mathematics and refined their ancient forebear's physical model to account for any anomalies.

The revolution against Ptolemy began quietly enough in the fifteenth century, when the Austrian astronomer and mathematician Georg Purbach (1423–1461) noted the inaccuracies of existing astronomical tables and the need for better translations of Greek texts. Purbach attempted to produce a revised and corrected version of the *Almagest,* but he died before completing it. The job fell to his student, Johann Müller, who was known as Regiomontanus (1436–1476).

The *Epitome of the Almagest* (1463), begun by Purbach and completed by Regiomontanus, proved to be a turning point in astronomy. Like their medieval predecessors, the two men started out working in the Ptolemaic tradition, but by showing the errors in Ptolemy's work, they actually were criticizing him. Their discoveries were not lost on a young Polish astronomer named Nicolaus Copernicus (1473–1543).

THE COPERNICAN REVOLUTION. The story of the Copernican Revolution, the opening chapter in a larger movement known as the Scientific Revolution, is among the greatest sagas in the history of thought. It was a watershed event, marking the birth of modern science as such, but the change in thought patterns created by this revolution was not so much the work of Copernicus as it was of the Italian astronomer Galileo Galilei (1564–1642). Although he often is given less attention than Copernicus and the other most noted figure of the Scientific Revolution, the English natural philosopher Isaac Newton (1642–1727), Galileo was a thinker of the first order who took Copernicus's discoveries much further.

Copernicus had been concerned with how the planets move as they do, and in the course of his work he showed that all of them (Earth included) move around the Sun. Galileo, on the other hand, set out to discover *why* the planets revolve around the Sun, and in so doing he discovered the principles of inertia and gravitational acceleration that would influence Newton. He made numerous other contributions, such as the discovery that Jupiter had moons, but by far his greatest gift to science was his introduction of the scientific method.

Thanks to Galileo and others who later refined the method, thinkers would no longer be content to let mere conjecture guide their work. Before his time, scientists generally had followed a pattern of absorbing the received wisdom of the ancients and then seeking evidence that confirmed those suppositions. The new scientific method, on the other hand, required rigorous work: detailed observation, the formation of hypotheses, testing of hypotheses, formation of theories, testing of theories, formation of laws, testing of laws—and always more observation and testing.

REAL-LIFE APPLICATIONS

GRAVITY, THE SUN, AND EARTH

A fourth key figure in the Scientific Revolution was the German astronomer Johannes Kepler (1571–1630), whose laws of planetary motion directly influenced Newton's laws of gravitation and motion. Thanks to Kepler, we know that planets do not make circular orbits around the Sun; rather, those orbits are elliptical. As Newton later showed, the reason for this is the gravitational pull exerted by the Sun.

Gravitational force explains why Earth, the Sun, and all celestial bodies larger than asteroids are round—but also why they cannot be *perfectly* round. As to the latter issue—the fact that Earth bulges near the equator—it is a consequence of its motion around its axis. Because it is spinning rapidly, the mass of the planet's interior responds to the centripetal (inward) force of its motion, producing a centrifugal, or outward, component. If Earth were standing still, it would be much nearer to the shape of a sphere.

MASS AND SPHERICITY. Now to the larger question: Why is Earth round? The answer is that the gravitational pull of its interior forces a planetary body to assume a more or less uniform shape. Furthermore, the larger the mass of an object, the greater its tendency toward roundness. Earth's surface has a relatively small vertical differential: between the lowest point and the highest point is just 12.28 mi. (19.6 km), which is not a great distance, considering that Earth's radius is about 4,000 mi. (6,400 km).

JOHANNES KEPLER *(The Bettmann Archive. Reproduced by permission.)*

An object of less mass is more likely to retain a shape that is less than spherical. This can be shown by reference to the Martian moons Phobos and Deimos, both of which are oblong, and both of which are tiny, in terms of size and mass, compared with Earth's Moon. Mars itself has a radius half that of Earth, yet its mass is only about 10% of Earth's, and therefore it is capable of retaining a less perfectly spherical shape.

There is also the possibility of more pronounced differences in elevation, and thus it should not be surprising to learn that Mars is also home to the tallest mountain in the solar system. Standing 15 mi. (24 km) high, the volcano Olympus Mons is not only much taller than Earth's tallest peak, Mount Everest (29,028 ft., or 8,848 m), it is also 22% taller than the distance from the top of Mount Everest to the lowest spot on Earth, the Mariana Trench in the Pacific Ocean (–36,198 ft., or –10,911 m).

WHY EARTH IS SPECIAL

With regard to gravitation, a spherical object behaves as though its mass were concentrated near its center. Indeed, 33% of Earth's mass is at is core (as opposed to the crust or mantle), even though the core accounts for only about 20% of the planet's volume. Geologists believe that the composition of Earth's core must be molten iron, which creates the planet's vast electromagnetic field.

Certain particulars of Earth's core lead us to answering another great question about our home planet: Why is it alone capable of sustaining life—as far as we can tell—while the other planets of our solar system are either hellish worlds of fire or frigid, forbidding realms of ice crystals and liquefied gas?

DENSITY OF EARTH'S INTERIOR. At first glance, Earth seems to have few distinctions other than its ability to support life: it is neither the largest nor the smallest planet in the solar system, positions held by Jupiter and Pluto, respectively. (Earth ranks fifth.) Earth has a moon, but that is hardly a distinction: Saturn has 18 moons. And not only does Olympus Mons tower over Everest, but the gaseous oceans of Jupiter also are much deeper than the Mariana Trench. In the lists of planetary superlatives, Earth has only one: it is the most dense.

The only bodies that come close are Mercury and Venus, which along with Earth and Mars are designated as terrestrial planets. (Earth's Moon often is considered along with the terrestrial planets because its composition is similar to them and because it is a relatively large satellite.) The terrestrial planets are small, rocky, and dense; have relatively small amounts of gaseous elements; and are composed primarily of metals and silicates. This is in contrast to the Jovian planets, which are large, low in density, and composed primarily of gases. (The Jovian planets usually are designated as the four giants Jupiter, Saturn, Uranus, and Neptune. Pluto, the smallest of all nine planets, has a density higher than any Jovian planet.)

Density is simply the ratio of mass to volume, meaning that Earth packs more mass into a given volume than any other body in the solar system. Saturn, least dense among the planets, has a mass 95.16 times as great as that of Earth, yet its volume is 764 times greater, meaning that its density is only about 12% of Earth's. But whereas Saturn and other Jovian planets are composed primarily of gases surrounding small, dense cores, Earth—beneath its atmospheric layer and it waters—is a hard little ball. Its core, composed of iron, nickel, and traces of other elements, including uranium, is relatively heavy.

That gives it a strong gravitational pull and, in combination with the comparatively high speed of the planet's rotation, causes Earth to have a powerful magnetic field. It is also important to note the significance of planetary mass in making possible the formation of an atmosphere. Because of their mass, larger planetary bodies exert enough gravitational pull to retain gases around their surfaces; by contrast, the Moon and Mercury are too small and have no atmosphere. Of course, Earth is the only planet whose atmosphere is capable of sustaining life as we know it, and this is a result of activity beneath the planet's surface.

A VOLATILE PLANET. Earth is the only terrestrial planet on which the processes of plate tectonics, or the shifting of plates beneath the planetary surface, take place. The other terrestrial planets have crusts of fairly uniform thickness, suggesting that they have never experienced the internal shifting that has helped give our planet its unique topography. Earth also has a relatively thin lithosphere—the upper layer of the planetary surface, including the crust and the brittle portion at the top of the mantle—which helps make it a particularly volatile body.

Of the terrestrial planets, the only ones still given to volcanic activity are Earth and Venus. Mars seems to have experienced volcanic activity at some point in the past billion years, while Mercury and the Moon have not had volcanoes for several billion years. This is also an important factor in determining Earth's capacity to support living things, because volcanoes—which transport gases from the planet's interior to its atmosphere—have been crucial to the creation of the conditions necessary for sustaining life.

The heat generated by internal volatility is also a component influencing the sustainability of life on Earth. At the time Earth and other planets were formed, some 4.5 billion years ago, the planets experienced such heat that they melted, causing a separation of chemical compounds. The heavier compounds, mostly containing iron, sank to the core of the planet, where they remain today, while the lighter ones rose to the surface. Included in these lighter substances were oxygen and other elements essential to the sustenance of life. Even now Earth and Venus, because of their volcanic activity, are cooling at rates slower than those of the other planets, and this has facilitated the separation of elements.

THE SILHOUETTE OF A STARGAZER AGAINST THE MILKY WAY. *(© F. Zullo/Photo Researchers. Reproduced by permission.)*

THE CREATION AND SUSTENANCE OF LIFE. Aside from the distinctive features of its core, Earth's position relative to the Sun has helped make it possible for life to take root on this planet. For decades scientists believed that Earth is unique in possessing that life-sustaining compound of hydrogen and oxygen, H_2O or water; but now we know that even Jupiter—not to mention Venus and Mars—have water on their surfaces. The problem is that Venus's water is too hot, existing as vapor in the upper atmosphere, while the water on Mars and Jupiter takes the form of ice crystals. Earth is uniquely placed to sustain liquid water.

The existence of liquid water made it possible for the first microorganisms to form on Earth, leading over hundreds of millions of years to the development of the complex biosphere known today. The existence of life in simple forms promoted the development of the atmosphere and geosphere, because these life-forms took in carbon dioxide and water, processed them, and returned them to the environment as oxygen and organic materials.

COMPUTER IMAGE OF THE NINE PLANETS OF OUR SOLAR SYSTEM. *(© Photo Researchers. Reproduced by permission.)*

THE SOLAR SYSTEM AND BEYOND

The reader may have noticed that earlier in this essay, we ceased discussing progress in cosmology after about 1650. This is not because nothing happened after that time; on the contrary, the centuries that have elapsed since then have seen the greatest progress in astronomical study since the dawn of civilization. To give this topic the coverage it warrants, however, would require a lengthy discussion—one that would take us away from the earth sciences and toward the sister science of astronomy.

Up until the Scientific Revolution, the earth sciences hardly existed, except inasmuch as various people over the millennia had recorded data concerning Earth and made sometimes unscientific speculations regarding its origin and composition. As the oldest of the physical sciences, astronomy was much more mature, but even it could progress only so far under the restrictions of the Ptolemaic system. Unfettered, it began to progress rapidly, and the result has been an unfolding vision of the universe that is at once more clear and more complex.

THE SIZE OF THE UNIVERSE. One of the dominant themes in astronomy from Galileo's time to the present day is astronomers' quite literally expanding vision of the universe. Up until 1781, when the German-born English astronomer William Herschel (1738–1822) discovered Uranus, scientists had known only of the

KEY TERMS

ATMOSPHERE: In general, an atmosphere is a blanket of gases surrounding a planet. Unless otherwise identified, however, the term refers to the atmosphere of Earth, which consists of nitrogen (78%), oxygen (21%), argon (0.93%), and other substances that include water vapor, carbon dioxide, ozone, and noble gases such as neon, which together comprise 0.07%.

BIOSPHERE: A combination of all living things on Earth—plants, mammals, birds, reptiles, amphibians, marine life, insects, viruses, single-cell organisms, and so on—as well as all formerly living things that have not yet decomposed. Typically, after decomposing, a formerly living organism becomes part of the geosphere.

CENTRIFUGAL: A term describing the tendency of objects in uniform circular motion to move outward, away from the center of the circle. Though the term *centrifugal force* often is used, it is inertia, rather than force, that causes the object to move outward.

CENTRIPETAL FORCE: The force that causes an object in uniform circular motion to move inward, toward the center of the circle.

COMPOUND: A substance made up of atoms of more than one element chemically bonded to one another.

COSMOLOGY: The study of the origin, structure, and evolution of the universe.

COSMOS: The universe.

ELECTROMAGNETIC ENERGY: A form of energy with electric and magnetic components that travels in waves and which, depending on the frequency and energy level, can take the form of long-wave and short-wave radio; microwaves; infrared, visible, and ultraviolet light; x rays, and gamma rays.

ELEMENT: A substance made up of only one kind of atom. Unlike compounds, elements cannot be chemically broken into other substances.

GEOCENTRIC: Earth-centered.

GEOCHEMISTRY: A branch of the earth sciences, combining aspects of geology and chemistry, that is concerned with the chemical properties and processes of Earth.

GEOLOGY: The study of the solid earth, in particular, its rocks, minerals, fossils, and land formations.

GEOPHYSICS: A branch of the earth sciences that combines aspects of geology and physics. Geophysics addresses the planet's physical processes as well as its magnetic and electric properties and the means by which energy is transmitted through its interior.

GEOSPHERE: The upper part of Earth's continental crust, or that portion of the solid earth on which human beings live and which provides them with most of their food and natural resources.

HELIOCENTRIC: Sun-centered.

HYDROSPHERE: The entirety of Earth's water, excluding water vapor in the atmosphere but including all oceans, lakes, streams, groundwater, snow, and ice.

HYPOTHESIS: An unproven statement regarding an observed phenomenon.

KEY TERMS CONTINUED

INERTIA: The tendency of an object in motion to remain in motion and of an object at rest to remain at rest.

JOVIAN PLANETS: The planets between Mars (the last terrestrial planet) and Pluto, all of which are large, low in density, and composed primarily of gases.

LAW: A scientific principle that is shown always to be the case and for which no exceptions are deemed possible.

LITHOSPHERE: The upper layer of Earth's interior, including the crust and the brittle portion at the top of the mantle.

MANTLE: The layer, approximately 1,429 mi. (2,300 km) thick, between Earth's crust and its core.

MASS: A measure of inertia, indicating the resistance of an object to a change in its motion. (By contrast, weight—which people tend to think of as similar to mass—is a measure of gravitational force, or mass multiplied by the acceleration due to gravity.)

PLANETARY SCIENCE: The branch of the earth sciences, sometimes known as planetology or planetary studies, that focuses on the study of other planetary bodies. This discipline, or set of disciplines, is concerned with the geologic, geophysical, and geochemical properties of other planets but also draws on aspects of astronomy, such as cosmology.

PROTON: A positively charged particle in the nucleus of an atom.

SCIENTIFIC METHOD: A set of principles and procedures for systematic study that includes observation; the formation of hypotheses, theories, and laws; and continual testing and reexamination.

SCIENTIFIC REVOLUTION: A period of accelerated scientific discovery that completely reshaped the world. Usually dated from about 1550 to 1700, the Scientific Revolution saw the origination of the scientific method and the introduction of such ideas as the heliocentric (Sun-centered) universe and gravity.

TERRESTRIAL PLANETS: The four inner planets of the solar system: Mercury, Venus, Earth, and Mars. These planets are all small, rocky, and dense; have relatively modest amounts of gaseous elements; and are composed primarily of metals and silicates. Compare with Jovian planets.

THEORY: A general statement derived from a hypothesis that has withstood sufficient testing.

UNIFORM CIRCULAR MOTION: The motion of an object around the center of a circle in such a manner that speed is constant or unchanging.

five other planets visible to the naked eye, all of which had been discovered in prehistoric times. (Neptune was discovered in 1846 and Pluto not until 1930.)

In the seventeenth century, astronomers still regarded what we call the solar system as the entire universe, but Herschel was instrumental in ascertaining that Earth is part of a bright band of stars called the Milky Way. Just as Earth once had been believed to be the center of the "universe," or solar system, astronomers then came to believe it was at the center of the Milky Way. Only since 1920 has it been known that our solar system is, in fact, somewhere between the center and the edge of the vast galaxy. Even the Milky Way, composed of several hundred billion stars and about 120,000 light-years in diameter, is not the entire universe; it is only one of many hundreds of galaxies or "island universes."

As discussed at the beginning of this essay, such a scale is almost too much for the human mind to comprehend, particularly inasmuch as Earth is the only planet known to sustain intelligent life. As the British science-fiction writer Arthur C. Clarke (1917–) has observed, either there are other intelligent life-forms out there in the universe, or there are not—and either possibility is mind-boggling.

THE BIG BANG AND THE SOLAR SYSTEM. Not only has astronomers' understanding of the universe expanded, along with their idea of its size; it also appears that the universe itself is expanding. Today the most widely accepted model regarding the formation of the universe is the big bang theory, first put forward by the Belgian astrophysicist Georges Édouard Lemaître (1894–1966) in 1927. According to this theory, an explosion 10-20 billion years ago resulted in the rapid creation of all matter in the universe, and that matter is continuing to move outward, expanding the frontiers of the universe.

Our own solar system appears to be about five billion years old, meaning that the Sun is a relatively young star. It seems that the future solar system was just one of many great balls of gas, rotating as they moved outward, that were scattered around the universe as a result of the big bang. Just as these balls of gas exploded from the center, the material of the various stars emerged from the center of the ball that became our solar system.

FORMATION OF THE PLANETS. The proto–solar system we have described here was a great rotating cloud, and though it has long since ceased to be a cloud, it continues to rotate—only now it is in the form of planets turning around a sun at the center. The hottest portion of the cloud, at the center, became the Sun, while cooler portions at the fringes became planets. The Sun itself is composed primarily of hydrogen and helium, the two most plentiful elements in the universe. In the extraordinarily high temperatures on the Sun, atoms of hydrogen (which has one proton in its nucleus) experience nuclear fusion, becoming atoms of helium, which has two protons. It appears that continued fusion resulted in the creation of the heavier elements (for instance, nitrogen, carbon, oxygen, and silicon) of which the planets—in particular, our own—are composed.

Earth's elemental makeup is discussed elsewhere in this book, as is the structure of its interior. So, too, is the Sun's effect on Earth. These matters are not unrelated. In studying the solar system and the planets that make it up, one is confronted again and again with the fact that a planet's destiny is governed by its position relative to the Sun. Ultimately, the planets in our solar system are ruled by the same principle that drives the sale of real estate: location, location, location!

This is true not only of the atmosphere and temperature of planets but also of their relative density. It is no mistake that the terrestrial planets are closer to the Sun: their internal composition is as it is because these bodies became the destination of most of the heavier elements that emanated from it. Many of the lighter elements continued to move outward, where they gathered around rocky centers to become the mostly gaseous Jovian planets.

WHERE TO LEARN MORE

Astronomy and Cosmology (Web site). <http://www.netlabs.net/hp/tremor/>.

Beatty, J. Kelly. *The New Solar System.* New York: Sky, 1999.

Cambridge Cosmology (Web site). <http://www.damtp.cam.ac.uk/user/gr/public/cos_home.html>.

Canup, Robin M., and Kevin Righter. *Origin of the Earth and Moon.* Tucson: University of Arizona Press, 2000.

The Cosmology: Explore the Largest Mystery (Web site). <http://library.thinkquest.org/28181/>.

Gallant, Roy A. *Earth's Place in Space.* New York: Benchmark Books, 1999.

Lambert, David, and Martin Redfern. *The Kingfisher Book of the Universe.* New York: Kingfisher, 2001.

Llamas Ruiz, Andrés. *The Origin of the Universe.* Illus. Luis Rizo. New York: Sterling Publishers, 1997.

Ned Wright's Cosmology Tutorial (Web site). <http://www.astro.ucla.edu/~wright/cosmolog.htm>.

Skinner, Brian J., Stephen C. Porter, and Daniel B. Botkin. *The Blue Planet: An Introduction to Earth System Science.* 2d ed. New York: John Wiley and Sons, 1999.

SUN, MOON, AND EARTH

CONCEPT

Earth is intimately tied to the star around which it revolves, the Sun, and the satellite that revolves around Earth itself, the Moon. Without the Sun, of course, life on Earth simply could not exist, not just because of the need for light but to an even greater degree because of the energy it supplies. For that matter, Earth itself would not exist: our planet appears to have developed from the same cosmic cloud that formed the Sun, and without the Sun to hold it in place with its gravitational pull, Earth would go spinning off into space. For all its influence on human life, however, the Sun has less impact on the tides than the Moon, which is smaller but much closer. Together, these two bodies literally define time in human experience, which has been marked by the movements of the Sun and Moon from a time before civilization began.

HOW IT WORKS

FORMATION OF THE SUN AND PLANETS

Scientists today believe that the universe began between 10 and 20 billion years ago, with an event nicknamed the "big bang," The galaxies continued (and still continue) to move outward from that center, and among them was our Milky Way. Somewhere between the center and the rim of the Milky Way, a rotating cloud of cosmic gas formed about six billion years ago,

The center of the cloud, where the greatest amount of gases gathered, was naturally the densest and most massive portion as well as the hottest. There, hydrogen—the lightest of all elements—experienced extraordinary amounts of compression, owing to the density of the clouded gases around it, and underwent nuclear fusion, or the bonding of atomic nuclei. This hot center became the Sun about five billion years ago, but there remained a vast nebula of gas surrounding it. As the fringes of this nebula began to cool, the gases condensed, forming solids around which particles began to accumulate. These were the future planets.

THE PLANETS AND THE ELEMENTS. Closest to the Sun, the planets and other satellites were formed of elements that could condense at high temperatures: iron, silicon, aluminum, calcium, and magnesium. These elements, along with oxygen-containing compounds, constituted the material foundation around which other particles accumulated to form the terrestrial planets: Mercury, Venus, Earth, and Mars. Further from the Sun, where temperatures were lower, gaseous compounds—including methane, ammonia, and even water—could condense. These compounds became the basis around which the five outer planets (the four Jovian planets and Pluto) formed.

The process of planetary formation involved additional steps and took place over a period of about 500 million years. Some of the factors that played a part in forming Earth are discussed in Planetary Science, but in the present context, let us consider the source of those elements mentioned here. Did they just magically form? Were they always there? In answer to the first question, there is nothing magical about the formation of "new" elements, though it almost seems so. As for the second question, the answer is yes and no.

EARTH WITH THE MOON IN ECLIPSE. *(© Chris Bjornberg/Photo Researchers. Reproduced by permission.)*

ELEMENTS AND THE SUN

The elements themselves were not always there in the universe or on Earth itself; however, the subatomic building blocks that make them up have indeed existed from the beginning of the universe. The basic atomic structure is as follows. There is a nucleus in which one or more protons (positively charged subatomic particles) may reside along with one or more neutrons, which have no charge. Spinning around the nucleus are one or more electrons, or negatively charged subatomic particles.

The number of protons and electrons in an atom is always the same, meaning that the atom has no electric charge. Atoms that have lost or gained electrons (in which case they would acquire a positive or negative charge, respectively) are called ions. Electrons, which move fast and possess very small mass compared with protons and neutrons, are very easy to dislodge from an atom; on the other hand, it takes an extraordinary event to change the number of protons in the nucleus.

This fact is significant, because it points toward the defining characteristic of an element: the number of protons in the nucleus. Atoms of a particular element *always* have the same number of protons, called their atomic number. The atomic number of an element can be determined by consulting the periodic table of elements: for instance, iron, with an atomic number of 26, *must* have 26 protons in its nucleus. If an atom has 25, it is manganese, and if it is has 27, it is cobalt.

FORMING NEW ELEMENTS. The number of neutrons in the nucleus may vary for atoms within a given element. Atoms that have the same number of protons (and are thus of the same element) but differ in their number of neutrons are called isotopes. Most isotopes are stable, meaning that their chemical composition will remain as it is; however, some isotopes are radioactive, meaning that they experience the spontaneous emission of particles or energy over a given period of time.

Radioactive decay is one of two ways that one element can become another. When a radioactive isotope emits an alpha particle, for example, its nucleus expels a positively charged nucleus consisting of two protons and two neutrons, which is the same thing as a helium atom stripped of its electrons. This obviously changes the number of protons in the nucleus of the isotope and may result in its stabilization. The other means of forming a new element is by nuclear fusion, in which two atomic nuclei fuse or bond.

NUCLEAR FUSION. Note that the first of these means by which elements are formed is subtractive; in other words, with radioactive decay, a different element is formed by the expulsion of protons. Nuclear fusion, on the other hand, is additive, resulting in the creation of different elements by the addition of protons to an atomic nucleus. Radioactive decay takes place inside Earth (among other places), while nuclear fusion is the source of the Sun's power.

Nuclear fusion involves the release of huge amounts of energy. On Earth, scientists have been able to bring about uncontrolled nuclear fusion in the form of the so-called hydrogen bomb, which is actually a "fusion bomb." They have yet to succeed in creating controlled nuclear fusion. If and when they do, it would provide a safe, clean source of almost limitless power and probably would constitute the greatest scientific or technological discovery since fire.

On the Sun, nuclear fusion has been taking place, and will continue to do so, for a long, long time. The 92 naturally occurring elements of the universe are the result of fusion reactions, meaning that all that we see around us was once part of a star. This represents a major break with the ancient belief that Earth is made of fundamentally different substances than are the bodies of space (see Earth, Science, and Nonscience). In fact, our world and everything in it—including our own bodies—is truly "the stuff of stars."

THE MOON AND EARTH

In comparison to the Sun, the Moon is altogether less remarkable. Below, we review some statistics about the sizes of each, but as every elementary-school student today knows, the Sun is much, much larger and exerts far more impact on the fate of the solar system—including Earth. The Moon does not even have its own energy sources: its light comes from the Sun, and the absence of an atmosphere, of volcanic activity, or even of a significant magnetic field makes it a very dull place indeed.

Yet the Moon has inspired at least as much fascination among humans over the ages as has the Sun. There is its physical beauty, though comparisons with the Sun are hardly fair, since we cannot look at the Sun without damage to our eyes. There is also its influence on earthly cycles ranging from the tides to the months themselves, though some claims of lunar influence have little basis in fact. During the Middle Ages, many believed that the Moon caused madness, a superstition still reflected in our word lunacy.

Still, humans have long associated a spirit of mystery with the Moon, in part because of its ever-changing appearance and in part because it has always showed just one side to Earth. (We discuss the reason why later.) Only in 1959, when the Soviet space probe *Luna* traveled to the "dark side of the Moon," did scientists gain a glimpse of it. Unmanned and later manned journeys, which culminated with the U.S. Moon landing in 1969, also changed astronomers' understanding of the Moon's origins.

THE "BIG WHACK." One of the curious things about the Moon is its size in relation to Earth. Nowhere in the universe is there such a small size differential between a satellite and the planet around which it orbits, the only possible exception being Pluto and its moon, Charon. Because our Moon is so close in size to Earth, scientists once speculated that they might have shared origins, and this speculation informed several theories concerning the formation of the Moon.

According to the fission theory, the Moon was a piece of Earth that had been torn away, perhaps from the Pacific basin. The simultaneous creation theory likewise depicted Earth and the Moon as sharing origins, but in this case they literally had been formed together from the same materials. Finally, there was the capture theory, which, in contrast to the others, assumed quite different origins for the two bodies: the Moon had formed somewhere else in the solar system and had been captured by Earth's gravitational field after it wandered too close to the planet.

As it turned out, the capture theory was closest to the theory accepted today, though it was discarded along with the other two on the basis of data brought back from the Apollo Moon landings. According to the giant impact theory, sometimes called the Big Whack model or the ring ejection theory, at a young age Earth was sideswiped by a celestial object as large or larger than Mars. As a result of that collision, a ring of crustal matter was spewed into space, and over time the matter in this ring agglomerated to form the Moon.

VITAL STATISTICS OF THE SUN AND MOON

The Moon is about 240,000 mi. (385,000 km) from Earth, meaning that it takes about 1.25 seconds for its light to reach Earth. By contrast, the Sun's distance from Earth is so great that it takes eight minutes for sunlight to reach our planet, even though light travels through space at the speed of about 186,000 mi. (299,339 km) per second.

The distance between Earth and the Sun is the basis for the astronomical unit (AU), a figure used for measuring the distance between bodies in the solar system. Equal to the average distance from Earth's center to the center of the Sun, an AU is designated $1.49597870691 \times 10^8$ km, or approximately 92,955,807 mi. Usually we think of the solar system as the area encompassed by the orbit of the most remote planet, Pluto, but that is only 39.44 AU, a tiny figure compared with the diameter of the realm within the Sun's gravitational pull, which is a staggering 100,000 AU.

VOLUME AND MASS. The Sun itself has a diameter of about 856,000 mi. (1,392,000 km), meaning that the distance across it is about 109 times that of Earth's diameter. Another way to consider that figure is this: if one were to draw a circle as big as the Sun around Earth, the edge of that circle would be about twice as far away as the Moon. The Sun's volume is so great that about 1.3 million Earths could fit inside it, and its mass is about 300,000 times that of Earth. In fact, it accounts for about 99.8% of the mass of the entire solar system.

By contrast, the Moon has a diameter of only about 2,160 mi. (3,475 km), a little less than the distance from New York to Los Angeles. Its mass is a little more than 1% of Earth's, and as a result, its gravitational pull is too small to retain the gases that make up the atmosphere. That small mass, combined with an imbalance in its distribution, explains why the Moon shows only one face to Earth. The side of the Moon facing Earth is of greater mass than the other side and is therefore more strongly attracted by Earth's gravitational force. The result is a phenomenon called gravitational locking, whereby the Moon rotates on its axis at exactly the same rate as it travels around Earth—once ever 29.5 days.

Clearly, the Moon is tiny in both volume and mass, and were it as far away as the Sun, we would hardly pay it notice. Yet it should be pointed that even the Sun itself, while remarkable in our own solar system, is far from a standout in the universe as a whole. It is a youngish star, of average size, and not all that different from billions of other stars throughout the cosmos. It is not even unique in being the only star with its own solar system. In 1999 astronomers discovered an entire solar system some 44 light-years from Earth, in which three large planets were found to be circling the star Upsilon Andromedae.

REAL-LIFE APPLICATIONS

ENERGY, TEMPERATURE, AND COMPOSITION OF THE MOON

Of course, the area in which the Sun and Moon differ most significantly is in terms of the energy they possess, which, in turn, affects both their

THE MOON'S SURFACE IS POCKMARKED WITH CRATERS, SHOWING ITS VULNERABILITY TO DAMAGE CAUSED BY PARTICLES FROM SPACE. *(© NASA/Photo Researchers. Reproduced by permission.)*

temperature and the composition of each body. With regard to the Moon's temperature, little need be said, since it is entirely a function of the Moon's exposure (or lack of exposure) to heat from the Sun. On the Moon figures range from 280°F (138°C) to –148°F (–100°C), with a mean temperature of –10°F (–23.33°C).

As for the Moon's composition, it has similarities to and differences from that of Earth, and both these similarities and differences are instructive. The internal composition of the Moon is such that it often is treated along with the terrestrial planets. In addition, it is not much smaller than the smallest terrestrial planet, Mercury. But whereas Mercury has a proportionally large, extremely hot core, the Moon's core is much smaller in proportion to its total size and much cooler. In all likelihood it is not even molten, or only a portion of it is.

It appears that the Moon has an internal structure not dissimilar to that of Earth—that is, a crust, mantle, and core. The materials that make up the Moon, however, are quite different. Aluminum, calcium, iron, magnesium, titanium, potassium, and phosphorus have been found on the Moon, but it seems bereft of organic compounds, indicating that life never existed there. Moon rocks are primarily of basalt, or hardened lava, and breccia, soil, and rock fragments that have melted together.

The formation of these rocks must have occurred a long, long time ago, because the Moon has not had volcanic activity for several billion years. Thus, it is almost entirely "dead," lacking even a magnetic field of any significance. Moon rocks have only the faintest magnetic field, suggesting the absence of a significant molten core, which is what gives Earth its own magnetism. From the traces of magnetism found in these rocks, it is possible that the Moon had a magnetic field of some significance at one time, but that time is long past.

THE VULNERABLE MOON. Owing to the Moon's lack of sufficient gravitational force, it retains no atmosphere, and this means that it is completely vulnerable to particles reaching it from space. Early in its life, the Moon was subjected to a 700-million-year-long meteor shower that formed its craters. The damage from these showers was so great that it melted the Moon's crust, and eventually lava from the lunar interior surfaced to fill in cracks made by the meteorites.

In 1609 the Italian astronomer Galileo Galilei (1564–1642), the first scientist to gaze at the Moon through a telescope, observed dark patches that looked to him like bodies of water. He named these patches *seas,* a term that has remained in use among lunar cartographers, though Galileo's "seas" long ago were identified as dark spots made by cooling lava in the cracked surface.

As it is vulnerable to particles from space, the Moon also is susceptible to dramatic temperature changes brought about by the presence or absence of sunlight. For this reason, parts of the lunar surface are extremely hot at certain times, while other portions have never been penetrated by sunlight and are almost inconceivably cold.

HUMANS ON THE MOON

One such cold spot is at the Moon's south pole, in a basin carved out by an asteroid. There, temperatures are as low as –387°F (–233°C), within 40°C of absolute zero, or the temperature at which all molecular motion virtually ceases. By contrast, Pluto, which is so far from the Sun that the great star appears merely as a bright dot from there, has an average surface temperature that is warmer by 8°C. Yet in this inhospitable spot on the lunar surface, the U.S. probe *Lunar Prospector* in 1998 found something quite surprising: water.

Mixed in with dirt in the South Pole–Aitken Basin are ice crystals, which scientists speculate make up about 10% of the material in the surrounding area. Apparently moisture residue from comets that struck the Moon over the past three billion years, the ice offers intriguing possibilities. If a cost-effective method for extracting ice from the soil can be developed, colonization and exploration of the Moon may become a reality.

MOON LANDINGS. For now, however, human presence on the Moon is more a thing of the past than of the future. The first man-made spacecraft to visit the Moon was the Soviet *Luna 2* in 1959, and seven years later, *Luna 9* became the first such craft to land on the lunar surface. Its landing dispelled long-held fears that the Moon was awash in thick layers of dust; in fact, the lunar surface is a grayish soil, primarily composed of rock fragments, from 5 ft. to 20 ft. (1.5 m to 6 m) deep.

Though the Soviets were the first to reach the Moon in unmanned craft, the United States was the first (and so far the only) nation to put a man on the Moon. On July 20, 1969, in one of the most dramatic events of human history, the U.S. astronauts Neil Armstrong (1930–) and Edwin ("Buzz") Aldrin (1930–) walked on the lunar surface while Michael Collins (1930–) piloted *Apollo 11* in its orbit around the Moon.

Photographs from the Moon landing show the American flag extended, as though waving in a breeze as it would on Earth, but it actually was held in place mechanically, since there is no wind on the Moon. Aldrin and Armstrong demonstrated one of the most dramatic differences between the Moon and Earth: a decrease in weight. A man who weighs 200 lb. (90.72 kg) on Earth would weigh 33.04 lb. on the Moon and would therefore be much easier to lift. But he would be no easier to push from side to side, because his mass would not have changed.

Weight is the product of mass multiplied by the rate of gravitational acceleration and is therefore dependent on the gravitational force of the celestial body on which it is measured. Mass, on the other hand, does not vary anywhere in the universe. Thus, although 33.04 lb. is equal to 14.99 kg on Earth, the hypothetical astronaut described here would still have a mass of 90.72 kg.

Physicists had long known these facts about weight and mass, but footage of Armstrong and Aldrin bouncing around on the lunar surface provided a much more vivid demonstration. As it turned out, however, their foray to the Moon was the beginning of an all-too-brief chapter. Over the next three years, the United States conducted five more lunar landings as well as the failed 1970 mission designated *Apollo 13* (portrayed in the 1995 movie of that name), which never landed due to an onboard explosion. The last Moon mission took place in 1972, and soon afterward America was thrown into a recession spawned by the 1973 oil crisis. The next sustained effort at space flight, which began with the launch of the space shuttle in 1981, had an entirely different mission and destination—one that lay well within Earth's gravitational field.

SOLAR ENERGY, TEMPERATURE, AND COMPOSITION

Needless to say, there never will be any space missions, manned or otherwise, to the Sun. Spacecraft have passed close to Mercury, but as for the Sun, even its corona, or outermost atmospheric layer, has a temperature of 2,000,000 on the Kelvin scale of absolute temperature, or 3,599,541°F.

The Sun could not be more different from the Moon, most notably in terms of energy and temperature. It should be noted, however, that the Sun's temperature does not increase uniformly from the corona to the core. The "coolest part," at 5,800K (9,981°F), is the photosphere, a layer some 300 mi. (480 km) thick that constitutes the visible surface of the Sun. Above it, temperatures rise through the chromosphere, which is about 1,600 mi. (2,560 km) thick, becoming hotter still in the corona. Scientists do not understand the reasons for this temperature rise at the outer surface of the Sun.

Less surprising is the continued increase of temperature beneath the photosphere. The deeper inside the Sun, the higher the temperatures, until it reaches a staggering 15,000,000K (27,000,000°F) at the core. It is so hot there that not even atoms can exist; instead, the Sun's core is made up of subatomic particles—specifically, protons and electrons. Heat and movement are directly related in physics, and these particles are moving very fast—so fast that nuclear fusion can and does occur when these particles smash together.

NUCLEAR FUSION IN THE SOLAR CORE. When four protons fuse and absorb two electrons, amazing things happen. The result is the creation of a helium nucleus, whose mass is slightly smaller than the combined mass of the separate particles. Where did that "missing" mass go? It was converted to energy—an amount of energy that is staggering when multiplied by the large numbers of hydrogen atoms being converted to helium at any given moment.

That, in essence, is how the Sun creates energy—by converting hydrogen to helium. This energy is radiating from the Sun in electromagnetic waves at an amazing rate every second, and eventually it will be used up. Long before that happens, however, the Sun will begin to expand and cool. "Cooling," of course, is a relative term; this expanding Sun, a red giant, will burn up Earth even as it absorbs Mercury entirely. Then, when it has used up all its hydrogen, nothing will remain but a glowing core called a white dwarf.

There is no need to worry, however, because the events described will not happen anytime soon. The Sun has about as much life left in it as the amount of time it has lasted so far—approximately five billion years, or longer than Earth has existed. Given the fact that about 4.96 million tons (4.5×10^6 metric tons) of hydrogen are converted to helium every second, this gives some idea of the vast energy reserves on the Sun.

THE SUN'S ENERGY AND EARTH. Sunlight is more than just light, though that alone is a marvelous thing. The rays projected by the Sun are electromagnetic energy, of which visible light is only a small part. Also contained in the electromagnetic spectrum are long waves, short waves, and microwaves (including those used for transmitting radio and television signals); infrared and ultraviolet light; and x rays and gamma rays. Solar energy travels to Earth by means of radiation, a form of heat transfer that, unlike conduction or convection, requires no physical medium such as air. It can move through the vacuum of space to Earth's atmosphere, where a portion of it is absorbed and becomes the fuel that powers spaceship Earth.

It is a measure of the Sun's vast energy that Earth receives only 0.00000005% of its total output at any given moment. Of that small fraction (equal to one part in two billion), a much smaller portion makes it through Earth's atmosphere—yet that is enough to light the world and to facilitate the myriad other functions, such as photosynthesis, for which we depend on the Sun. Many effects of the Sun's light and energy are less than desirable, of course: sunburns and sometimes even tans, bleaching of materials exposed to light for too long, temporary blindness caused by gazing at the Sun or even at something reflecting it, and so on. These effects, too, attest to the Sun's awesome power.

The Sun makes possible the operations of three of four Earth systems and indirectly affects the fourth (see Earth Systems). The hydrosphere and atmosphere are both affected, for instance, by the evaporation of water for eventual precipitation, a process powered by the Sun. Likewise, the biosphere could not exist without photosynthesis and the other biological processes dependent on sunlight. Even the geosphere, though not directly powered by the Sun, is influenced by Sun-powered phenomena from the other three spheres.

CURIOUS SOLAR PHENOMENA

Even traveling at the speed of light, a photon takes nearly 30,000 years to travel from the center of the Sun to the corona. If it were in a vacuum, it could make the journey in about four seconds, but it continually bumps into other particles, and this slows it down (to put it mildly). Once it finally escapes the solar surface, it travels rather quickly, transmitting the Sun's light to the solar system.

When that light reaches Earth's atmosphere, it creates a number of strange optical effects. One of the best known is the rainbow, produced by the refraction, or bending, of light inside a raindrop. Because the colors of the visible spectrum are bent at different angles, they disperse as though in a prism; after being refracted a second time by the surrounding air, the billions of raindrops that fill the atmosphere after a storm produce a brilliant band of light.

When sunlight strikes oxygen or nitrogen, the two most significant elements in our atmosphere, the shorter wavelengths of light—green, blue, and violet—are the ones reflected. When

THE AURORA BOREALIS, OR NORTHERN LIGHTS. THIS PHENOMENON IS CAUSED BY SOLAR PARTICLES DRAWN TO EARTH'S ATMOSPHERE, WHERE THEY COLLIDE WITH OXYGEN AND NITROGEN MOLECULES AND BECOME ELECTRICALLY CHARGED, TAKING ON A COLORED GLOW. *(© Michael Giannechini/Photo Researchers. Reproduced by permission.)*

these wavelenths combine, they appear bluish. The combination of light and differing temperatures on the ground produces a mirage, while ice crystals in the air may bring about haloes around the Sun or Moon and, in some cases, a sun dog, a reflected image of the Sun.

THE NORTHERN LIGHTS. One of the most breathtaking displays of atmospheric optical phenomena are auroras (or aurorae), natural fireworks that appear in the night sky. Visible in the northern United States and Canada, the auroras there are known as aurora borealis, or northern lights. A similar phenomenon occurs in the Southern Hemisphere, where it is called the southern lights, or aurora australis.

The source of the auroras is a stream of particles, called the *solar wind,* that emanates from the Sun. The solar wind is a consequence of the extreme heat on the Sun's surface, which causes atoms there to move so rapidly that even the enormous gravitational force of the Sun itself cannot hold them. Solar particles pass through the entire solar system and out to space beyond, but few of them reach Earth's atmosphere, because they are deflected by the planet's magnetic field.

The particles that cause auroras are electrons that instead of being deflected, are drawn toward Earth's polar regions. Two great rings of charged particles, called the Van Allen belts, are located high above each polar region, and they catch the majority of charged particles flowing toward them from the northern and southern hemispheres, respectively. The Van Allen belts cannot trap all the charged particles, however, and some enter the atmosphere, where they collide with oxygen and nitrogen molecules at about 50 mi. to 600 mi. (80 km to 1,000 km) above sea level. Some of these molecules become electrically charged and take on a colored glow. At higher levels oxygen glows red, while oxygen at lower levels is yellowish-green. Nitrogen glows blue.

SUNSPOTS AND THE SOLAR ACTIVITY CYCLE. Though auroras are visually remarkable, a more significant solar phenomenon is the sunspot. Sunspots are cooler regions—many of them vast in size—on the photosphere. They tend to be about 2,700°F (1,500°C) cooler than the surrounding areas, and for this reason they appear darker.

Sunspots seem to occur in 11-year cycles—a period known as a solar activity cycle—and are the result of strong magnetic fields. The latter, in turn, result from something called differential rotation: the Sun's equator rotates once every 26 days, whereas its poles rotate every 36 days, resulting in massive twisting of magnetic fields. This can produce anomalies such as sunspots, which disrupt radio communications on Earth.

Other solar anomalies, such as prominences, or hot spots formed by magnetic loops in the Sun's atmosphere, also follow the 11-year solar activity cycle. Actually, the cycle as it is known and measured *today* is about 11 years long. Its length seems to have varied over time, producing past changes in global temperature: the shorter the cycle, the warmer the temperatures on Earth.

THE SUN, MOON, AND EVENTS ON EARTH

As noted earlier, the Sun and Moon have long inspired awe in humans, and in this the Moon has been a more than equal partner. Prehistoric and ancient humans regarded both lunar eclipses, in which Earth's shadow covers the Moon, and solar eclipses, in which the Moon comes between Earth and the Sun, as portents from the gods. From at least the era of the Babylonians onward, the Sun and Moon played complementary roles in marking time.

The marking of time according to the Sun relies on an objective reality, rather than a mere human construct. The year, which involves a complete cycle of seasons, is based on the amount of time Earth takes to revolve around the Sun. During this time, the planet also moves on its axis, causing the changes in orientation that bring about the seasons. Likewise a day is an objective reality, being the amount of time it takes Earth to revolve on its axis.

The month and week, on the other hand, relate to the phases of the Moon, which are themselves dependent on the Moon's position relative to Earth and the Sun. A lunar cycle lasts about 29.5 days, and this became the basis for the month, which lasts from 28 to 31 days. Within that lunar cycle are four phases—new, first quarter, full, and last quarter—which eventually became the basis for the idea that a month has about four seven-day weeks.

DAYS OF THE WEEK. Until about 1500, people thought of the Sun and Moon as two of the seven "planets" (including the five planets visible with the naked eye) that supposedly revolved around Earth. They further related these "planets" to the days of the week. As a result, virtually every culture that speaks a Euro-

KEY TERMS

ABSOLUTE ZERO: The temperature at which all molecular motion virtually ceases.

ASTRONOMICAL UNIT (AU): A figure equal to the average distance from Earth's center to the center of the Sun. The SI figure for an AU, adopted in 1996, is equal to $1.49597870691 \times 10^8$ km, or approximately 92,955,807 mi.

ATMOSPHERE: In general, an atmosphere is a blanket of gases surrounding a planet. Unless otherwise identified, however, the term refers to the atmosphere of Earth, which consists of nitrogen (78%), oxygen (21%), argon (0.93%), and other substances that include water vapor, carbon dioxide, ozone, and noble gases such as neon, which together comprise 0.07%.

ATOM: The smallest particle of an element, consisting of protons, neutrons, and electrons. An atom can exist either alone or in combination with other atoms in a molecule.

ATOMIC MASS UNIT: An SI unit (abbreviated amu), equal to 1.66×10^{-24} g, for measuring the mass of atoms.

ATOMIC NUMBER: The number of protons in the nucleus of an atom. Since this number is different for each element, elements are listed on the periodic table in order of atomic number.

AVERAGE ATOMIC MASS: A figure used by chemists to specify the mass—in atomic mass units—of the average atom in a large sample.

BIOSPHERE: A combination of all living things on Earth—plants, mammals, birds, reptiles, amphibians, aquatic life, insects, viruses, single-cell organisms, and so on—as well as all formerly living things that have not yet decomposed. Typically, after decomposing, a formerly living organism becomes part of the geosphere.

CORE: The center of Earth, which appears to be of molten iron. For terrestrial planets in general, *core* refers to the center, which in most cases is probably molten metal of some kind.

COSMOLOGY: The study of the origin, structure, and evolution of the universe.

COSMOS: The universe.

ELECTROMAGNETIC ENERGY: A form of energy with electric and magnetic components, which travels in waves.

ELECTROMAGNETIC SPECTRUM: The complete range of electromagnetic waves on a continuous distribution from a very low range of frequencies and energy levels, with a correspondingly long wavelength, to a very high range of frequencies and energy levels, with a correspondingly short wavelength. Included on the electromagnetic spectrum are long-wave and short-wave radio; microwaves; infrared, visible, and ultraviolet light; x rays, and gamma rays.

ELECTRON: A negatively charged particle in an atom, which spins around the nucleus.

ELEMENT: A substance made up of only one kind of atom. Unlike compounds, elements cannot be broken chemically into other substances.

GEOSPHERE: The upper part of Earth's continental crust, or that portion of the solid earth on which human beings live

KEY TERMS CONTINUED

and which provides them with most of their food and natural resources.

HYDROSPHERE: The entirety of Earth's water, excluding water vapor in the atmosphere but including all oceans, lakes, streams, groundwater, snow, and ice.

INERTIA: The tendency of an object in motion to remain in motion and of an object at rest to remain at rest.

ION: An atom that has lost or gained one or more electrons and thus has a net electric charge.

ISOTOPES: Atoms that have an equal number of protons, and hence are of the same element, but differ in their number of neutrons. This results in a difference of mass. An isotope may be either stable or radioactive.

JOVIAN PLANETS: The planets between Mars (the last terrestrial planet) and Pluto, all of which are large, low in density, and composed primarily of gases.

KELVIN SCALE: Established by William Thomson, Baron Kelvin (1824–1907), the Kelvin scale measures temperature in relation to absolute zero, or 0K. (Note that units in the Kelvin system, known as Kelvins, do not include the word or symbol for "degree.") The Kelvin scale, which is the system usually favored by scientists, is directly related to the Celsius scale; hence Celsius temperatures can be converted to Kelvin by adding 273.15.

LIGHT-YEAR: A unit of distance used by astronomers for measuring the extremely large expanses of space. Equal to the distance light travels in a year, a light-year is $9.460528405 \times 10^{12}$ km, or approximately 5.88 trillion mi.

LITHOSPHERE: The upper layer of Earth's interior, including the crust and the brittle portion at the top of the mantle.

MANTLE: The layer, approximately 1,429 mi. (2,300 km) thick, between Earth's crust and its core. In reference to the other terrestrial planets, *mantle* simply means the area of dense rock between the crust and core.

MASS: A measure of inertia, indicating the resistance of an object to a change in its motion. (By contrast, weight—which people tend to think of as analogous to mass—is a measure of gravitational force, or mass multiplied by the acceleration due to gravity.)

NEUTRON: A subatomic particle that has no electric charge. Neutrons are found at the nucleus of an atom, alongside protons.

NUCLEAR FUSION: A nuclear reaction that involves the joining of atomic nuclei.

pean language—not only in Europe itself but also among former European colonies in the Americas, Africa, Asia, and the Pacific—refers to the first day of the week as the Sun's day and the second as the Moon's day.

Today, in countries that speak Romance languages, or languages derived from Latin (most notably Italian, French, Spanish, and Portuguese), variations on the original Latin names remain in use: Domingo and Lunes in Spanish, for instance, or Dimanche and Lundi in French. Germanic languages adopted the ideas of the Latin day names, for the most part, but translated them into their own tongues: hence, Sonntag

KEY TERMS CONTINUED

NUCLEUS: The center of an atom, a region where protons and neutrons are located and around which electrons spin.

ORGANIC: At one time, chemists used the term "organic" only in reference to living things. Now the word is applied to most compounds containing carbon, with the exception of calcium carbonate (limestone) and oxides, such as carbon dioxide.

PERIODIC TABLE OF ELEMENTS: A chart that shows the elements arranged in order of atomic number, along with chemical symbol and the average atomic mass for that particular element.

PHOTON: A particle of electromagnetic radiation carrying a specific amount of energy.

PLANETARY SCIENCE: The branch of the earth sciences, sometimes known as planetology or planetary studies, that focuses on the study of other planetary bodies. This discipline, or set of disciplines, is concerned with the geologic, geophysical, and geochemical properties of other planets but also draws on aspects of astronomy, such as cosmology.

PROTON: A positively charged particle in an atom.

RADIATION: The transfer of energy by means of electromagnetic waves, which require no physical medium (for example, water or air) for the transfer. Earth receives the Sun's energy, via the electromagnetic spectrum, by means of radiation.

RADIOACTIVITY: A term describing a phenomenon whereby certain materials are subject to a form of decay brought about by the emission of high-energy particles or radiation. Forms of particles or energy include alpha particles (positively charged helium nuclei); beta particles (either electrons or subatomic particles called positrons); or gamma rays, which occupy the highest energy level in the electromagnetic spectrum.

REFRACTION: The bending of light as it passes at an angle from one transparent material into a second transparent material. Refraction accounts for the fact that objects under water appear to have a different size and location than they have in air.

SI: An abbreviation of the French term *Système International d'Unités,* or "International System of Units." Based on the metric system, SI is the system of measurement units in use by scientists worldwide.

TERRESTRIAL PLANETS: The four inner planets of the solar system: Mercury, Venus, Earth, and Mars. They are all small, rocky, dense, have relatively small amounts of gaseous elements, and are composed primarily of metals and silicates. Compare with Jovian planets.

and Montag in German, or Sunday and Monday in English.

CYCLES AND TIDES. One reason the ancients respected the Moon as much as the Sun is that it seemed to affect many aspects of human life—as indeed it does. Medieval ideas about "lunacy" were themselves more than a little off-kilter, and the belief that the Moon affects female menstrual cycles has come under significant challenge. Nevertheless, the Moon does seem to have some effect on human biological cycles.

The human circadian rhythm, or cycle of sleep and wakefulness, stretches over a period of

25 hours, meaning that a person living in a cave without any exposure to sunlight would eventually assume a 25-hour-a-day schedule. The reason for this disparity between the human circadian cycle and the length of a solar day is not known, but it is possible that the circadian cycle is based on the length of a lunar day, or the interval between periods of time when the Moon appears in the sky over a given spot on Earth. Because the Moon is moving even as Earth is rotating, this span of time is not 24 hours, the interval it takes Earth to rotate on its axis, but 24 hours and 50 minutes.

It is also more than an old wives' tale that people behave strangely around the time of a full moon: in fact, more deaths and accidents do occur at that point in the monthly cycle. One of the most significant lunar-influenced cycles, however, is not monthly but semidiurnal, or twice daily. These are the cycles of Earth's tides, brought about by the gravitational attraction exerted on Earth by the nearest significant celestial object. (The Sun also affects tides, but much less so because of its greater distance from Earth.) Though the Moon's effect on tides has been known since ancient times, the ancients did not understand the gravitational nature of its attraction, which actually causes a bulge to appear in the oceans.

In fact, two bulges appear, one in the oceans on the side of Earth nearest the Moon and one on the opposite side. The latter is also a result of the Moon's gravitation, which pulls Earth in the other direction, thus drawing the solid earth away from the water. These bulges are known as high tide, and the resulting displacement of water causes a low tide. In most places on Earth, tides are semidiurnal, meaning that there are two full cycles of high and low tide per day. Sometimes these variations can be very great, as at the Bay of Fundy in Canada, where the tidal range is as large as 46 ft. (14 m).

The tides are related closely to phases of the Moon, or, to put it another way, they are affected by the alignment between the Moon and the Sun. At the new moon, the Moon is between Earth and the Sun, and during a full moon, the Moon is on the other side of Earth from the Sun. (Because of a 20° angle between Earth's orbit and that of the Moon, Earth rarely casts a shadow on a full moon, except in the case of a lunar eclipse.) In the new-moon and full-moon phases, the Moon and Sun are aligned, and this combination of lunar and solar gravitation produces strong spring tides every 14 days. On the other hand, when the Moon and Sun are at angles to each other—during the first quarter and last quarter, sometimes known as half-moons—it produces a much weaker neap tide.

WHERE TO LEARN MORE

Beatty, J. Kelly, Carolyn Collins Petersen, and Andrew Chaikin. *The New Solar System.* 4th ed. New York: Sky Publishing, 1999.

Earth and Moon Viewer (Web site). <http://www.fourmilab.ch/earthview/vplanet.html>.

Lewis, John S. *Physics and Chemistry of the Solar System.* San Diego: Academic Press, 1997.

Lodders, Katharina, and Bruce Fegley. *The Planetary Scientist's Companion.* New York: Oxford University Press, 1998.

The Moon (Web site). <http://seds.lpl.arizona.edu/nineplanets/nineplanets/luna.html>.

Moon Phases (Web site). <http://www.astro.wisc.edu/~dolan/java/MoonPhase.html>

Moore, Patrick. *The Data Book of Astronomy.* Philadelphia: Institute of Physics, 2000.

Pérez, Miguel. *The Earth and the Universe: How the Sun, Moon, and Stars Cause Changes on Earth.* Illus. María Rius. Happage, NY: Barron's Educational Series, 1998.

Skinner, Brian J., Stephen C. Porter, and Daniel B. Botkin. *The Blue Planet: An Introduction to Earth System Science.* 2d ed. New York: John Wiley and Sons, 1999.

The Solar and Heliospheric Observatory (Web site). <http://sohowww.nascom.nasa.gov/>.

The Sun (Web site). <http://www.solarviews.com/eng/sun.htm>.

The Sun (Web site). <http://www.seds.org/nineplanets/nineplanets/sol.html>.

SCIENCE OF EVERYDAY THINGS

REAL-LIFE EARTH SCIENCE

GEOLOGY

HISTORICAL GEOLOGY

CONCEPT

Geologists are concerned primarily with two subjects: Earth's physical features and the study of the planet's history. These two principal branches of geology are known, appropriately enough, as physical geology and historical geology. Today they are of equal importance, but in the early modern era, geologists were most focused on topics related to historical geology, in particular, Earth's age and the means by which Earth was formed. This debate pitted adherents of religion, which seemed to require a very young Earth, against adherents of science. A breakthrough came with the introduction of uniformitarianism, a still-influential principle based on the idea that the geologic processes at work today have always been at work. Opposing uniformitarianism was catastrophism, or the idea that Earth was formed in a short time by a series of cataclysmic events. Discredited at the time, catastrophism later gained acceptance, though this did not lead to support for the concept of a young Earth. In fact, the planet is very old—so old that all of human history is almost inconceivably short in comparison.

HOW IT WORKS

Explaining Origins

For thousands of years, humans were content to rely on religiously inspired stories, rather than scientific research, to provide an explanation regarding Earth's origins. This topic, along with the scientific challenge to those early accounts of Earth's formation, is discussed in considerable detail within the essay Earth, Science, and Nonscience. Other aspects of the subject, particularly the challenge to mythological explanations put forward by earth scientists in modern times, are examined here.

All religious explanations of the planet's origins can be called myths, which is not necessarily a pejorative term: a myth is simply a story to explain how something came into being. So pervasive are myths about geology that a term, *geomythology,* has been coined to identify such myths. Geomythology in particular and mythology in general stand in sharp contrast to scientific explanations derived by using the scientific method of observation, hypothesis formation, testing of hypotheses, and the development and testing of theories.

CONTRASTING SCIENCE AND RELIGIOUS GEOMYTHOLOGY. Science and myth have in common the aim of explaining how things came to be, but the means by which they reach that explanation are quite different. So, too, are the reasons that drive science on the one hand and religion or myth on the other in seeking to develop such an explanation.

The most famous of all religious explanations of Earth's origins, of course, is that found in the biblical book of Genesis. Probably written in the latter part of the second millennium B.C., it offered a compelling story of creation that virtually defined the Western view of Earth's origins for more than 1,500 years, from about A.D. 300 to the beginning of the nineteenth century. Its purpose, of course, was not scientific, and it was not written as the result of research; rather, the Genesis account depicts nature as a vast stage on which a cosmic drama of love, sin, redemption, and salvation has been played out over the ages.

By contrast, the scientific search for Earth's origins is driven merely, or at least primarily, by curiosity to explain how things came to be as they are. Scientists certainly have their biases and are just as capable of error as anyone, yet at least they have a standard in the form of the scientific method. If a scientist's findings, and the resulting theory, withstand the rigorous testing required by the scientific method, the theory is rewarded with increasing acceptance and new research designed to test further its ability to explain the world. If the theory fails those tests, its adherents may hold on to it for a time, but eventually they die off, and the theory is discarded. On the other hand, adherence to the religious explanation of Earth's origins has proved more intractable, as we shall see.

THE RELIGIOUS WAR CONCERNING EARTH'S ORIGINS

The great Italian artist and scientist Leonardo da Vinci (1452–1519) was among the first Western thinkers to speculate that fossils might have been made by the remains of long-dead animals. This was a daring supposition to make in the Renaissance, and it would become even more daring to uphold such an idea in the centuries that followed. The concept of fossils seemed to imply an Earth older than the biblical account suggested, and with the Catholic Church under attack by the forces of religious reformation (i.e., Protestantism) and other forms of "heresy," church leaders became less and less inclined to tolerate any deviation from orthodoxy.

The ecclesiastical view of Earth's history reached a sort of extreme in the seventeenth century, with the Irish bishop James Ussher (1581–1656). The New Testament contains a thorough accounting of Jesus' lineage, both through his mother and his earthly father, Joseph, all the way back to the time of Adam. Jesus was descended from David, whose lineage is provided in the Old Testament, complete with each ancestor's life span and the age at which he fathered a successor in the Davidic line of descent. From these figures, Ussher concluded that God finished making Earth at 9:00 A.M. on Sunday, October 23, 4004 B.C. Accepted by the Church, Ussher's calculation gave the idea of a very young Earth an aura of "scientific" justification.

Ironically, one of the first scientists to discover evidence that pointed toward an extremely old Earth was also a minister, the English astronomer Henry Gellibrand (1597–1636). While researching Earth's magnetic field, Gellibrand discovered that the field had changed over time (as indeed it has—for more on this subject, see Geomagnetism). This was one of the first indications that the planet's history can be studied scientifically, even though humans have no direct information regarding the origins of Earth.

EARLY STRATIGRAPHIC STUDIES

After Gellibrand came the Danish geologist Nicolaus Steno (1638–1687), who studied the age of rock beds. Thus was born the concept of stratigraphy, or the study of rock layers beneath Earth's surface, which revealed a great deal about the planet's age. (See Stratigraphy, which discusses many topics related to historical geology.) Along with the English physicist Robert Hooke (1635–1703) and others, Steno also became one of the first thinkers to confront the possibility that Earth must be much more than 6,000 years old.

During the eighteenth century, the German geologist Johann Gottlob Lehmann (1719–1767) built on ideas introduced by Steno concerning the formation of rock beds. Lehmann put forward the theory that certain groups of rocks tend to be associated with each other and that each layer of rock is a sort of chapter in the history of Earth. Aspects of Lehmann's theory were incorrect, but the general principle marked an advancement over previous ideas in geology and helped point the way toward a new view of the earth sciences.

Previously, geologic studies had tended to be qualitative and descriptive, meaning that earth scientists used very generalized terminology and failed to possess a grasp of larger issues. Thanks to Lehmann and others who followed him, the earth sciences became more truly quantitative and predictive, offering explanations of what had happened in the past, along with justifiable theories concerning what might happen in the future.

The German geologist Abraham Gottlob Werner (1750–1817) put forward a theory that was largely incorrect, yet one that nevertheless advanced the earth sciences. His "neptunist" theory was based on the idea that water had been the main force in shaping Earth's surface. Though this theory was not accurate, his idea was significant, because it constituted the first well-ordered geologic theory of Earth's origins and early history. At the same time, that history was turning out to be very long indeed.

LEONARDO DA VINCI WAS THE FIRST TO SUGGEST THAT FOSSILS ARE LONG-DEAD ANIMAL REMAINS, IMPLYING THAT EARTH IS OLDER THAN THE BIBLICAL ACCOUNT IN GENESIS WOULD SUGGEST. *(AP/Wide World Photos. Reproduced by permission.)*

RELIGION AND EARTH'S AGE

In 1774 the French mathematician Georges-Louis Leclerc, Comte de Buffon (1707–1788), applied new scientific ideas to the study of Earth and estimated its age at 75,000 years. Privately he admitted that he actually thought Earth was billions of years old but did not think that such a figure would be understood. At the time, after all, the concept of a "billion" was hardly a familiar one, as it is today. More important, the idea of

Earth being that old was shocking and downright frightening to people who accepted a strict interpretation of the biblical account.

In an earlier century, the Italian astronomer Galileo Galilei (1564–1642) had been forced, on pain of death, to recant his support for the Polish astronomer Nicolaus Copernicus's (1473–1543) discovery that Earth is not the center of the universe. In Buffon's day, by contrast, few Europeans faced such dire threats for endorsing apparently unbiblical ideas. A scientist could still lose his job for supporting the wrong principles, and thus Buffon had to renounce his position on threat of losing his post at the University of Paris.

AN ATHEISTIC REACTION. Other forces were at work in the sciences during the eighteenth century, and some were openly hostile to religious belief. An extreme example was the French physician and philosopher Julien de La Mettrie (1709–1751), a leading figure in the mechanist school of the biological sciences. La Mettrie maintained that humans are essentially a variety of monkey, to whom they were superior only by virtue of possessing the power of language. Moving far beyond the territory of science itself, he also taught that atheism is the only road to happiness and that the purpose of human life is to experience pleasure.

In the physical sciences, an interesting example of reaction to religious belief can be found in the case of the French mathematician Pierre Simon de Laplace (1749–1827). Like those of La Mettrie, Laplace's aims were not purely scientific; instead, he envisioned himself as a warrior against religious belief. Correctly enough, Laplace maintained that the origins of the universe as well as its workings could be explained fully without any reference to God. He also introduced a highly influential theory, widely accepted today, that the solar system originated from a cloud of gas. (See the entries Planetary Science and Sun, Moon, and Earth.)

Like La Mettrie, Laplace took his ideas far beyond their justifiable purview in the realm of science, however, wielding them as a sword in a religious war. Laplace maintained that because it was possible to discuss the origins of the cosmos without reference to God, there must be no God—which is far from a logically necessary conclusion. Misguided as La Mettrie's and Laplace's atheistic crusades may have been, they are historically understandable: in France, far more than anywhere in western Europe except perhaps Spain, the Church had come to be seen as a force of political oppression, allied as it was with the French royalty. It is no wonder, then, that the French Revolution of 1789 was directed as much against the Church as against the king.

REAL-LIFE APPLICATIONS

Uniformitarianism

Late in the eighteenth century, the Scottish geologist James Hutton (1726–1797) put forward an idea that transcended the debate over Earth's origins. Rather than speculate as to how Earth had come into being, Hutton analyzed the processes at work on the planet in his time and reasoned that they must be a key to understanding the means by which Earth was shaped. This was the principle of uniformitarianism, which is still a key concept in the study of Earth. Thanks to his introduction of this influential idea, Hutton today is regarded as the father of modern scientific geology.

Uniformitarianism, in general, is the idea that the geologic processes at work today provide a key to understanding the geologic past. This means that the laws of nature have always been the same. The uniformitarianism promoted by Hutton and his fellow Scottish geologist Charles Lyell (1797–1875), however, has undergone some modification, namely, by the addition of the qualifying statement that the speed and intensity of those processes may not always be the same at any juncture in geologic history. For instance, land does not erode today at the same rate that it did before plants existed to hold rocks and soil in place.

GOULD'S FOUR UNIFORMITIES. In the late twentieth century, the American paleontologist Stephen Jay Gould (1941–2002) identified four different meanings of uniformity in science, not all of which are equally valid. Gould's listing and analysis of these four meanings is as follows:

- Uniformity of law: The assumption that natural laws do not change over time. This idea governs all sciences.
- Uniformity of process: The idea embodied in the most well-known definition of geo-

METEOR CRATER, ARIZONA. MOST NOTABLE AMONG CATASTROPHE THEORIES OF EARTH'S FORMATION IS THE COLLISION OF METEORITES. *(© Francois Gohier/Photo Researchers. Reproduced by permission.)*

logic uniformitarianism, "The present is key to the past."

- Uniformity of rate: The incorrect assumption that the rate at which processes occur presently is the same as the rate at which they occurred in the past.
- Uniformity of state: The incorrect assumption that the state of the universe always has been as it is today.

As noted, uniformity of law is essential to all sciences. For instance, there is every reason to believe that the conservation of energy (a law stating that the total amount of energy in the universe remains constant) always has been the case. If the contrary were true, the conservation of energy could no longer properly be called a law, because it might cease to be the case at some time in the future.

The statement "The present is key to the past" was formulated by yet another Scottish geologist, Sir Archibald Geikie (1835–1924). Geikie's statement often has been criticized as an oversimplification, because processes that occurred in the past may not necessarily be occurring now, or vice versa, even though they could occur again. This idea has required modification of uniformitarianism, as noted earlier, to take into account the fact that the speed and intensity of processes may not always be the same. Part of this modification has involved acceptance of a form of catastrophism, discussed later in this essay.

Variations in the speed and intensity of processes also were addressed by Gould, with his observation that "uniformity of rate" is a fallacy. So, too, is "uniformity of state," which is one of the few areas on which adherents of creationism (a strict interpretation of the Genesis account) would agree with their opponents. Even the Bible, after all, says "In the beginning ... the earth was without form, and void."

CATASTROPHISM

In *Theory of the Earth* (1795), Hutton suggested that the weathering effects of water produced the sedimentary layers of Earth. Based on observation of river flow and mud content, he realized that this process would require much longer than 6,000 years. So, too, did Lyell, author of the highly influential *Principles of Geology,* which appeared in 12 editions from 1830 to 1875 and which presented a strict version of uniformitarianism.

Aqueducts and other structures erected by the Romans had stood for a good one-fourth to one-third of the entire history of Earth, assuming that it was as young as Ussher's biblical interpretation implied. Yet these Roman constructions had experienced very little weathering and certainly much less than mountains would have had to experience to leave behind the sediments observed by geologists. Surely, then, Earth must be millions upon millions of years old, not just a few thousand.

CUVIER'S CATASTROPHIC THEORY. Not so, countered adherents of a movement known as catastrophism, which arose in opposition to uniformitarianism during the late eighteenth and early nineteenth centuries. Catastrophism associates geologic phenomena with sudden, dramatic changes rather than ongoing and long-term processes, as in uniformitarianism. The leading proponent of catastrophism was the French geologist Baron Georges Cuvier (1769–1832), who used this theory to explain unconformities. These apparent gaps in the geologic record, revealed by observing rock layers, or strata, are discussed in the essay, Stratigraphy.

Whereas Cuvier's countryman (and fellow French noble) Buffon had asserted that Earth was 75,000 years old while actually believing that it was much older, Cuvier maintained that the planet is *just* 75,000 years old. The formation of mountains and other landforms, which should have taken millions of years, could be explained by sudden, violent changes, an example of which was Noah's Flood in the Book of Genesis. As the ocean waters receded, they moved rocks far from their sources, carved out valleys, and left behind lakes and other bodies of fresh water.

CATASTROPHISM TODAY. As more and more evidence for a very old Earth began to accumulate during the nineteenth century, catastrophism fell into disfavor. Discoveries from the 1970s onward, however, influenced a new look at catastrophism, and, as a result, the idea has received new attention in later years.

This has not led to a wholesale endorsement of creationism; rather, scientists have come to understand that the generally steady pace of processes on Earth periodically is broken by catastrophic events. Most notable among types of catastrophe is the collision of a meteorite with Earth, a remarkable example of which apparently occurred some 65 million years ago. That dramatic event seems to have forced so much dust and gas into the atmosphere that it blocked out the Sun, leading to the ultimate extinction of the dinosaurs.

UNDERSTANDING GEOLOGIC TIME

So just how old *is* Earth? Modern earth scientists working in the realm of historical geology, and specifically geochronology, estimate its age at about 4.6 billion years. (The dating techniques used to determine the age of the planet are discussed in the essay Stratigraphy.) Such a vast span of time is more than a little difficult for humans to comprehend, given the fact that our lives last 70–80 years, on average, and the entire history of human civilization is only about 5,500 years long.

For this reason, it is helpful to use scales of comparison, such as that offered at the Web site listed under the title *Comprehending Geologic Time.* Suppose that the entire geologic history of Earth were likened to a single year of 365.25 days, starting with the formation of the planet from a cloud of dust and ending with the present. More than two months would have been required simply for the accretion of Earth from a gas cloud to a planetesimal to something like its present form, but by about March 5 this evolution would have been accomplished.

The entire spring would be analogous to a long, long period of time in which Earth was pounded by meteor showers and the oceans began to form. Not even the oldest known rocks date back this far, and many of our ideas about this phase in Earth's history are based on conjecture. Much more is known about the second half of geologic history, beginning with the origins of the first single-cell life-forms on June 16.

FROM SINGLE CELLS TO DINOSAURS. We are now almost halfway through the year and still a long, long way from any sort of complex living beings. This is not surprising, given the fact that the formation of the continental plates and the development of oxygen in the atmosphere would have occurred only by about August 26. Even in the week after Thanksgiving, the most complex organisms would have been snails. Finally, a few days before the beginning of December, creatures would have begun to invade the land.

We tend to associate the dinosaurs with the early phases of Earth's history, but this only illustrates our distorted view of geologic time. In fact the Jurassic period, when dinosaurs roamed

KEY TERMS

CATASTROPHISM: The idea that geologic phenomena are brought about by sudden dramatic changes rather than ongoing and long-term processes, as in uniformitarianism. Although it was once used to promote the idea of a very young Earth, catastrophism today is accepted, in a very modified form, by many earth scientists.

GEOCHRONOLOGY: The study of Earth's age and the dating of specific formations in terms of geologic time.

GEOLOGY: The study of the solid earth, in particular, its rocks, minerals, fossils, and land formations.

GEOMYTHOLOGY: Folklore inspired by geologic phenomena.

HISTORICAL GEOLOGY: The study of Earth's physical history. Historical geology is one of two principal branches of geology, the other being physical geology.

QUALITATIVE: Involving a comparison between qualities that are not defined precisely, such as "fast" and "slow" or "warm" and "cold."

QUANTITATIVE: Involving a comparison between precise quantities—for instance, 10 lb. versus 100 lb. or 50 mi. per hour versus 120 mi. per hour.

SCIENTIFIC METHOD: A set of principles and procedures for systematic study that includes observation; the formation of hypotheses, theories, and laws; and continual testing and reexamination.

SEDIMENT: Material deposited at or near Earth's surface from a number of sources, most notably preexisting rock.

STRATIGRAPHY: The study of rock layers, or strata, beneath Earth's surface.

UNCONFORMITY: An apparent gap in the geologic record, as revealed by observing rock layers, or strata.

UNIFORMITARIANISM: The idea that the geologic processes at work today provide a key to understanding the geologic past. The speed and intensity of those processes, however, may not always be the same at any juncture in geologic history. Uniformitarianism usually is contrasted with catastrophism.

WEATHERING: The breakdown of rocks and minerals at or near the surface of Earth due to physical or chemical processes or both.

Earth, would be parallel to a period of about five days, from December 15 to 20. By Christmas Day, the meteorite referred to earlier would have hit Earth, and the dinosaurs would be headed toward extinction, their dead bodies eventually forming the fossil fuels that have powered much of human civilization.

THE SHORT SPAN OF HUMANITY'S EXISTENCE. By this point, we are within a few days of the year's end, and yet nothing remotely resembling a human has appeared. Our own species, *Homo sapiens,* would not have come on the scene until the last 0.16 days of the year—that is, at a few minutes after 8:00 p.m. on December 31. The New Year's Eve countdown would be nearing by the time human civilization began, at about 42 seconds before midnight.

Now we have come to a period about 6,000 years ago, or the point at which, according to Bishop Ussher, Earth was created. No wonder many people wanted to believe in a young Earth, and some even hold on to that belief today: when viewed against the backdrop of the planet's true age, humanity seems very insignificant indeed.

Christ's birth would have occurred at about 14 seconds before midnight, and the final 10-second countdown would begin about the time the Roman Empire fell. The life span of the average person would correspond to about half a second or less.

HOW DO WE KNOW EARTH'S AGE?

What we know about Earth's age comes, of course, not from direct observation but from the study of materials. One of the most important techniques for determining the age of samples taken from the earth is radiometric dating, discussed in more detail in Geologic Time. Radiometric dating involves ratios between two different kinds of atoms for a given element: stable and radioactive isotopes. Because chemists know how long it takes for half the isotopes in a given sample to stabilize (a half-life), they can judge the age of the sample by examining the ratio of stable to radioactive isotopes. In the case of uranium, one isotopic form, uranium-238, has a half-life of 4,470 million years, which is very close to the age of Earth itself. Use of uranium dating has detected rocks of an age between 3.8 and 3.9 billion years old, as well as even older crystal formations that suggest the earth had solid ground as early as 4.2 billion years ago.

A rock discovered in the Australian desert during the early 1980s appears to be the oldest rock sample in the world, according to data originally reported in *Nature* and included on the *Scientific American* Web site in early 2001. This zircon crystal, according to Simon Wilde of Curtin University in Western Australia, is 4.4 billion years old. Wilde and associates reported that extensive study of the sample suggested that at the time of its formation, Earth was already covered in water—something that had supposedly happened many millions of years later. If this was the case, it could suggest the possibility that life appeared much earlier than has previously been supposed, and perhaps even that life disappeared and reappeared several times before finally taking hold.

WHERE TO LEARN MORE

Bishop, A. C., A. Woolley, and A. Hamilton. *Cambridge Guide to Minerals, Rocks, and Fossils.* New York: Cambridge University Press, 1992.

Boggy's Links to Stratigraphy and Geochronology (Web site). <http://geologylinks.freeyellow.com/stratigraphy.html>.

Comprehending Geologic Time (Web site). <http://www.athro.com/geo/hgfr1.html>.

Harris, Nicholas, Alessandro Rabatti, and Andrea Ricciardi. *The Incredible Journey to the Beginning of Time.* New York: Peter Bedrick Books, 1998.

"Historical Geology." Georgia Perimeter College (Web site). <http://www.dc.peachnet.edu/~pgore/geology/geo102.htm>.

Lamb, Simon, and David Sington. *Earth Story: The Shaping of Our World.* Princeton, NJ: Princeton University Press, 1998.

MacRae, Andrew. *Radiometric Dating and the Geological Time Scale* (Web site). <http://www.talkorigins.org/faqs/dating.html>.

Reeves, Hubert. *Origins: Cosmos, Earth, and Mankind.* New York: Arcade, 1998.

Spickert, Diane Nelson, and Marianne D. Wallace. *Earthsteps: A Rock's Journey Through Time.* Golden, CO: Fulcrum Kids, 2000.

UCMP (University of California, Berkeley, Museum of Paleontology) Web Time Machine (Web site). <http://www.ucmp.berkeley.edu/help/timeform.html>.

GEOLOGIC TIME

CONCEPT

The expression geologic time refers to the vast span from Earth's beginnings to the present, about 4.6 billion years. To examine the history of Earth, one must discard most familiar ideas about time. Instead of thinking in terms of years, centuries, or even millennia, the most basic unit is a million years, and even that is rather small when compared with the four eons into which geologic time is divided. Earth scientists' knowledge of the first three eons is fairly limited. What they do know comes from a combination of absolute dating, mostly by the study of radioactive decay, and relative dating through the stratigraphic record of rock layers.

HOW IT WORKS

Historical Geology

The study of geologic time is encompassed within the larger subject of historical geology. The latter, the study of Earth's physical history, is one of the two principal branches of geology, the other being physical geology, or the study of Earth's physical components and the forces that have shaped them.

The background of historical geology is discussed in some detail within the Historical Geology essay. Its principal subdisciplines include stratigraphy, the study of rock layers, or strata, beneath Earth's surface; geochronology, the study of Earth's age and the dating of specific formations in terms of geologic time; sedimentology, the study and interpretation of sediments, including sedimentary processes and formations; paleontology, the study of fossilized plants and animals; and paleoecology, the study of the relationship between prehistoric plants and animals and their environments. Several of these subjects are examined in essays within this book.

Divisions of Geologic Time

Geologic time is divided according to two scales. The more well-known of these is the geologic scale, which divides time into named groupings according to six basic units: eon, era, period, epoch, age, and chron. In addition, the chronostratigraphic scale identifies successive layers of rock with specific units of time.

As noted earlier, stratigraphy is the study of rock layers, or strata, beneath Earth's surface, while chronostratigraphy is a subdiscipline devoted to studying the ages of rocks and what they reveal about geologic time. The chronostratigraphic scale likewise has six time units, analogous to those of the geologic scale: eonothem, erathem, system, series, stage, and chronozone. For the most part, we will not be concerned with the chronostratigraphic terms in the present context.

RELATIVE AND ABSOLUTE TIME. To discuss the divisions of geologic time, it is necessary first to discuss the concepts of relative and absolute time. The term *relative* refers to a quality or quantity that is comparative, or dependent on something else. Its opposite is *absolute,* a term designating a quality or quantity that is independent and not defined in relation to another quality or quantity.

If we say that Abraham Lincoln was born in 1809, it is an absolute designation of his birth year, whereas if we say that he was born 10 years after the death of George Washington (which

SUBATOMIC PARTICLE TRACKS. STUDY OF RADIOACTIVE DECAY, OR THE RELEASE OF SUBATOMIC PARTICLES, IS A METHOD OF ABSOLUTE DATING. *(© John Giannicchi/Photo Researchers. Reproduced by permission.)*

occurred in 1799), that is an example of a relative time measurement. In actuality, of course, there is no truly absolute measure of time. For example, the reference to 1809 as Lincoln's birth year is based on the system of time measurement developed in the West, which, in turn, is based on early ideas regarding the date of Christ's birth. (As it turns out, Christ likely was born in about 6 B.C.)

Since the B.C./A.D. system of dating is widely accepted and used, or at least recognized, by most of the non-Western world, a date rendered according to this system constitutes the closest possible approximation to an absolute measure of time. In any case, one knows the difference between absolute and relative when one sees it: thus, to say that Lincoln was born ten years after Washington died is obviously and unmistakably a relative statement.

In terms of geology, the absolute age of a geologic phenomenon is its age in Earth years. On the other hand, its relative age is its age in comparison with other geologic phenomena, particularly the stratigraphic record of rock layers. Thus, references to relative age are given in terms of chronostratigraphic time divisions rather than millions of years.

RELATIVE DATING

Given the meaning of relative age, it is easy enough to guess what relative dating would be, once one knows that dating, in a scientific context, usually refers to any effort directed toward finding the age of a particular item or phenomenon. Relative dating, then, assigns an age relative to that of other items, whereas absolute dating determines the age in actual years or millions of years.

One of the principal means of relative dating is through stratigraphy, which is based on the assumption that the deeper a layer of rock lies beneath Earth's surface, the earlier it was deposited. This holds true, however, for only one of the three major types of rock: sedimentary rock, which is formed by compression and deposition (i.e., formation of deposits) on the part of rock and mineral particles. (The other types of rock are igneous and metamorphic.)

Aside from stratigraphy, discussed in a separate essay, other relative dating techniques include seriation, faunal dating, and pollen dating, or palynology. Used, for instance, in archaeological studies, seriation analyzes the abundance of a particular item (for instance, pieces of pottery) and assigns relative dates based on this abundance. The term faunal dating refers to fauna, or animal life, and faunal dating is the use of animal bones to determine age. Finally, pollen dating, or palynology, involves analysis of pollen deposits.

ABSOLUTE DATING

As dating technology has progressed, it has become increasingly possible for scientists to provide absolute dates for specimens. One such method, introduced in the 1960s, is amino-acid racimization. Amino acids exist in two forms, designated *L*-forms and *D*-forms, which are stereoisomers, or mirror images of each other. Virtually all living organisms (except some microbes) incorporate only the L-forms, but once the organism dies the L-amino acids gradually convert to D-amino acids. Several factors influence the rate of conversion, and though amino-acid racimization was popular in the 1970s, these uncertainties have led scientists to treat it with increasing disfavor.

The principles that undergird amino-acid racimization, however, are essential to most forms of absolute dating. Generally, absolute dating uses ratios between the quantities of a particular substance (let us call it *Substance A*) and the quantities of a mirror substance (*Substance B*) to which it is converted over a period of time. The greater the ratio of Substance B to Substance A, the longer the time that has elapsed. The scale of time for various substances, however, differs greatly. Carbon-14 decay, for instance, takes place over a few thousand years, making it useful for measuring the age of human artifacts. On the other hand, uranium decay takes billions of years, and thus it is used for dating rocks.

Cation-ratio dating, for instance, measures the amount of cations, or positively charged ions, that have formed on an exposed rock surface. (An ion is an atom or group of atoms that have lost or gained electrons, thus acquiring a net electric charge. Electron loss creates a cation, as opposed to a negatively charged anion, created when an atom or atoms gain electrons.) Cation-ratio dating is based on the idea that the ratio of potassium and calcium cations to titanium cations decreases with age. It is applicable only to rocks in desert areas, where the dry air stabilizes the cation "varnish."

RADIOACTIVE DECAY. Various forms of radiometric dating employ ratios as well. Every element has a particular number of protons, or positively charged particles, in its nucleus, but it may have varying numbers of neutrons, particles with a neutral electric charge but relatively great mass. (Neutrons and protons have approximately the same mass, which is more than 1,800 times greater than that of an electron.) When two or more atoms of the same element have a differing number of neutrons, they are called isotopes.

Some types of isotopes "fit" better with a particular element and tend to be most abundant. For instance, carbon has six protons, and it so happens that the most abundant carbon isotope has six neutrons. Because there are six protons and six neutrons, totaling 12, this carbon isotope is designated *carbon-12,* which accounts for 98.9% of the carbon in nature. Generally speaking, the most abundant isotope is also the most stable one, or the one least likely to release particles and thus change into something else.

This release of particles is known as radioactive decay. In the context of radioactivity, "to decay" does not mean "to rot" rather, the isotope expels alpha particles (positively charged helium nuclei), beta particles (either electrons or subatomic particles called positrons), or gamma rays, which occupy the highest energy level in the electromagnetic spectrum. In so doing, it eventually will become another isotope, either of the same element or of a different element, and will stabilize. The amount of time it takes for half the isotopes in a sample to stabilize is called its half-life. This half-life varies greatly between isotopes, some of which have a half-life that runs into the billions of years.

DETERMINING ABSOLUTE AGE

When an organism is alive, it incorporates a certain ratio of carbon-12 in proportion to the amount of the radioisotope (that is, radioactive isotope) carbon-14 that it receives from the atmosphere. As soon as the organism dies, however, it stops incorporating new carbon, and the ratio between carbon-12 and carbon-14 will begin to change as the carbon-14 decays to form nitrogen-14. A scientist can use the ratios of carbon-12, carbon-14, and nitrogen-14 to ascertain the age of an organic sample.

Carbon-14, known as radiocarbon, has a half-life of 5,730 years, meaning that it takes that long for half the isotopes in a sample to decay to nitrogen-14. Note that half-life is *not* half the amount of time it takes for the entire sample to decay, especially because the first half of the sample usually decays faster than the second half. Imagine, for instance, that you had 100 units and wanted to reduce it to zero units by continually halving it. At first, the results would be dramatic, as 100 became 50, then 25, then 12.5, and so on. Eventually you would be down to smaller and smaller fractions of 1, and each division by 2 would yield a smaller number—but never zero.

Radioactive decay works that way as well, and, thus, while carbon-14 has a half-life of less than 6,000 years, it takes much longer than 6,000 years for the other half of the isotopes in a carbon-14 sample to decay. For this reason, the use of proper instrumentation makes it possible to judge the age of charcoal, wood, and other biological materials over a span of as long as 70,000 years. While this may be useful for archaeologists, it is not very helpful for measuring the vast spans of time encompassed in the earth sciences. Furthermore, there is a good likelihood that the sample will become contaminated by additional carbon from the soil. Moreover, it cannot be said with certainty that the ratio of carbon-12 to carbon-14 in the atmosphere has been constant throughout time.

POTASSIUM-ARGON DATING. Much more useful, from the standpoint of geology, is potassium-argon dating. When volcanic rocks are subjected to extremely high temperatures, they release the element argon, a noble gas. As the rocks cool, the stable isotope argon-40 accumulates. Because argon-40 is formed by the radioactive decay of a potassium isotope, potassium-40, the amount of argon-40 that forms is proportional to the rate of decay for potassium-40.

Potassium-40 has a half-life of 1.3 billion years, and with the help of argon-40, geologists have been able to estimate the age of volcanic layers above and below fossil and artifact remains in eastern Africa. Potassium-argon dating is most effective for rocks that are at least three million years old, because it takes about that long to accumulate enough argon-40 to make accurate measurements possible.

This brings up a notable aspect of radiometric dating techniques. No one technique is most effective; rather, each technique is suited to a particular span of time. Thus, potassium-argon dating would be virtually useless for measuring the relatively short time scales for which radiocarbon dating is ideally suited. The converse is also true: as we have noted, radiocarbon dating simply does not cover a wide enough span of time to be useful in most geologic studies.

URANIUM-SERIES DATING. We now come to the element most useful for dating the age of material samples over a broad chronological spectrum: uranium, which has an atomic number of 92. This means that it has 92 protons in its nucleus, making uranium atoms typically the heaviest atoms that occur in nature. (There are about 20 elements with atomic numbers higher than 92, but all of them have been created artificially, either in laboratories or as the result of nuclear testing.)

Both uranium and thorium, with an atomic number of 90, have unstable "parent" isotopes that decay into even more unstable "daughter" isotopes before eventually stabilizing as isotopes of lead. These daughter isotopes have half-lives that range from just a few years to a few hundred thousand years, whereas the half-lives of the parent isotopes are much longer. That of uranium-235, for instance, is 7.038×10^8 years, or more than 700 million years. On the other hand, the daughter isotope protactinium-231 has a half-life of 32,760 years.

When uranium-235 is deposited in an area, over time it will decay to form daughter isotopes. Assuming that the sample has been left undisturbed (isotopes have neither entered nor exited the deposit since its initial formation), the age of certain types of sample may thus be determined. For mollusks and corals, for instance, the amount of protactinium-231, a daughter isotope that begins to accumulate only after the organism dies, makes it possible to date a sample. In some cases, large amounts of a daughter isotope may be deposited initially alongside samples of a parent, and if these are present in water, the quantities of each can be judged according to the amount that has dissolved. For example, the daughter isotope uranium-234 dissolves more readily in water than the parent, uranium-238.

REAL-LIFE APPLICATIONS

How Do We Know Earth's Age?

We can now begin to answer a question almost inevitably raised when discussing geologic time: just how do we know that Earth is about 4.6 billion years old? A clue lies in the half-life of uranium-238, which is 4.47×10^9 years, or 4,470 million years. Geologists typically would abbreviate this as 4.47 Ga, the latter referring to "gigayears," a unit of a billion years.

As uranium atoms undergo fission, or splitting, this process releases energy that causes marks, called tracks, to form on the surface of volcanic minerals. In splitting, two daughter atoms shoot away from each other, forming tracks, and thus the rate of track formation is proportional to the rate of decay on the part of the parent isotope.

Incidentally, fission-track dating with uranium-238 defies the statement made earlier that certain types of dating are more suited to long periods of time, while others are best for shorter periods. When heated, the tracks disappear from

Relative dating uses the stratigraphic record of rock layers, as seen in the wall of this gorge, to determine the age of a geologic formation compared with others. *(© B. Bachman/Photo Researchers. Reproduced by permission.)*

a sample containing uranium-238, thus resetting the dating clock. As a result, if an object was heated just a few decades ago, it can be dated; so, too, can meteorites billions of years old.

Most meteorites found in the solar system tend to be about 4.56 Ga; hence, the rough figure of 4.6 Ga is used for the point at which the solar system, including Earth, began to form. The oldest known materials on Earth are zircon crystals from western Australia, dated about 4.3 Ga. Small samples of gneiss in Canada's Northwest Territories have been dated to about 4.0 Ga, but the oldest large-scale sample is a belt of 3.8 Ga gneiss in western Greenland.

A Question of Scale

Having discussed at least the rudiments of the dating system used by geologists, it is possible to examine geologic time itself. This requires a mental adjustment of monumental proportions, because one must discard all notions used in studying the history of human civilization. Concepts such as medieval, ancient, and prehistoric are practically useless when discussing geologic time, which dwarfs the scale of human events.

Human civilization has existed for about 5,500 years, the blink of an eye in geologic terms. Even the span of time that the human species, *Homo sapiens,* has existed—about two million years—is negligible in the grand scheme of Earth's history. The latter stretches back some 4,600 million years, meaning that human beings have existed on this Earth for just 0.043% of the planet's history. As discussed in the essay Historical Geology, if the entire history of Earth were likened to a single year, humans would have appeared on the scene at a few minutes after 8:00 p.m. on December 31. Human civilization would date only from about 42 seconds before midnight, and the age of machinery and industrialization would not fill up even the final two seconds of the year.

ANOTHER ANALOGY: LOS ANGELES TO NEW YORK. When discussing distances in space, astronomers dispense with miles, because they would be useless, given the vastness of the scale involved. The same is true of geologic time, in which the concept of years is hardly relevant. Instead, geologists speak in terms of millions of years, or megayears, abbreviated "Ma." (Geologists also use the much larger unit of a gigayear, to which we have already referred.) To discuss the age of Earth in terms of years, in fact, would be rather like measuring the distance from Los Angeles to New York in feet; instead, of course, we use miles. Now let us consider geologic time in terms of the 2,462 mi. between Los Angeles and New York, with 1 mi. equal to 1.8684 Ma, or 1,868,400 years.

Suppose we have left Los Angeles and driven a good deal of the distance to New York—46% of the way, in fact, to western Nebraska, a spot analogous to the beginning of the Proterozoic era. In the preceding miles, a duration equivalent to about 1,133 Ma, Earth was formed from a cloud of gas, pounded by meteors, and gradually became the home to oceans—but no atmosphere resembling the one we know now. The end of the Proterozoic era (about 545 Ma, or 545 million years ago) would be at about 88% of the distance from Los Angeles to New York—somewhere around Pittsburgh, Pennsylvania. By this point, the continental plates have been formed, oxygen has entered the atmosphere, and soft-bodied organisms have appeared.

We are a long way from Los Angeles, and yet almost the entire history of life on Earth, at least in terms of relatively complex organisms, lies ahead of us. If we skip ahead by about 339 Ma (a huge leap in terms of biological development), we come to the time when the dinosaurs appeared. We are now 95% of the way from the beginning of Earth's history to the present, and if measured against the distance from Los Angeles to New York, this would put us at a longitude equivalent to that of Baltimore, Maryland. Another 89 mi. would put us at about 65 million years ago, or the point when the dinosaurs became extinct.

We would then have only 33.7 mi. to drive to reach the point where humans appeared, by which time we would be in the middle of Manhattan. Compared with the distance from Los Angeles to New York, the span of human existence would be much smaller than the cab ride from Central Park to the Empire State Building. The entire sweep of written human history, from about a thousand years before the building of the pyramids to the beginning of the third millennium A.D., would be much smaller than a city block. In fact, it would be about the width of a modest storefront, or 15.54 ft.

The Very, Very Distant Past

So what happened for all those hundreds of millions of years before humans appeared on the scene? We will attempt to answer that question in an extremely cursory, abbreviated fashion, but for further clarification, the reader is strongly encouraged to consult a chart of geologic time. Such a chart can be found in virtually any earth sciences textbook; indeed, several versions (including a chronostratigraphic chart) may appear in a single book.

In addition to showing geologic time in both absolute and relative terms, these charts typically provide information about the magnetic polarity over a given span, since that has changed many times since Earth came into existence. In other words, what is today the magnetic North Pole was once the magnetic South Pole, and vice versa. (For more on this subject, see the discussion of paleomagnetism in the entries Plate Tectonics and Geomagnetism.)

As one might expect, disagreement between earth scientists is greatest with regard to the most distant phases of Earth's geologic history. This encompasses nearly 90% of all geologic time, dating back to about 545 Ma, thus showing how little geologists know, even today, about the geologic events of the very distant past. For this reason, when discussing Precambrian time, it is usually necessary to consider only the three eons that composed it. Discussion of era and period, on the other hand, is reserved for the three eras, and 11 periods, of the Phanerozoic eon. The smaller division of epoch is generally only of concern with regard to the most recent era, the Cenozoic. As for divisions smaller than an epoch, these will not concern us here.

THE PRECAMBRIAN EONS. The last paragraph of the preceding section encompasses a number of ideas, which now need to be explained, in at least general terms. The term *Precambrian* encompasses about four billion years of Earth's history, including three of the four eons (Hadean or Priscoan, Archaean, and Proterozoic) of the planet's existence. The names of these eons are derived from Greek, with the first being taken from the name of the deity who ruled over the Underworld. The latter two are derived, respectively, from the Greek words for *beginning* and *new life.*

The Hadean eon (sometimes called the Priscoan) lasted from about 4,560 Ma to 4,000

Ma ago, when the planet was being formed, or accreted, as pieces of solid matter floating around in the young solar system began to join one another. Meteorites showered the planet, bringing both solid matter and water, and thus forming the basis of the oceans. There was no atmosphere as such, but by the end of the eon, volcanic activity had ejected enough carbon dioxide and other substances into the air to form the beginnings of one. The oceans began to cool, making possible the beginnings of life—that is, molecules of carbon-based matter that were capable of replicating themselves. These appeared at the end of the Hadean eon, perhaps arriving from space in a meteorite.

The boundaries of the Precambrian eons are far from certain, so it is possible only to say that the Archaean eon lasted from about 4,000 Ma to 2,500 Ma ago. The earliest known datable materials, described earlier, all come from this time; in fact, outcrops of Archaean rock have been found on all seven continents. The rocks of this eon contain the first clear evidence of life, in the form of microorganisms. Over the course of the Archaean eon, prokaryotes, or cells without a nucleus, made their appearance, and later they were followed by eukaryotes, or cells with a nucleus.

During this great span of time, more than 20% of Earth's history, the atmosphere and hydrosphere developed considerably, even as the biosphere had its true beginnings. As for the geosphere, it also matured enormously in the course of the Archaean eon. During the Hadean eon, Earth's interior had begun to differentiate into core, mantle, and crust, and cooling in the two upper layers influenced the beginnings of the earliest plate-tectonic activity (see Plate Tectonics).

Even longer was the Proterozoic eon, which appears to have lasted from about 2,500 Ma to 545 Ma. This phase saw the beginnings of very basic forms of plant life, such that photosynthesis (the biological conversion of electromagnetic energy from the Sun into chemical energy in plants) began to take place. Plate-tectonic processes accelerated as well, with continents moving about over Earth's surface and smashing against one another. Oxygen in the atmosphere assumed about 4% of its present levels, but animal life still consisted primarily of eukaryotes.

THE PHANEROZOIC ERAS AND PERIODS. The end of the Proterozoic eon, once again, is not sharply defined in the stratigraphic record, such that there is considerable dispute as to the time periods involved. In any case, it is clear that the pace of development in the biosphere increased dramatically in the Phanerozoic, the eon in which we are now living. During the beginning of the Phanerozoic eon, algae appeared, and there followed an acceleration in the development of living organisms that ultimately produced the varied biosphere we know today.

As noted earlier, the only eras and periods that need concern most students of the earth sciences are those of the Phanerozoic eon. The three eras are as follows:

Eras of the Phanerozoic Eon

- Paleozoic (about 545 to 248.2 Ma)
- Mesozoic (about 248.2 to 65 Ma)
- Cenozoic (about 65 Ma to the present)

Within these eras are the following periods:

Periods of the Paleozoic Era

- Cambrian (about 545 to 495 Ma)
- Ordovician (about 495 to 443 Ma)
- Silurian (about 443 to 417 Ma)
- Devonian (about 417 to 354 Ma)
- Carboniferous (about 354 to 290 Ma)
- Permian (about 290 to 248.2 Ma)

Periods of the Mesozoic Era

- Triassic (about 248.2 to 205.7 Ma)
- Jurassic (about 205.7 to 142 Ma)
- Cretaceous (about 142 to 65 Ma)

Periods of the Cenozoic Era

- Palaeogene (about 65 to 23.8 Ma)
- Neogene (about 23.8 to 1.8 Ma)
- Quaternary (about 1.8 Ma to present)

These divisions, as well as the two most recent epochs of the Quaternary period (Pleistocene and Holocene), are discussed elsewhere in this book. It should be noted that there are variations for many of the eon, era, and period names given here; also, the Palaeogene and Neogene are often grouped together as a subera called the Tertiary. The latter nomenclature fits with a mnemonic device used by geology students memorizing the names of the 11 Phanerozoic periods: "Camels Ordinarily Sit Down Carefully; Perhaps Their Joints Creak Tremendously Quietly."

KEY TERMS

ABSOLUTE AGE: The absolute age of a geologic phenomenon is its age in Earth years. Compare with *relative age.*

ATMOSPHERE: In general, an atmosphere is a blanket of gases surrounding a planet. Unless otherwise identified, however, the term refers to the atmosphere of Earth, which consists of nitrogen (78%), oxygen (21%), argon (0.93%), and other substances that include water vapor, carbon dioxide, ozone, and noble gases such as neon, which together comprise 0.07%.

ATOM: The smallest particle of an element, consisting of protons, neutrons, and electrons. An atom can exist either alone or in combination with other atoms in a molecule.

BIOSPHERE: A combination of all living things on Earth—plants, animals, birds, marine life, insects, viruses, single-cell organisms, and so on—as well as all formerly living things that have not yet decomposed.

CHRONOSTRATIGRAPHY: A subdiscipline of stratigraphy devoted to studying the ages of rocks and what they reveal about geologic time.

DATING: Any effort directed toward finding the age of a particular item or phenomenon. Methods of geologic dating are either relative (i.e., comparative and usually based on rock strata) or absolute. The latter, based on such methods as the study of radioactive isotopes, usually is given in terms of actual years or millions of years.

ELECTRON: A negatively charged particle in an atom, which spins around the nucleus.

ELEMENT: A substance made up of only one kind of atom. Unlike compounds, elements cannot be chemically broken into other substances.

EON: The longest phase of geologic time. Earth's history has consisted of four eons, the Hadean or Priscoan, Archaean, Proterozoic, and Phanerozoic. The next-smallest subdivision of geologic time is the era.

EPOCH: The fourth-longest phase of geologic time, shorter than an era and longer than an age and a chron. The current epoch is the Holocene, which began about 0.01 Ma (10,000 years) ago.

ERA: The second-longest phase of geologic time, after an eon. The current eon, the Phanerozoic, has had three eras, the Paleozoic, Mesozoic, and Cenozoic, which is the current era. The next-smallest subdivision of geologic time is the period.

GA: An abbreviation meaning "gigayears," or "billion years." The age of Earth is about 4.6 Ga.

GEOCHRONOLOGY: The study of Earth's age and the dating of specific formations in terms of geologic time.

GEOLOGIC TIME: The vast stretch of time over which Earth's geologic development has occurred. This span (about 4.6 billion years) dwarfs the history of human existence, which is only about two million years. Much smaller still is the span of human civilization, only about 5,500 years.

GEOSPHERE: The upper part of Earth's continental crust, or that portion of the solid earth on which human beings live

KEY TERMS CONTINUED

and which provides them with most of their food and natural resources.

HISTORICAL GEOLOGY: The study of Earth's physical history. Historical geology is one of two principal branches of geology, the other being physical geology.

HYDROSPHERE: The entirety of Earth's water, excluding water vapor in the atmosphere but including all oceans, lakes, streams, groundwater, snow, and ice.

ISOTOPES: Atoms that have an equal number of protons and hence are of the same element but differ in their number of neutrons. This results in a difference of mass. An isotope may be either stable or radioactive.

MA: An abbreviation used by earth scientists, meaning "million years" or "megayears." When an event is dated to, for instance, 160 Ma, it usually means that it took place 160 million years ago.

NEUTRON: A subatomic particle that has no electric charge. Neutrons are found at the nucleus of an atom, alongside protons.

NUCLEUS: The center of an atom, a region where protons and neutrons are located and around which electrons spin.

PERIOD: The third-longest phase of geologic time, after an era. The current eon, the Phanerozoic, has had 11 periods, and the current era, the Cenozoic, has consisted of three periods, of which the most recent is the Quaternary. The next-smallest subdivision of geologic time is the epoch.

PRECAMBRIAN TIME: A term that refers to the first three of four eons in Earth's history, which lasted from about 4,Ma to about 545 Ma ago.

PROTON: A positively charged particle in an atom.

RADIOACTIVITY: A term describing a phenomenon whereby certain materials are subject to a form of decay brought about by the emission of high-energy particles or radiation. Forms of particles or energy include alpha particles (positively charged helium nuclei), beta particles (either electrons or subatomic particles called *positrons*, or gamma rays, which occupy the highest energy level in the electromagnetic spectrum.

RADIOMETRIC DATING: A method of absolute dating using ratios between "parent" isotopes and "daughter" isotopes, which are formed by the radioactive decay of parent isotopes.

RELATIVE AGE: The relative age of a geologic phenomenon is its age in comparison with other geologic phenomena, particularly the stratigraphic record of rock layers. Compare with *absolute age.*

SEDIMENT: Material deposited at or near Earth's surface from a number of sources, most notably preexisting rock.

SEDIMENTARY ROCK: Rock formed by compression and deposition (i.e., formation of deposits) on the part of other rock and mineral particles.

SEDIMENTOLOGY: The study and interpretation of sediments, including sedimentary processes and formations.

STRATIGRAPHY: The study of rock layers, or strata, beneath Earth's surface.

WHERE TO LEARN MORE

Boggy's Links to Stratigraphy and Geochronology (Web site). <http://geologylinks.freeyellow.com/stratigraphy.html>.

Comprehending Geologic Time (Web site). <http://www.athro.com/geo/hgfr1.html>.

Hancock, Paul L., and Brian J. Skinner. *The Oxford Companion to the Earth.* New York: Oxford University Press, 2000.

Harris, Nicholas, Alessandro Rabatti, and Andrea Ricciardi. *The Incredible Journey to the Beginning of Time.* New York: Peter Bedrick Books, 1998.

"Historical Geology." Georgia Perimeter College (Web site). <http://www.dc.peachnet.edu/~pgore/geology/geo102.htm>.

Lamb, Simon, and David Sington. *Earth Story: The Shaping of Our World.* Princeton, NJ: Princeton University Press, 1998.

MacRae, Andrew. *Radiometric Dating and the Geological Time Scale* (Web site). <http://www.talkorigins.org/faqs/dating.html>.

Reeves, Hubert. *Origins: Cosmos, Earth, and Mankind.* New York: Arcade, 1998.

Spickert, Diane Nelson, and Marianne D. Wallace. *Earthsteps: A Rock's Journey Through Time.* Golden, Colo.: Fulcrum Kids, 2000.

UCMP (University of California, Berkeley, Museum of Paleontology) Web Time Machine (Web site). <http://www.ucmp.berkeley.edu/help/timeform.html>.

STRATIGRAPHY

CONCEPT

Stratigraphy is the study of rock layers (strata) deposited in the earth. It is one of the most challenging of geologic subdisciplines, comparable to an exacting form of detective work, yet it is also one of the most important branches of study in the geologic sciences. Earth's history, quite literally, is written on the strata of its rocks, and from observing these layers, geologists have been able to form an idea of the various phases in that long history. Naturally, information is more readily discernible about the more recent phases, though even in studying these phases, it is possible to be misled by gaps in the rock record, known as unconformities.

HOW IT WORKS

THE FOUNDATIONS OF STRATIGRAPHY

Historical geology, the study of Earth's physical history, is one of the two principal branches of geology, the other being physical geology, or the study of Earth's physical components and the forces that have shaped them. Among the principal subdisciplines of historical geology is stratigraphy, the study of rock layers, which are called strata or, in the singular form, a *stratum.*

Other important subdisciplines include geochronology, the study of Earth's age and the dating of specific formations in terms of geologic time; sedimentology, the study and interpretation of sediments, including sedimentary processes and formations; paleontology, the study of fossilized plants and animals; and paleoecology, the study of the relationship between prehistoric plants and animals and their environments. Several of these subjects are examined in other essays within this book.

EARLY WORK IN STRATIGRAPHY. Among the earliest contributions to what could be called historical geology came from the Italian scientist and artist Leonardo da Vinci (1452–1519), who speculated that fossils might have come from the remains of long-dead animals. Nearly two centuries later, stratigraphy itself had its beginnings when the Danish geologist Nicolaus Steno (1638–1687) studied the age of rock strata.

Steno formulated what came to be known as the law of superposition, or the idea that strata are deposited in a sequence such that the deeper the layer, the older the rock. This, of course, assumes that the rock has been undisturbed, and it is applicable only for one of the three major types of rock, sedimentary (as opposed to igneous or metamorphic). Later, the German geologist Johann Gottlob Lehmann (1719–1767) put forward the theory that certain groups of rocks tend to be associated with each other and that each layer of rock is a sort of chapter in the history of Earth.

Thus, along with Steno, Lehmann helped pioneer the idea of the stratigraphic column, discussed later in this essay. The man credited as the "father of stratigraphy," however, was the English engineer and geologist William Smith (1769–1839). In 1815 Smith produced the first modern geologic map, showing rock strata in England and Wales. Smith's achievement, discussed in Measuring and Mapping Earth, influenced all of geology to the present day by introducing the idea of geologic, as opposed to geographic, map-

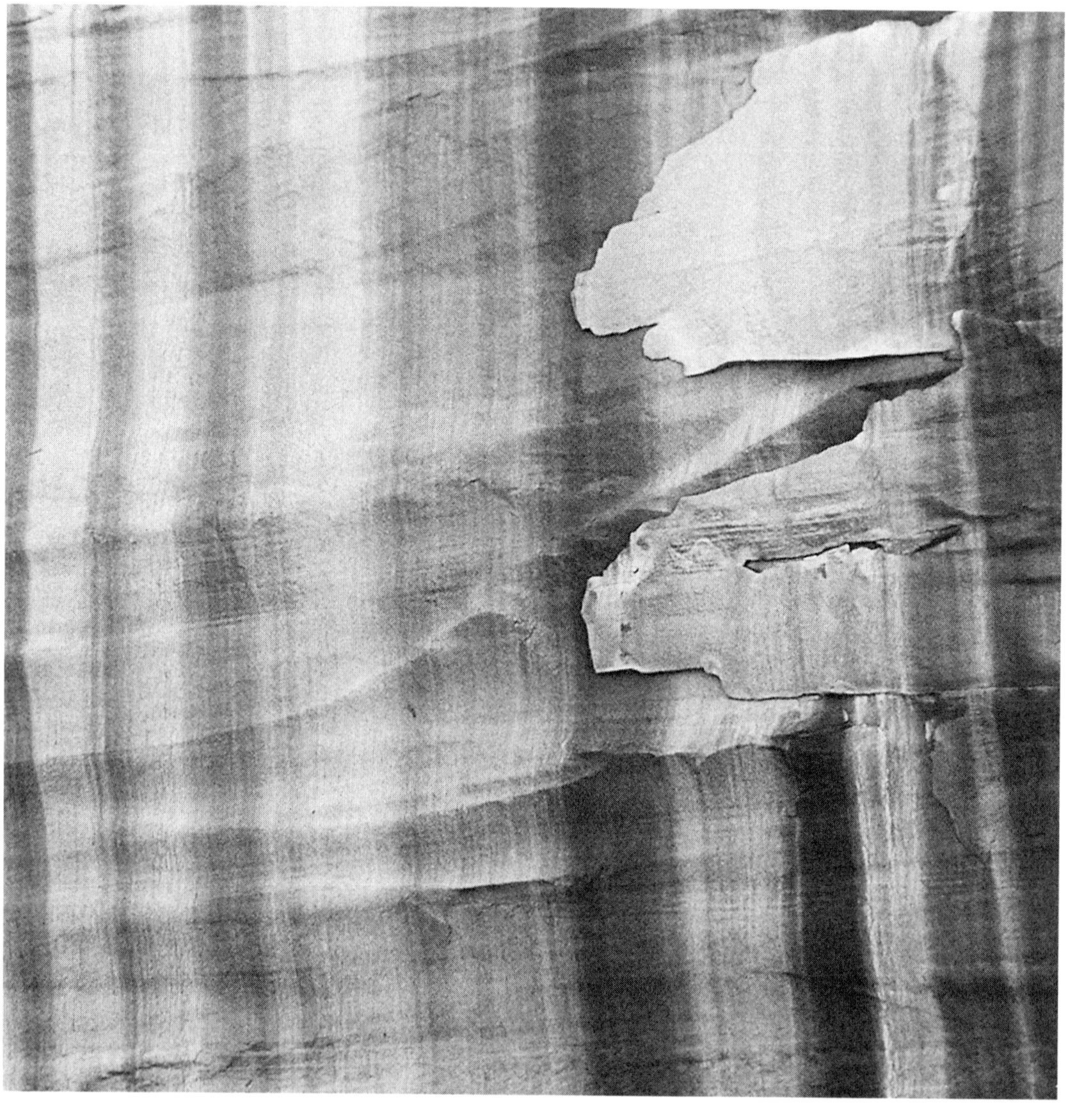

STRIATIONS VISIBLE IN SANDSTONE FROM NEON CANYON, UTAH. *(© Rod Planck/Photo Researchers. Reproduced by permission.)*

ping. Furthermore, by linking stratigraphy with paleontology, he formulated an important division of stratigraphy, known as biostratigraphy.

AREAS OF STRATIGRAPHIC STUDY

Along with biostratigraphy, the major areas of stratigraphy include lithostratigraphy, chronostratigraphy, geochronometry, and magnetostratigraphy. The most basic type of stratigraphy, and the first to emerge, was lithostratigraphy, which is simply the study and description of rock layers. Earth scientists working in the area of lithostratigraphy identify various types of layers, which include (from the most specific to the most general), formations, members, beds, groups, and supergroups.

Biostratigraphy involves the study of fossilized plants and animals to establish dates for and correlate relations between stratigraphic layers. Scientists in this field also identify categories of biostratigraphic units, the most basic being a biozone. Magnetostratigraphy is based on the investigation of geomagnetism and the reversals in Earth's magnetic field that have occurred over time. (See Geomagnetism as well as the discussion of paleomagnetism in Plate Tectonics.)

Chronostratigraphy is devoted to studying the ages of rocks and what they reveal about geologic time, or the vast stretch of history (approximately 4.6 billion years, abbreviated 4.6 Ga) over which Earth's geologic development has occurred. It is concerned primarily with relative dating, whereas geochronometry includes the determination of absolute dates and time intervals. This typically calls for the use of radiometric dating.

THE STRATIGRAPHIC COLUMN

The stratigraphic column is the succession of rock strata laid down over the course of time, each of which correlates to specific phases in Earth's geologic history. The record provided by the stratigraphic column is most reliable for studying the Phanerozoic, the current eon of geologic history, as opposed to the Precambrian, which constituted the first three eons and hence the vast majority of Earth's geologic history. The relatively brief span of time since the Phanerozoic began (about 545 million years, or Ma) has seen by far the most dramatic changes in plant and animal life. It was in this eon that the fossil record emerged, giving us far more detailed information about comparatively recent events than about a much longer span of time in the more distant past.

RELATIVE AND ABSOLUTE DATING. Precambrian time is so designated because it precedes the Cambrian period, one of 11 periods in the Phanerozoic eon. The Cambrian period extended for about 50 million years, from approximately 545 Ma to 495 Ma ago. This statement in terms of years, however inexact, is an example of absolute age. By contrast, if we say that the Cambrian period occurred at the beginning of the Paleozoic era, after the end of the Proterozoic eon and before the beginning of the Ordovician period, this is a statement of relative age. Both statements are true, and though it is obviously preferable to measure time in absolute terms, sometimes relative terms are the only ones available.

Dating, in scientific terms, is any effort directed toward finding the age of a particular item or phenomenon. Relative dating methods assign an age relative to that of other items, whereas absolute dating determines age in actual years or millions of years. When geologists first embarked on stratigraphic studies, the only means of dating available to them were relative. Using Steno's law of superposition, they reasoned that a deeper layer of sedimentary rock was necessarily older than a shallower layer.

Advances in our understanding of atomic structure during the twentieth century, however, made possible a particularly useful absolute form of dating through the study of radioactive decay. Radiometric dating, which is explained in more detail in Geologic Time, uses ratios between "parent" and "daughter" isotopes. Radioactive isotopes decay, or emit particles, until they become stable, and as this takes place, parent isotopes spawn daughters. The amount of time that it takes for half the isotopes in a sample to stabilize is termed a *half-life.* Elements such as uranium, which has isotopes with half-lives that extend into the billions of years, make possible the determination of absolute dates for extremely old geologic materials.

DIVISIONS OF THE STRATIGRAPHIC COLUMN. Geologic time is divided into named groupings according to six basic units, which are (in order of size from longest to shortest) eon, era, period, epoch, age, and chron. There is no absolute standard for the length of any unit; rather, it takes at least two ages to make an epoch, at least two epochs to compose a period, and so on. The dates for specific eons, eras, periods, and so on are usually given in relative terms, however; an example is the designation of the Cambrian period given earlier.

Chronostratigraphy also uses six time units: the eonothem, erathem, system, series, stage, and chronozone. These time units are analogous to the terms in the geologic time scale, the major difference being that chronostratigraphic units are conceived in terms of relative time and are not assigned dates. The more distant in time a particular unit is, the more controversy exists regarding its boundary with preceding and successive units. This is true both of the geologic and the chronostratigraphic scales.

For this reason, the International Union of Geological Sciences, the leading worldwide body of geologic scientists, has established a Commission on Stratigraphy to determine such boundaries. The commission selects and defines what are called Global Stratotype Sections and Points (GSGPs), which are typically marine fossil formations. Because it is believed that life has existed longest on Earth in its oceans, samples from the water provide the most reliable stratigraphic record.

NAMING OF CHRONOSTRATIGRAPHIC UNITS

As noted, the chronostratigraphic divisions correspond to units of geologic time, even though chronostratigraphic units are based on relative dating methods and geologic ones use absolute time measures. Because attempts at relative dating have been taking place since the late eighteenth century, today's geologic units originated as what would be called *stratigraphic* or *chronostratigraphic* units. Even today the names of the phases are the same, with the only difference being the units in which they are expressed. Thus, when speaking in terms of geologic time, one would refer to the Jurassic period, whereas in stratigraphic terms, this would be the Jurassic *system.*

In 1759 the Italian geologist Giovanni Arduino (1714–1795) developed the idea of primary, secondary, and tertiary groups of rocks. Though the use of the terms *primary* and *secondary* has been discarded, vestiges of Arduino's nomenclature survive in the modern designation of the Tertiary subera of the Cenozoic era (erathem in stratigraphic terminology) as well as in the name of the present period or system, the Quaternary. (Just as primary, secondary, and tertiary refer to a first, second, and third level, respectively, the term *quaternary* indicates a fourth level.)

We are living in the fourth of four eons, or eonothems, the Phanerozoic, which is divided into three eras, or erathems: Paleozoic, Mesozoic, and Cenozoic. These eras, in turn, are divided into 11 periods, or systems, whose names (except for Tertiary and Quaternary) refer to the locations in which the respective stratigraphic systems were first observed. The names of these systems, along with their dates in millions of years before the present and the origin of their names, are as follows (from the most distant to the most recent):

<u>Periods/Systems of the Paleozoic Era/Erathem</u>

- Cambrian (about 545 to 495 Ma): Cambria, the Roman name for the province of Wales
- Ordovician (about 495 to 443 Ma): Ordovices, the name of a Celtic tribe in ancient Wales
- Silurian (about 443 to 417 Ma): Silures, another ancient Welsh Celtic tribe
- Devonian (about 417 to 354 Ma): Devonshire, a county in southwest England
- Mississippian (a subperiod of the Carboniferous period, about 354 to 323 Ma): the Mississippi River
- Pennsylvanian (a subperiod of the Carboniferous, about 323 to 290 Ma): the state of Pennsylvania
- Permian (about 290 to 248.2 Ma): Perm, a province in Russia

<u>Periods/Systems of the Mesozoic Era/Erathem</u>

- Triassic (about 248.2 to 205.7 Ma): a tripartite, or threefold, division of rocks in Germany
- Jurassic (about 205.7 to 142 Ma): the Jura Mountains of Switzerland and France
- Cretaceous (about 142 to 65 Ma): from a Latin word for "chalk," a reference to the chalky cliffs of southern England and France

Within the more recent Cenozoic era, or erathem, names of epochs (or "series" in stratigraphic terminology) become important. They are all derived from Greek words, whose meanings are given below:

<u>Epochs/Series of the Cenozoic Era/Erathem</u>

- Paleocene (about 65 to 54.8 Ma): "early dawn of the recent"
- Eocene (about 54.8 to 33.7 Ma): "dawn of the recent"
- Oligocene (about 33.7 to 23.8 Ma): "slightly recent"
- Miocene (about 23.8 to 5.3 Ma): "less recent"
- Pliocene (about 5.3 to 1.8 Ma): "more recent"
- Pleistocene (about 1.8 to 0.01 Ma): "most recent"
- Holocene (about 0.01 Ma to present): "wholly recent"

REAL-LIFE APPLICATIONS

CORRELATION

The geologist studying the stratigraphic record is a sort of detective, looking for clues. Just as detectives have their methods for solving crimes, geologists rely on correlation, or methods of establishing age relationships between various strata. There are two basic types of correlation: physical correlation, which requires comparison of the physical characteristics of the strata, and fossil correlation, the comparison of fossil types.

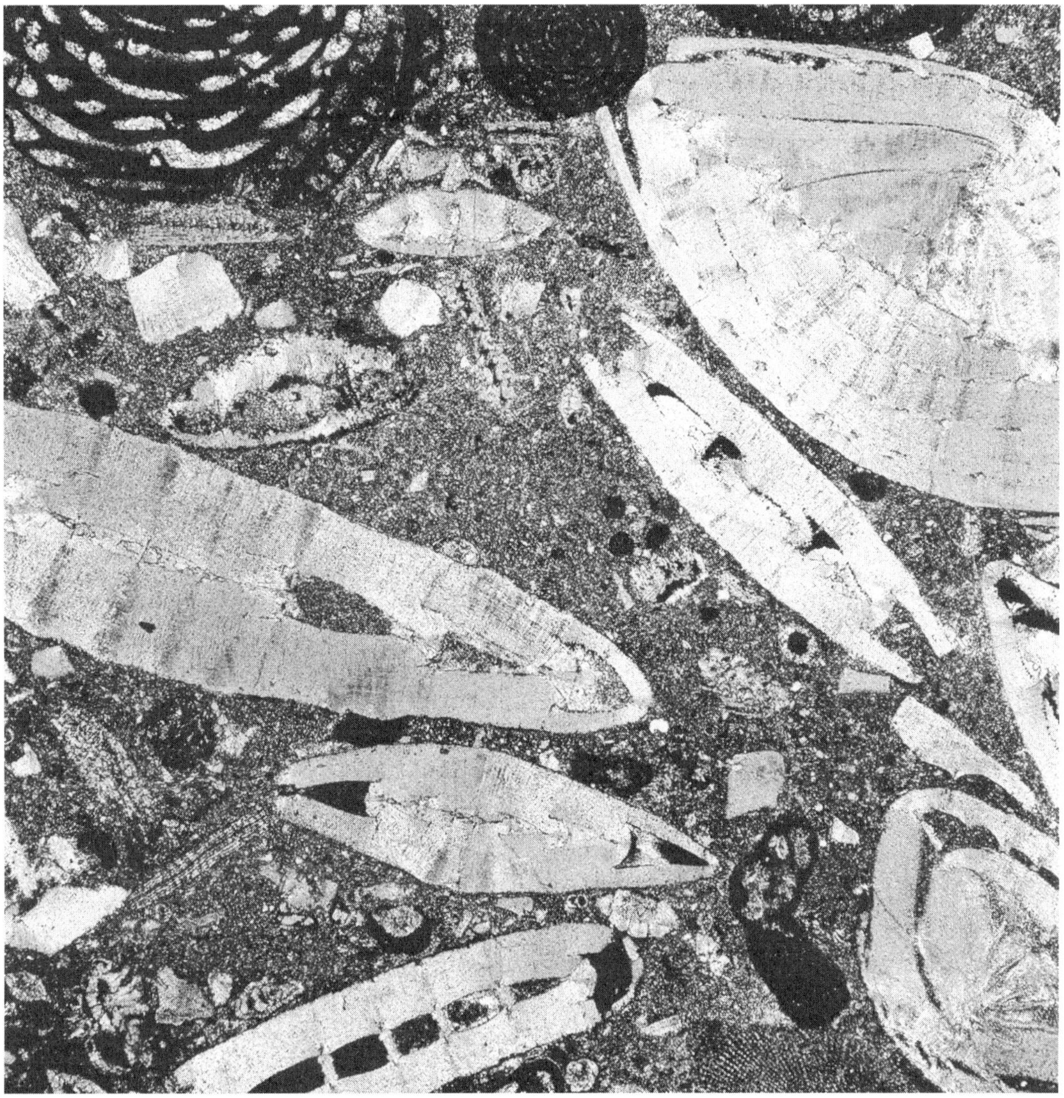

A POLARIZED-LIGHT MICROGRAPH SHOWS FOSSILS IN LIMESTONE DATING TO THE EOCENE AND OLIGOCENE EPOCHS. *(© A. Pasieka/Photo Researchers. Reproduced by permission.)*

Actually, chronostratigraphic work is very similar some of the toughest cases confronted by police detectives, because more often than not the geologic detective has little evidence on which to operate. First of all, as noted earlier, only sedimentary rock can be used in making such determinations: for instance, igneous rock in its molten form, as when it is expelled from a volcano, could force itself underneath a rock stratum, thus confusing the stratigraphic record.

POTENTIAL PITFALLS. Even when the rock is sedimentary, there is still plenty of room for error. The layers may be many feet or less than an inch deep, and it is up to the geologist to determine whether the stratum has been affected by such geologic forces as erosion. If erosion has occurred, it can cause a disturbance, or unconformity (discussed later), which tends to render inaccurate any reading of the stratigraphic record.

Another possible source of disturbance is an earthquake, which could cause one part of Earth's crust to shift over an adjacent section, making the stratigraphic record difficult, if not impossible, to read. Under the best of conditions, after all, the strata are hardly neat, easily defined lines. If one observes a horizontal section, there is likely to be a change in thickness, because as the stratum extends outward, it merges with the edges of adjacent deposits.

Yet another potential pitfall in stratigraphic correlation involves one of the most useful tools

KEY TERMS

ABSOLUTE AGE: The absolute age of a geologic phenomenon is its age in Earth years. Compare with *relative age.*

BIOSTRATIGRAPHY: An area of stratigraphy involving the study of fossilized plants and animals in order to establish dates for and correlations between stratigraphic layers.

CHRONOSTRATIGRAPHY: A subdiscipline of stratigraphy devoted to studying the relative ages of rocks. Compare with *geochronometry.*

CORRELATION: A method of establishing age relationships between various rock strata. There are two basic types of correlation: physical correlation, which requires comparison of the physical characteristics of the strata, and fossil correlation, the comparison of fossil types.

DATING: Any effort directed toward finding the age of a particular item or phenomenon. Methods of geologic dating are either relative (i.e., comparative and usually based on rock strata) or absolute. The latter, based on such methods as the study of radioactive isotopes, usually is given in terms of actual years or millions of years.

EON: The longest phase of geologic time, equivalent to an eonothem in the stratigraphic time scale. Earth's history has consisted of four eons, the Hadean or Priscoan, Archaean, Proterozoic, and Phanerozoic. The next-smallest subdivision of geologic time is the era.

EPOCH: The fourth-longest phase of geologic time, shorter than an era and longer than an age and a chron. An epoch is equivalent to a series in the stratigraphic time scale. The current epoch is the Holocene, which began about 0.01 Ma (10,000 years) ago.

ERA: The second-longest phase of geologic time, after an eon, and equivalent to an erathem in the stratigraphic time scale. The current eon, the Phanerozoic, has had three eras, the Paleozoic, Mesozoic, and Cenozoic, which is the current era. The next-smallest subdivision of geologic time is the period.

EROSION: The movement of soil and rock due to forces produced by water, wind, glaciers, gravity, and other influences.

GA: An abbreviation meaning "giga-years" or "billion years." The age of Earth is about 4.6 Ga.

GEOCHRONOMETRY: An area of stratigraphy devoted to determining absolute dates and time intervals. Compare with *chronostratigraphy.*

GEOLOGIC MAP: A map showing the rocks beneath Earth's surface, including their distribution according to type as well as their ages, relationships, and structural features.

GEOLOGIC TIME: The vast stretch of time over which Earth's geologic development has occurred. This span (about 4.6 billion years) dwarfs the history of human existence, which is only about two million years. Much smaller still is the span of human civilization, only about 5,500 years.

KEY TERMS CONTINUED

HISTORICAL GEOLOGY: The study of Earth's physical history. Historical geology is one of two principal branches of geology, the other being physical geology.

ISOTOPES: Atoms that have an equal number of protons, and hence are of the same element, but differ in their number of neutrons. This results in a difference of mass. An isotope may be either stable or radioactive.

LAW OF FAUNAL SUCCESSION: The principle that all samples of any given fossil species were deposited on Earth, regardless of location, at more or less the same time. This makes it possible to correlate widely separated strata.

LAW OF SUPERPOSITION: The principle that strata are deposited in a sequence such that the deeper the layer, the older the rock. This is applicable only or sedimentary rock, as opposed to igneous or metamorphic rock.

LITHOSTRATIGRAPHY: An area of stratigraphy devoted to the study and description (but not the dating) of rock layers.

MA: An abbreviation used by earth scientists, meaning "million years" or "megayears." When an event is designated as, for instance, 160 Ma, it usually means 160 million years ago.

PALEONTOLOGY: The study of fossilized plants and animals, or flora and fauna.

PERIOD: The third-longest phase of geologic time, after an era; it is equivalent to a system in the stratigraphic time scale. The current eon, the Phanerozoic, has had 11 periods, and the current era, the Cenozoic, has consisted of three periods, of which the most recent is the Quaternary. The next-smallest subdivision of geologic time is the epoch.

PRECAMBRIAN TIME: A term that refers to the first three of four eons in Earth's history, which lasted from about 4,560 Ma to about 545 Ma ago.

RADIOACTIVITY: A term describing a phenomenon whereby certain materials are subject to a form of decay brought about by the emission of high-energy particles or radiation. Forms of particles or energy include alpha particles (positively charged helium nuclei), beta particles (either electrons or subatomic particles called positrons), or gamma rays, which occupy the highest energy level in the electromagnetic spectrum.

RADIOMETRIC DATING: A method of absolute dating using ratios between "parent" isotopes and "daughter" isotopes, which are formed by the radioactive decay of parent isotopes.

RELATIVE AGE: The relative age of a geologic phenomenon is its age compared with the ages of other geologic phenomena, particularly the stratigraphic record of rock layers. Compare with *absolute age.*

KEY TERMS CONTINUED

SEDIMENT: Material deposited at or near Earth's surface from a number of sources, most notably preexisting rock.

SEDIMENTARY ROCK: Rock formed by compression and deposition (i.e., formation of deposits) on the part of other rock and mineral particles. Sedimentary rock is one of the three major types of rock, along with igneous and metamorphic.

SEDIMENTOLOGY: The study and interpretation of sediments, including sedimentary processes and formations.

STRATA: Layers, or beds, of rocks beneath Earth's surface. The singular form is *stratum*.

STRATIGRAPHIC COLUMN: The succession of rock strata laid down over the course of time, each of which correlates to specific junctures in Earth's geologic history.

STRATIGRAPHY: The study of rock layers, or strata, beneath Earth's surface.

UNCONFORMITY: An apparent gap in the geologic record, as revealed by observing rock layers or strata.

WEATHERING: The breakdown of rocks and minerals at or near the surface of Earth due to physical or chemical processes, or both.

available to a geologist attempting to find an absolute age for the materials he or she is studying: radiometric dating. Though this method can provide accurate absolute dates, it is quite possible that the age thus determined will be the age of the parent rock from which a sample is taken, not the age of the sample itself. The grains of sand in a piece of sandstone, for instance, are much older than the larger unit of sandstone, and for this reason, radiometric dating is useful only in specific circumstances.

PHYSICAL AND FOSSIL CORRELATION. Given all these challenges, it is a wonder that geologists manage to correlate strata successfully, yet they do. Physical correlations are achieved on the basis of several criteria, including color, the size of grains, and the varieties of minerals found within a stratum. By such means, it is sometimes possible to correlate widely separated strata.

Particularly impressive feats of correlation can result from the study of fossils, whose stratigraphic implications, as we have noted, were first discovered by William Smith. Smith hit upon the idea of biostratigraphy while excavating land for a set of canals near London. As he discovered, any given stratum contains the same types of fossils, and strata in two different areas thus can be correlated.

Long before his countryman Charles Darwin (1809–1882) developed the theory of evolution, Smith conceived his own law of faunal succession, which hints at the idea that species developed and disappeared over given phases in Earth's past. According to the law of faunal succession, all samples of any given fossil species were deposited on Earth, regardless of location, at more or less the same time. As a result, if a geologist finds a stratum in one area that contains a particular fossil and another in a distant area containing the same fossil, it is possible to conclude that the strata are the same.

UNCONFORMITIES

In discussing the many challenges facing a geologist studying stratigraphic data, the role of erosion was noted. Let us return to that subject, because erosion is a source of what are known as unconformities, or gaps in the rock record. Unconformities are of three types: angular unconformities, disconformities, and nonconformities.

Angular unconformities involve a tilting of the layers, such that an upper layer does not lie perfectly parallel to a lower one. Disconformities are more deceptive, because the layers are parallel, yet there is still an unconformity between them, and only a study of the fossil record can reveal the unconformity. Finally, a nonconformity arises when sedimentary rocks are divided from a type of igneous rock known as *intrusive* (meaning "cooled within Earth").

ANGULAR UNCONFORMITIES. Angular unconformities emerge as a by-product of the dramatic shifts and collisions that take place in plate tectonics (see Plate Tectonics). Sediment accumulates and then, as a result of plate movement, is moved about and eventually experiences weathering and erosion. Layers are tilted and then flattened by more erosion, and as the solid earth rises or sinks, they are shifted further. Such is the case, for instance, along the Colorado River at the Grand Canyon, where angular unconformities reveal a series of movements over the years.

Another famous angular unconformity can be found at Siccar Point in Scotland, where nearly horizontal deposits of sandstone rest atop nearly vertical ones of graywacke, another sedimentary rock. Observations of this unconformity led the great geologist James Hutton (1726–1797) to the realization that Earth is much, much older than the 6,000 years claimed by theologians in his day (see Historical Geology).

WHERE TO LEARN MORE

Bishop, A. C., A. Woolley, and A. Hamilton. *Cambridge Guide to Minerals, Rocks, and Fossils.* New York: Cambridge University Press, 1992.

Boggy's Links to Stratigraphy and Geochronology (Web site). <http://geologylinks.freeyellow.com/stratigraphy.html>.

Harris, Nicholas, Alessandro Rabatti, and Andrea Ricciardi. *The Incredible Journey to the Beginning of Time.* New York: Peter Bedrick Books, 1998.

Lamb, Simon, and David Sington. *Earth Story: The Shaping of Our World.* Princeton, NJ: Princeton University Press, 1998.

MacRae, Andrew. *Radiometric Dating and the Geological Time Scale* (Web site). <http://www.talkorigins.org/faqs/dating.html>.

Reeves, Hubert. *Origins: Cosmos, Earth, and Mankind.* New York: Arcade, 1998.

Spickert, Diane Nelson, and Marianne D. Wallace. *Earthsteps: A Rock's Journey Through Time.* Golden, CO: Fulcrum Kids, 2000.

Stratigraphy and Earth History—West's Geology Directory (Web site). <http://www.soton.ac.uk/~imw/stratig.htm>.

University of Georgia Stratigraphy Lab (Web site). <http://www.uga.edu/~strata/home.html>.

Web Time Machine. UCMP (University of California, Berkeley, Museum of Paleontology) (Web site). <http://www.ucmp.berkeley.edu/help/timeform.html>.

PALEONTOLOGY

CONCEPT

Thanks to a certain 1993 blockbuster, most people know the name of at least one period in geologic history. *Jurassic Park* spurred widespread interest in dinosaurs and, despite its fantastic plot, encouraged popular admiration and respect for the work of paleontologists. Paleontology is the study of life-forms from the distant past, as revealed primarily through the record of fossils left on and in the earth. It is a complex and varied subdiscipline of historical geology that is tied closely to the biological sciences. As with other types of historical geology, the work of a paleontologist is similar to that of a detective investigating a case with few and deceptive clues. Reliable fossil samples are more rare, compared with the vast number of species that have lived on Earth, than one might imagine. Furthermore, several factors pose challenges for paleontologists attempting to interpret the fossil record. Nonetheless, paleontologic research has led to a growing understanding of how life emerged, how Earth has changed, and how vast animal populations became extinct over relatively short periods of time.

HOW IT WORKS

Thinking in Terms of Geologic Time

The term geologic time refers to the great sweep of Earth's history, a timescale that dwarfs the span of human existence. The essays Historical Geology and Geologic Time offer several comparisons to emphasize the proportions involved and to illustrate the very short period during which human life has existed on this planet.

As one example shows, if all of geologic time were compressed into a single year, the first *Homo sapiens* would have appeared on the scene at about 8:00 p.m. on December 31. Human civilization, which dates back about 5,500 years (a millennium before the building of Egypt's great pyramids) would have emerged within the last minute of the year.

In another example, geologic time is compared to the distance from Los Angeles to New York City. On this scale, the period of time in which humans have existed on the planet would be equivalent to the distance from New York's Central Park to the Empire State Building, or less than 2 mi. (3.2 km). The history of human civilization, on the other hand, would be less than 16 ft. (4.9 m) long.

THE "ABYSS OF TIME." Needless to say, the scope of geologic time compared with the units with which we are accustomed to measuring our lives (or even the history of our civilization) is more than a little intimidating. This fact perhaps was best expressed by the Scottish geologist John Playfair (1748–1819), friend and countryman of the "father of geology," James Hutton (1726–1797). At a time when many people were content to believe that the Earth had been around no more than 6,000 years (see Historical Geology), Hutton suggested that to undergo the complex processes that had shaped its landforms, the planet had to be much, much older. Commenting on Hutton's discoveries, Playfair said, "The mind seemed to grow giddy by looking so far into the abyss of time."

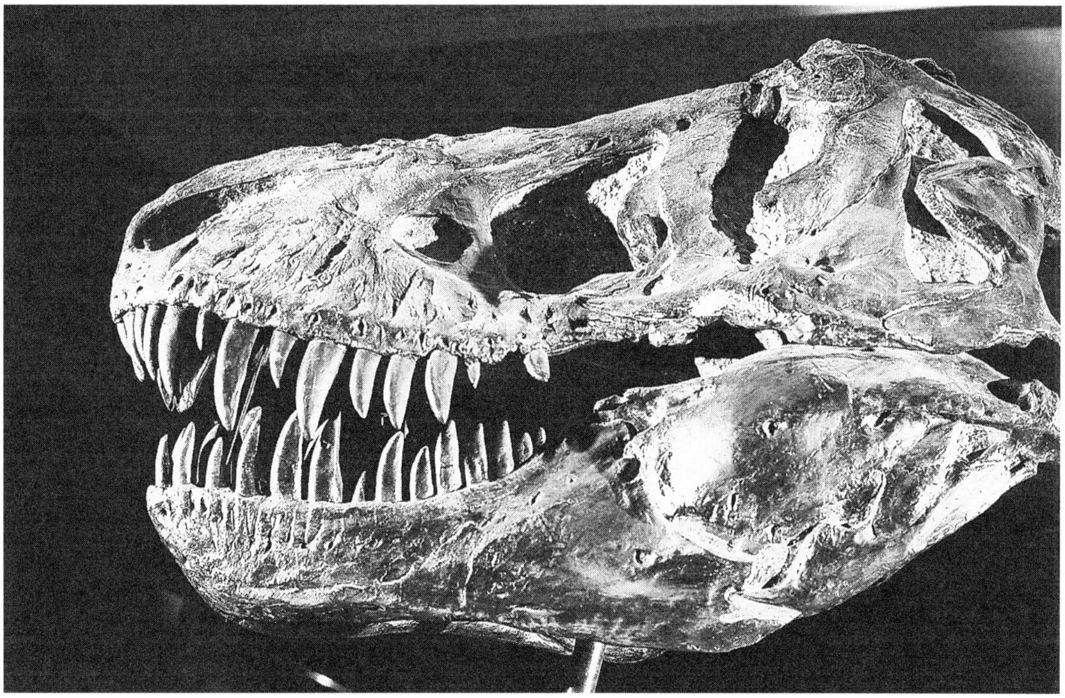

THE HEAD OF A TYRANNOSAURUS REX ("KING OF THE TERRIBLE LIZARDS") FOUND IN MONTANA, DATING TO THE MESOZOIC PERIOD. *(© Tom McHugh/Photo Researchers. Reproduced by permission.)*

CARBON: THE MEANING OF "LIFE"

A discussion of life on Earth requires us to go deep into this "abyss," though not nearly as far back as the planet's origins. It does appear that life on Earth existed at a very early point, but in this context "life" refers merely to molecules of carbon-based matter capable of replicating themselves. Knowledge of these very early forms is extremely limited.

Carbon appears in all living things, in things that were once living, and in materials produced by living things (for example, sap, blood, and urine). Hence, the term *organic,* which once meant only living matter, refers to almost all types of material containing carbon. The only carbon-containing materials that are not considered organic are oxides, such as carbon dioxide and carbon monoxide, and carbonates, a class of minerals that is extremely abundant on Earth.

PRECAMBRIAN TIME

We will return to the subject of carbon, which plays a role in one technique for dating relatively recent items or phenomena. For the present, however, let us set our bearings for a discussion of the Phanerozoic eon, the fourth and last of the major divisions of geologic time. Though extremely primitive life-forms existed before the Phanerozoic eon, the vast majority of species have evolved since it began, and consequently paleontological work is concerned primarily with the Phanerozoic eon.

The divisions of geologic time are not arranged in terms of strict mathematical relationships of the type to which we are accustomed, for example, ten years in a decade, ten decades in a century, and so on. Instead, each era consists of two or more periods, each period consists of two or more epochs, and so on. The first 4,000 million years or so of Earth's existence (abbreviated as 4,000 Ma, or 4 Ga) are known as Precambrian time. In discussing this period of time, the vast majority of the planet's history, it is seldom necessary to speak of geologic time divisions smaller than the largest unit, the eon. Precambrian time consisted of three eons, the Hadean or Priscoan, Archaean, and Proterozoic.

THE FIRST THREE EONS. The Hadean (sometimes called the Priscoan and dating to about 4,560 Ma to 4,000 Ma ago) saw the formation of the planet and the beginnings of the oceans and an early form of atmosphere that consisted primarily of carbon dioxide. It was during this eon that the carbon-based matter referred earlier made its appearance, perhaps by means of the meteorites that bombarded the planet during that long-ago time.

In the Archaean eon (about 4,000 Ma to 2,500 Ma ago) the first clear evidence of life appeared in the form of microorganisms. These were prokaryotes, or cells without a nucleus, which eventually were followed by eukaryotes, or cells with a nucleus. Many of the prerequisites for life as we know it were established during this time, though our present oxygen-containing atmosphere still lay far in the future.

Longest of the four eons was the Proterozoic eon (about 2,500 Ma to 545 Ma). This phase saw the beginnings of very basic forms of plant life, while oxygen in the atmosphere assumed about 4% of its present levels. Animal life, meanwhile, still consisted primarily of eukaryotes.

THE PHANEROZOIC EON

The majority of paleontologic history has taken place during the Phanerozoic eon. In the course of this essay, we discuss its eras and periods (the second- and third-longest spans of geologic time, respectively) as they relate to life on Earth. The three Phanerozoic eras are as follows:

Eras of the Phanerozoic Eon

- Paleozoic (about 545–248.2 Ma)
- Mesozoic (about 248.2–65 Ma)
- Cenozoic (about 65 Ma–present)

Within these eras are the following periods:

Periods of the Paleozoic Era

- Cambrian (about 545–495 Ma)
- Ordovician (about 495–443 Ma)
- Silurian (about 443–417 Ma)
- Devonian (about 417–354 Ma)
- Carboniferous (about 354–290 Ma)
- Permian (about 290–248.2 Ma)

Periods of the Mesozoic Era

- Triassic (about 248.2–205.7 Ma)
- Jurassic (about 205.7–142 Ma)
- Cretaceous (about 142–65 Ma)

Periods of the Cenozoic Era

- Palaeogene (about 65–23.8 Ma)
- Neogene (about 23.8–1.8 Ma)
- Quaternary (about 1.8 Ma to the present)

The Carboniferous period of the Paleozoic era usually is divided into two subperiods, the Mississippian (about 354 to 323 Ma) and the Pennsylvanian (about 323–290 Ma). In addition, the Palaeogene and Neogene periods of the Cenozoic era often are lumped together as a subera called the Tertiary. By substituting that name for those of the two periods, it is possible to use a time-honored mnemonic device by which geology students have memorized the names of the 11 Phanerozoic periods: "Camels Ordinarily Sit Down Carefully; Perhaps Their Joints Creak Tremendously Quietly."

EPOCHS OF THE CENOZOIC ERA. An epoch is the fourth-largest division of geologic time and is, for the most part, the smallest one with which we will be concerned. (There are two smaller categories, the age and the chron.) Listed here are the epochs of the Cenozoic era from the most distant to the Holocene, in which we are now living. Their names are derived from Greek words whose meanings are provided:

Epochs of the Cenozoic Era

- Paleocene (about 65–54.8 Ma): "early dawn of the recent"
- Eocene (about 54.8–33.7 Ma): "dawn of the recent"
- Oligocene (about 33.7–23.8 Ma): "slightly recent"
- Miocene (about 23.8–5.3 Ma): "less recent"
- Pliocene (about 5.3–1.8 Ma): "more recent"
- Pleistocene (about 1.8–0.01 Ma): "most recent"
- Holocene (about 0.01 Ma to present): "wholly recent"

A BRIEF OVERVIEW OF PALEONTOLOGIC HISTORY

The title "A Brief Overview of Paleontologic History" is almost a contradiction in terms, since virtually nothing about the history of Earth has been brief. Moreover, the history of *life* on Earth is so filled with detail and complexity that it could fill many books, as indeed it has. Owing to that complexity, anything approaching an exhaustive treatment of the subject would burden the reader with so much technical terminology that it would obscure the larger overview of paleontology and the materials of the paleontologist's work. Therefore, only the most cursory of treatments is possible, or indeed warranted, in the present context. For additional detail, the reader is invited to consult other texts, including

A DINOSAUR IS EXCAVATED AT DINOSAUR NATIONAL MONUMENT IN COLORADO. *(© James L. Amos/Photo Researchers. Reproduced by permission.)*

those listed in the suggested reading section at the end of this essay.

As with many another process, the evolution of organisms was exceedingly slow in the beginning (and here the comparative term *slow* refers even to the standards of geologic time), but it sped up considerably over the course of Earth's history. This is not to suggest that the development of life-forms has been a steady process; on the contrary, it has been punctuated by mass extinctions, discussed at the conclusion of this essay. Nonetheless, it is correct to say that during the first 80%–90% of Earth's history, the few existing life-forms underwent an extremely slow process of change.

PRECAMBRIAN AND PALEOZOIC LIFE-FORMS. Life existed in Precambrian time, as noted, but over the course of those four billion years, it evolved only to the level of single-cell microorganisms. Samples of these organisms have been found in the fossil record, but the fossilized history of life on Earth really began in earnest only with the Cambrian period at the beginning of the Paleozoic era and the Phanerozoic eon. The early Cambrian period saw an explosion of invertebrate (without an internal skeleton) marine forms, which dominated from about 545 Ma–417 Ma ago. By about 420 Ma–410 Ma, life had appeared on land, in the form of algae and primitive insects.

The beginning of the Devonian period (approximately 417 Ma) saw the appearance of the first vertebrates (animals with an internal skeleton), which were jawless fish. Plant life on land consisted of ferns and mosses. By the late Devonian (about 360 Ma), fish had evolved jaws, and amphibians had appeared on land. Reptiles emerged between about 320 Ma and 300 Ma, in the Pennsylvanian subperiod of the Carboniferous. In the last period of the Paleozoic era, the Permian (about 290–248.2 Ma), reptiles became the dominant land creatures.

MESOZOIC AND CENOZOIC LIFE-FORMS. The next era, the Mesozoic (about 248.2–65 Ma), belonged to a particularly impressive form of reptile, known as the terrible lizard: the dinosaur. These creatures are divided into groups based on the shape of their hips, which were either lizardlike or birdlike. Though the lizardlike Saurischia emerged first, they lived alongside the birdlike Ornithischia throughout the late Triassic, Jurassic, and Cretaceous periods. Ornithischia were all herbivores, or plant eaters, whereas Saurischia included both herbivores and carnivores, or meat eaters. Naturally, the most fierce of the dinosaurs were carnivores, a group that included the largest carnivore ever to walk the earth, *Tyrannosaurus.*

Though dinosaurs receive the most attention, the Mesozoic world was alive with varied forms, including flying reptiles and birds. (In fact, dinosaurs may have been related to birds, and, in the opinion of some paleontologists, they may have been warm-blooded, like birds and mammals, rather than cold-blooded, like other reptiles.) Botanical life included grasses, flowering plants, and trees of both the deciduous (leaf-shedding) and coniferous (cone-bearing) varieties.

A violent event, discussed in the context of mass extinction later in this essay, brought an end to the Mesozoic era. This cleared the way for the emergence of mammalian forms at the beginning of the Cenozoic, though it still would be a long time before anything approaching an ape, let alone a human, appeared on the scene. The earliest hominid, or humanlike creature, dates back to about four million years ago, in the Pliocene epoch of the Neogene period.

HISTORICAL GEOLOGY AND PALEONTOLOGY

One of the two principal divisions of geology (along with physical geology) is historical geology, the study of Earth's physical history. Other subdisciplines of historical geology are stratigraphy, the study of rock layers, or strata, beneath Earth's surface; geochronology, the study of Earth's age and the dating of specific formations in terms of geologic time; and sedimentology, the study and interpretation of sediments, including sedimentary processes and formations.

Paleontology, the investigation of life-forms from the distant past (primarily through the study of fossilized plants and animals), is another subdiscipline of historical geology. Though it is rooted in the physical sciences, it obviously crosses boundaries into the biological or life sciences as well. Related or subordinate fields include paleozoology, which focuses on the study of prehistoric animal life; paleobotany, the study of past plant life; and paleoecology, the study of the relationship between prehistoric plants and animals and their environments.

CLASSIFYING PLANTS AND ANIMALS

Given the close relationship between paleontology and the biological sciences, it is necessary to discuss briefly the taxonomic system applied in biology, botany, zoology, and related fields. Taxonomy is an area of biology devoted to the identification, classification, and naming of organisms. Devised in the eighteenth century by the Swedish botanist Carolus Linnaeus (1707–1778) and improved in succeeding years by many others, the taxonomic system revolutionized biology.

Linnaeus's taxonomy provided a framework for classifying known species not simply by superficial similarities but also by systemic characteristics. For example, worms and snakes have something in common on a surface level, because they are both without appendages and move by writhing on the ground. A worm is an invertebrate, however, whereas a snake is a vertebrate. The Linnaean system therefore would classify them in widely separated categories: they are not siblings or even first cousins but more like fourth cousins.

Moreover, the system created by Linnaeus gave scientists a means for classifying and thereby potentially understanding much about the history and characteristics of species as yet undiscovered. Thus, it would prove of immeasurable significance to the English naturalist Charles Darwin (1809–1882) in formulating his theory of evolution. As Darwin showed, the varieties of different organisms have increased over time, as those organisms developed characteristics that made them more adaptable to their environments. Plants and animals that failed to adapt simply became extinct, though failure to adapt is only one of several causes for extinction, as we shall see.

A BRIEF OVERVIEW. The Linnaean system uses binomial nomenclature, or a two-part naming scheme (in Latin), to identify each separate type of organism. If a man is named John Smith, then "Smith" identifies his family, while John identifies him singularly. Likewise each variety of organism is identified by genus, equivalent to Smith, and species, analogous to John. In the Linnaean system, there are eight levels of classification, which, from most general to most specific, are kingdom, phylum, subphylum, class, order, family, genus, and species.

These levels can be illustrated by identifying a species near and dear to all of us: *Homo sapiens,* commonly known as *humans.* We belong to the animal kingdom (Animalia), the Chordata (i.e., possessing some form of central nervous system) phylum, and the Vertebrata subphylum, indicating the existence of a backbone. Within the mammal (Mammalia) class we are part of the primate (Primata) order, along with apes. Humans are distinguished further as members of the hominid (Hominoidea), or "human-like" family; the genus *Homo* ("man"); and the species *sapiens* ("wise").

DATING MATERIALS FROM THE PAST

In studying the past, paleontologists and other earth scientists working in the field of historical geology rely on a variety of dating techniques. "Dating," in a scientific context, usually refers to any effort directed toward finding the age of a particular item or phenomenon. It may be relative, devoted to finding an item's age in relation to that of other items; or absolute, involving the determination of age in actual years or millions of years.

Among the methods of relative dating are stratigraphic dating, discussed in the essay Stratigraphy, as well as seriation, faunal dating, and pollen dating. Seriation entails analyzing the abundance of a particular item and assigning relative dates based on that abundance. Faunal dating is the use of bones from animals (fauna) to determine age, and pollen dating, or palynology, analyzes pollen deposits.

FAUNAL DATING AND PALYNOLOGY. The concept of faunal dating emerged from early work by the English engineer and geologist William Smith (1769–1839), widely credited as the "father of stratigraphy." In particular, Smith established an important division of stratigraphy, known as biostratigraphy, that is closely tied to paleontology. While excavating land for a set of canals near London, he discovered that any given stratum, or rock layer, contains the same types of fossils, and therefore strata in two different areas can be correlated.

Smith stated this in what became known as the law of faunal succession: all samples of any given fossil species were deposited on Earth, regardless of location, at more or less the same time. As a result, if a geologist finds a stratum in one area that contains a particular fossil and another in a distant area containing the same fossil, it is possible to conclude that the strata are the same.

Pollen dating, or palynology, is based on the fact that seed-bearing plants release large numbers of pollen grains each year. As a result, pollen spreads over the surrounding area, and in many cases pollen from the distant past has been preserved. This has occurred primarily in lake beds, peat bogs, and, occasionally, in areas with cool or acidic soil. By observing the species of pollen deposited in an area, scientists are able to develop a sort of "pollen calendar," which provides information about such details as changes in climate.

DENDROCHRONOLOGY. Scientists use relative dating when they must, but they would prefer to determine dates in an absolute sense wherever possible. Most methods of absolute dating rely on processes that are not immediately comprehensible to the average person, but there is one exception: dendrochronology, or the dating of tree rings. As almost everyone knows, trees produce one growth ring per year. There is nothing magical about this, since a year is not an abstract unit of time; rather, it is based on Earth's revolution around the Sun, during which time the planet undergoes changes in orientation that result in the four seasons, which, in turn, affect the tree's growth.

Though dendrochronology makes use of a principle familiar to most people, the work of the dendrochronologist requires detailed, often complex study. Just as the layers of rock beneath Earth's surface reveal information about past geologic events (a matter discussed in the essay Stratigraphy), tree rings can tell us much about environmental changes. Thin rings, for instance, suggest climatic anomalies and may provide clues about cataclysmic events that were understood only vaguely by the ancient humans who experienced them. (An example of this is the apparent cataclysm of A.D. 535, which is discussed Earth Systems.)

AMINO-ACID RACIMIZATION. Dendrochronology is useful only for studying the relatively recent past, up to about 10,000 years—a span equivalent to the Holocene epoch, which began with the end of the last ice age. To investigate more distant phases of Earth's history, it is necessary to use forms of radiometric dating, which we will discuss shortly. The principles of radiometric dating, however, are illustrated by another method, amino-acid racimization.

With the exception of some microbes, living organisms incorporate only one of two forms of amino acids, known as L-forms. Once the organism dies, the L-amino acids gradually convert to D-amino acids. In the 1960s, scientists discovered that by comparing the ratios between the L- and D-forms, it was possible to date organisms that were several thousand years old. Unfortunately, it has since come to light that because of the many factors affecting the rate of amino-acid conversion, this method is less reliable than once was believed. Moisture, temperature, and pH (the relative acidity and alkalinity of a substance) all play a part, and because these factors vary so widely, amino-acid racimization no longer is used commonly.

Nonetheless, the basic principle behind amino-acid racimization plays a part in other, more reliable forms of absolute dating. Many of them are based on the fact that over time, a particular substance converts to another, mirror substance. By comparing the ratios between them, it is possible to arrive at some estimate of the amount of time that has elapsed since the organism died.

RADIOCARBON DATING. The most significant method of absolute dating available to scientists today is radiometric dating, which is explained in detail in the essay Geologic Time. Each chemical element is distinguished by the number of protons (positively charged particles) in its atomic nucleus, but atoms of a particular element may have differing numbers of neutrons, or neutrally charged particles, in their nuclei. Such atoms are referred to as isotopes.

Certain isotopes are stable, whereas others are radioactive, meaning that they are likely to eject particles from the nucleus over time. The amount of time it takes for half the isotopes in a sample to stabilize is called its half-life. By analyzing the quantity of radioactive isotopes in a given sample that have converted to stable isotopes, it is possible to determine the age of the sample. In other situations, it is necessary to compare ratios of unstable "parent" isotopes to even more unstable "daughter" isotopes produced by the parent.

As we noted earlier, carbon is present in all living things, and thus an important means of dating available to paleontologists uses a radioactive form of carbon. All atoms of carbon have six protons, and the most stable and abundant carbon isotope is carbon-12, so designated because it has six neutrons. On the other hand, carbon-14, with eight neutrons, is unstable.

When an organism is alive, it incorporates a certain ratio of carbon-12 in proportion to the (very small) amount of carbon-14 that it receives from the atmosphere. Once the organism dies, however, it stops incorporating new carbon, and the ratio between carbon-12 and carbon-14 begins to change as the carbon-14 decays to form nitrogen-14. Therefore, a scientist can use the ratios of carbon-12, carbon-14, and nitrogen-14 to estimate the age of an organic sample. This method is known as radiocarbon dating.

Carbon-14, or radiocarbon, has a half-life of 5,730 years, meaning that it is useful for analyzing only fairly recent samples. Nonetheless, it takes much longer than 5,730 years for the other half of the radiocarbon isotopes in a given sample to stabilize, and for this reason radiocarbon dating can be used with considerable accuracy for 30,000–40,000 years. Sophisticated instrumentation can extend this range even further, up to 70,000 years.

THE LIMITS OF ABSOLUTE DATING. While 70,000 years, or 0.07 Ma, may be a long time in human terms, from the standpoint of the earth scientist, 0.07 Ma is only yesterday—the latter part of the last epoch, the Pleistocene. Other forms of radiometric dating, such as potassium-argon dating and uranium-series dating, can be used to measure truly long spans of times, in the billions of years. These methods are discussed in Geologic Time.

Potassium-argon dating and uranium-series dating can be useful to the paleontologist, inasmuch as they aid in determining the age of the geologic samples in which the remains of life-forms are found. Nothing is simple, however, when it comes to dating specimens from the distant past. After all, geologists working in the realm of stratigraphy face numerous challenges in judging the age of samples, even with these sophisticated forms of radiometric dating. (For more on this subject, see Stratigraphy.)

In fact, the work of a paleontologist is much like that of the stratigrapher. The success of either type of scientist relies more on detailed and painstaking detective work than it does on sophisticated technology. Both must analyze lay-

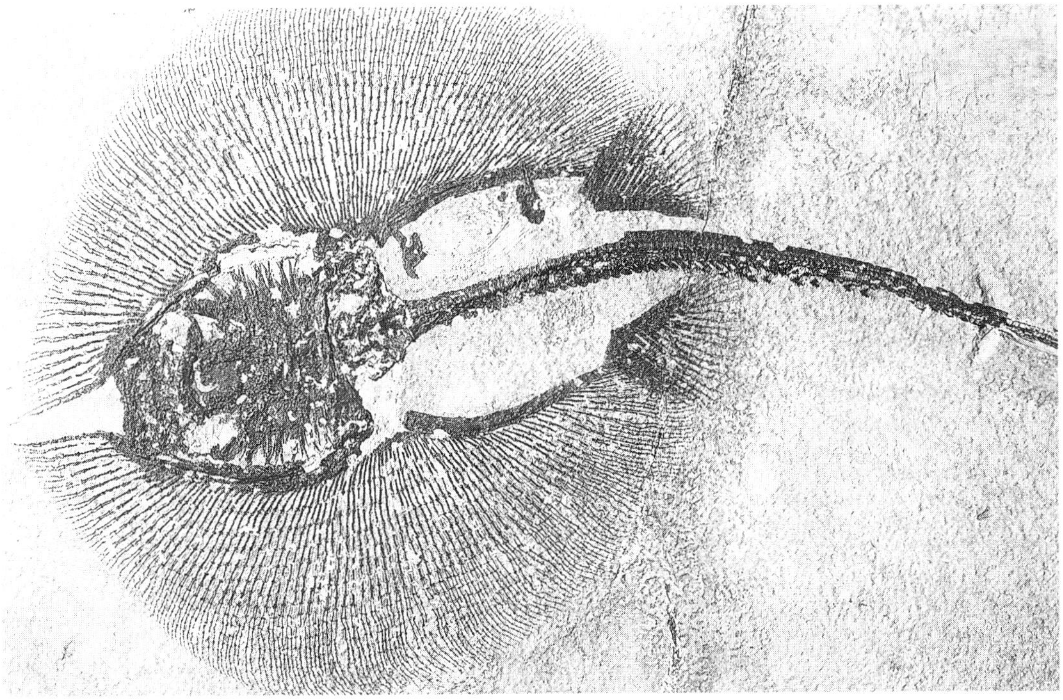

STINGRAY FOSSIL, POSSIBLY DATING TO THE JURASSIC PERIOD. THIS IS A RARE FIND, BECAUSE STINGRAYS HAVE NO BONE, ONLY CARTILAGE, WHICH MAKES IT HARDER FOR THEM TO UNDERGO MINERALIZATION AND BE PRESERVED AS FOSSILS. *(© Gary Retherford/Photo Researchers. Reproduced by permission.)*

ered samples from the past to form a picture of the chronology in which the samples evolved, and oftentimes the apparent evidence can be deceptive. The principal difference between stratigraphic and paleontologic study relates to the materials: rocks in the first instance and fossils in the second.

REAL-LIFE APPLICATIONS

FOSSILS AND FOSSILIZATION

The term fossil refers to the remains of any prehistoric life-form, especially those preserved in rock before the end of the last ice age. The process by which a once-living thing becomes a fossil is known as fossilization. Generally, fossilization refers to changes in the hard portions, including bones, teeth, shells, and so on. This series of changes, in which minerals are replaced by different minerals, is known as mineralization. Sometimes, soft parts may experience mineralization and thus be preserved as fossils. A deceased organism in the process of becoming a fossil is known as a subfossil.

The majority of fossils come from invertebrates, such as mussels, that possess hard parts. Generally speaking, the older and smaller the organism, the more likely it is to have experienced fossilization, though other factors (which we will discuss later) also play a part. One of the most important factors involves location: for the most part, the lower the altitude, the greater the likelihood that a region will contain fossils. The best place of all is in the ocean, particularly the ocean floor. Nonetheless, fossils have been found on every continent of Earth, and the great distances that sometimes separate samples of the same species have aided earth scientists of many fields in understanding the processes that shaped our planet.

FOSSILS AND GEOLOGIC HISTORY. The earth beneath our feet is not standing still; rather, it is constantly moving, and over the great stretches of geologic time, the positions of the continents have shifted considerably. The details of these shifts are discussed in Plate Tectonics, an area of geologic study that explains much about the earth, from earthquakes and volcanoes to continental drift.

Paleontology has contributed to the study of plate tectonics by revealing apparent anomalies, such as fossilized dinosaur parts in Antarctica. No dinosaur could have lived on that forbidding continent, so there must be some other explanation: the continental plates themselves have moved. Long ago, the present continents were united in a "supercontinent" called Pangaea. When Pangaea split apart to form the present continents, the remains of various species were separated from one another and from the latitudes to which they were accustomed in life.

Fossilized remains of single-cell organisms have been found in rock samples as old as 3.5 Ga, and animal fossils have been located in rocks that date to the latter part of Precambrian time, as old as 1 Ga. Just as paleontologists have benefited from studies in chronostratigraphy and geochronometry, realms of stratigraphy concerned with the dating of rock samples, stratigraphers and other geologists have used fossil samples to date the rock strata in which they were found. Not all fossilized life-forms are equally suited to this purpose. Certain ones, known as index fossils or indicator species, have been associated strongly with particular intervals of geologic time. An example is the ammonoid, a mollusk that proliferated for about 350 Ma from the late Devonian to the early Cretaceous before experiencing mass extinction.

MAKING THE GRADE AS A FOSSIL. Everything that is living eventually dies, but not nearly all living things will become fossils. And even if they do, there are numerous reasons why fossils might not be preserved in such a way as to provide meaningful evidence for a paleontologist many millions of years later. In the potential pool of candidates for fossilization, as we have noted, organisms without hard structural portions are unlikely to become fossilized. Fossilization of soft-bodied creatures sometimes occurs, however, as, for instance, at Burgess Shale in British Columbia, where environmental conditions made possible the preservation of a wide range of samples.

Furthermore, location is a powerful factor. Sedimentary rock, formed by compression and deposition (i.e., formation of deposits) on the part of other rock and mineral particles, provides the setting for many fossils. Best of all is sediment, such as sand or mud, that has not yet consolidated into harder sandstone, limestone, or other rocks. Organisms that die in upland locations are more likely to be disturbed either by wind or by scavengers, creatures that feed on the remains of living things. On the other hand, an organism at the bottom of an ocean is out of reach from most scavengers. Even at lesser depths, if the organism is in a calm, relatively scavenger-free marine environment, there is a good chance that it will be preserved.

Assuming that all the conditions are right and the dead organism is capable of undergoing fossilization, it will experience mineralization of one type or another. Living things already contain minerals, which is the reason why people take mineral supplements to augment the substances nature has placed in their bodies to preserve and extend life. In the mineralization of a fossil, the minerals in the organism's body may be replaced by other ones, or other minerals may be added to existing ones. It is also possible that both the hard and soft parts will dissolve and be replaced by a mineral cement that forms a mold that preserves the shape of the organism.

FINDING AND STUDYING FOSSILS. Only about 30% of species are ever fossilized, a fact that scientists must take into account, because it could skew their reading of the paleontologic record. If a paleontologist judges the past only from the fossils that have been found in an area, it will result in a picture of a past environment that contained only certain species, when, in fact, others were present. Furthermore, there are many factors that contribute to the loss of fossils. For instance, if the area has been subjected to violent tectonic activity, it is likely that the sample will be destroyed partially or wholly.

The removal of a fossil from its home in the rock is a painstaking process akin to restoring a valuable piece of art. Before removing it, the paleontologist photographs the fossil and surrounding strata and records details about the environment. Only when these steps have been taken is the fossil removed. This is done with a rock saw, which is used to cut out carefully a large area surrounding the fossil. The sample is then jacketed, or wrapped in muslin with an additional layer of wet plaster, and taken to a laboratory for study.

Fossil research can reveal a great deal about the history of life on Earth, including the relationships between species or between species and their habitats. Studies of dinosaur bones have brought to light proteins that existed in the bodies of these long-gone creatures, while research on certain oxygen isotopes has aided attempts to discover whether dinosaurs were warm-blooded creatures. Thanks to advances in the understanding of DNA (deoxyribonucleic acid), which provides the genetic codes for all living things, it may be possible to make even more detailed studies in the future.

Mass Extinction

The remains of dinosaurs, of course, have an importance aside from their significance to paleontology. The bodies of these giant lizards have been deposited in the earth, where over time they became coal, peat, petroleum, and other fossil fuels. The latter are discussed in Economic Geology, but the fact that the dinosaurs disappeared at all is of particular interest to paleontology. Why are there no dinosaurs roaming the earth today? The answer appears to be that they were wiped out in a dramatic event, perhaps brought about as the result of a meteorite impact.

Numerous species have become extinct, typically as a result of their inability to adapt to changes in their natural environment. More recently, some extinctions or endangerments of species have been attributed to human activities, including hunting and the disruption of natural habitats. For the most part, however, extinction is simply a part of Earth's history, a result of the fact that nature has a way of destroying organisms that do not adapt (the "survival of the fittest"). But there have been occasions in the course of the planet's past in which vast numbers of individuals and species perished at once. A natural catastrophe may destroy a large population of individuals within a locality, a phenomenon known as mass mortality. Or mass mortality may take place on a global scale, destroying many species, in which case it is known as mass extinction.

CAUSES OF MASS EXTINCTION. The Bible depicts an example of near mass extinction, in the form of Noah's flood, and, indeed, several instances of mass extinction have resulted from sudden and dramatic changes in ocean levels. Others have been caused by tectonic events, most notably vast volcanic eruptions that filled the atmosphere with so much dust that they caused a violent change in temperature. Scientific speculation concerning other such extinctions has pointed to events in or from space—either the explosion of a star or the impact of a meteorite on Earth—as the cause of atmospheric changes and hence mass extinction.

Even though scientists have a reasonable idea of the immediate causes of mass extinction in some cases, their understanding of the ultimate or root causes is still limited. This fact was expressed by the University of Chicago paleobiologist David M. Raup, who wrote: "The disturbing reality is that for none of the thousands of well-documented extinctions in the geologic past do we have a solid explanation of why the extinction occurred."

EXAMPLES OF MASS EXTINCTION. The five largest known mass extinctions occurred at intervals of 50 Ma to 100 Ma over a span of time from about 435 to 65 million years ago. Most occurred at the end of a period, which is no accident, since geologists have used mass extinction as a factor in determining the parameters of a specific period.

In the late Ordovician period, about 435 Ma ago, a drop in the ocean level wiped out one-fourth of all marine families. Similarly, changes in sea level, along with climate changes, appear to have caused the destruction of one-fifth of existing marine families during the late Devonian period (about 357 Ma ago). Worst of all was the "great dying," as the extinction at the end of the Permian period (about 250 Ma) is known. Perhaps caused by a volcanic eruption in Siberia, it eliminated a staggering 96% of all species over a period of about a million years.

During the late Triassic period, about 198 million years ago, another catastrophe eliminated a quarter of marine families. Paleontologists know this, as they know about other mass extinctions, by the inordinate numbers of fossilized samples found in rock strata dating to that period. This reliance on the fossil record is also reflected in the fact that the scope of early mass extinctions usually is expressed in terms of marine life. As we have seen, the ocean environment provides the most reliable fossil record. Creatures died on land as well, but the terrestrial record is simply less reliable or less complete.

Scientific disagreement over the late-Triassic mass extinction exemplifies the fact that our knowledge of these distant events is not firmly established, but rather is subject to much scientific conjecture and dispute. (This does not mean that just any old idea can compete on an equal

KEY TERMS

ABSOLUTE AGE: The absolute age of a geologic phenomenon is its age in Earth years. Compare with *relative age.*

ATMOSPHERE: In general, an atmosphere is a blanket of gases surrounding a planet. Unless otherwise identified, however, the term refers to the atmosphere of Earth, which consists of nitrogen (78%), oxygen (21%), argon (0.93%), and other substances that include water vapor, carbon dioxide, ozone, and noble gases such as neon, which together comprise 0.07%.

BIOSPHERE: A combination of all living things on Earth—plants, mammals, birds, reptiles, amphibians, aquatic life, insects, viruses, single-cell organisms, and so on—as well as all formerly living things that have not yet decomposed.

BIOSTRATIGRAPHY: An area of stratigraphy involving the study of fossilized plants and animals to establish dates for and correlations between stratigraphic layers.

CHRONOSTRATIGRAPHY: A subdiscipline of stratigraphy devoted to studying the relative ages of rocks. Compare with *geochronometry.*

CONTINENTAL DRIFT: The theory that the configuration of Earth's continents was once different than it is today; that some of the individual landmasses of today once were joined in other continental forms; and that these landmasses later separated and moved to their present locations.

CORRELATION: A method of establishing age relationships between various rock strata. There are two basic types of correlation: physical correlation, which requires comparison of the physical characteristics of the strata, and fossil correlation, the comparison of fossil types.

DATING: Any effort directed toward finding the age of a particular item or phenomenon. Methods of geologic dating are either relative (i.e., comparative and usually based on rock strata) or absolute. The latter, based on such methods as the study of radioactive isotopes, typically is given in terms of actual years or millions of years.

EON: The longest phase of geologic time. Earth's history has consisted of four eons, the Hadean or Priscoan, Archaean, Proterozoic, and Phanerozoic. The next-smallest subdivision of geologic time is the era.

EPOCH: The fourth-longest phase of geologic time, shorter than an era and longer than an age or a chron. The current epoch is the Holocene, which began about 0.01 Ma (10,000 years) ago.

ERA: The second-longest phase of geologic time, after an eon. The current eon, the Phanerozoic, has had three eras, the Paleozoic, Mesozoic, and Cenozoic, which is the current era. The next-smallest subdivision of geologic time is the period.

FOSSIL: The mineralized remains of any prehistoric life-form, especially those preserved in rock before the end of the last ice age.

FOSSILIZATION: The process by which a once-living organism becomes a fossil. Generally, fossilization involves mineralization of the organism's hard portions, such as bones, teeth, and shells.

KEY TERMS CONTINUED

GA: An abbreviation meaning "gigayears," or "billion years." The age of Earth is about 4.6 Ga.

GEOCHRONOMETRY: An area of stratigraphy devoted to determining absolute dates and time intervals. Compare with *chronostratigraphy.*

GEOLOGIC TIME: The vast stretch of time over which Earth's geologic development has occurred. This span (about 4.6 billion years) dwarfs the history of human existence, which is only about two million years. Much smaller still is the span of human civilization, only about 5,500 years.

HISTORICAL GEOLOGY: The study of Earth's physical history. Historical geology is one of two principal branches of geology, the other being physical geology.

INVERTEBRATE: An animal without an internal skeleton.

ISOTOPES: Atoms that have an equal number of protons, and hence are of the same element, but differ in their number of neutrons. This results in a difference of mass. An isotope may be either stable or radioactive.

LAW OF FAUNAL SUCCESSION: The principle that all samples of any given fossil species were deposited on Earth, regardless of location, at more or less the same time. This makes it possible to correlate widely separated strata.

MA: An abbreviation used by earth scientists, meaning "million years," or "megayears." When an event is designated as, for instance, 160 Ma, it usually means 160 million years ago.

MASS EXTINCTION: A phenomenon in which numerous species cease to exist at or around the same time, usually as the result of a natural calamity.

MINERALIZATION: A series of changes experienced by a once-living organism during fossilization. In mineralization, minerals in the organism are either replaced or augmented by different minerals, or the hard portions of the organism dissolve completely.

ORGANIC: At one time, chemists used the term *organic* only in reference to living things. Now the word is applied to most compounds containing carbon, with the exception of carbonates (which are minerals), and oxides, such as carbon dioxide.

PALEOBOTANY: An area of paleontology involving the study of past plant life.

PALEOECOLOGY: An area of paleontology devoted to studying the relationship between prehistoric plants and animals and their environments.

PALEONTOLOGY: The study of life-forms from the distant past, primarily as revealed through the fossilized remains of plants and animals.

PALEOZOOLOGY: An area of paleontology devoted to the study of prehistoric animal life.

PERIOD: The third-longest phase of geologic time, after an era. The current eon, the Phanerozoic, has had 11 periods, and the current era, the Cenozoic, has consisted of three periods, of which the most recent is the Quaternary. The next-smallest subdivision of geologic time is the epoch.

KEY TERMS CONTINUED

PRECAMBRIAN TIME: A term that refers to the first three of four eons in Earth's history, which lasted from about 4,560 to about 545 Ma ago.

RADIOACTIVITY: A term describing a phenomenon whereby certain materials are subject to a form of decay brought about by the emission of high-energy particles or radiation. Forms of particles or energy include alpha particles (positively charged helium nuclei), beta particles (either electrons or subatomic particles called positrons, or gamma rays, which occupy the highest energy level in the electromagnetic spectrum.

RADIOMETRIC DATING: A method of absolute dating using ratios between "parent" isotopes and "daughter" isotopes, which are formed by the radioactive decay of parent isotopes. Radiometric dating also may involve ratios between radioactive isotopes and stable isotopes.

RELATIVE AGE: The relative age of a geologic phenomenon is its age compared with other geologic phenomena, particularly the stratigraphic record of rock layers. Compare with *absolute age.*

SEDIMENT: Material deposited at or near Earth's surface from a number of sources, most notably preexisting rock.

SEDIMENTARY ROCK: Rock formed by compression and deposition (i.e., formation of deposits) on the part of other rock and mineral particles. Sedimentary rock is one of the three major types of rock, along with igneous and metamorphic.

SEDIMENTOLOGY: The study and interpretation of sediments, including sedimentary processes and formations.

STRATA: Layers, or beds, of rocks beneath Earth's surface. The singular form is *stratum.*

STRATIGRAPHY: The study of rock layers, or strata, beneath Earth's surface.

VERTEBRATE: An animal with an internal skeleton.

footing: we are talking here about differences of opinion among highly trained specialists.) At any rate, some scientists refer to the late-Triassic mass extinction as being one of the less exciting or eventful mass extinctions. Of course, it is hard to see how a mass extinction could be unexciting or uneventful, but they mean this in comparative terms; on the other hand, some paleontologists maintain that the late-Triassic was among the most devastating.

As to the cause, some theorists point to a group of impact sites spread across Canada, the northern United States, and Ukraine, places that would have been more or less contiguous at the time of the mass extinction. Difficulties in analyzing the "signatures" left by the projectiles that made these impressions have prevented theorists from saying with any degree of certainty whether it was a comet or an asteroid that caused the impact. Others, in particular a team from the University of California at Berkeley led by geologist Paul R. Renne, cite a volcanic eruption as either the cause of the mass extinction, or at least a major abetting factor to an extinction already in progress. According to Renne and his team, basalt outcroppings scattered from New Jersey to Brazil to west Africa (again, areas that would have been contiguous then) suggest that a volcanic eruption of almost inconceivable magnitude occurred about 200 million years ago. Such an eruption would surely have destroyed vast quantities of living things.

The last and best known mass extinction occurred about 65 million years ago, marking the end of the Cretaceous period—and the end of the dinosaurs. As to what happened, paleontologists and other scientists have proposed a number of theories: a rapid climate change; the emergence of new poisonous botanical species, eaten by herbivorous dinosaurs, that resulted in the passing of toxins along the food web (see Ecosystems); an inability to compete successfully with the rapidly evolving mammals; and even an epidemic disease to which the dinosaurs possessed no immunity.

Interesting as many of these theories are, none has gained anything like the widespread acceptance achieved by another scenario. According to this highly credible theory, an asteroid hit Earth, hurtling vast quantities of debris into the atmosphere, blocking out the sunlight, and greatly lowering Earth's surface temperature. Around the world, geologists have found traces of iridium deposited at a layer equivalent to the boundary between the Cretaceous and Tertiary periods, the Tertiary being the beginning of the present Cenozoic era. This is significant, because iridium seldom appears on Earth's surface—but it *is* found in asteroids.

HUMANS AND MASS EXTINCTION. There have been much more recent, if less dramatic, examples of mass extinction, including those caused by the most highly developed of all life-forms: humans. Among these examples are the well-documented (and *very* recent) mass extinctions brought on by destruction of tropical rainforests. Such activities are killing off a vast array of organisms: according to the highly respected Harvard biologist Edward O. Wilson, some 17,500 species are disappearing each year. But cases of mass extinction are not limited to modern times.

When prehistoric hunters (the ancestors of today's Native Americans) crossed the Bering land bridge from Siberia to Alaska some 12,000 years ago, they found an array of species unknown in the Americas today. These species included mammoths and mastodons; giant bears, beaver, and bison; and even saber-toothed tigers, camels, and lions. Perhaps most remarkable of all, it appears that prehistoric America was once home to a creature that would prove to be of enormous benefit to humans until the beginning of the automotive age: the horse. Horses did not reappear in the Americas until Europeans arrived to conquer those lands after A.D. 1500.

WERE DINOSAURS WARM-BLOODED?

One of the most significant scientific debates of the later twentieth and early twenty-first centuries, not only in paleontology but in the earth sciences or even science itself, is the question of whether or not the dinosaurs were warm-blooded. In other words, were they like modern reptiles, which must adjust their temperature by moving into the sunlight when they are cold, and into the shade when they are too hot? Or were they more like modern birds and mammals, whose bodies generate their own heat?

A warm-blooded animal always has a more or less constant body temperature, regardless of the temperature of its environment. This is due to the fact that it produces heat by the burning of food, as well as by physical activity, and stores that heat under a layer of fat just beneath the skin. Warm-blooded animals are also capable of cooling down their bodies by perspiring and panting. Birds and mammals are the only warm-blooded animals; all others are cold-blooded. A cold-blooded creature, on the other hand, lacks control over its body temperature and therefore is warm when its environment is warm, and cold when its environment is cold.

The difference between warm- and cold-blooded animals is partly one of metabolic rate, or the rate at which nutrients are broken down and converted into energy. Cold-blooded creatures have slow metabolic rates; think of a python that swallows a medium-sized mammal whole and takes several days to digest it. The dinosaur debate is therefore often framed as a question of whether the dinosaurs' bodies had a relatively high or relatively low metabolic rate.

THE DEBATE. Until the 1960s, there was no debate: dinosaurs, whose existence had been known for about a century, were assumed to be big, dumb, slow, cold-blooded creatures. Then, in 1968, Robert T. Bakker—an undergraduate at Yale University, not a professor or a full-fledged paleontologist—revolutionized the world of paleontology with a paper called "The Superiority of Dinosaurs."

In his article, Bakker described dinosaurs as "fast, agile, energetic creatures" whose physiology was so advanced that even the biggest and heaviest of them could outrun a human. Just a year later, John H. Ostrom, a professor of paleontology who also happened to be at Yale, wrote that a recently identified species of theropod dinosaur must have been "an active and very agile predator."

Thus began the great dinosaur debate, which rages even today. *Jurassic Park* reinforced the Bakker-Ostrom position, portraying *Velociraptor* as a cunning, fast-moving predator with clear links to birds. And indeed there are many arguments for endothermy (warm-bloodedness) in dinosaurs—arguments that relate to everything from brain size to rate of growth to the latitudes at which dinosaur fossils have been located. On the other hand, there is plenty of evidence for ectothermy (cold-bloodedness), based on the dinosaurs' size, scaliness, the climate in the Mesozoic era, and so on.

To explore, compare, and judge these many arguments, the reader is encouraged to consult the "Were Dinosaurs Warm-Blooded?" Web site listed in the "Where to Learn More" section at the conclusion of this essay. However, a word of warning, as noted on that site: "The issue is a tangled, complex one. There are not just two sides to the issue; there are numerous competing hypotheses. If you're looking for a major controversy in science, look no further!"

WHERE TO LEARN MORE

Cadbury, Deborah. *Terrible Lizard: The First Dinosaur Hunters and the Birth of a New Science.* New York: Holt, 2001.

K-12: Paleontology: Dinos (Web site). <http://www.ceismc.gatech.edu/busyt/paleo.html>.

Morris, S. Conway. *The Crucible of Creation: The Burgess Shale and the Rise of Animals.* New York: Oxford University Press, 1998.

Munro, Margaret, and Karen Reczuch. *The Story of Life on Earth.* Toronto: Douglas and McIntyre, 2000.

Oceans of Kansas Paleontology (Web site). <http://www.oceansofkansas.com/>.

Paleontology and Fossils Resources—University of Arizona Library, Tucson (Web site). <http://www.library.arizona.edu/users/mount/paleont.html>.

Palmer, Douglas. *Atlas of the Prehistoric World.* Bethesda, MD: Discovery Communications, 1999.

Sanders, Robert. "New Evidence Links Mass Extinction with Massive Eruptions That Split Pangea Supercontinent and Created the Atlantic 200 Million Years Ago." University of California, Berkeley. (Web site). http://www.berkeley.edu/news/media/releases/99legacy/4-22-1999b.html>.

Singer, Ronald. *Encyclopedia of Paleontology.* Chicago: Fitzroy Dearborn Publishers, 1999.

University of California, Berkeley Museum of Paleontology (Web site). <http://www.ucmp.berkeley.edu/>.

USGS (United States Geological Survey) Paleontology Home Page (Web site). <http://geology.er.usgs.gov/paleo/>.

"Were Dinosaurs Warm-Blooded?" (Web site). <http://pubs.usgs.gov/gip/dinosaurs/warmblood.html>.

MINERALS

CONCEPT

A mineral is a naturally occurring, typically inorganic substance with a specific chemical composition and structure. An unknown mineral usually can be identified according to known characteristics of specific minerals in terms of certain parameters that include its appearance, its hardness, and the ways it breaks apart when fractured. Minerals are not to be confused with rocks, which are typically aggregates of minerals. There are some 3,700 varieties of mineral, a handful of which are abundant and wide-ranging in their application. Many more occur less frequently but are extremely important within a more limited field of uses.

HOW IT WORKS

INTRODUCTION TO MINERALS

The particulars of the mineral definition deserve some expansion, especially inasmuch as mineral has an everyday definition somewhat broader than its scientific definition. In everyday usage, minerals would be the natural, nonliving materials that make up rocks and are mined from the earth. According to this definition, minerals would include all metals, gemstones, clays, and ores. The scientific definition, on the other hand, is much narrower, as we shall see.

The fact that a mineral must be inorganic brings up another term that has a broader meaning in everyday life than in the world of science. At one time, the scientific definition of organic was more or less like the meaning assigned to it by nonscientists today, as describing all living or formerly living things, their parts, and substances that come from them. Today, however, chemists use the word organic to refer to any compound that contains carbon bonded to hydrogen, thus excluding carbonates (which are a type of mineral) and oxides such as carbon dioxide or carbon monoxide. Because a mineral must be inorganic, this definition eliminates coal and peat, both of which come from a wide-ranging group of organic substances known as hydrocarbons.

A mineral also occurs naturally, meaning that even though there are artificial substances that might be described as "mineral-like," they are not minerals. In this sense, the definition of a mineral is even more restricted than that of an element, discussed later in this essay, even though there are nearly 4,000 minerals and more than 92 elements. The number 92, of course, is not arbitrary: that is the number of elements that occur in nature. But there are additional elements, numbering 20 at the end of the twentieth century, that have been created artificially.

PHYSICAL AND CHEMICAL PROPERTIES OF MINERALS. The specific characteristics of minerals can be discussed both in physical and in chemical terms. From the standpoint of physics, which is concerned with matter, energy, and the interactions between the two, minerals would be described as crystalline solids. The definition of a mineral is narrowed further in terms of its chemistry, or its atomic characteristics, since a mineral must be of unvarying composition.

A mineral, then, must be solid under ordinary conditions of pressure and temperature. This excludes petroleum, for instance (which, in any case, would have been disqualified owing to its organic origins), as well as all other liquids and gases. Moreover, a mineral cannot be just any type of solid but must be a crystalline one—that is, a solid in which the constituent parts have a simple and definite geometric arrangement that is repeated in all directions. This rule, for instance, eliminates clay, an example of an amorphous solid.

Chemically, a mineral must be of unvarying composition, a stipulation that effectively limits minerals to elements and compounds. Neither sand nor glass, for instance, is a mineral, because the composition of both can vary. Another way of putting this is to say that all minerals must have a definite chemical formula, which is not true of sand, dirt, glass, or any other mixture. Let us now look a bit more deeply into the nature of elements and compounds, which are collectively known as pure substances, so as to understand the minerals that are a subset of this larger grouping.

ELEMENTS

The periodic table of elements is a chart that appears in most classrooms where any of the physical sciences are taught. It lists all elements in order of atomic number, or the number of protons (positively charged subatomic particles) in the atomic nucleus. The highest atomic number of any naturally occurring element is 92, for uranium, though it should be noted that a very few elements with an atomic number lower than 92 have never actually been found on Earth. On the other hand, *all* elements with an atomic number higher than 92 are artificial, created either in laboratories or as the result of atomic testing.

An element is a substance made of only one type of atom, meaning that it cannot be broken down chemically to create a simpler substance. In the sense that each is a fundamental building block in the chemistry of the universe, all elements are, as it were, "created equal." They are not equal, however, in terms of their abundance. The first two elements on the periodic table, hydrogen and helium, represent 99.9% of the matter in the entire universe. Though Earth contains little of either, our planet is only a tiny dot within the vastness of space; by contrast, stars such as our Sun are composed almost entirely of those elements (see Sun, Moon, and Earth).

ABUNDANCE ON EARTH. Of all elements, oxygen is by far the most plentiful on Earth, representing nearly half—49.2%—of the total mass of atoms found on this planet. (Here the term mass refers to the known elemental mass of the planet's atmosphere, waters, and crust; below the crust, scientists can only speculate, though it is likely that much of Earth's interior consists of iron.)

Together with silicon (25.7%), oxygen accounts for almost exactly three-fourths of the elemental mass of Earth. If we add in aluminum (7.5%), iron (4.71%), calcium (3.39%), sodium (2.63%), potassium (2.4%), and magnesium (1.93%), these eight elements make up about 97.46% of Earth's material. Hydrogen, so plentiful in the universe at large, ranks ninth on Earth, accounting for only 0.87% of the planet's known elemental mass. Nine other elements account for a total of 2% of Earth's composition: titanium (0.58%), chlorine (0.19%), phosphorus (0.11%), manganese (0.09%), carbon (0.08%), sulfur (0.06%), barium (0.04%), nitrogen (0.03%), and fluorine (0.03%). The remaining 0.49% is made up of various other elements.

Looking only at Earth's crust, the numbers change somewhat, especially at the lower end of the list. Listed below are the 12 most abundant elements in the planet's crust, known to earth scientists simply as "the abundant elements." These 12, which make up 99.23% of the known crustal mass, together form approximately 40 different minerals that account for the vast majority of that 99.23%. Following the name and chemical symbol of each element is the percentage of the crustal mass it composes.

Abundance of Elements in Earth's Crust

- Oxygen (O): 45.2%
- Silicon (Si): 27.2%
- Aluminum (Al): 8.0%
- Iron (Fe): 5.8%
- Calcium (Ca): 5.06%
- Magnesium (Mg): 2.77%
- Sodium (Na): 2.32%
- Potassium (K): 1.68%
- Titanium (Ti): 0.86%
- Hydrogen (H): 0.14%
- Manganese (Mn): 0.1%
- Phosphorus (P): 0.1%

ATOMS, MOLECULES, AND BONDING

As noted earlier, an element is identified by the number of protons in its nucleus, such that any atom with six protons *must* be carbon, since carbon has an atomic number of 6. The number of electrons, or negatively charged subatomic particles, is the same as the number of protons, giving an atom no net electric charge.

An atom may lose or gain electrons, however, in which case it becomes an ion, an atom or group of atoms with a net electric charge. An atom that has gained electrons, and thus has a negative charge, is called an anion. On the other hand, an atom that has lost electrons, thus becoming positive in charge, is a cation.

In addition to protons and electrons, an atom has neutrons, or neutrally charged particles, in its nucleus. Neutrons have a mass close to that of a proton, which is much larger than that of an electron, and thus the number of neutrons in an atom has a significant effect on its mass. Atoms that have the same number of protons (and therefore are of the same element), but differ in their number of neutrons, are called isotopes.

COMPOUNDS AND MIXTURES. Whereas there are only a very few elements, there are millions of compounds, or substances made of more than one atom. A simple example is water, formed by the bonding of two hydrogen atoms with one oxygen atom; hence the chemical formula for water, which is H_2O. Note that this is quite different from a mere mixture of hydrogen and oxygen, which would be something else entirely. Given the gaseous composition of the two elements, combined with the fact that both are extremely flammable, the result could hardly be more different from liquid water, which, of course, is used for putting out fires.

The difference between water and the hydrogen-oxygen mixture described is that whereas the latter is the result of mere physical mixing, water is created by chemical bonding. Chemical bonding is the joining, through electromagnetic attraction, of two or more atoms to create a compound. Of the three principal subatomic particles, only electrons are involved in chemical bonding—and only a small portion of those, known as valence electrons, which occupy the outer shell of an atom. Each element has a characteristic pattern of valence electrons, which determines the ways in which the atom bonds.

CHEMICAL BONDING. Noble gases, of which helium is an example, are noted for their lack of chemical reactivity, or their resistance to bonding. While studying these elements, the German chemist Richard Abegg (1869–1910) discovered that they all have eight valence electrons. His observation led to one of the most important principles of chemical bonding: atoms bond in such a way that they achieve the electron configuration of a noble gas. This concept, known as the octet rule, has been shown to be the case in most stable chemical compounds.

Abegg hypothesized that atoms combine with one another because they exchange electrons in such a way that both end up with eight valence electrons. This was an early model of ionic bonding, which results from attractions between ions with opposite electric charges: when they bond, these ions "complete" each other. Metals tend to form cations and bond with nonmetals that have formed anions. The bond between anions and cations is known as an ionic bond, and is extremely strong.

The other principal type of bond is a covalent bond. The result, once again, is eight valence electrons for each atom, but in this case, the nuclei of the two atoms share electrons. Neither atom "owns" them; rather, they share electrons. Today, chemists understand that most bonds are neither purely ionic nor purely covalent; instead, there is a wide range of hybrids between the two extremes, which are a function of the respective elements' electronegativity, or the relative ability of an atom to attract valence electrons. If one element has a much higher electronegativity value than the other one, the bond will be purely ionic, but if two elements have equal electronegativity values, the bond is purely covalent. Most bonds, however, fall somewhere between these two extremes.

INTERMOLECULAR BONDING. Chemical bonds exist between atoms and within a molecule. But there are also bonds *between* molecules, which affect the physical composition of a substance. The strength of intermolecular bonds is affected by the characteristics of the interatomic, or chemical, bond.

For example, the difference in electronegativity values between hydrogen and oxygen is great enough that the bond between them is not purely covalent, but instead is described as a polar covalent bond. Oxygen has a much higher electronegativity (3.5) than hydrogen (2.1), and therefore the electrons tend to gravitate toward the oxygen atom. As a result, water molecules have a strong negative charge on the side occupied by the oxygen atom, with a resulting positive charge on the hydrogen side.

By contrast, molecules of petroleum, a combination of carbon and hydrogen, tend to be nonpolar, because carbon (with an electronegativity value of 2.5) and hydrogen have very similar electronegativity values. Therefore the electric charges are more or less evenly distributed in the molecule. As a result, water molecules form strong attractions, known as dipole-dipole attractions, to each other. Molecules of petroleum, on the other hand, have little attraction to each other, and the differences in charge distribution account for the fact that water and oil do not mix.

Even weaker than the bonds between nonpolar molecules, however, are those between highly reactive elements, such as the noble gases and the "noble metals"—gold, silver, and copper, which resist bonding with other elements. The type of intermolecular attraction that exists in such a situation is described by the term London dispersion forces, a reference to the German-born American physicist Fritz Wolfgang London (1900–1954).

The bonding between molecules of most other metals, however, is described by the electron sea model, which depicts metal atoms as floating in a "sea" of valence electrons. These valence electrons are highly mobile within the crystalline structure of the metal, and this mobility helps explain metals' high electric conductivity. The ease with which metal crystals allow themselves to be rearranged explains not only metals' ductility (their ability to be shaped) but also their ability to form alloys, a mixture containing two or more metals.

THE CRYSTALLINE STRUCTURE OF MINERALS

By definition, a solid is a type of matter whose particles resist attempts at compression. Because of their close proximity, solid particles are fixed in an orderly and definite pattern. Within the larger category of solids are crystalline solids, or those in which the constituent parts are arranged in a simple, definite geometric pattern that is repeated in all directions.

The term crystal is popularly associated with glass and with quartz, but only one of these is a crystalline solid. Quartz is a member of the silicates, a large group of minerals that we will discuss later in this essay. Glass, on the other hand, is an amorphous solid, meaning that its molecules are not arranged in an orderly pattern.

CRYSTAL SYSTEMS. Elsewhere in this book (Earth, Science, and Nonscience and Planetary Science), there is considerable discussion of misconceptions originating with Aristotle (384–322 B.C.). Despite his many achievements, including significant contributions to the biological sciences, the great Greek philosopher spawned a number of erroneous concepts, which prevailed in the physical sciences until the dawn of the modern era. At least Aristotle made an attempt at scientific study, however; for instance, he dissected dead animals to observe their anatomic structures. His teacher, Plato (427?–347 B.C.), on the other hand, is hardly ever placed among the ranks of those who contributed, even ever so slightly, to progress in the sciences.

There is a reason for this. Plato, in contrast to his pupil, made virtually no attempt to draw his ideas about the universe from an actual study of it. Within Plato's worldview, the specific qualities of any item, including those in the physical world, reflected the existence of perfect and pure ideas that were more "real" than the physical objects themselves. Typical of his philosophy was his idea of the five Platonic solids, or "perfect" geometric shapes that, he claimed, formed the atomic substructure of the world.

The "perfection" of the Platonic solids lay in the fact that they are the only five three-dimensional objects in which the faces constitute a single type of polygon (a closed shape with three or more sides, all straight), while the vertices (edges) are all alike. These five are the tetrahedron, octahedron, and icosahedron, composed of equilateral triangles (four, eight, and twenty, respectively); the cube, which, of course, is made of six squares; and the dodecahedron, made up of twelve pentagons. Plato associated the latter solid with the shape of atoms in outer space, while the other four corresponded to what the Greeks believed were the elements on Earth: fire (tetrahedron), earth (cube), air (octahedron), and water (icosahedron).

All of this, of course, is nonsense from the standpoint of science, though the Platonic solids are of interest within the realm of mathematics. Yet amazingly, Plato in his unscientific way actually touched on something close to the truth, as applied to the crystalline structure of minerals.

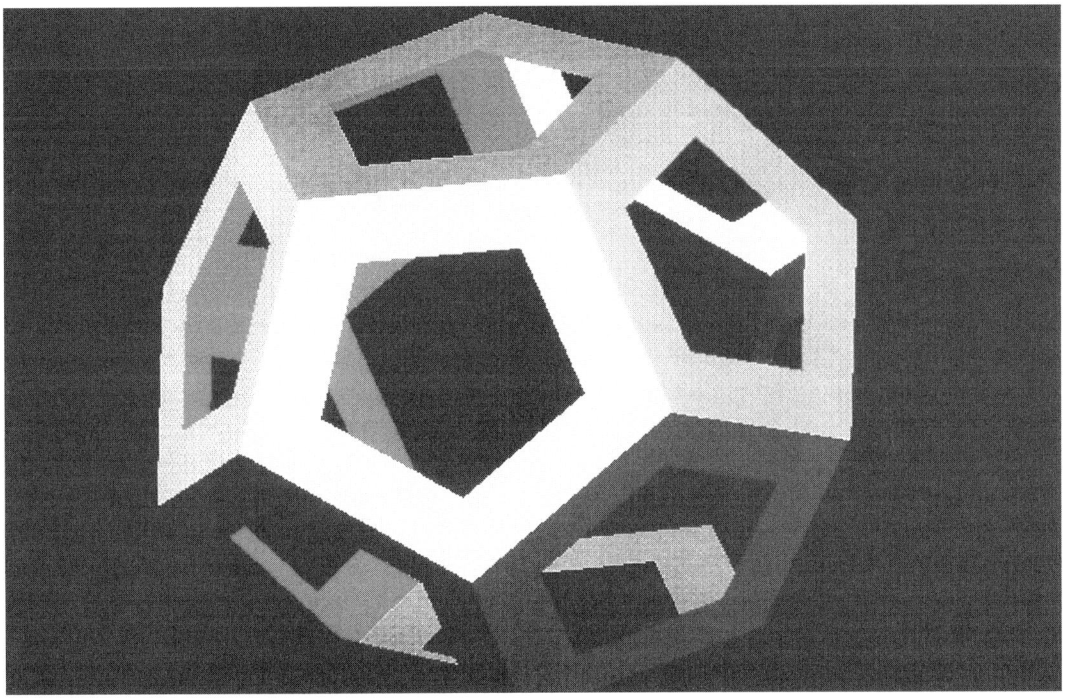

A DODECAHEDRON, ONE OF THE PLATONIC SOLIDS. *(© Richard Duncan/Photo Researchers. Reproduced by permission.)*

Despite the large number of minerals, there are just six crystal systems, or geometric shapes formed by crystals. For any given mineral, it is possible for a crystallographer (a type of mineralogist concerned with the study of crystal structures) to identify its crystal system by studying a good, well-formed specimen, observing the faces of the crystal and the angles at which they meet.

An isometric crystal system is the most symmetrical of all, with faces and angles that are most clearly uniform. Because of differing types of polygon that make up the faces, as well as differing numbers of vertices, these crystals appear in 15 forms, several of which are almost eerily reminiscent of Plato's solids: not just the cube (exemplified by halite crystals) but also the octahedron (typical of spinels) and even the dodecahedron (garnets).

REAL-LIFE APPLICATIONS

MINERAL GROUPS

Before the time of the great German mineralogist Georgius Agricola (1494–1555), attempts to classify minerals were almost entirely overshadowed by the mysticism of alchemy, by other nonscientific preoccupations, or by simple lack of knowledge. Agricola's *De re metallica* (On minerals, 1556), published after his death, constituted the first attempt at scientific mineralogy and mineral classification, but it would be two and a half centuries before the Swedish chemist Jöns Berzelius (1779–1848) developed the basics of the classification system used today.

Berzelius's classification system was refined later in the nineteenth century by the American mineralogist James Dwight Dana (1813–1895) and simplified by the American geologists Brian Mason (1917–) and L. G. Berry (1914–). In general terms, the classification system accepted by mineralogists today is as follows:

- Class 1: Native elements
- Class 2: Sulfides
- Class 3: Oxides and hydroxides
- Class 4: Halides
- Class 5: Carbonates, nitrates, borates, iodates
- Class 6: Sulfates, chromates, molybdates, tungstates
- Class 7: Phosphates, arsenates, vanadates
- Class 8: Silicates

NATIVE ELEMENTS. The first group, native elements, includes (among other things) metallic elements that appear in pure form somewhere on Earth: aluminum, cadmium, chromium, copper, gold, indium, iron, lead, mercury, nickel, platinum, silver, tellurium, tin, titanium, and zinc. This may seem like a great number of elements, but it is only a small portion of the 87 metallic elements listed on the periodic table.

The native elements also include certain metallic alloys, a fact that might seem strange for several reasons. First of all, an alloy is a mixture, not a compound, and, second, people tend to think of alloys as being man-made, not natural. The list of metallic alloys included among the native elements, however, is very small, and they meet certain very specific mineralogic criteria regarding consistency of composition.

The native elements class also includes native nonmetals such as carbon, in the form of graphite or its considerably more valuable alter ego, diamond, as well as elemental silicon (an extremely important building block for minerals, as we shall see) and sulfur. For a full list of native elements and an explanation of criteria for inclusion, as well as similar data for the other classes of mineral, the reader is encouraged to consult the *Minerals by Name* Web site, the address of which is provided in "Where to Learn More" at the end of this essay.

SULFIDES AND HALIDES. Most important ores (a rock or mineral possessing economic value)—copper, lead, and silver—belong to the sulfides class, as does a mineral that often has been mistaken for a precious metal—iron sulfide, or pyrite. Better known by the colloquial term fool's gold, pyrite has proved valuable primarily to con artists who passed it off as the genuine article. During World War II, however, pyrite deposits near Ducktown, Tennessee, became valuable owing to the content of sulfur, which was extracted for use in defense applications.

Whereas the sulfides fit the common notion of a mineral as a hard substance, halides, which are typically soft and transparent, do not. Yet they are indeed a class of minerals, and they include one of the best-known minerals on Earth: halite, known chemically as NaCl or sodium chloride—or, in everyday language, table salt.

OXIDES. Oxides, as their name suggests, are minerals containing oxygen; however, if all oxygen-containing minerals were lumped into just one group, that group would take up almost the entire list. For instance, under the present system, silicates account for the vast majority of minerals, but since *those* contain oxygen as well, a list that grouped all oxygen-based minerals together would consist of only four classes: native elements, sulfides, halides, and a swollen oxide category that would include 90% of all known minerals.

Instead, the oxides class is limited only to noncomplex minerals that contain either oxygen or hydroxide (OH). Examples of oxides include magnetite (iron oxide) and corundum (aluminum oxide.) It should be pointed out that a single chemical name, such as iron oxide or aluminum oxide, is not limited to a single mineral; for example, anatase and brookite are both titanium oxide, but they represent different combinations.

OTHER NONSILICATES. All the mineral classes discussed to this point, as well as several others to follow, are called nonsilicates, a term that stresses the importance of silicates among mineral classes.

Like the oxides, the carbonates, or carbon-based minerals, are a varied group. This class also contains a large number of minerals, making it the most extensive group aside from silicates and phosphates. Among these are limestones and dolostones, some of the most abundant rocks on Earth.

The phosphates, despite their name, may or may not include phosphorus; in some cases, arsenic, vanadium, or antimony may appear in its place. The same is true of the sulfates, which may or may not involve sulfur; some include chromium, tungsten, selenium, tellurium, or molybdenum instead.

TWO QUESTIONABLE CLASSES. In addition to the seven formal classes just described, there are two other somewhat questionable classes of nonsilicate that might be included in a listing of minerals. They would be included, if at all, only with major reservations, since they do not strictly fit the fourfold definition of a mineral as crystalline in structure, natural, inorganic, and identifiable by a precise chemical formula. These two questionable groups are organics and mineraloids.

Organics, as their name suggests, have organic components, but as we have observed, "organic" is not the same as "biological." This class excludes hard substances created in a biological setting—for example, bone or pearl—and includes only minerals that develop in a geologic setting yet have organic chemicals in their composition. By far the best-known example of this class, which includes only a half-dozen minerals, is amber, which is fossilized tree sap.

Amber is also among the mineraloids, which are not really "questionable" at all—they are clearly *not* minerals, since they do not have the necessary crystalline structure. Nevertheless, they often are listed among minerals in reference books and are likely to be sold by mineral dealers. The other four mineraloids include two other well-known substances, opal and obsidian.

SILICATES

Where minerals are concerned, the silicates are the "stars of the show": the most abundant and most widely used class of minerals. That being said, it should be pointed out that there are a handful of abundant nonsilicates, most notably the iron oxides hematite, magnetite, and goethite. A few other nonsilicates, while they are less abundant, are important to the makeup of Earth's crust, examples being the carbonates calcite and dolomite; the sulfides pyrite, sphalerite, galena, and chalcopyrite; and the sulfate gypsum. Yet the nonsilicates are not nearly as important as the class of minerals built around the element silicon.

Though it was discovered by Jöns Berzelius in 1823, owing to its abundance in the planet's minerals, silicon has been in use by humans for thousands of years. Indeed, silicon may have been one of the first elements formed in the Precambrian eons (see Geologic Time). Geologists believe that Earth once was composed primarily of molten iron, oxygen, silicon, and aluminum, which, of course, are still the predominant elements in the planet's crust. But because iron has a greater atomic mass, it settled toward the center, while the more lightweight elements rose to the surface. After oxygen, silicon is the most abundant of all elements on the planet, and compounds involving the two make up about 90% of the mass of Earth's crust.

SILICON, CARBON, AND OXYGEN. On the periodic table, silicon lies just below carbon, with which it shares an ability to form long strings of atoms. Because of this and other chemical characteristics, silicon, like carbon, is at the center of a vast array of compounds—organic in the case of carbon and inorganic in the case of silicon. Silicates, which, as noted earlier, account for nine-tenths of the mass of Earth's crust (and 30% of all minerals), are to silicon and mineralogy what hydrocarbons are to carbon and organic chemistry.

Whereas carbon forms it most important compounds with hydrogen—hydrocarbons such as petroleum—the most important silicon-containing compounds are those formed by bonds with oxygen. There is silica (SiO_2), for instance, commonly known as sand. Aside from its many applications on the beaches of the world, silica, when mixed with lime and soda (sodium carbonate) and other substances, makes glass. Like carbon, silicon has the ability to form polymers, or long, chainlike molecules. And whereas carbon polymers are built of hydrocarbons (plastics are an example), silicon polymers are made of silicon and oxygen in monomers, or strings of atoms, that form ribbons or sheets many millions of units long.

SILICATE SUBCLASSES. There are six subclasses of silicate, differentiated by structure. Nesosilicates include some the garnet group; gadolinite, which played a significant role in the isolation of the lanthanide series of elements during the nineteenth century; and zircon. The latter may seem to be associated with the cheap diamond simulant, or substitute, called cubic zirconium, or CZ. CZ, however, is an artificial "mineral," whereas zircon is the real thing—yet it, too, has been applied as a diamond simulant.

Just as silicon's close relative, carbon, can form sheets (this is the basic composition of graphite), so silicon can appear in sheets as the phyllosilicate subclass. Included among this group are minerals known for their softness: kaolinite, talc, and various types of mica. These are used in everything from countertops to talcum powder. The kaolinite derivative known as kaolin is applied, for instance, in the manufacture of porcelain, while some people in parts of Georgia, a state noted for its kaolinite deposits, claim that it can and should be chewed as an antacid stomach remedy. (One can even find little bags of kaolin sold for this purpose at convenience stores around Columbus in southern Georgia.)

CALCITE WITH QUARTZ. *(© Mark A. Schneider/Photo Researchers. Reproduced by permission.)*

Included in another subclass, the tectosilicates, are the feldspar and quartz groups, which are the two most abundant types of mineral in Earth's crust. Note that these are both groups: to a mineralogist, feldspar and quartz refer not to single minerals but to several within a larger grouping. Feldspar, whose name comes from the Swedish words for "field" and "mineral" (a reference to the fact that miners and farmers found the same rocks in their respective areas of labor), includes a number of varieties, such as albite (sodium aluminum silicate) or sanidine (potassium aluminum silicate).

Other, more obscure silicate subclasses include sorosilicates and inosilicates. Finally, there are cyclosilicates, such as beryl or beryllium aluminum silicate.

IDENTIFYING MINERALS

Mineralogists identify unknown minerals by judging them in terms of various physical properties, including hardness, color and streak, luster, cleavage and fracture, density and specific gravity, and other factors, such as crystal form. Hardness, or the ability of one mineral to scratch another, may be measured against the Mohs scale, introduced in 1812 by the German mineralogist Friedrich Mohs (1773–1839). The scale rates minerals from 1 to 10, with 10 being equivalent to the hardness of a diamond and 1 that of talc, the softest mineral. (See Economic Geology for other scales, some of which are more applicable to specific types of minerals.)

Minerals sometimes can be identified by color, but this property can be so affected by the presence of impurities that mineralogists rely instead on streak. The latter term refers to the color of the powder produced when one mineral is scratched by another, harder one. Another visual property is luster, or the appearance of a mineral when light reflects off its surface. Among the terms used in identifying luster are metallic, vitreous (glassy), and dull.

The term cleavage refers to the way in which a mineral breaks—that is, the planes across which the mineral splits into pieces. For instance, muscovite tends to cleave only in one direction, forming thin sheets, while halite cleaves in three directions, which are all perpendicular to one another, forming cubes. The cleavage of a mineral reveals its crystal system; however, minerals are more likely to fracture (break along something other than a flat surface) than they are to cleave.

DENSITY, SPECIFIC GRAVITY, AND OTHER PROPERTIES. Density is the ratio of mass to volume, and specific gravity is the ratio between the density of a particular substance and that of water. Specific gravity almost always is measured according to the metric system, because of the convenience: since the density of water is 1 g per cubic centimeter (g/cm^3), the specific gravity of a substance is identical to its density, except that specific gravity involves no units.

For example, gold has a density of 19.3 g/cm^3 and a specific gravity of 19.3. Its specific gravity, incidentally, is extremely high, and, indeed, one of the few metals that comes close is lead, which has a specific gravity of 11. By comparing specific gravity values and measuring the displacement of water when an object is set down in it, it is possible to determine whether an item purported to be gold actually is gold.

In addition to these more common parameters for identifying minerals, it may be possible to identify certain ones according to other specifics. There are minerals that exhibit fluorescent or phosphorescent characteristics, for instance. The first term refers to objects that glow when viewed under ultraviolet light, while the second term describes those that continue to glow after being exposed to visible light for a short period of time. Some minerals are magnetic, while others are radioactive.

NAMING MINERALS. Chemists long ago adopted a system for naming compounds so as to avoid the confusion of proliferating common names. The only compounds routinely referred to by their common names in the world of chemistry are water and ammonia; all others are known according to chemical nomenclature that is governed by specific rules. Thus, for instance, NaCl is never "salt," but "sodium chloride."

Geologists have not been able to develop such a consistent means of naming minerals. For one thing, as noted earlier, two minerals may be different from each other yet include the same elements. Furthermore, it is difficult (unlike the case of chemical compounds) to give minerals names that provide a great deal of information regarding their makeup. Instead, most minerals are simply named after people (usually scientists) or the locale in which they were found.

ABRASIVES

The physical properties of minerals, including many of the characteristics we have just discussed, have an enormous impact on their usefulness and commercial value. Some minerals, such as diamonds and corundum, are prized for their hardness, while others, ranging from marble to the "mineral" alabaster, are useful precisely because they are soft. Others, among them copper and gold, are not just soft but highly malleable, and this property makes them particularly useful in making products such as electrical wiring.

Diamonds, corundum, and other minerals valued for their hardness belong to a larger class of materials called abrasives. The latter includes sandpaper, which of course is made from one of the leading silicate derivatives, sand. Sandstone and quartz are abrasives, as are numerous variants of corundum, such as sapphire and garnets.

In 1891, American inventor Edward G. Acheson (1856–1931) created silicon carbide, later sold under the trade name Carborundum, by heating a mixture of clay and coke (almost pure carbon). For 50 years, Carborundum was the second-hardest substance known, diamonds being the hardest. Today other synthetic abrasives, made from aluminum oxide, boron carbide, and boron nitride, have supplanted Carborundum in importance.

Corundum, from the oxides class of mineral, can have numerous uses. Extremely hard, corundum, in the form of an unconsolidated rock commonly called emery, has been used as an abrasive since ancient times. Owing to its very high melting point—even higher than that of iron—corundum also is employed in making alumina, a fireproof product used in furnaces and fireplaces. Though pure corundum is colorless, when combined with trace amounts of certain elements, it can yield brilliant colors: hence, corundum with traces of chromium becomes a red ruby, while traces of iron, titanium, and other elements yield varieties of sapphire in yellow, green, and violet as well as the familiar blue.

This brings up an important point: many of the minerals named here are valued for much more than their abrasive qualities. Many of the 16 minerals used as gemstones, including corundum (source of both rubies and sapphires, as we have noted), garnet, quartz, and of course diamond, happen to be abrasives as well. (See Economic Geology for the full list of precious gems.)

DIAMONDS. Diamonds, in fact, are so greatly prized for their beauty and their application in jewelry that their role as "working" minerals—not just decorations—should be emphasized. The diamonds used in industry look quite different from the ones that appear in jewelry. Industrial diamonds are small, dark, and cloudy in appearance, and though they have the same chemical properties as gem-quality diamonds, they are cut with functionality (rather than beauty) in mind. A diamond is hard, but brittle: in other words, it can be broken, but it is very difficult to scratch or cut a diamond—except with another diamond.

On the other hand, the cutting of fine diamonds for jewelry is an art, exemplified in the alluring qualities of such famous gems as the jewels in the British Crown or the infamous Hope Diamond in Washington, D.C.'s Smithsonian Institution. Such diamonds—as well as the diamonds on an engagement ring—are cut to refract or bend light rays and to disperse the colors of visible light.

SOFT AND DUCTILE MINERALS

At the other end of the Mohs scale are an array of minerals valued not for their hardness, but for opposite qualities. Calcite, for example, is often used in cleansers because, unlike an abrasive (also used for cleaning in some situations), it will not scratch a surface to which it is applied. Calcite takes another significant form, that of marble, which is used in sculpture, flooring, and ornamentation because of its softness and ease in carving—not to mention its great beauty.

Gypsum, used in plaster of paris and wallboard, is another soft mineral with applications in building. Though, obviously, soft minerals are not much value as structural materials, when stud walls of wood provide the structural stability for gypsum sheet wall coverings, the softness of the latter can be an advantage. Gypsum wallboard makes it easy to put in tacks or nails for pictures and other decorations, or to cut out a hole for a new door, yet it is plenty sturdy if bumped. Furthermore, it is much less expensive than most materials, such as wood paneling, that might be used to cover interior walls.

GOLD. Quite different sorts of minerals are valued not only for their softness but also their ductility or malleability. There is gold, for instance, the most ductile of all metals. A single troy ounce (31.1 g) can be hammered into a sheet just 0.00025 in. (0.00064 cm) thick, covering 68 sq. ft. (6.3 sq m), while a piece of gold weighing about as much as a raisin (0.0022 lb., or 1 g) can be pulled into the shape of a wire 1.5 mi. (2.4 km) long. This, along with its qualities as a conductor of heat and electricity, would give it a number of other applications, were it not for the high cost of gold.

Therefore, gold, if it were a person, would have to be content with being only the most prized and admired of all metallic minerals, an element for which men and whole armies have fought and sometimes died. Gold is one of the few metals that is not silver, gray, or white in appearance, and its beautifully distinctive color caught the eyes of metalsmiths and royalty from the beginning of civilization. Hence it was one of the first widely used metals.

Records from India dating back to 5000 B.C. suggest a familiarity with gold, and jewelry found in Egyptian tombs indicates the use of sophisticated techniques among the goldsmiths of Egypt as early as 2600 B.C. Likewise, the Bible mentions gold in several passages, and the Romans called it *aurum* ("shining dawn"), which explains its chemical symbol, Au.

COPPER. Copper, gold, and silver are together known as coinage metals. They have all been used for making coins, a reflection not only of their attractiveness and malleability, but also of their resistance to oxidation. (Oxygen has a highly corrosive influence on metals, causing rust, tarnishing, and other effects normally associated with aging but in fact resulting from the reaction of metal and oxygen.) Of the three coinage metals, copper is by far the most versatile, widely used for electrical wiring and in making cookware. Due to the high conductivity of copper, a heated copper pan has a uniform temperature, but copper pots must be coated with tin because too much copper in food is toxic.

Its resistance to corrosion makes copper ideal for plumbing. Likewise, its use in making coins resulted from its anticorrosive qualities, combined with its beauty. These qualities led to the use of copper in decorative applications for which gold would have been much too expensive: many old buildings used copper roofs, and the Statue of Liberty is covered in 300 thick copper plates. As for why the statue and many old copper roofs are green rather than copper-colored, the reason is that copper does eventually corrode when exposed to air for long periods of time. It develops a thin layer of black copper oxide, and as the years pass, the reaction with carbon dioxide in the air leads to the formation of copper carbonate, which imparts a greenish color.

Unlike silver and gold, copper is still used as a coinage metal, though it, too, has been increasingly taken off the market for this purpose due to the high expense involved. Ironically, though most people think of pennies as containing copper, in fact the penny is the only American coin that contains *no* copper alloys. Because the amount of copper necessary to make a penny today costs more than one cent, a penny is actually made of zinc with a thin copper coating.

INSULATION AND OTHER APPLICATIONS

Whereas copper is useful because it conducts heat and electricity well, other minerals (e.g., kyanite, andalusite, muscovite, and silimanite) are valuable for their ability *not* to conduct heat or electricity. Muscovite is often used for insulation in electrical devices, though its many qualities make it a mineral prized for a number of reasons.

Its cleavage and lustrous appearance, combined with its transparency and almost complete lack of color, made it useful for glass in the windowpanes of homes owned by noblemen and other wealthy Europeans of the Middle Ages. Today, muscovite is the material in furnace and stove doors: like ordinary glass, it makes it possible for one to look inside without opening the door, but unlike glass, it is an excellent insulator. The glass-like quality of muscovite also makes it a popular material in wallpaper, where ground muscovite provides a glassy sheen.

In the same vein, asbestos—which may be made of chrysotile, crocidolite, or other minerals—has been prized for a number of qualities, including its flexibility and fiber-like cleavage. These factors, combined with its great heat resistance and its resistance to flame, have made it useful for fireproofing applications, as for instance in roofing materials, insulation for heating and electrical devices, brake linings, and suits for firefighters and others who must work around flames and great heat. However, information linking asbestos and certain forms of cancer, which began to circulate in the 1970s, led to a sharp decline in the asbestos industry.

MINERALS FOR HEALTH OR OTHERWISE. All sorts of other properties give minerals value. Halite, or table salt, is an important—perhaps too important!—part of the American diet. Nor is it the only consumable mineral; people also take minerals in dietary supplements, which is appropriate since the human body itself contains numerous minerals. In addition to a very high proportion of carbon, the body also contains a significant amount of iron, a critical component in red blood cells, as well as smaller amounts of minerals such as zinc. Additionally, there are trace minerals, so called because only traces of them are present in the body, that include cobalt, copper, manganese, molybdenum, nickel, selenium, silicon, and vanadium.

One mineral that does not belong in the human body is lead, which has been linked with a number of health risks. The human body can only excrete very small quantities of lead a day, and this is particularly true of children. Even in small concentrations, lead can cause elevation of blood pressure, and higher concentrations can effect the central nervous system, resulting in decreased mental functioning, hearing damage, coma, and possibly even death.

The ancient Romans, however, did not know this, and used what they called *plumbum* in making water pipes. (The Latin word is the root of our own term *plumber.*) Many historians believe that *plumbum* in the Romans' water supply was one of the reasons behind the decline and fall of the Roman Empire.

Even in the early twentieth century, people did not know about the hazards associated with lead, and therefore it was applied as an ingredient in paint. In addition, it was used in water pipes, and as an antiknock agent in gasolines. Increased awareness of the health hazards involved have led to a discontinuation of these practices.

GRAPHITE. Pencil "lead," on the other hand, is actually a mixture of clay with graphite, a form of carbon that is also useful as a dry lubricant because of its unusual cleavage. It is slippery because it is actually a series of atomic sheets, rather like a big, thick stack of carbon paper: if the stack is heavy, the sheets are likely to slide against one another.

Actually, people born after about 1980 may have little experience with carbon paper, which was gradually phased out as photocopiers became cheaper and more readily available. Today, carbon paper is most often encountered

KEY TERMS

ALLOY: A mixture of two or more metals.

ANION: The negative ion that results when an atom or group of atoms gains one or more electrons.

ATOM: The smallest particle of an element, consisting of protons, neutrons, and electrons. An atom can exist either alone or in combination with other atoms in a molecule.

ATOMIC NUMBER: The number of protons in the nucleus of an atom. Since this number is different for each element, elements are listed on the periodic table in order of atomic number.

CATION: The positive ion that results when an atom or group of atoms loses one or more electrons.

CHEMICAL BONDING: The joining, through electromagnetic forces, of atoms representing different elements. The principal methods of combining are through covalent and ionic bonding, though few bonds are purely one or the other.

CLEAVAGE: A term referring to the characteristic patterns by which a mineral breaks and specifically to the planes across which breaking occurs.

COMPOUND: A substance made up of atoms of more than one element, chemically bonded to one another.

COVALENT BONDING: A type of chemical bonding in which two atoms share valence electrons.

CRUST: The uppermost division of the solid earth, representing less than 1% of its volume and varying in depth from 3 mi. to 37 mi. (5–60 km).

CRYSTALLINE SOLID: A type of solid in which the constituent parts have a simple and definite geometric arrangement that is repeated in all directions.

ELECTRON: A negatively charged particle in an atom, which spins around the nucleus.

ELECTRONEGATIVITY: The relative ability of an atom to attract valence electrons.

ELEMENT: A substance made up of only one kind of atom. Unlike compounds, elements cannot be broken chemically into other substances.

HARDNESS: In mineralogy, the ability of one mineral to scratch another. This is measured by the Mohs scale.

HYDROCARBON: Any chemical compound whose molecules are made up of nothing but carbon and hydrogen atoms.

ION: An atom or group of atoms that has lost or gained one or more electrons and thus has a net electric charge. Positively charged ions are called *cations,* and negatively charged ones are called *anions.*

IONIC BONDING: A form of chemical bonding that results from attractions between ions with opposite electric charges.

when signing a credit-card receipt: the signature goes through the graphite-based backing of the receipt onto a customer copy.

In such a situation, one might notice that the copied image of the signature looks as though it were signed in pencil, which of course is fitting

KEY TERMS CONTINUED

LUSTER: The appearance of a mineral when light reflects off its surface. Among the terms used in identifying luster are metallic, vitreous (glassy), and dull.

MINERAL: A naturally occurring, typically inorganic substance with a specific chemical composition and a crystalline structure. Unknown minerals usually can be identified in terms of specific parameters, such as hardness or luster.

MINERALOGY: The study of minerals, which includes a number of smaller subdisciplines, such as crystallography.

MIXTURE: A substance with a variable composition, meaning that it is composed of molecules or atoms of differing types and in variable proportions.

MOHS SCALE: A scale, introduced in 1812 by the German mineralogist Friedrich Mohs (1773–1839), that rates the hardness of minerals from 1 to 10. Ten is equivalent to the hardness of a diamond and 1 that of talc, an extremely soft mineral.

MONOMERS: Small, individual subunits that join together to form polymers.

NUCLEUS: The center of an atom, a region where protons and neutrons are located and around which electrons spin.

ORE: A rock or mineral possessing economic value.

ORGANIC: At one time, chemists used the term *organic* only in reference to living things. Now the word is applied to most compounds containing carbon and hydrogen, thus excluding carbonates (which are minerals) and oxides such as carbon dioxide.

PERIODIC TABLE OF ELEMENTS: A chart that shows the elements arranged in order of atomic number, along with the chemical symbol and the average atomic mass for that particular element.

POLYMERS: Large, typically chainlike molecules composed of numerous smaller, repeating units known as monomers.

PROTON: A positively charged particle in an atom.

PURE SUBSTANCE: A substance, whether an element or compound, that has the same chemical composition throughout. Compare with *mixture.*

REACTIVITY: A term referring to the ability of one element to bond with others. The higher the reactivity (and, hence, the electronegativity value), the greater the tendency to bond.

ROCK: An aggregate of minerals.

SPECIFIC GRAVITY: The ratio between the density of a particular substance and that of water.

STREAK: The color of the powder produced when one mineral is scratched by another, harder one.

VALENCE ELECTRONS: Electrons that occupy the highest principal energy level in an atom. These are the electrons involved in chemical bonding.

due to the application of graphite in pencil "lead." In ancient times, people did indeed use lead—which is part of the "carbon family" of elements, along with carbon and silicon—for writing, because it left gray marks on a surface. Even today, people still use the word "lead" in reference to pencils, much as they still refer to a galvanized steel roof with a zinc coating as a "tin roof."

(For more about minerals, see Rocks. The economic applications of both minerals and rocks are discussed in Economic Geology. In addition, Paleontology contains a discussion of fossilization, a process in which minerals eventually replace organic material in long-dead organisms.)

WHERE TO LEARN MORE

Atlas of Rocks, Minerals, and Textures (Web site). <http://www.geosci.unc.edu/Petunia/IgMetAtlas/mainmenu.html>.

Hurlbut, Cornelius, W. Edwin Sharp, and Edward Salisbury Dana. *Dana's Minerals and How to Study Them.* New York: John Wiley and Sons, 1997.

The Mineral and Gemstone Kingdom: Minerals A–Z (Web site). <http://www.minerals.net/mineral/>.

Minerals and Metals: A World to Discover. Natural Resources Canada (Web site). <http://www.nrcan.gc.ca/mms/school/e_mine.htm>.

Minerals by Name (Web site). <http://mineral.galleries.com/minerals/by_name.htm>.

Pough, Frederick H. *A Field Guide to Rocks and Minerals.* Boston: Houghton Mifflin, 1996.

Roberts, Willard Lincoln, Thomas J. Campbell, and George Robert Rapp. *Encyclopedia of Minerals.* New York: Van Nostrand Reinhold, 1990.

Sorrell, Charles A., and George F. Sandström. *A Field Guide and Introduction to the Geology and Chemistry of Rocks and Minerals.* New York: St. Martin's, 2001.

Symes, R. F. *Rocks and Minerals.* Illus. Colin Keates and Andreas Einsiedel. New York: Dorling Kindersley, 2000.

"USGS Minerals Statistics and Information." United States Geological Survey (Web site). <http://minerals.usgs.gov/minerals/>.

ROCKS

CONCEPT

It might come as a surprise to learn that geologists regularly use an unscientific-sounding term, rocks. Yet as is almost always the case with a word used both in everyday language and within the realm of a scientific discipline, the meanings are not the same. For one thing, *rock* and *stone* are not interchangeable, as they are in ordinary discussion. The second of these two terms is used only occasionally, primarily as a suffix in the names of various rocks, such as limestone or sandstone. On the other hand, a rock is an aggregate of minerals or organic material. Rocks are of three different types: igneous, formed by crystallization of molten minerals, as in a volcano; sedimentary, usually formed by deposition, compaction, or cementation of weathered rock; and metamorphic, formed by alteration of preexisting rock.

HOW IT WORKS

AN INTRODUCTION TO ROCKS

To expand somewhat on the definition of *rock,* the term may be said to describe an aggregate of minerals or organic material, which may or may not appear in consolidated form. Consolidation, which we will explore further within the context of sedimentary rock, is a process whereby materials become compacted, or experience an increase in density. It is likely that the image that comes to mind when the word rock is mentioned is that of a consolidated one, but it is important to remember that the term also can apply to loose particles.

The role of organic material in forming rocks also belongs primarily within the context of sedimentary, as opposed to igneous or metamorphic, rocks. There are, indeed, a handful of rocks that include organic material, an example being coal, but the vast majority are purely inorganic in origin. The inorganic materials that make up rocks are minerals, discussed in the next section. Rocks and minerals of economic value are called ores, which are examined in greater depth elsewhere, within the context of Economic Geology.

MINERALS DEFINED

The definition of a mineral includes four components: it must appear in nature and therefore not be artificial, it must be inorganic in origin, it must have a definite chemical composition, and it must have a crystalline internal structure. The first of these stipulations clearly indicates that there is no such thing as a man-made mineral; as for the other three parts of the definition, they deserve a bit of clarification.

At one time, the term *organic,* even within the realm of chemistry, referred to all living or formerly living things, their parts, and substances that come from them. Today, however, chemists use the word to describe any compound that contains carbon and hydrogen, thus excluding carbonates (which are a type of mineral) and oxides such as carbon dioxide or carbon monoxide.

NONVARYING COMPOSITION. The third stipulation, that a mineral must be of nonvarying composition, limits minerals almost exclusively to elements and compounds—that is, either to substances that cannot be chemically broken down to yield simpler substances or to substances formed by the chemical bonding of elements. The chemical bonding of elements is a process quite different from mixing, and a compound is not to be confused with a mixture, whose composition is highly variable.

Another way of putting this is to say that all minerals must have a definite chemical formula, which is not possible with a mixture such as dirt or glass. The Minerals essay, which the reader is encouraged to consult for further information, makes reference to certain alloys, or mixtures of metals, that are classified as minerals. These alloys, however, are exceptional and fit certain specific characteristics of interest to mineralogists. The vast majority of the more than 3,700 known varieties of mineral constitute either a single element or a single compound.

CRYSTALLINE STRUCTURE. The fact that a mineral must have a crystalline structure implies that it must be a solid, since all crystalline substances are solids. A solid, of course, is a type of matter whose particles, in contrast to those of a gas or liquid, maintain an orderly and definite arrangement and resist attempts at compression. Thus, petroleum cannot be a mineral, nor is "mineral spirits," a liquid paint thinner made from petroleum (and further disqualified by the fact that it is artificial in origin).

Crystalline solids are those in which the constituent parts are arranged in a simple, definite geometric pattern that is repeated in all directions. These solids are contrasted with amorphous solids, such as clay. Metals are crystalline in structure; indeed, several metallic elements that appear on Earth in pure form (for example, gold, copper, and silver) also are classified as minerals.

Identifying Minerals

The type of crystal that appears in a mineral is one of several characteristics that make it possible for a mineralogist to identify an unidentified mineral. Although, as noted earlier, there are nearly 4,000 known varieties of mineral, there are just six crystal systems, or geometric shapes formed by crystals. Crystallographers, or mineralogists concerned with the study of crystal structures, are able to identify the crystal system by studying a good, well-formed specimen of a mineral, observing the faces of the crystal and the angles at which they meet.

Other characteristics by which minerals can be studied and identified visually are color, streak, and luster. The first of these features is not particularly reliable, because impurities in the mineral may greatly affect its hue. Therefore, mineralogists are much more likely to rely on streak, or the color of the powder produced when one mineral is scratched by a harder one. Luster, the appearance of a mineral when light reflects off its surface, is described by such terms as vitreous (glassy), dull, or metallic.

HARDNESS. Minerals also can be identified according to what might be called tactile properties, or characteristics best discerned through the sense of touch. One of the most important among such properties is hardness, defined as the ability of one mineral to scratch another. Hardness is measured by the Mohs scale, introduced in 1812 by the German mineralogist Friedrich Mohs (1773–1839).

The scale rates minerals from 1 to 10, with 1 being equivalent to the hardness of talc, a mineral so soft that it is used for making talcum powder. A 2 on the Mohs scale is the hardness of gypsum, which is still so soft that it can be scratched by a human fingernail. Above a 5 on the scale, roughly equal to the hardness of a pocketknife or glass, are potassium feldspar (6), quartz (7), topaz (8), corundum (9), and diamond (10).

OTHER PROPERTIES. Other tactile parameters are cleavage, the planes across which the mineral breaks, and fracture, the tendency to break along something other than a flat surface. Minerals also can be evaluated by their density (ratio of mass to volume) or specific gravity (ratio between the mineral's density and that of water). Density and specific-gravity measures are particularly important for extremely dense materials, such as lead or gold.

In addition to these specifics, others may be used for identifying some kinds of minerals. Magnetite and a few other minerals, for instance, are magnetic, while minerals containing uranium and other elements with a high atomic number may be radioactive, or subject to the spontaneous emission of high-energy particles. Still others are fluorescent, meaning that they glow when viewed under ultraviolet light, or phosphorescent, meaning that they continue to glow after being exposed to visible light for a short period of time.

Mineral Groups

Minerals are classified into eight basic groups:

- Class 1: Native elements
- Class 2: Sulfides
- Class 3: Oxides and hydroxides
- Class 4: Halides
- Class 5: Carbonates, nitrates, borates, iodates
- Class 6: Sulfates, chromates, molybdates, tungstates
- Class 7: Phosphates, arsenates, vanadates
- Class 8: Silicates

The first group, native elements, includes metallic elements that appear in pure form somewhere on Earth; certain metallic alloys, alluded to earlier; and native nonmetals, semimetals, and minerals with metallic and nonmetallic elements. Sulfides include the most important ores of copper, lead, and silver, while halides are typically soft and transparent minerals containing at least one element from the halogens family: fluorine, chlorine, iodine, and bromine. (The most well known halide, table salt, is a good example of an unconsolidated mineral.)

Oxides are noncomplex minerals that contain either oxygen or hydroxide (OH). Included in the oxide class are such well-known materials as magnetite and corundum, widely used in industry. Other nonsilicates (a term that stresses the importance of silicates among mineral classes) include carbonates, or carbon-based minerals, as well as phosphates and sulfates. The latter are distinguished from sulfides by virtue of the fact that they include a complex anion (a negatively charged atom or group of atoms) in which an atom of sulfur, chromium, tungsten, selenium, tellurium, or molybdenum (or a combination of these) is attached to four oxygen atoms.

There are two other somewhat questionable classes of nonsilicate that might be included in a listing of minerals—organics and mineraloids. Though they have organic components, organics—for example, amber—originated in a geologic and not a biological setting. Mineraloids, among them, opal and obsidian, are not minerals because they lack the necessary crystalline structure, but they can be listed under the more loosely defined heading of "rocks."

SILICATES. Only a few abundant or important minerals are nonsilicates, for example, the iron oxides hematite, magnetite, and goethite; the carbonates calcite and dolomite; the sulfides pyrite, sphalerite, galena, and chalcopyrite; and the sulfate gypsum. The vast majority of minerals, including the most abundant ones, belong to a single class, that of silicates, which accounts for 30% of all minerals. As their name implies, they are built around the element silicon, which bonds to four oxygen atoms to form what are called silica tetrahedra.

Silicon, which lies just below carbon on the periodic table of elements, is noted, like carbon, for its ability to form long strings of atoms. Carbon-hydrogen formations, or hydrocarbons, are the foundation of organic chemistry, while formations of oxygen and silicon—the two most abundant elements on Earth—provide the basis for a vast array of geologic materials. There is silica, for instance, better known as sand, which consists of silicon bonded to two carbon atoms.

Then there are the silicates, which are grouped according to structure into six subclasses. Among these subclasses, discussed in the Minerals essay, are smaller groupings that include a number of well-known mineral types: garnet, zircon, kaolinite, talc, mica, and the two most abundant minerals on Earth, feldspar and quartz. The name *feldspar* comes from the Swedish words *feld* ("field") and *spar* ("mineral"), because Swedish miners tended to come across the same rocks that Swedish farmers found themselves extracting from their fields.

REAL-LIFE APPLICATIONS

Rocks and Human Existence

Rocks are all around us, especially in our building materials but also in everything from jewelry to chalk. Then, of course, there are the rocks that exist in nature, whether in our backyards or in some more dramatic setting, such as a national park or along a rugged coastline. Indeed, humans have a long history of involvement with rocks—a history that goes far back to the aptly named Stone Age.

CHICHÉN-ITZÁ, A MAYAN STONE PYRAMID IN THE STATE OF YUCATÁN, MEXICO. *(© Ulrike Welsch/Photo Researchers. Reproduced by permission.)*

The latter term refers to a period in which the most sophisticated human tools were those made of rock—that is, before the development of the first important alloy used in making tools, bronze. The Bronze Age began in the Near East in about 3300 B.C. and lasted until about 1200 B.C., when the development of iron-making technology introduced still more advanced varieties of tools.

These dates apply to the Near East, specifically to such areas as Mesopotamia and Egypt, which took the lead in ancient technology, followed much later by China and the Indus Valley civilization of what is now Pakistan. The rest of the world was even slower in adopting the use of metal: for instance, the civilizations of the Americas did not enter the Bronze Age for almost 4,000 years, in about A.D. 1100. Nor did they ever develop iron tools before the arrival of the Europeans in about 1500.

THE STONE AGE. In any case, the Stone Age, which practically began with the species *Homo sapiens* itself, was unquestionably the longest of the three ages. The Stone Age is divided into two periods: Paleolithic and Neolithic, sometimes called Old and New Stone Age, respectively. (There was also a middle phase, called the Mesolithic, but this term is not used as widely as Paleolithic or Neolithic.) Throughout much of this time, humans lived in rock caves and used rock tools, including arrowheads for killing animals and (relatively late in prehistory) flint for creating fire.

The Paleolithic, characterized by the use of crude tools chipped from pieces of stone, began sometime between 2.5 and 1.8 million years ago and lasted until last ice age ended (and the present Holocene epoch began), about 10,000 years ago. The Neolithic period that followed saw enormous advances in technology, so many advances that historians speak of a "Neolithic Revolution" that included the development of much more sophisticated, polished tools. The mining of gold, copper, and various other ores began long before the development of the first alloys (bronze is formed by the mixture of copper and tin). Yet even after humans discovered metals, they continued to use stone tools.

THE PYRAMIDS AND OTHER STONE STRUCTURES. Indeed, the great pyramids of Egypt, built during the period from about 2600-2400 B.C., were constructed primarily with the use of stone rather than metal tools. The structures themselves, of course, also reflect the tight connection between humans and rocks. Built of limestone, the pyramids are still standing some 4,500 years later, even as structures of clay and mud built at about the same time in Mesopotamia (a region poor in stone resources) have long since dwindled to dust.

Incidentally, the great pyramids once had surfaces of polished limestone, such that they gleamed in the desert sun. Centuries later, Arab invaders in the seventh century A.D. stripped this limestone facing to use it in other structures, and the only part of the facing that remains today is high atop the pyramid of Khafre. For this reason, Khafre's pyramid is slightly taller than the structure known as *the* Great Pyramid, that of Cheops, or Khufu, which was originally the largest pyramid.

The centuries that have followed the building of those great structures likewise are defined, at least in part, by their buildings of stone. The Bible is full of references to stones, whether those used in building Solomon's temple or the precious gemstones said to form the gates of the New Jerusalem described in the Book of Revelation. Greece and Rome, too, are known for their structures of stone, ranging from marble (limestone that has undergone metamorphism) to unconsolidated stones in early forms of concrete, pioneered by the Egyptians.

Still later, medieval Europe built its cathedrals and castles of stone, though it should be noted that the idea of the castle came from the Middle East, where the absence of lumber for fortresses caused Syrian castle builders to make use of abundant sandstone instead. Other societies left behind their own great stone monuments: the Great Wall of China, Angkor Wat in southeast Asia, the pyramids of Central America and Machu Picchu in South America, the great cliffside dwellings of what is now the southwestern United States, and the stone churches of medieval Ethiopia.

Certainly there were civilizations that created great structures of wood, but these structures were simply not as durable. The oldest wood building, a Buddhist temple at Horyuji in Japan, dates back only to A.D. 607, which, of course, is quite impressive for a wooden structure. But it hardly compares to what may well be the oldest known human structure, a windbreak discovered by the paleobiologist Mary Leakey (1913–1996) in Tanzania in 1960. Consisting of a group of lava blocks that form a rough circle, it is believed to be 1.75 million years old.

MINERALOGY AND PETROLOGY

Not surprisingly, mineralogy is concerned with minerals—their physical properties, chemical makeup, crystalline structures, occurrence, distribution, and physical origins. Researchers whose work focuses on the physical origins of minerals study data and draw on the principles of physics and chemistry to develop hypotheses regarding the ways minerals form. Other mineralogical studies may involve the identification of a newly discovered mineral or the synthesis of mineral-like materials for industrial purposes.

The study of rocks is called petrology, from a Greek root meaning "rock." (Hence also the words *petroleum* and *petrify.*) Its areas of interest with regard to rocks are much the same as those of mineralogy as they relate to minerals: physical properties, distribution, and origins. It includes two major subdisciplines, experimental petrology, or the synthesis of rocks in a laboratory as a means of learning the conditions under which rocks are formed in the natural world, and petrography, or the study of rocks observed in thin sections through a petrographic microscope, which uses polarized light.

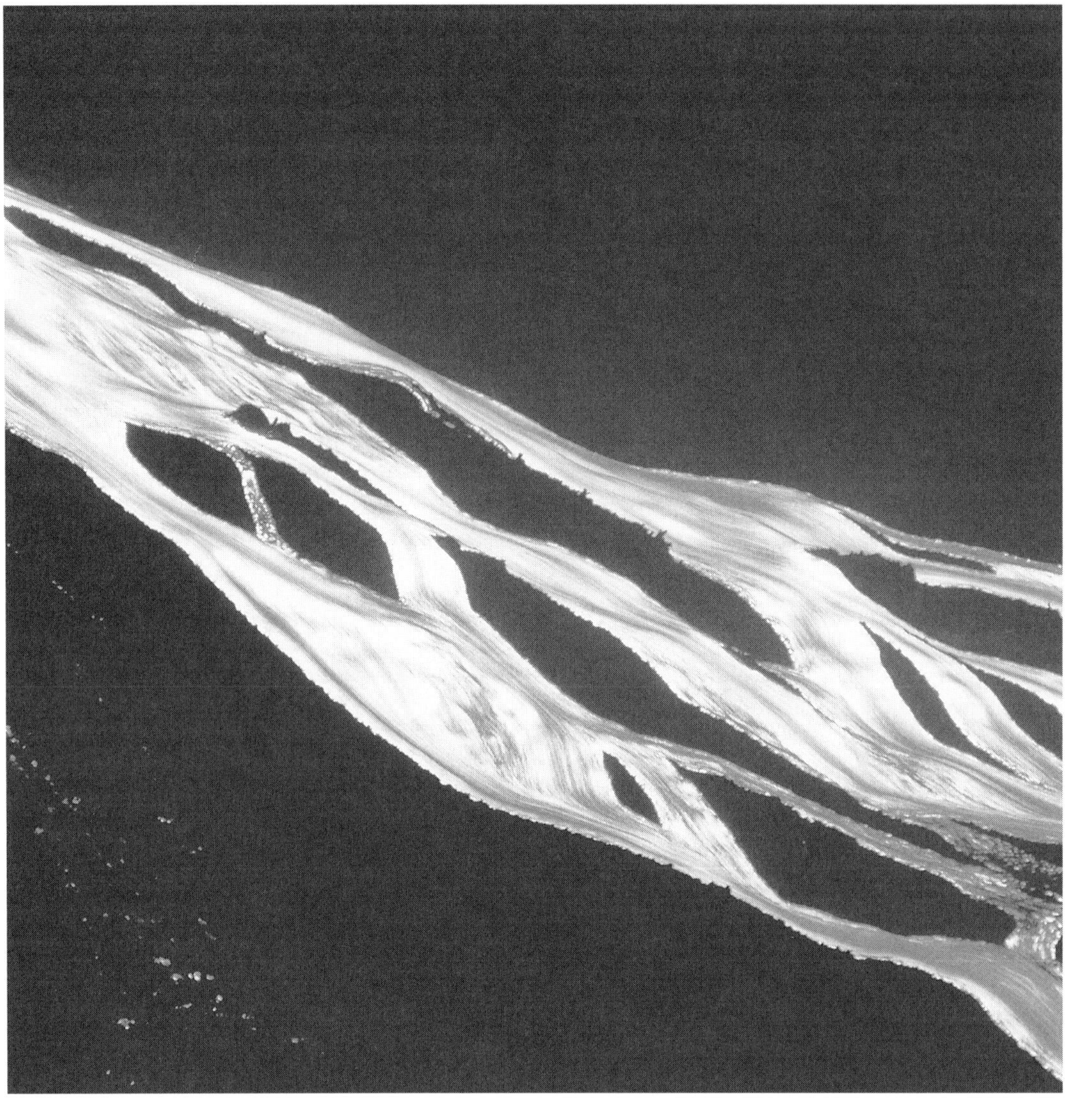

LAVA FLOW AFTER THE 1992 ERUPTION OF MOUNT ETNA IN SICILY. WHEN IT COOLS, LAVA BECOMES IGNEOUS ROCK. *(© B. Edmaier/Photo Researchers. Reproduced by permission.)*

Owing to the fact that most rocks contain minerals, petrology draws on and overlaps with mineralogical studies to a great extent. At the same time, it goes beyond mineralogy, inasmuch as it is concerned with materials that contain organic substances, which are most likely to appear within the realm of sedimentary rock. Petrologists also are concerned with the other two principal types of rock, igneous and metamorphic.

IGNEOUS ROCKS

Igneous rock is rock formed by the crystallization of molten materials. It most commonly is associated with volcanoes, though, in fact, it comes into play in the context of numerous plate tectonic processes, such as seafloor spreading (see Plate Tectonics). The molten rock that becomes igneous rock is known as magma when it is below the surface of the earth and lava when it is at or near the earth's surface. Its most notable characteristic is its interlocking crystals. For the most part, igneous rocks do not have a layered texture.

When igneous rocks form deep within the Earth, they are likely to have large crystals, an indication of the fact that a longer period of time elapsed while the magma was cooling. On the other hand, volcanic rocks and others that form at or near Earth's surface are apt to have very small crystals. Obsidian (which, as we have noted, is not truly a mineral owing to its lack of crystals) is formed when hot lava comes into contact with water; as a result, it cools so quickly that crystals never have time to develop. Sometimes called volcanic glass, it once was used by prehistoric peoples as a cutting tool.

CLASSIFYING AND IDENTIFYING IGNEOUS ROCKS. Igneous rocks can be classified in several ways, referring to the means by which they were formed, the size of their crystals, and their mineral content. Extrusive igneous rocks, ejected by volcanoes to crystallize at or near Earth's surface, have small crystals, whereas intrusive igneous rocks, which cooled slowly beneath the surface, have larger crystals. Sometimes the terms *plutonic* and *volcanic,* which roughly correspond to *intrusive* and *extrusive,* respectively, are used.

Igneous rocks made of fragments from volcanic explosions are known as *pyroclastic,* or "fire-broken," rocks. Those that consist of dense, dark materials are known as *mafic* igneous rocks. On the other hand, those made of lightly colored, less-dense minerals, such as quartz, mica, and feldspar, are called *felsic* igneous rocks. Among the most well known varieties of igneous rock is granite, an intrusive, felsic rock that includes quartz, feldspar, mica, and amphibole in its makeup. Also notable is basalt, which is mafic and extrusive.

Sedimentary Rocks

Earlier, we touched on the subject of consolidation, which can be explained in more depth within the context of sedimentary rock. Consolidation is the compacting of loose materials by any number of processes, including recrystallization and cementation. The first of these processes is the formation of new mineral grains as a result of changes in temperature, pressure, or other factors. In cementation, particles of sediment (material deposited at or near Earth's surface from several sources, most notably preexisting rock) are cemented together, usually with mud.

Compaction, recrystallization, and other processes, such as dehydration (which also may contribute to compaction), are collectively known as diagenesis. The latter term refers to all the changes experienced by a sediment sample under conditions of low temperature and low pressure following deposition. If the temperature and pressure increase, diagenesis may turn into metamorphism, discussed later in the context of metamorphic rock.

FORMATION OF SEDIMENTARY ROCKS. Sedimentary rock is formed by the deposition, compaction, and cementation of rock that has experienced weathering (breakdown of rock due to physical, chemical, or biological processes) or as a result of chemical precipitation. The latter term refers not to "precipitation" in terms of weather but to the formation of a solid from a liquid, by chemical rather than physical means. (The freezing of water, a physical process, is not an example of precipitation.)

Sedimentary rock usually forms at or near the surface of the earth, as the erosive action of wind, water, ice, gravity, or a combination of these forces moves sediment. Yet this formation also may occur when chemicals precipitate from seawater or when organic material, such as plant debris or animal shells, accumulate. Evaporation of saltwater, for instance, produces gypsum, a mineral noted for its lack of thermal conductivity; hence its use in drywall, the material that covers walls in most modern homes. (Ancient peoples made alabaster, a fine-grained ornamental stone, from gypsum.)

CLASSIFICATION AND SIZES. Sedimentary rock is classified with reference to the size of the particles from which the rock is made as well as the origin of those particles. Clastic rock comes from fragments of preexisting rock (whether igneous, sedimentary, or metamorphic) and organic matter, while nonclastic sedimentary rock is formed either by precipitation or by organic means. Examples include gypsum, salts, and other rocks formed by precipitation of saltwater as well as those created from organic material or organic activity—coal, for example.

Ranging in size from fine clay (less than 0.00015 in., or 0.004 mm) to boulders (defined as any rock larger than 10 in., or 0.254 m), sedimentary rock bears a record of the environment in which the original sediments were deposited. This record lies in the sediment itself. For example, rocks containing conglomerate, material ranging in size from clay to boulders (including the intermediate categories of silt, sand, gravel, pebbles, and cobble), come from sediment that was deposited rapidly as the result of slides or slumps. (Slides and slumps are discussed in Mass Wasting.)

METAMORPHIC ROCK IS FORMED BY THE ALTERATION OF PREEXISTING ROCK. THE PRESENCE OF MICA, SHOWN HERE, IS A SIGN THAT ROCK MIGHT BE METAPHORPHIC. *(© C. D. Winters/Photo Researchers. Reproduced by permission.)*

Sedimentary rocks are of particular interest to paleontologists, stratigraphers, and others working in the field of historical geology, because they are the only kinds of rock in which fossils are preserved. The pressure and temperature levels that produce igneous and metamorphic rock would destroy the organic remnants that produce fossils; on the other hand, sedimentary rock—created by much less destructive processes—permits the formation of fossils. Thus, the study of these formations has contributed greatly to geologists' understanding of the distant past. (See the essays Historical Geology, Stratigraphy, and Paleontology. For more about sedimentary rock, see Sediment and Sedimentation.)

METAMORPHIC ROCKS

Metamorphic rock is formed through the alteration of preexisting rock as a result of changes in temperature, pressure, or the activity of fluids (usually gas or water). These changes in temperature must be extreme (figures are given later), such that the preexisting rock—whether igneous, sedimentary, or metamorphic—is no longer stable.

Often formed in mountain environments, metamorphic rocks include such well-known varieties as marble, slate, and gneiss—metamorphosed forms of limestone, shale, and granite, respectively. Also notable is schist, composed of various minerals, such as talc, mica, and muscovite. There is not always a one-to-one correspondence between precursor rocks and metamorphic ones: increasing temperature and pressure can turn shale progressively into slate, phyllite, schist, and gneiss.

The presence of mica in a rock—or of other minerals, including amphibole, staurolite, and garnet—is a sign that the rock might be metamorphic. These minerals, typical of metamorphic rocks, are known as metamorphic facies. Also indicative of metamorphism are layers in the rock, more or less parallel lines along which minerals are laid as a result of the high pressures applied to the rock in its formation. Metamorphism, the process whereby metamorphic rock is created, also may produce characteristic formations, such as an alignment of elongate crystals or the separation of minerals into layers.

METAMORPHISM. Given the conditions described for metamorphism, one might conclude that in terms of violence, drama, and stress, it is a process somewhere between sedimentation and the formation of igneous rock. That, in fact, is precisely the case: the temperature and pressure conditions necessary for metamorphism lie between those of diagenesis, on the one hand, and the extreme conditions necessary for the production of igneous rock, on the other hand. Specifically, metamorphism occurs at temperatures between 392°F (200°C) and 1,472°F (800°C) and under levels of pressure between 1,000 and 10,000 bars. (A bar is slightly less than the standard atmospheric pressure at sea level. The latter, equal to 14.7 lb. per square inch, or 101,325 Pa, is equal to 1.01325 bars.)

There are several types of metamorphism: regional, contact, dynamic, and hydrothermal. Regional metamorphism results from a major tectonic event or events, producing widespread changes in rocks. Contact or thermal metamorphism results from contact between igneous intrusions and cooler rocks above them, which recrystallize as a result of heating. Dynamic metamorphism takes place in the high-pressure conditions along faults. Finally, hydrothermal metamorphism ensues from contact with fluids

KEY TERMS

ALLOY: A mixture of two or more metals.

CEMENTATION: A process of consolidation whereby particles of sediment are cemented together, usually with mud.

COMPOUND: A substance made up of atoms of more than one element, chemically bonded to one another.

CONGLOMERATE: Unconsolidated rock material containing rocks ranging in size from very small clay (less than 0.00015 in., or 0.004 mm) to boulders (defined as any rock larger than 10 in., or 0.254 m). Sedimentary rock often appears in the form of conglomerate.

CONSOLIDATION: A process whereby materials become compacted, or experience an increase in density. This takes place through several processes, including recrystallization and cementation.

CRYSTALLINE SOLID: A type of solid in which the constituent parts have a simple and definite geometric arrangement that is repeated in all directions.

DEPOSITION: The process whereby sediment is laid down on the Earth's surface.

DIAGENESIS: A term referring to all the changes experienced by a sediment sample under conditions of low temperature and low pressure following deposition. Higher temperature and pressure conditions may lead to metamorphism.

ELEMENT: A substance made up of only one kind of atom. Unlike compounds, elements cannot be broken chemically into other substances.

EROSION: The movement of soil and rock as the result of forces produced by water, wind, glaciers, gravity, and other influences. In most cases, a fluid medium, such as air or water, is involved.

IGNEOUS ROCK: One of the three principal types of rock, along with sedimentary and metamorphic rock. Igneous rock is formed by the crystallization of molten materials, for instance, in a volcano or other setting where plate tectonic processes take place.

LAVA: Molten rock at or near the surface of the earth that becomes igneous rock. Below the surface, lava is known as *magma.*

MAGMA: Molten rock beneath the surface of the earth that becomes igneous rock. Once it is at or near the surface, magma is known as *lava.*

MINERAL: A naturally occurring, typically inorganic substance with a specific chemical composition and a crystalline structure.

MINERALOGY: An area of geology devoted to the study of minerals. Mineralogy includes several subdisciplines, such as crystallography, the study of crystal formations within minerals.

MIXTURE: A substance with a variable composition, meaning that it is composed of molecules or atoms of differing types in varying proportions.

ORE: A rock or mineral possessing economic value.

ORGANIC: At one time, chemists used the term *organic* only in reference to living things. Now the word is applied to most

KEY TERMS CONTINUED

compounds containing carbon and hydrogen, thus excluding carbonates (which are minerals), and oxides such as carbon dioxide.

PETROLOGY: An area of geology devoted to the study of rocks, including their physical properties, distribution, and origins.

PRECIPITATION: In the context of chemistry, precipitation refers to the formation of a solid from a liquid.

RECRYSTALLIZATION: The formation of new mineral grains as a result of changes in temperature, pressure, or other factors.

ROCK: An aggregate of minerals or organic matter, which may be consolidated or unconsolidated.

ROCK CYCLE: The ongoing process whereby rocks continually change from one type to another, typically through melting, metamorphism, uplift, weathering, burial, or other processes.

SAND: A term that can have several meanings. The sand at a beach could be a variety of unconsolidated materials, though most likely it is silica (SiO_2). Sand is also a term used for a size of rock ranging from very fine to very coarse.

SEDIMENT: Material deposited at or near Earth's surface from a number of sources, most notably preexisting rock.

SEDIMENTARY ROCK: One of the three major types of rock, along with igneous and metamorphic rock. Sedimentary rock usually is formed by the deposition, compaction, and cementation of rock that has experienced weathering. It also may be formed as a result of chemical precipitation.

UNCONSOLIDATED ROCK: Rock that appears in the form of loose particles, such as sand.

UPLIFT: A process whereby the surface of Earth rises, owing to either a decrease in downward force or an increase in upward force.

WEATHERING: The breakdown of rocks and minerals at or near the surface of Earth as the result of physical, chemical, or biological processes.

heated by igneous rock. Reacting with minerals in the surrounding rock, the fluids produce different minerals, which, in turn, yield metamorphic rocks.

TYPES OF METAMORPHIC ROCKS. Metamorphic rocks that contain elongate or platy minerals, such as mica and amphibole, are called foliated *rocks.* These rocks have a layered texture, which may manifest as the almost perfect arrangement of materials in slate or as the alternating patterns of light and dark found in some other varieties of rock. Metamorphic rocks without visible layers are referred to as unfoliated rocks. As a foliated metamorphic rock, slate is particularly good for splitting into thin layers—hence one of its most important applications is in making shingles for roofing. By contrast, marble, which is unfoliated, is valued precisely for its lack of tendency to split.

Petrologists attempting to determine exactly which rocks or combinations of rocks metamorphosed to produce a particular sample often face a challenge. Many metamorphic rocks are stubborn about giving up their secrets; on the other hand, it is possible to match up precursor rocks with certain varieties. For example, as noted earlier, marble comes from limestone, while gneiss usually (but not always) comes from granite. Quartzite is metamorphosed sandstone. Nonetheless, it is not as easy to trace the history of a metamorphic rock as it is to say that a raisin was once a grape or that a pickle was once a cucumber.

Where to Find Rocks

In general, one might find igneous rocks such as basalt in any place known for volcanic activity either in the recent or distant past. This would include such well-known areas of volcanism as Hawaii, the Philippines, and Italy, but also places where volcanic activity occurred in the distant past. (See, for instance, the discussion in the essay titled "Paleontology" regarding possible volcanic activity in what is now the continental United States at the conclusion of the Triassic period.)

The best place for metamorphic rock would be in areas of mountain-building and powerful tectonic activity, as for instance in the Himalayas or the Alps of central Europe. Sedimentary rock is basically everywhere, but a good place to find large samples of it would include areas with large oil deposits, which are always found in sedimentary rock.

Closer to home, a wide array of sedimentary rocks can be located in the plains and lowlands of the United States, particularly in the West and Midwest, where large samples are exposed. Igneous and metamorphic rocks can be found, predictably, in regions where mountains provide evidence of past tectonic activity: New England, the Appalachians, and the various mountain ranges of the western United States such as the Rockies, Cascades, and Sierra Nevada.

The Rock Cycle

Given what we have seen about the characteristics of the three rock varieties—igneous, sedimentary, and metamorphic—it should be clear that there is no such thing as a rock that simply is what it is, without any possibility of changing. Rocks, in fact, are constantly changing, as is Earth itself. This process whereby rocks continually change from one type to another—typically through melting, metamorphism, uplift, weathering, burial, or other processes—is known as the rock cycle.

The rock cycle can go something like this: Exposed to surface conditions such as wind and the activity of water, rocks experience weathering. The result is the formation of sediments that are eventually compacted to make sedimentary rocks. As the latter are buried deeper and deeper beneath greater amounts of sediment, the pressure and temperature builds. This process ultimately can result in the creation of metamorphic rock. On the other hand, the rock may undergo such extreme conditions of temperature that it recrystallizes to form igneous rock. Whatever the variety—igneous, sedimentary, or metamorphic—the rock likely will be in a position eventually to experience erosion, in which case the rock cycle begins all over again.

Where to Learn More

Atlas of Igneous and Metamorphic Rocks, Minerals, and Textures (Web site). <http://www.geosci.unc.edu/Petunia/IgMetAtlas/mainmenu.html>.

Bishop, A. C., Alan Robert Woolley, and William Roger Hamilton. *Cambridge Guide to Minerals, Rocks, and Fossils.* New York: Cambridge University Press, 1999.

Busbey, Arthur Bresnahan. *Rocks and Fossils.* Alexandria, VA: Time-Life Books, 1996.

Discovery Channel Rocks and Minerals: An Explore Your World Handbook. New York: Discovery Books, 1999.

"The Essential Guide to Rocks." BBC Education (Web site). <http://www.bbc.co.uk/education/rocks/>.

"Igneous, Sedimentary, and Metamorphic Rock Info." University of British Columbia (Web site). <http://www.science.ubc.ca/~geol202/petrology/rock.html>.

RocksForKids.com (Web site). <http://www.rocksforkids.com/>.

Rocks and Minerals (Web site). <http://www.fi.edu/tfi/units/rocks/rocks.html>.

Symes, R. F., Colin Keates, and Andreas Einsiedel. *Rocks and Minerals.* New York: Dorling Kindersley, 2000.

Vernon, R. H. *Beneath Our Feet: The Rocks of Planet Earth.* New York: Cambridge University Press, 2000.

ECONOMIC GEOLOGY

CONCEPT

Economic geology is the study of fuels, metals, and other materials from the earth that are of interest to industry or the economy in general. It is concerned with the distribution of resources, the costs and benefits of their recovery, and the value and availability of existing materials. These materials include ore (rocks or minerals possessing economic value) as well as fossil fuels, which embrace a range of products from petroleum to coal. Rooted in several subdisciplines of the geologic sciences—particularly geophysics, structural geology, and stratigraphy—economic geology affects daily life in myriad ways. Masonry stones and gasoline, gypsum wallboard (sometimes known by the brand name Sheetrock) and jewelry, natural gas, and table salt—these and many more products are the result of efforts in the broad field known as economic geology.

HOW IT WORKS

Background of Economic Geology

Some sources of information in the geologic sciences use a definition of "economic geology" narrower than the one applied here. Rather than including nonmineral resources that develop in and are recovered from a geologic environment—a category that consists primarily of fossil fuels—this more limited definition restricts the scope of economic geology to minerals and ores. Given the obvious economic importance of fossil fuels such as petroleum and its many byproducts as well as coal and peat, however, it seems only appropriate to discuss these valuable organic resources alongside valuable inorganic ones.

The concept of economic geology as such is a relatively new one, even though humans have been extracting metals and minerals of value from the ground since prehistoric times. For all their ability to appreciate the worth of such resources, however, premodern peoples possessed little in the way of scientific theories regarding either their formation or the means of extracting them.

The Greeks, for instance, believed that veins of metallic materials in the earth indicated that those materials were living things putting down roots after the manner of trees. Astrologers of medieval times maintained that each of the "seven planets" (Sun, Moon, and the five planets, besides Earth, known at the time) ruled one of the seven known metals—gold, copper, silver, lead, tin, iron, and mercury—which supposedly had been created under the influence of their respective "planets."

AGRICOLA'S CONTRIBUTION. The first thinker who attempted to go beyond such unscientific (if imaginative) ideas was a German physician writing under the Latinized name Georgius Agricola (1494–1555). As a result of treating miners for various conditions, Agricola, whose real name was Georg Bauer, became fascinated with minerals. The result was a series of written works, culminating with *De re metallica* (On the nature of minerals, 1556, released posthumusly), that collectively initiated the modern subdiscipline of physical geology. (It is worth noting that the first translators of *Metallica* into English were Lou [d. 1944] and Herbert Clark Hoover [1874–1964]. The couple published their translation in 1912 in London's *Mining Magazine,* and the husband went on to become the thirty-first president of the United States in 1929.)

Rejecting the works of the ancients and all manner of fanciful explanations for geologic phenomena, Agricola instead favored careful observation, on the basis of which he formed verifiable hypotheses. Regarded as the father of both mineralogy and economic geology, Agricola introduced several ideas that provided a scientific foundation for the study of Earth and its products. In *De ortu et causis subterraneorum* (1546), he critiqued all preceding ideas regarding the formation of ores, including the Greek and astrological notions mentioned earlier as well as the alchemical belief that all metals are composed of mercury and sulfur. Instead, he maintained that subterranean fluids carry dissolved minerals, which, when cooled, leave deposits in the cracks of rocks and thus give rise to mineral veins. Agricola's ideas later helped form the basis for modern theories regarding the formation of ore deposits.

In *De natura fossilium* (On the nature of fossils, 1546), Agricola also introduced a method for the classification of "fossils," as minerals were then known. Agricola's system, which categorizes minerals according to such properties as color, texture, weight, and transparency, is the basis for the system of mineral classification in use today. Of all his works, however, the most important was *De re metallica,* which would remain the leading textbook for miners and mineralogists during the two centuries that followed. In this monumental work, he introduced many new ideas, including the concept that rocks contain ores that are older than the rocks themselves. He also explored in detail the mining practices in use during his time, itself an extraordinary feat in that miners of the sixteenth century tended to guard their trade secrets closely.

METALS, MINERALS, AND ROCKS

Of all known chemical elements, 87, or about 80%, are metals. The latter group is identified as being lustrous or shiny in appearance and malleable or ductile, meaning that they can be molded into different shapes without breaking. Despite their ductility, metals are extremely durable, have high melting and boiling points, and are excellent conductors of heat and electricity. Some, though far from all, register high on the Mohs hardness scale, discussed later in the context of minerals.

The bonds that metals form with each other, or with nonmetals, are known as ionic bonds, the strongest type of chemical bond. Even within a metal, however, there are extremely strong, nondirectional bonds. Therefore, though it is easy to shape metals, it is very difficult to separate metal atoms. Obviously, most metal are solids at room temperature, though this is not true of all: mercury is liquid at ordinary temperatures, and gallium melts at just 85.6°F (29.76°C). Generally, however, metals would be described as crystalline solids, meaning that their constituent parts have a simple and definite geometric arrangement that is repeated in all directions. Crystalline structure is important also within the context of minerals as well as the rocks that contain them.

MINERALS. Whereas there are only 87 varieties of metal, there are some 3,700 types of mineral. There is considerable overlap between metals and minerals, but that overlap is far from complete: many minerals include nonmetallic elements, such as oxygen and silicon. A mineral is a substance that appears in nature and therefore cannot be created artificially, is inorganic in origin, has a definite chemical composition, and possesses a crystalline internal structure.

The term *organic* does not refer simply to substances with a biological origin; rather, it describes any compound that contains carbon, with the exception of carbonates (which are a type of mineral) and oxides, such as carbon dioxide or carbon monoxide. The fact that a mineral must be of nonvarying composition limits minerals almost exclusively to elements and compounds—that is, either to substances that cannot be broken down chemically to yield simpler substances or to substances formed by the chemical bonding of elements. Only in a few highly specific circumstances are naturally occurring alloys, or mixtures of metals, considered minerals.

MINERAL GROUPS. Minerals are classified into eight basic groups:

- Class 1: Native elements
- Class 2: Sulfides
- Class 3: Oxides and hydroxides
- Class 4: Halides

- Class 5: Carbonates, nitrates, borates, iodates
- Class 6: Sulfates, chromates, molybdates, tungstates
- Class 7: Phosphates, arsenates, vanadates
- Class 8: Silicates

The first group, native elements, includes metallic elements that appear in pure form somewhere on Earth; certain metallic alloys, alluded to earlier; as well as native nonmetals, semimetals, and minerals with metallic and nonmetallic elements. The native elements, along with the six classes that follow them in this list, are collectively known as nonsilicates, a term that emphasizes the importance of the eighth group. (For more about the nonsilicates, as well as other subjects covered in the present context, see Minerals.)

The vast majority of minerals, including the most abundant ones, belong to the silicates class, which is built around the element silicon. Just as carbon can form long strings of atoms, particularly in combination with hydrogen (as we discuss in the context of fossil fuels later in this essay), silicon also forms long strings, though its "partner of choice" is typically oxygen rather than hydrogen. Together with oxygen, silicon—known as a metalloid because it exhibits characteristics of both metals and nonmetals—forms the basis for an astonishing array of products, both natural and man-made, which we examine in brief later.

CHARACTERISTICS OF MINERALS. From the list of parameters first developed by Agricola has grown a whole array of characteristics by which minerals are classified. These characteristics also can be used to evaluate an unknown mineral and thus to determine the mineral class within which it fits. One such parameter is the type of crystal of which a mineral is composed. Though there are thousands of minerals, there are just six crystal systems, or basic geometric shapes formed by crystals. Crystallographers, mineralogists concerned with the study of crystal structures, are able to identify the crystal system (the simplest being isometric, or cubic) by studying a good specimen of a mineral and observing the faces of the crystal and the angles at which they meet.

Minerals also can be identified by their hardness, defined as the ability of one mineral to scratch another. Hardness can be measured by the Mohs scale, introduced in 1812 by the German mineralogist Friedrich Mohs (1773–1839), which rates minerals from 1 (talc) to 10 (diamond.) Though it is useful for geologists attempting to identify a mineral in the field, the Mohs scale is not considered helpful for the industrial testing of fine-grained materials, such as steel or ceramics. For such purposes, the Vickers or Knoop scales are applied. These scales (named, respectively, after a British company and an American official) also have an advantage over Mohs in that they offer a precise, proportional scale in which each increase of number indicates the same increase in hardness. By contrast, on the Mohs scale, an increase from 3 to 4 (calcite to fluorite) indicates an additional 25% in hardness, whereas a shift from 9 to 10 (corundum to diamond) marks an increase of 300%.

Other properties significant in identifying minerals are color; streak, or the appearance of the powder produced when one mineral is scratched by a harder one; luster, the appearance of a mineral when light reflects off its surface; cleavage, the planes across which a mineral breaks; fracture, the tendency to break along something other than a flat surface; density, or ratio of mass to volume; and specific gravity, or the ratio between the mineral's density and that of water. Sometimes minerals can be identified in terms of qualities unique to a specific mineral group or groups: magnetism, radioactivity, fluorescence, phosphorescence, and so on. (For more about mineral characteristics, see Minerals.)

ROCKS. A rock is an aggregate of minerals or organic material, which can appear in consolidated or unconsolidated form. Rocks are of three different types: igneous, formed by crystallization of molten minerals, as in a volcano; sedimentary, usually formed by deposition, compaction, or cementation of weathered rock; and metamorphic, formed by alteration of preexisting rock. Rocks made from organic material are typically sedimentary, an example being coal.

Rocks have possessed economic importance from a time long before "economics" as we know it existed—a time when there was nothing to buy and nothing to sell. That time, of course, would be the Stone Age, which dates back practically to the beginnings of the human species and overlapped with the beginnings of civilization some 5,500 years ago. In the hundreds of thousands of years when stone constituted the most advanced toolmaking material, humans developed an array of stone devices for making fire, sharpening knives, killing animals (and other humans), cutting food or animal skins, and so on.

The Stone Age, both in the popular imagination and (with some qualifications) in actual archaeological fact, was a time when people lived in caves. Since that time, of course, humans have generally departed from the caves, though exceptions exist, as the United States military found in 2001 when attempting to hunt for terrorists in the caves of Afghanistan. In any case, the human attachment to stone dwellings has taken other forms, beginning with the pyramids and continuing through today's masonry homes. Nor is rock simply a structural material for building, as the use of gypsum wallboard, slate countertops, marble finishes, and graveled walkways attests. And, of course, construction is only one of many applications to which rocks and minerals are directed, as we shall see.

HYDROCARBONS

As noted earlier, the focus of economic geology is on both rocks and minerals, on the one hand, and fossil fuels, on the other. The latter may be defined as fuel (specifically, coal, oil, and gas) derived from deposits of organic material that have experienced decomposition and chemical alteration under conditions of high pressure. Given this derivation from organic material, by definition all fossil fuels are carbon-based, and, specifically, they are built around hydrocarbons—chemical compounds whose molecules are made up of nothing but carbon and hydrogen atoms.

Theoretically, there is no limit to the number of possible hydrocarbons. Carbon forms itself into apparently limitless molecular shapes, and hydrogen is a particularly versatile chemical partner. Hydrocarbons may form straight chains, branched chains, or rings, and the result is a variety of compounds distinguished not by the elements in their makeup or even (in some cases) by the numbers of different atoms in each molecule, but rather by the structure of a given molecule.

VARIETIES OF HYDROCARBON. Among the various groups of hydrocarbons are alkanes or saturated hydrocarbons, so designated because all the chemical bonds are filled to their capacity (that is, "saturated") with hydrogen atoms. Included among them are such familiar names as methane (CH_4), ethane (C_2H_6), propane (C_3H_8), and butane (C_4H_{10}). The first four, being the lowest in molecular mass, are gases at room temperature, while the heavier ones—including octane (C_8H_{18})—are oily liquids. Alkanes even heavier than octane tend to be waxy solids, an example being paraffin wax, for making candles.

With regard to octane, incidentally, there is a reason why its name is so familiar, while that of heptane (C_7H_{16}) is not. Heptane does not fire smoothly in an internal-combustion engine and therefore disrupts the engine's rhythm. For this reason, it has a rating of zero on a scale of desirability, while octane has a rating of 100. This is why gas stations list octane ratings at the pump: the higher the content of octane, the better the gas is for one's automobile.

In a hydrocarbon chain, if one or more hydrogen atoms is removed, a new bond may be formed. The hydrocarbon chain is then named by adding the suffix *yl*—hence such names as methyl, ethyl, and so on. This indicates that the substance is an alkane, and that something other than hydrogen can be attached to the chain; for example, the attachment of a chlorine atom could yield methyl chloride. Two other large structural groups of hydrocarbons are alkenes and alkynes, which contain double or triple bonds between carbon atoms. Such hydrocarbons are unsaturated—in other words, if the double or triple bond is broken, some of the carbon atoms are then free to form other bonds. Among the products of these groups is the alkene known as acetylene, or C_2H_2, used for welding steel. In addition to alkanes, alkenes, and alkynes, all of which tend to form carbon chains, there are the aromatic hydrocarbons, a traditional name that actually has nothing to do with smell.

All aromatic hydrocarbons contain what is known as a benzene ring, which has the chemical formula C_6H_6 and appears in characteristic ring shapes. In this group are such products as naphthalene, toluene, and dimethyl benzene. These last two are used as solvents as well as in the synthesis of drugs, dyes, and plastics. One of the more famous (or infamous) products in this part of the vast hydrocarbon network is trinitrotoluene, or TNT. Naphthalene is derived from coal tar and used in the synthesis of other compounds. A crystalline solid with a powerful odor, it is found in mothballs and various deodorant disinfectants.

REAL-LIFE APPLICATIONS

FOSSIL FUELS

The organic material that has decomposed to create the hydrocarbons in fossil fuels comes primarily from dinosaurs and prehistoric plants, though it just as easily could have come from any other organisms that died in large numbers a long, long time ago. To form petroleum, there must be very large quantities of organic material deposited along with sediments and buried under more sediment. The accumulated sediments and organic material are called *source rock.*

What happens after accumulation of this material is critical and depends a great deal on the nature of the source rock. It is important that the organic material—for example, the vast numbers of dinosaurs that died in a mass extinction about 65 million years ago (see Paleontology)—not be allowed simply to rot, as would happen in an aerobic, or oxygen-containing, environment. Instead, the organic material undergoes transformation into hydrocarbons as a result of anaerobic chemical activity, or activity that takes place in the absence of oxygen.

Good source rocks for this transformation are shale or limestone, provided the particular rocks are composed of between 1% and 5% organic carbon. The source rocks should be deep enough that the pressure heats the organic material, yet not so deep that the pressure and temperature cause the rocks to undergo metamorphism or transform them into graphite or other non-hydrocarbon versions of carbon. Temperatures of up to 302°F (150°C) are considered optimal for petroleum generation.

Once generated, petroleum gradually moves from the source rock to a reservoir rock, or a rock that stores petroleum in its pores. A good reservoir rock is one in which the pore space constitutes more than 30% of the rock volume. Yet the rock must be sealed by another rock that is much less porous; indeed, for a seal or cap rock, as it is called, a virtually impermeable rock is preferred. Thus, the best kind of seal-forming rock is one made of very small, closely fitting pieces of sediment, for instance, shale. Such a rock is capable of holding petroleum in place for millions of years until it is ready to be discovered and used.

HUMANS AND PETROLEUM. People have known about petroleum from prehistory, simply because there were places on Earth where it literally seeped from the ground. The modern era of petroleum drilling, however, began in 1853, when an American lawyer named George Bissell (1821–1884) recognized its potential for use as a lamp fuel. He hired "Colonel" Edwin Drake (1819–1880) to oversee the drilling of an oil well at Titusville, Pennsylvania, and in 1859 Drake struck oil. The legend of "black gold," of fortunes to be made by drilling holes in the ground, was born.

In the wake of the development and widespread application of the internal-combustion engine during the latter part of the nineteenth and the early part of the twentieth centuries, interest in oil became much more intense, and wells sprouted up around the world. Sumatra, Indonesia, yielded oil from its first wells in 1885, and in 1901, successful drilling began in Texas—the source of many a Texas-sized fortune. An early form of the company known today as British Petroleum (BP) discovered the first Middle Eastern oil in Persia (now Iran) in 1908. Over the next 50 years, the economic importance and prospects of that region changed considerably.

With the vast expansion in automobile ownership that began following World War I (1914–1918) and reached even greater heights after World War II (1939–1945), the value and importance of petroleum soared. The oil industry boomed, and, as a result, many geologists found employment in a sector that offered far more in the way of financial benefits than university or government positions ever could. Today geologists assist their employers in locating oil reserves, not an easy task because so many variables must line up to produce a viable oil source. Given the cost of drilling a new oil well, which may run to $30 million or more, it is clearly important to exercise good judgment in assessing the possibilities of finding oil.

The oil industry has been fraught with environmental concerns over the impact of drilling (much of which takes place offshore, on rigs placed in the ocean); possible biohazards associated with spills, such as the one involving the Exxon *Valdez* in 1989; and the effect on the

BLOCKS OF SHALE AT THE PARAHO OIL SHALE FACILITY IN COLORADO. SHALE IS KNOWN AS A SEAL OR CAP ROCK, A VIRTUALLY IMPERMEABLE ROCK MADE OF SMALL, CLOSELY FITTING PIECES OF SEDIMENT THAT COVERS A MORE POROUS ROCK HOLDING PETROLEUM. *(© U.S. Department of Energy/Photo Researchers. Reproduced by permission.)*

atmosphere of carbon monoxide and other greenhouse gases produced by petroleum-burning internal-combustion engines. There is even more wide-ranging concern over United States dependence on oil sources in foreign countries (some of which are openly hostile to the United States) as well as the possible dwindling of resources.

OFFSHORE OIL RIG AT SABINE PASS, TEXAS. *(© Garry D. McMichael/Photo Researchers. Reproduced by permission.)*

At the present rate of consumption, oil reserves will be exhausted by about the year 2040, but this takes into account only reserves that are considered viable today. As exploration continues, the tapping of United States reserves, such as those in Alaska, will become more and more profitable, leading to increased exploitation of U.S. resources and decreased dependence on oil produced by Middle Eastern states, many of which openly or covertly support terrorist attacks against the United States. In the long run, however, it will be necessary to develop new means of fueling the industrialized world, because petroleum is a nonrenewable resource: there is only so much of it underground, and when it is gone, it will not be replaced for millions of years (if at all).

PETROCHEMICALS. In the meantime, however, petroleum—a mixture of alkanes, alkenes, and aromatic hydrocarbons—makes the world (or at least the industrialized world) go 'round. Petroleum itself is a raw material from which numerous products, collectively known as petrochemicals or petroleum derivatives, are obtained. Through a process termed fractional distillation, the petrochemicals of the lowest molecular mass boil off first, and those having higher mass separate at higher temperatures.

Natural gas separates from petroleum at temperatures below 96.8°F (36°C)—far lower than the boiling point of water. At somewhat higher temperatures, petroleum ether and naphtha, both solvents (naphtha is used in paint thinner), separate; then, in the region between 156.2°F and 165.2°F (69–74°C), gasoline separates. Still higher temperatures yield other substances, each thicker than the one before it: kerosene; fuel for heating and the operation of diesel engines; lubricating oils; petroleum jelly; paraffin wax; and pitch, or tar. A host of other organic chemicals, including various drugs, plastics, paints, adhesives, fibers, detergents, synthetic rubber, and agricultural chemicals, owe their existence to petrochemicals.

SILICON, SILICATES, AND OTHER COMPOUNDS

It was stated earlier that both carbon and silicon have the tendency to produce long strings of atoms, usually in combination with hydrogen in the first case and oxygen in the second. This is no accident, since silicon lies just below carbon on the periodic table of elements and they share certain chemical features (see Minerals). Just as carbon is at the center of a vast world of hydrocarbons, so silicon is equally important to inorganic substances ranging from sand or silica (SiO_2) to silicone (a highly versatile set of silicon-based products), to the rocks known as silicates.

Silicates are the basis for several well-known mineral types, including garnet, topaz, zircon, kaolinite, talc, mica, and the two most abundant minerals on Earth, feldspar and quartz. (Note that most of the terms used here refer to a group of minerals, not to a single mineral.) Made of compounds formed around silicon and oxygen and comprising various metals, such as aluminum, iron, sodium, and potassium, the silicates account for 30% of all minerals. As such, they appear in everything from gemstones to building materials; yet they are far from the only notable products centered around silicon.

SILICONE AND OTHER COMPOUNDS. Silicone is not a mineral; rather, it is a synthetic product often used as a substitute for organic oils, greases, and rubber. Instead of attaching to oxygen atoms, as in a silicate, silicon atoms in silicone attach to organic groups, that is, molecules containing carbon. Silicone oils frequently are used in place of organic petroleum as a lubricant because they can withstand greater variations in temperature. And because the body tolerates the introduction of silicone implants better than it does organic ones, silicones are used in surgical implants as well. Silicone rubbers appear in everything from bouncing balls to space vehicles, and silicones are also present in electrical insulators, rust preventives, fabric softeners, hair sprays, hand creams, furniture and automobile polishes, paints, adhesives, and even chewing gum.

Even this list does not exhaust the many applications of silicon, which (together with oxygen) accounts for the vast majority of the mass in Earth's crust. Owing to its semimetallic qualities, silicon is used as a semiconductor of electricity. Computer chips are tiny slices of ultrapure silicon, etched with as many as half a million microscopic and intricately connected electronic circuits. These chips manipulate voltages using binary codes, for which 1 means "voltage on" and 0 means "voltage off." By means of these pulses, silicon chips perform multitudes of calculations in seconds—calculations that would take humans hours or months or even years.

A porous form of silica known as silica gel absorbs water vapor from the air and is often packed alongside moisture-sensitive products,

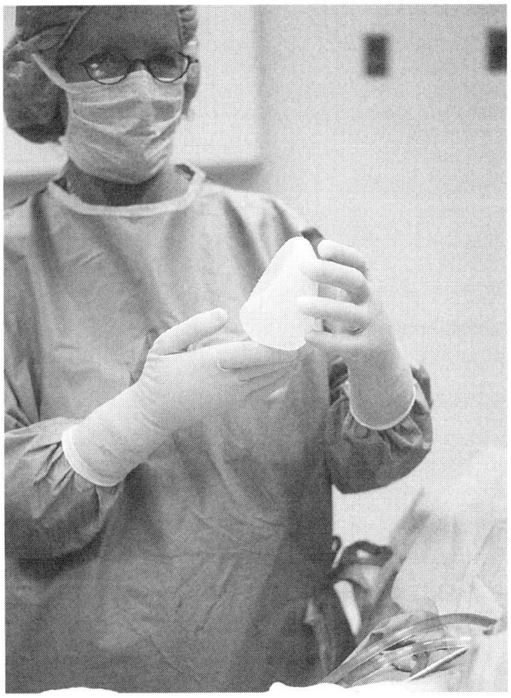

SILICONE, A SYNTHETIC PRODUCT, HAS A VARIETY OF USES, FROM ELECTRICAL INSULATORS TO FABRIC SOFTENERS TO PAINTS AND ADHESIVES. IT IS TOLERATED BETTER BY THE HUMAN BODY THAN ORGANIC COMPOUNDS AND OFTEN IS USED FOR SURGICAL IMPLANTS. *(© Michelle Del Guercio/Photo Researchers. Reproduced by permission.)*

such as electronics components, to keep them dry. Silicon carbine, an extremely hard crystalline material manufactured by fusing sand with coke (almost pure carbon) at high temperatures, has applications as an abrasive.

ORES

Earlier, it was stated that an ore is a rock or mineral that possesses economic value. This is true, but a more targeted definition would include the adjective metalliferous, since economically valuable minerals that contain no metals usually are treated as a separate category, industrial minerals. Indeed, it can be said that the interests of economic geology are divided into three areas: ores, industrial minerals, and fuels, which we have discussed already.

The very word ore seems to call to mind one of the oldest-known metals in the world and probably the first material worked by prehistoric metallurgists: gold. Even the Spanish word for gold, *oro,* suggests a connection. When conquistadors from Spain arrived in the New World after about 1500, *oro* was their obsession, and it was said that the Spanish invaders of Mexico found every bit of gold or silver ore located at the surface of the earth. However, miners of the sixteenth century lacked much of the knowledge that helps geologists today find ore deposits that are *not* at the surface.

LOCATING AND EXTRACTING ORES. The modern approach uses knowledge gained from experience. As in Agricola's day, much of the wealth possessed by a mining company is in the form of information regarding the means of best seeking out and retrieving materials from the solid earth. Certain surface geochemical and geophysical indicators help direct the steps of geologists and miners searching for ore. Thus, by the time a company in search of ore begins drilling, a great deal of exploratory work has been done. Only at that point is it possible to determine the value of the deposits, which may simply be minerals of little economic interest.

It is estimated that a cubic mile (1.6 km^3) of average rock contains about $1 trillion worth of metals, which at first sounds promising—until one does the math. A trillion dollars is a lot of money, but 1 cu. mi. (equal to 5,280 $\times$ 5,280 $\times$ 5,280 ft., or 1,609 km^3) is a lot of space too. The result is that 1 cu. ft. (0.028 m^3) is worth only about $6.79. But that is an average cubic foot in an average cubic mile of rock, and no mining company would even consider attempting to extract metals from an average piece of ground. Rather, viable ore appears only in regions that have been subjected to geologic processes that concentrate metals in such a way that their abundance is usually many hundreds of times greater than it would be on Earth as a whole.

Ore contains other minerals, known as *gangue,* which are of no economic value but which serve as a telltale sign that ore is to be found in that region. The presence of quartz, for example, may suggest deposits of gold. Ore may appear in igneous, metamorphic, or sedimentary deposits as well as in hydrothermal fluids. The latter are emanations from igneous rock, in the form of gas or water, that dissolve metals from rocks through which they pass and later deposit the ore in other locations.

CONFRONTING THE HAZARDS OF MINING. Mining, a means of extracting not only ores but many industrial minerals and fuels, such as coal, is difficult work fraught with numerous hazards. There are short-term dangers to the miners, such as cave-ins, flooding, or the release of gases in the mines, as well as long-term dangers that include such mining-related diseases as black lung (typically a hazard of coal miners). Then there is the sheer mental and emotional stress that comes from spending eight or more hours a day away from the sunlight, in claustrophobic surroundings.

And, of course, there is the environmental stress created by mining—not just by the immediate impact of cutting a gash in Earth's surface, which may disrupt ecosystems on the surface, but myriad additional problems, such as the seepage of pollutants into the water table. Abandoned mines present further dangers, including the threat of subsidence, which make these locations unsafe for the long term.

Higher environmental and occupational safety standards, established in the United States during the last third of the twentieth century, have led to changes in the way mining is performed as well as in the way mines are left when the work is completed. For example, mining companies have experimented with the use of chemicals or even bacteria, which can dissolve a metal underground and allow it to be pumped to the surface without the need to create actual underground shafts and tunnels or to send human miners to work them.

Industrial Minerals and Other Products

Industrial minerals, as noted earlier, are non-metal-containing mineral resources of interest to economic geology. Examples include asbestos, a generic term for a large group of minerals that are highly resistant to heat and flame; boron compounds, which are used for making heat-resistant glass, enamels, and ceramics; phosphates and potassium salts, used in making fertilizers; and sulfur, applied in a range of products, from refrigerants to explosives to purifiers used in the production of sugar.

Just one industrial mineral, corundum (from the oxides class of mineral), can have numerous uses. Extremely hard, corundum in the form of an unconsolidated rock commonly called emery has been used as an abrasive since ancient times. Owing to its very high melting point—even higher than that of iron—corundum also is employed in making alumina, a fire-

KEY TERMS

ALLOY: A mixture of two or more metals.

ATOM: The smallest particle of an element, consisting of protons, neutrons, and electrons. An atom can exist either alone or in combination with other atoms in a molecule.

CHEMICAL BONDING: The joining through electromagnetic force of atoms that sometimes, but not always, represent more than one chemical element.

COMPOUND: A substance made up of atoms of more than one element, chemically bonded to one another.

CONSOLIDATION: A process whereby materials become compacted, or experience an increase in density.

CRYSTALLINE SOLID: A type of solid in which the constituent parts have a simple and definite geometric arrangement that is repeated in all directions.

DEPOSITION: The process whereby sediment is laid down on the Earth's surface.

DUCTILE: Capable of being bent or molded into various shapes without breaking.

ECONOMIC GEOLOGY: The study of fuels, metals, and other materials from the Earth that are of interest to industry or the economy in general.

ELECTRON: A negatively charged particle in an atom, which spins around the nucleus.

FOSSIL FUELS: Fuel derived from deposits of organic material that have experienced decomposition and chemical alteration under conditions of high pressure. These nonrenewable forms of bioenergy include petroleum, coal, peat, natural gas, and their derivatives.

GANGUE: Minerals of no economic value, which appear in nature with ore. Recognition of certain characteristic combinations can help geologists find ore on the basis of its attendant gangue. (The *ue* is silent, as in *tongue.*)

HARDNESS: In mineralogy, the ability of one mineral to scratch another. This can be measured by the Mohs scale.

HYDROCARBON: Any organic chemical compound whose molecules are made up of nothing but carbon and hydrogen atoms.

IGNEOUS ROCK: One of the three principal types of rock, along with sedimentary and metamorphic rock. Igneous rock is formed by the crystallization of molten materials, for instance, in a volcano or other setting where plate tectonic processes take place.

INDUSTRIAL MINERALS: Nonmetallic minerals with uses for industry.

LUSTER: The appearance of a mineral when light reflects off its surface. Among the terms used in identifying luster are metallic, vitreous (glassy), and dull.

METALS: Substances that are ductile, lustrous or shiny in appearance, extremely durable, and excellent conductors of heat and electricity. Metals have very high melting and boiling points, and some (though far from all) have a high degree of hardness.

KEY TERMS CONTINUED

METAMORPHIC ROCK: One of the three principal varieties of rock, along with sedimentary and igneous rock. Metamorphic rock is formed through the alteration of preexisting rock as a result of changes in temperature, pressure, or the activity of fluids. These changes are known as metamorphism.

MINERAL: A naturally occurring, typically inorganic substance with a specific chemical composition and a crystalline structure. Unknown minerals usually can be identified in terms of specific parameters, such as hardness or luster.

MINERALOGY: An area of geology devoted to the study of minerals. Mineralogy includes a number of subdisciplines, such as crystallography, or the study of crystal formations within minerals.

MOHS SCALE: A scale introduced in 1812 by the German mineralogist Friedrich Mohs (1773–1839) that rates the hardness of minerals from 1 to 10. Ten is equivalent to the hardness of a diamond and 1 that of talc, an extremely soft mineral.

MOLECULE: A group of atoms, usually but not always representing more than one element, joined in a structure. Compounds are typically made up of molecules.

NUCLEUS: The center of an atom, a region where protons and neutrons are located and around which electrons spin.

ORE: A metalliferous rock or mineral possessing economic value.

ORGANIC: At one time chemists used the term *organic* only in reference to living things. Now the word is applied to most compounds containing carbon and hydrogen, thus excluding carbonates (which are minerals), and oxides such as carbon dioxide.

PHYSICAL GEOLOGY: The study of the material components of Earth and of the forces that have shaped the planet. Physical geology is one of two principal branches of geology, the other being historical geology.

PROTON: A positively charged particle in an atom.

ROCK: An aggregate of minerals or organic matter, which may be consolidated or unconsolidated.

SEDIMENT: Material deposited at or near Earth's surface from a number of sources, most notably preexisting rock.

SEDIMENTARY ROCK: One of the three major types of rock, along with igneous and metamorphic rock. Sedimentary rock usually is formed by the deposition, compaction, and cementation of rock that has experienced weathering. It also may be formed as a result of chemical precipitation.

STREAK: The color of the powder produced when one mineral is scratched by another, harder one.

UNCONSOLIDATED ROCK: Rock that appears in the form of loose particles, such as sand.

proof product used in furnaces and fireplaces. Though pure corundum is colorless, trace amounts of certain elements can yield brilliant colors: hence, corundum with traces of chromium becomes a red ruby, while traces of iron, titanium, and other elements yield varieties of sapphire in yellow, green, and violet as well as the familiar blue.

AN ARRAY OF APPLICATIONS. We have only begun to scratch the surface, as it were, of the uses to which minerals can be put: after all, everything—literally, every solid object—that people use is either organic in origin or a mineral. The wide array of applications of minerals is clear from the following list of mineral categories, classified by application: abrasives (corundum, diamond), ceramics (feldspar, quartz), chemical minerals (halite, sulfur, borax), and natural pigments (hematite, limonite).

Lime, cement, and plaster comes from calcite and gypsum, while building materials—both structural and ornamental—are products of agate, as well as the two aforementioned minerals. Table salt is a mineral, and so is chalk, as are countless other products. There are rocks, such as granite and marble, used in building, decoration, or artwork, and then there are "rocks"—to use a word that is at once a geologic term and a slang expression—that appear in the form of jewelry.

JEWELRY. Out of all minerals, 16 are important for their use as gems: beryl, chrysoberyl, corundum, diamond, feldspar, garnet, jade, lazurite, olivine, opal, quartz, spinel, topaz, tourmaline, turquoise, and zircon. Not all forms of these minerals, of course, are precious. Furthermore, some minerals provide more than one type of gem: corundum, as we have noted, is a source of rubies and sapphires, while beryl produces both emeralds and aquamarines.

Note that many of the precious gems familiar to most of us are not minerals in their own right but versions of minerals. At least one, the pearl, is not on this list because, with its organic origin, it is not a mineral. Certainly not all minerals are created equal: even in the list of 16 just provided, the name *diamond* stands out, representing a worldwide standard of value. Yet a diamond is nothing but pure carbon, which also appears in the form of graphite and (with a very few impurities) as coke for burning.

A diamond is unusual, however, in many respects, including the fact that it is basically a huge "molecule" composed of carbon atoms strung together by chemical bonds. The size of this formation corresponds to the size of the diamond, such that a diamond of 1 carat is simply a gargantuan "molecule" containing about 10^{22} (10,000,000,000,000,000,000,000, or 10 billion trillion) carbon atoms. Not only is a diamond rare and (when properly selected, cut, and polished) extremely beautiful, it is also extraordinarily hard. At the top of the Mohs scale, it can cut any other substance, but nothing can cut a diamond except another diamond.

WHERE TO LEARN MORE

Atlas of Rocks, Minerals, and Textures (Web site). <http://www.geosci.unc.edu/Petunia/IgMetAtlas/mainmenu.html>.

Bates, Robert Latimer. *Industrial Minerals: How They Are Found and Used.* Hillside, NJ: Enslow Publishers, 1988.

McGraw-Hill Encyclopedia of Science and Technology. 8th ed. New York: McGraw-Hill, 1997.

The Mineral and Gemstone Kingdom: Minerals A–Z (Web site). <http://www.minerals.net/mineral/>.

"Minerals and Metals: A World to Discover." Natural Resources Canada (Web site). <http://www.nrcan.gc.ca/mms/school/e_mine.htm>.

Ohio Department of Natural Resources Division of Mineral Resources Management (Web site). <http://www.dnr.state.oh.us/mineral/index.html>.

Spitz, Peter H. *Petrochemicals: The Rise of an Industry.* New York: John Wiley and Sons, 1988.

Stevens, Paul. *Oil and Gas Dictionary.* New York: Nichols, 1988.

Symes, R. F. *Rocks and Minerals.* Illus. Colin Keates and Andreas Einsiedel. New York: Dorling Kindersley, 2000.

Western Australia Department of Mineral and Petroleum Resources (Web site). <http://www.dme.wa.gov.au/index.html>.

GEOPHYSICS

GRAVITY AND GEODESY

CONCEPT

Thanks to the force known as gravity, Earth maintains its position in orbit around the Sun, and the Moon in orbit around Earth. Likewise, everything on and around Earth holds its place—the waters of the ocean, the gases of the atmosphere, and so on—owing to gravity, which is also the force that imparts to Earth its nearly spherical shape. Though no one can really say what gravity *is,* it can be quantified in terms of mass and the inverse of the distance between objects. Earth scientists working in the realm of geophysics known as geodesy measure gravitational fields, as well as anomalies within them, for a number of purposes, ranging from the prediction of tectonic processes to the location of oil reserves.

HOW IT WORKS

GRAVITATION

Not only does gravity keep Earth and all other planets in orbit around the Sun, it also makes it possible for our solar system to maintain its position in the Milky Way, rather than floating off through space. Likewise, the position of our galaxy within the larger universe is maintained because of gravity. As for the universe itself, though many questions remain about its size, mass, and boundaries, it seems clear that the cosmos is held together by gravity.

Thanks to gravity, all objects on Earth as well as those within its gravitational field remain fixed in place. These objects include man-made satellites, which have grown to number in the thousands since the first was launched in 1957, as well as the greatest satellite of them all: the Moon. Even though people are accustomed to thinking of gravity in these large terms, with regard to vast bodies such as Earth or the Moon, every object in the universe, in fact, exerts some gravitational pull on another.

This attraction is proportional to the product of the mass of the two bodies, and inversely related to the distance between them. Bodies have to be fairly large (i.e., larger than an asteroid) for this attraction to be appreciable, but it is there, and thus gravity acts as a sort of "glue" holding together the universe. As to what gravity really is or exactly why it works, both of which are legitimate questions, the answers have so far largely eluded scientists.

Present-day scientists are able to understand how gravity works, however, inasmuch as it can be described as a function of mass and the gravitational constant (discussed later) and an inverse function of distance. They also are able to measure gravitational fields and anomalies within them. That, in fact, is the focus of geodesy, an area of geophysics devoted to the measurement of Earth's shape and gravitational field.

DISCOVERING GRAVITY

As discussed in several places within this book (see the entries Earth, Science, and Nonscience and Studying Earth), the physical sciences made little progress until the early sixteenth century. For centuries, the writings of the Greek philosopher Aristotle (384–322 B.C.) and the Alexandrian astronomer Ptolemy (*ca.* A.D. 100–170) had remained dominant, reinforcing an almost entirely erroneous view of the universe. This Aristotelian/Ptolemaic universe had Earth at its

Galileo Galilei *(Corbis-Bettmann. Reproduced by permission.)*

center, with the Sun, Moon, and other planets orbiting it in perfect circles.

The discovery by the Polish astronomer Nicolaus Copernicus (1473–1543) that Earth rotates on its axis and revolves around the Sun ultimately led to the overturning of the Ptolemaic model. This breakthrough, which inaugurated the Scientific Revolution (*ca.* 1550–1700), opened the way for the birth of physics, chemistry, and geology as genuine sciences. Copernicus himself was a precursor to this revolution rather than its leader; by contrast, the Italian astronomer Galileo Galilei (1564–1642) introduced the principles of study, known as the scientific method, that govern the work of scientists to this day.

GALILEO AND GRAVITATIONAL ACCELERATION. Galileo applied his scientific method in his studies of falling objects and was able to show that objects fall as they do, not because of their weight (as Aristotle had claimed) but as a consequence of gravitational force. This meant that the acceleration of all falling bodies would have to be the same, regardless of weight.

Of course, everyone knows that a stone falls faster than a feather, but Galileo reasoned that this was a result of factors other than weight, and later investigations confirmed that air resistance, rather than weight, is responsible for this difference. In other words, a stone falls faster than a feather not because it is heavier but because the feather encounters greater air resistance. In a vacuum, or an area devoid of all matter, including air, they would fall at the same rate.

On the other hand, if one drops two objects that meet similar air resistance but differ in weight—say, a large stone and a smaller one—they fall at almost exactly the same rate. To test this hypothesis directly, however, would have been difficult for Galileo: stones fall so fast that even if dropped from a great height, they would hit the ground too soon for their rate of fall to be tested with the instruments then available.

Instead, Galileo used the motion of a pendulum and the behavior of objects rolling or sliding down inclined planes as his models. On the basis of his observations, he concluded that all bodies are subject to a uniform rate of gravitational acceleration, later calibrated at 32 ft. (9.8 m) per second per second. What this means is that for every 32 ft. an object falls, it is accelerating at a rate of 32 ft. per second as well; hence, after two seconds it falls at the rate of 64 ft. per second, after three seconds it falls at 96 ft. per second, and so on.

NEWTON'S BREAKTHROUGH. Building on the work of his distinguished forebear, Sir Isaac Newton (1642–1727), who was born the same year Galileo died, developed a paradigm for gravitation that even today explains the behavior of objects in virtually all situations throughout the universe. Indeed, the Newtonian model reigned supreme until the early twentieth century, when Albert Einstein (1879–1955) challenged it on certain specifics.

Even so, Einstein's relativity did not *disprove* the Newtonian system as Copernicus and Galileo had disproved Aristotle's and Ptolemy's theories; rather, it showed the limitations of Newtonian mechanics for describing the behavior of certain objects and phenomena. In the ordinary world of day-to-day experience, however, the Newtonian system still offers the key to how and why things work as they do. This is particularly the case with regard to gravity.

Understanding the Law of Universal Gravitation

Like Galileo, Newton began in part with the aim of testing hypotheses put forward by an astronomer—in this case, Johannes Kepler (1571–1630). In the early years of the seventeenth century, Kepler published his three laws of planetary motion, which together identified the elliptical (oval-shaped) path of the planets around the Sun. Kepler had discovered a mathematical relationship that connected the distances of the planets from the Sun to the period of their revolution around it. Like Galileo with Copernicus, Newton sought to generalize these principles to explain not only *how* the planets moved but also *why* they did so.

The result was Newton's *Philosophiae naturalis principia mathematica* (Mathematical principles of natural philosophy, 1687). Usually referred to simply as the *Principia,* the book proved to be one of the most influential works ever written. In it, Newton presented his law of universal gravitation, along with his three laws of motion. These principles offered a new model for understanding the mechanics of the universe.

THE THREE LAWS OF MOTION

Newton's three laws of motion may be summarized in this way:

- An object at rest will remain at rest, and an object in motion will remain in motion at a constant velocity unless and until outside forces act upon it.
- The net force acting upon an object is a product of its mass multiplied by its acceleration.
- When one object exerts a force on another, the second object exerts on the first a force equal in magnitude but opposite in direction.

The first law of motion identifies inertia, a concept introduced by Galileo to explain what kept the planets moving around the Sun. Inertia is the tendency of an object either to keep moving or to keep standing still, depending on what it is already doing. Note that the first law refers to an object moving at a constant velocity: velocity is speed in a certain direction, so a constant velocity would be the same speed in the same direction.

Inertia is measured by mass, which—as the second law states—is a component of force and is inversely related to acceleration. The latter, as defined by physics, has a much broader meaning than it usually is given in ordinary life. Acceleration does not mean simply an increase of speed for an object moving in a straight line; rather, it is a change in velocity—that is, a change of speed or direction or both.

By definition, then, rotational motion (such as that of Earth around the Sun) involves acceleration, because any movement other than motion in a straight line at a constant speed requires a change in velocity. This further means that an object experiencing rotational motion *must* be under the influence of some force. That force is gravity, and as the third law shows, every force exerted in one direction is matched by an equal force in the opposing direction. (This law is sometimes rendered "For every action there is an equal and opposite reaction.")

NEWTON'S GRAVITATIONAL FORMULA. The law of universal gravitation can be stated as a formula for calculating the gravitational attraction between two objects of a certain mass, m_1 and m_2: $F_{grav} = G \times (m_1 m_2)/r^2$. In this equation, F_{grav} is gravitational force, and r^2 is the square of the distance between the two objects.

As for *G,* in Newton's time the value of this number was unknown. Newton was aware simply that it represented a very small quantity: without it, $(m_1 m_2)/r^2$ could be quite sizable for objects of relatively great mass separated by a relatively small distance. When multiplied by this very small number, however, the gravitational attraction would be revealed to be very small as well. Only in 1798, more than a century after Newton's writing, did the English physicist Henry Cavendish (1731–1810) calculate the value of *G* using a precision instrument called a torsion balance.

The value of *G* is expressed in units of force multiplied by distance squared, and then divided by,mass squared; in other words, *G* is a certain value of $(N \times m^2)/kg^2$, where *N* stands for newtons, *m* for meters, and *kg* for kilograms. Nor is the numerical value of *G* a whole number such as 1. A figure as large as 1, in fact, is astronomically huge compared with *G,* whose value is 6.67×10^{-11}—in other words, 0.0000000000667.

PHYSICAL GEODESY

Within the realm of geodesy is that of physical geodesy, which is concerned specifically with the measurement of Earth's gravitational field as well as the geoid. The latter may be defined as a surface of uniform gravitational potential covering the entire Earth at a height equal to sea level. ("Potential" here is analogous to height or, more specifically, position in a field. For a discussion of potential in a gravitational field, see Energy and Earth.)

Thus, in areas that are above sea level, the geoid would be below ground—indeed, *far* below it in mountainous regions. Yet in some places (most notably the Dead Sea and its shores, the lowest point on Earth), it would be above the solid earth and waters. The geoid is also subject to deviations or anomalies, owing to the fact that the planet's mass is not distributed uniformly; in addition, small temporary disturbances in the geoid may occur on the seas as a result of wind, tides, and currents.

Generally speaking, however, the geoid is a stable reference platform from which to measure gravitational anomalies. It is a sort of imaginary gravitational "skin" covering the planet, and in the past, countries conducting geodetic surveys tended to choose a spot within their boundaries as the reference point for all measurements. With the development of satellites and their use for geodetic research, however, it has become more common for national geodetic societies to use global points of reference such as the planet's center of mass.

MEASUREMENTS FROM SPACE, LAND, AND SEA. The geoid can be determined by using such a satellite, equipped with a radar altimeter, but there is also the much older technique of terrestrial gravity measurement. The terrestrial method is much more difficult and prone to error, however, and calculations require detailed checking and correction to remove potential anomalies due to the presence of matter in areas above the points at which gravitational measurements were obtained.

Also highly subject to error are measurements made from a vessel at sea. This has to do not only with the effect of the ship's pitch and roll but also with something called the Eötvös effect. Named for the Hungarian physicist Baron Roland Eötvös (1848–1919), who conducted extensive studies on gravity, the effect is related to the Coriolis force, which causes the deflection of atmospheric and oceanic currents in response to Earth's rotation. Measurements of gravity from the air are also subject to the Eötvös effect, though the use of GPS (global positioning system) information, obtained from satellites, can improve greatly the accuracy of seaborne measurements.

HOW GRAVITY IS MEASURED. Scientists can obtain absolute terrestrial gravity measurements by measuring the amount of time it takes for a pellet to fall a certain distance within a vacuum—that is, a chamber from which all matter, including air, has been removed. This, of course, is the same technology Galileo used in making his observations more than 400 years ago. It is also possible to obtain relative gravity measurements with the use of mechanical balance instruments.

As noted earlier, the acceleration due to gravity is 9.8 m/s^2, or 9.8 m s^{-2}. (Scientists sometimes use the latter notation, in which the minus sign is not meant to indicate a negative but rather is used in place of "per".) This number is the measure of Earth's gravitational field. In measuring gravitational anomalies, scientists may use the Gal, named after Galileo, which is equal to 0.01 m/s^2. Typically, however, the milligal, equal to one-thousandth of a Gal, is used. Note that "Gal" sometimes is rendered in lowercase, but this can be confusing, because it looks like the abbreviation for "gallon. ")

WHY MEASURE GRAVITY? Why is it important to measure gravity and gravitational anomalies? One answer is that weight values can vary considerably, depending on one's position relative to Earth's gravitational field. A fairly heavy person might weigh as much as a pound less at the equator than at the poles and less still at the top of a high mountain. The value of the gravity field at sea level has a range from 9.78 to 9.83 m/s^2, a difference of about 50,000 g.u., and it is likely to be much lower than 9.78 m/s^2 at higher altitudes.

Indeed, the higher one goes, the weaker Earth's gravitational field becomes. At the same time, the gases of the atmosphere dissipate, which is the reason why it is hard to breathe on high mountains without an artificial air supply and impossible to do so in the stratosphere or above it. At the upper edge of the mesosphere, Earth's gravitational field is no longer strong enough to hold large quantities of hydrogen, lightest of all elements, which constitutes the atmosphere at that point. Beyond the mesosphere, the atmosphere simply fades away, because there is not sufficient gravitational force to hold its particles in place.

Back down on Earth, gravity measurements are of great importance to the petroleum industry, which uses them to locate oil-containing salt domes. Furthermore, geologists, in general, remain acutely interested in measurements of gravity, the force behind tectonics, or the deformation of Earth's crust. Thus, gravity, responsible for fashioning Earth's exterior into the nearly spherical shape it has, is key to the shaping of its interior as well.

REAL-LIFE APPLICATIONS

GRAVITY ON EARTH

Using Newton's gravitational formula, it is relatively easy to calculate the pull of gravity between two objects. It is also easy to see why the attraction is insignificant unless at least one of the objects has enormous mass. In addition, application of the formula makes it clear why *G* is such a tiny number.

Suppose two people each have a mass of 45.5 kg—equal to 100 lb. on Earth, though not on the Moon, a matter that will be explained later in this essay—and they stand 1 m (3.28 ft.) apart. Thus, m_1m_2 is equal to 2,070 kg (4,555 lb.), and r^2 is equal to 1 m squared. Applied to the gravitational formula, this figure is rendered as 2,070 kg^2/1 m^2. This number then is multiplied by the gravitational constant, and the result is a net gravitational force of 0.000000138 N (0.00000003 lb.)—about the weight of a single-cell organism!

WEIGHT. What about Earth's gravitational force on one of those people? To calculate this force, we could apply the formula for universal gravitation, substituting Earth for m_2, especially because the mass of Earth is known: 5.98×10^{24} kg, or 5.98 septillion (1 followed by 24 zeroes) kg. We know the value of that mass, in fact, through the application of Newton's laws and the formulas derived from them. But for measuring the gravitational force between something as massive as Earth and something as small as a human body, it makes more sense to apply instead the formula embodied in Newton's second law of motion: $F = ma$. (Force equals mass multiplied by acceleration.)

For a body of any mass on Earth, acceleration is figured in terms of *g*—the acceleration due to gravity, which, as noted earlier, is equal to 32 ft. (9.8 m) per second squared. (Note, also, that this is a lowercase *g*, as opposed to the uppercase *G* that represents the gravitational constant.) Using the metric system, by multiplying the appropriate mass figure in kilograms by 9.8 m/s^2, one would obtain a value in newtons (N). To perform the same calculation with the English system, used in America, it would be necessary first to calculate the value of mass in slugs (which, needless to say, is a little-known unit) and multiply it by 32 ft./s^2 to yield a value in pounds.

In both cases, the value obtained, whether in newtons or pounds, is a measure of weight rather than of mass, which is measured in kilograms or slugs. For this reason, it is not entirely accurate to say that 1 kg is equal to 2.2 lb. This is true on Earth, but it would not be true on the Moon. The kilogram is a unit of mass, and as such it would not change anywhere in the universe, whereas the pound is a unit of force (in this case, gravitational force) and therefore varies according to the rate of acceleration for the gravitational field in which it is measured. For this reason, scientists prefer to use figures for mass, which is one of the fundamental properties (along with length, time, and electric charge) of the universe.

WHY EARTH IS ROUND—AND NOT ROUND

Everyone knows that Earth, the Sun, and all other large bodies in space are "round" (i.e., spherical), but why is that true? The reason is that gravity will not allow them to be otherwise: for any large object, the gravitational pull of its interior forces the surface to assume a relatively uniform shape. The most uniform of three-dimensional shape is that of a sphere, and the larger the mass of an object, the greater its tendency toward sphericity.

Earth has a relatively small vertical differential between its highest and lowest surface points, Mount Everest (29,028 ft., or 8,848 m) on the Nepal-Tibet border and the Mariana Trench (–36,198 ft., –10,911 m) in the Pacific Ocean, respectively. The difference is just 12.28 mi. (19.6 km)—not a great distance, considering that Earth's radius is about 4,000 mi. (6,400 km).

On the other hand, an object of less mass is more likely to retain a shape that is far less than spherical. This can be shown by reference to the Martian moons Phobos and Deimos, both of which are oblong—and both of which are tiny, in

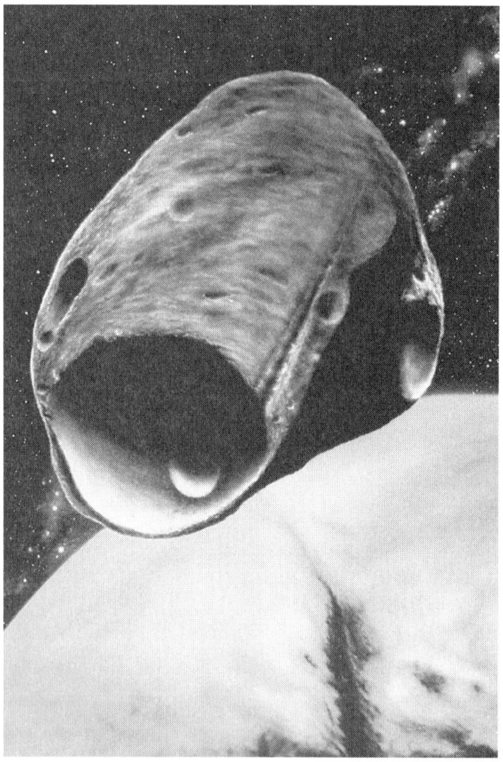

THE FORCE OF GRAVITY IMPARTS A SPHERICAL SHAPE TO EARTH BECAUSE OF ITS LARGE MASS. AN OBJECT OF LESS MASS WILL HAVE A FAR LESS SPHERICAL SHAPE. THE MARTIAN MOON PHOBOS, SHOWN HERE, IS OBLONG OWING TO ITS TINY MASS. *(© Julian Baum/Photo Researchers. Reproduced by permission.)*

terms of size and mass, compared with Earth's Moon. Mars itself has a radius half that of Earth, yet its mass is only about 10% of Earth's. In light of what has been said about mass, shape, and gravity, it should not be surprising to learn that Mars is also home to the tallest mountain in the solar system, the volcano Olympus Mons, which stands 16 mi. (27 km) high.

EARTH'S 'FLAT TOP' (AND BOTTOM). With regard to gravitation, a spherical object behaves as though its mass were concentrated near its center, and indeed, 33% of Earth's mass is at its core, even though the core accounts for only about 20% of the planet's volume. Geologists believe that the composition of Earth's core must be molten iron, which creates the planet's vast electromagnetic field.

It should be noted, however, that Earth is not really a perfect sphere, and the idea that its mass is concentrated at its center, while it works well in general, poses some problems in making exact gravitational measurements. If Earth were standing still, it would be much nearer to the shape of a sphere; however, it is not standing still but instead rotates on its axis, as does every other object of any significance in the solar system.

Incidentally, if Earth were suddenly to stop spinning, the gases in the atmosphere would keep moving at their current rate of about 1,000 MPH (1,600 km/h). They would sweep over the planet with the force of the greatest hurricane ever known, ripping up everything but the mountains. As to why Earth spins at all, scientists are not entirely sure. It may well be angular momentum (the momentum associated with rotational motion) imparted to it at some point in the very distant past, perhaps because it and the rest of the solar system were once part of a vast spinning cloud.

At any rate, the fact that Earth is spinning on its axis creates a certain centripetal, or inward-pulling, force, and this force produces a corresponding centrifugal (outward) component. To understand this concept, consider what happens to a sample of blood when it is rotated in a centrifuge. When the centrifuge spins, centripetal force pulls the material in the vial toward the center of the spin, but the material with greater mass has more inertia and therefore responds less to centripetal force. As a result, the heavier red blood cells tend to stay at the bottom of the vial (or, as it is spinning, on the outside), while the lighter plasma is pulled inward. The result is the separation between plasma and red blood cells.

Where Earth is concerned, this centrifugal component of centripetal force manifests as an equatorial bulging. Simply put, Earth's diameter around the equator is greater than at the poles, which are slightly flattened. The difference is small—the equatorial diameter of Earth is about 26.72 mi. (43 km) greater than the polar diameter—but it is not insignificant. In fact, as noted later, a person of fairly significant weight actually would notice a difference if he or she got on the scales at the equator (say, in Singapore) and then later weighed in near one of the poles (for instance, in the Norwegian possession of Svalbard, the northernmost human settlement on Earth).

Owing to this departure from a perfectly spherical shape, the Sun and Moon exert additional torques on Earth, and these torques cause shifts in the position of the planet's rotational axis in space. An imaginary line projected from

KEY TERMS

ACCELERATION: A change in velocity over time. The acceleration due to gravity, for instance, is 32 ft. (9.8 m) per second per second, meaning that for every second an object falls, its velocity is increasing as well.

ATMOSPHERE: In general, an atmosphere is a blanket of gases surrounding a planet. Unless otherwise identified, however, the term refers to the atmosphere of Earth, which consists of nitrogen (78%), oxygen (21%), argon (0.93%), and other substances that include water vapor, carbon dioxide, ozone, and noble gases such as neon, which together comprise 0.07%.

CENTRIFUGAL: A term describing the tendency of objects in uniform circular motion to move outward, away from the center of the circle. Though the term centrifugal force often is used, it is inertia, rather than force, that causes the object to move outward.

CENTRIPETAL FORCE: The force that causes an object in uniform circular motion to move inward, toward the center of the circle.

FORCE: The product of mass multiplied by acceleration.

GEODESY: An area of geophysics devoted to the measurement of Earth's shape and gravitational field.

GEOID: A surface of uniform gravitational potential covering the entire Earth at a height equal to sea level.

GEOPHYSICS: A branch of the earth sciences that combines aspects of geology and physics. Geophysics addresses the planet's physical processes as well as its gravitational, magnetic, and electric properties and the means by which energy is transmitted through its interior.

INERTIA: The tendency of an object at rest to remain at rest or an object in motion to remain in motion, at a uniform velocity, at a uniform velocity, unless acted upon by some outside force.

MASS: A measure of inertia, indicating the resistance of an object to a change in its motion.

POTENTIAL: Position in a field, such as a gravitational force field.

SCIENTIFIC METHOD: A set of principles and procedures for systematic study that includes observation; the formation of hypotheses, theories, and laws; and continual testing and reexamination.

SCIENTIFIC REVOLUTION: A period of accelerated scientific discovery that completely reshaped the world. Usually dated from about 1550 to 1700, the Scientific Revolution saw the origination of the scientific method and the introduction of ideas such as the heliocentric (Sun-centered) universe and gravity.

TORQUE: A force that produces, or tends to produce, rotational motion.

UNIFORM CIRCULAR MOTION: The motion of an object around the center of a circle in such a manner that speed is constant or unchanging.

VACUUM: An area devoid of matter, even air.

VELOCITY: Speed in a certain direction.

WEIGHT: A measure of the gravitational force on an object. Weight thus would change from planet to planet, whereas mass remains constant throughout the universe. A pound is a unit of weight, whereas a kilogram is a unit of mass.

the North Pole and into space therefore, over a period of time, would appear to move. In the course of about 25,800 years, this point (known as the celestial north pole) describes the shape of a cone, a movement known as Earth's *precession.*

SATELLITES

Why, then, does Earth move around the Sun, or the Moon around Earth? As should be clear from Newton's gravitational formula and the third law of motion, the force of gravity works both ways: not only does a stone fall toward Earth, but Earth also actually falls *toward it.* The mass of Earth is so great compared with that of the stone that the movement of Earth is imperceptible—but it does happen.

Furthermore, because Earth is round, when one hurls a projectile at a great distance, Earth curves away from the projectile. Eventually, gravity itself forces the projectile to the ground. If one were to fire a rocket at 17,700 mi. per hour (28,500 km per hour), however, something unusual would happen. At every instant of time, the projectile would be falling toward Earth with the force of gravity—*but* the curved Earth would be falling away from it at the same rate. Hence, the projectile would remain in constant motion around the planet—that is, it would be in orbit.

The same is true of an artificial satellite's orbit around Earth: even as the satellite falls toward Earth, Earth falls away from it. Change the names of the players, and this same relationship exists between Earth and its great natural satellite, the Moon. Furthermore, it is the same with the Sun and *its* many satellites, including Earth: Earth plunges toward the Sun with every instant of its movement, but at every instant, the Sun falls away.

WHERE TO LEARN MORE

Ardley, Neil. *The Science Book of Gravity.* San Diego, CA: Harcourt Brace Jovanovich, 1992.

Ask the Space Scientist (Web site). <http://image.gsfc.nasa.gov/poetry/ask/askmag.html>.

Beiser, Arthur. *Physics,* 5th ed. Reading, MA: Addison-Wesley, 1991.

Exploring Gravity (Web site). <http://www.curtin.edu.au/curtin/dept/phys-sci/gravity/>.

Geodesy for the Layman (Web site). <http://www.nima.mil/GandG/geolay/TR80003A.HTM>.

The Gravity Society (Web site). <http://www.gravity.org>.

Hancock, Paul L., and Brian J. Skinner. *The Oxford Companion to the Earth.* New York: Oxford University Press, 2000.

Riley, Peter D. *Earth.* Des Plaines, IL: Heinemann Interactive Library, 1998.

Smith, David G. *The Cambridge Encyclopedia of Earth Sciences.* New York: Cambridge University Press, 1981.

Wilford, John Noble. *The Mapmakers.* New York: Knopf, 2000.

GEOMAGNETISM

CONCEPT

Scientists have long recognized a connection between electricity and magnetism, but the specifics of this connection, along with the recognition that electromagnetism is one of the fundamental interactions in the universe, were worked out only in the mid–nineteenth century. By that time, geologists had come to an understanding of Earth as a giant magnet. This was the principle that made possible the operation of compasses, which greatly aided mariners in navigating the seas: magnetic materials, it so happened, point northward. As it turns out, however, Earth's magnetic North Pole is not the same as its geographic one, and even the pole's northerly location is not a permanent fact. Once upon a time and, in fact, at many times in Earth's history, the magnetic North Pole lay at the southern end of the planet.

HOW IT WORKS

Electromagnetism

The Greek philosopher Thales (640?–546 B.C.) was the first to observe that when amber is rubbed with certain types of materials, the friction imparts to it the ability to pick up light objects. The word electricity comes from the Greek word for amber, *elektron,* and, in fact, magnetism and electricity are simply manifestations of the same force. This concept of electric and magnetic interaction seems to have been established early in human history, though it would be almost 2,500 years before scientists came to a mature understanding of the relationship.

As in so much else, studies in electromagnetism made little progress from the time of the Romans to the late Renaissance, a span of nearly 1,500 years. Yet it is worth noting that thefirst ideas scientists had about studying Earth's history scientifically came from observing the planet's magnetic field. In the course of his work on that subject, the English astronomer Henry Gellibrand (1597–1636) showed that the field has changed over time. This suggested that it would be possible to form hypotheses about the planet's past, even though humans had no direct information regarding the origins of Earth. Thus, Gellibrand (who, ironically, was also a minister) helped make it possible for geologists to move beyond a strict interpretation of the Bible in studying the history of Earth. (See Earth, Science, and Nonscience for more on the Genesis account and its interpretations.)

ELECTROMAGNETIC STUDIES COME OF AGE. Beginning in the 1700s, a number of thinkers conducted experiments concerning the nature of electricity and magnetism and the relationship between them. Among these thinkers were several giants in physics and other disciplines, including one of America's greatest founding fathers, Benjamin Franklin (1706–1790). In addition to his famous (and highly dangerous) experiment with lightning, Franklin also contributed the names positive and negative to the differing electric charges discovered earlier by the French physicist Charles Du Fay (1698–1739).

In 1785 the French physicist and inventor Charles Coulomb (1736–1806) established the basic laws of electrostatics and magnetism. He maintained that there is an attractive force that, like gravitation, can be explained in terms of the

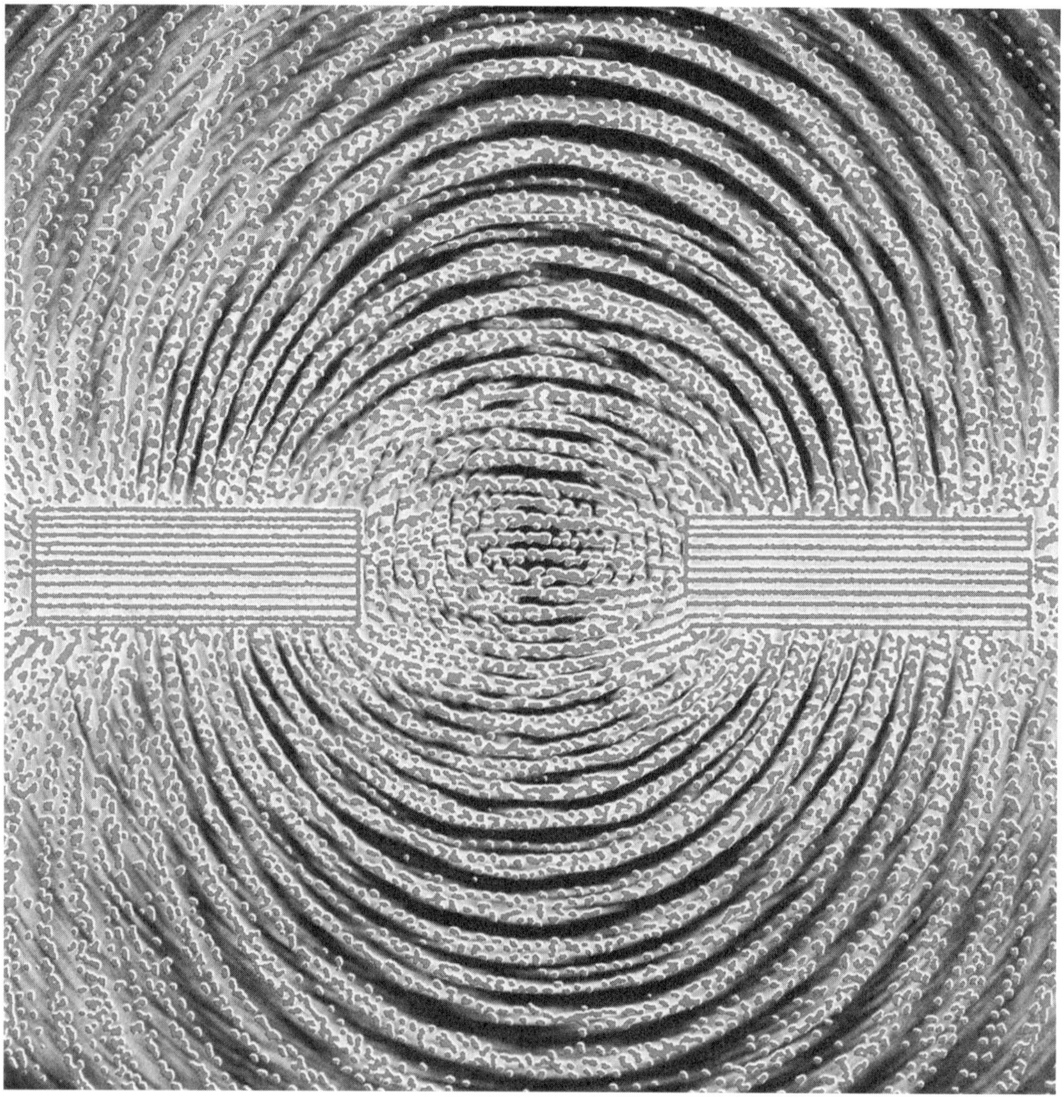

THE MAGNETIC FIELD AROUND TWO BAR MAGNETS. *(© A. Pasieka/Photo Researchers. Reproduced by permission.)*

inverse of the square of the distance between objects (see Gravity and Geodesy). That attraction itself, however, results not from gravity but from electric charge, according to Coulomb.

A few years later, the German mathematician Karl Friedrich Gauss (1777–1855) developed a mathematical theory for finding the magnetic potential of any point on Earth. His contemporary, the Danish physicist Hans Christian Oersted (1777–1851), became the first scientist to establish a clear relationship between electricity and magnetism. This led to the formalization of electromagnetism, the branch of physics devoted to the study of electric and magnetic phenomena.

The French mathematician and physicist André Marie Ampère (1775–1836) concluded that magnetism is the result of electricity in motion, and in 1831 the British physicist and chemist Michael Faraday (1791–1867) published his theory of electromagnetic induction. This theory shows how an electric current in one coil can set up a current in another through the development of a magnetic field, and it enabled Faraday's development of the first generator. For the first time in history humans were able to convert mechanical energy systematically into electric energy.

ELECTROMAGNETIC FORCE. By this point scientists were convinced that a relationship between electricity and magnetism existed, yet they did not know exactly *how* the two related. Then, in 1865, the Scottish physicist James Clerk Maxwell (1831–1879) published a groundbreaking paper, "On Faraday's Lines of Force," in which he outlined a theory of electromagnetic force. The latter may be defined as the total force on an electrically charged particle, which is a combination of forces due to electric and magnetic fields around the particle.

Maxwell thus had discovered a type of force in addition to gravity, and this reflected a "new" type of fundamental interaction, or a basic mode by which particles interact in nature. Nearly two centuries earlier Sir Isaac Newton (1642–1727) had identified the first, gravity, and in the twentieth century two other forms of fundamental interaction—strong nuclear and weak nuclear—were identified as well.

In his work Maxwell drew on the studies conducted by his predecessors but added a new statement. According to Maxwell, electric charge is conserved, meaning that the sum total of electric charge in the universe does not change, though it may be redistributed. This statement, which did not contradict any of the experimental work done by other physicists, was based on Maxwell's predictions regarding what should happen in situations of electromagnetism. Subsequent studies have supported his predictions.

MAGNETISM

What, then, is the difference between electricity and magnetism? It is primarily a matter of orientation. When two electric charges are at rest, it appears to an observer that the force between them is merely electric. If the charges are in motion relative to the observer, it appears as though a different sort of force, known as magnetism, exists between them.

An electromagnetic wave, such as that which is emitted by the Sun, carries both an electric and a magnetic component at mutually perpendicular angles. If you extend your hand, palm flat, with the fingers straight and the thumb pointing at a 90° angle to the fingers, the direction that the fingers are pointing would be that of the electromagnetic wave. Your thumb points in the direction of the electric field, and the flat of your palm indicates the direction of the magnetic field, which is perpendicular both to the electric field and to the direction of wave propagation.

A field, in this sense, is a region of space in which it is possible to define the physical properties of each point in the region at any given moment in time. Thus, an electric field and a magnetic field are simply regions in which the electric and magnetic components, respectively, of electromagnetic force are exerted.

MAGNETISM AT THE ATOMIC LEVEL. At the atomic level magnetism is the result of motion by electrons (negatively charged subatomic particles) in relation to one another. Rather like planets in a solar system, electrons revolve around the atom's nucleus and rotate on their own axes. (In fact, the exact nature of their movement is much more complex, but this analogy is accurate enough for the present purposes.) Both types of movement create a magnetic force field between electrons, and as a result the electron takes on the properties of a tiny bar magnet with a north pole and south pole. Surrounding this infinitesimal magnet are lines of magnetic force, which begin at the north pole and curve outward, describing an ellipse as they return to the south pole.

In most atoms, electrons are paired such that their magnetic fields cancel out one another. However, in certain cases, such as when there is an odd electron or when other factors become more significant, the fields line up to create what is known as a net magnetic dipole, or a unity of direction. These elements, among them, iron, cobalt, and nickel as well as various alloys or mixtures, are commonly known as magnetic metals, or natural magnets.

MAGNETIZATION. Magnetization occurs when an object is placed in a magnetic field. In this field magnetic force acts on a moving charged particle such that the particle would experience no force if it moved in the direction of the magnetic field. In other words, it would be "drawn," as a ten-penny nail is drawn to a common bar or horseshoe (U-shaped) magnet. An electric current is an example of a moving charge, and, indeed, one of the best ways to create a magnetic field is with a current. Often this is done by means of a solenoid, a current-carrying wire coil through which the material to be magnetized is passed, much as one would pass a straight wire up through the interior of a spring.

When a natural magnet becomes magnetized (that is, when a magnetic metal or alloy comes into contact with an external magnetic field), a change occurs at the level of the domain, a group of atoms equal in size to about 5×10^{-5} meters across—just large enough to be visible under a microscope.

In an unmagnetized sample, there may be an alignment of unpaired electron spins within a domain, but the direction of the various domains' magnetic forces in relation to one another is random. Once a natural magnet is placed within an external magnetic force field, however, one of two things happens to the domains. Either they all come into alignment with the field or, in certain types of material, those domains in alignment with the field grow, while the others shrink to nonexistence.

The first of these processes is called domain alignment, or ferromagnetism, and the second is termed domain growth, or ferrimagnetism. Both processes turn a natural magnet into what is known as a permanent magnet—or, more simply, a magnet. The magnet is then capable of temporarily magnetizing a ferromagnetic item, as, for instance, when one rubs a paper clip against a permanent magnet and then uses the magnetized clip to lift other paper clips. Of the two varieties, however, a ferromagnet is stronger, because it requires a more powerful magnetic force field to become magnetized. Most powerful of all is a saturated ferromagnetic metal, one in which all the unpaired electron spins are aligned.

REAL-LIFE APPLICATIONS

THE MAGNETIC COMPASS

A bar magnet placed in a magnetic field will rotate until it lines up with the field's direction. The same thing happens when one suspends a magnet from a string: it lines up with Earth's magnetic field and points in a north-south direction. The Chinese of the first century B.C. discovered that a strip of magnetic metal always tends to point toward geographic north, though they were unaware of the electromagnetic force that causes this to happen.

This led ultimately to the development of the magnetic compass, which typically consists of a magnetized iron needle suspended over a card marked with the four cardinal directions (north, south, east, and west). The needle is attached to a pivoting mechanism at its center, which allows it to move freely so that the tip of the needle will always point the user northward. The magnetic compass proved so important that it typically is ranked alongside paper, printing, and gunpowder as one of premodern China's four great gifts to the West. Before the compass, mariners had to depend purely on the position of the Sun and other, less reliable means of determining direction; hence, the invention quite literally helped open up the world.

ODD BEHAVIOR OF THE COMPASS. The compass, in fact, helped make possible the historic voyage of Christopher Columbus (1451–1506) in 1492. While sailing across the Atlantic, Columbus noticed something odd: his compass did not always point toward what he knew, based on the Sun's position, to be geographic north. The further he traveled, the more he noticed this phenomenon, which came to be known as magnetic declination.

When Columbus returned to Europe and reported on his observations of magnetic declination (along with the much bigger news of his landing in the New World, which he thought was Asia), his story perplexed mariners. Eventually, European scientists worked out tables of magnetic declination, showing the amount of deviation at various points on Earth, and this seemed to allay sailors' concerns.

Then, in 1544, the German astronomer Georg Hartmann (1489–1564) observed that a freely floating magnetized needle did not always stay perfectly horizontal and actually dipped more and more strongly as he traveled north. When he was moving south, on the other hand, the needle tended to become more closely horizontal. For many years, this phenomenon, along with magnetic declination, remained perplexing. Nor, for that matter, did scientists understand exactly how or why a compass works. Then, in 1600, the English physicist William Gilbert (1540–1603) became the first to suggest a reason.

EARTH AS A GIANT MAGNET

Gilbert coined the terms electric attraction, electric force, and magnetic pole. In De magnete (On the magnet), he became the first thinker to introduce the idea, now commonly accepted, that Earth itself is a giant magnet. Not only does it have north and south magnetic poles, but it also is surrounded by vast arcs of magnetic force, called the geomagnetic field. (The term geomagnetism, as opposed to magnetism, refers to the magnetic properties of Earth as a whole rather than those possessed by a single object or place on Earth.)

In the paragraphs that follow, we discuss the shape of this magnetic field, including the positions of the magnetic north and south poles; the origins of the field, primarily in terms of the known or suspected physical forces that sustain it (as opposed to the original cause of Earth's magnetic field, a more complicated and speculative subject); as well as changes in the magnetic field. Those changes, along with techniques for measuring the geomagnetic field, also are discussed at other places in this book.

EARTH'S MAGNETIC FIELD. Hartmann's compass phenomenon can be explained by the fact that Earth is a magnet and that its north and south magnetic poles are close to the geographic north and south. As for the phenomenon observed by Columbus, it is a result of the difference between magnetic and geographic north. If one continued to follow a compass northward, it would lead not to Earth's North Pole but to a point identified in 1984 as 77°N, 102°18' W—that is, in the Queen Elizabeth Islands of far northern Canada.

Earlier we described the lines that make up the magnetic force field around a bar magnet. A field of similar shape, though, of course, of much larger size (yet still invisible), also surrounds Earth. From the magnetic north and south poles, lines of magnetic force rise into space and form giant curves that come back around and reenter Earth at the opposing pole, so that the planet is surrounded by a vast series of concentric loops. If one could draw a straight line through the center of all these loops, it would reach Earth at a point 11° from the equator. Likewise the north and south magnetic poles—which are on a plane perpendicular to that of Earth's magnetic field—are 11° off the planet's axis.

THE MAGNETOSPHERE. Surrounding the planet is a vast region called the magnetosphere, an area in which ionized particles (i.e., ones that have lost or gained electrons so as to acquire a net electric charge) are affected by Earth's magnetic field. The magnetosphere is formed by the interaction between our planet's magnetic field and the solar wind, a stream of particles from the Sun. (See Sun, Moon, and Earth for more about the solar wind.) Its shape would be akin to that of Earth's magnetic field, as described earlier, were it not for the Sun's influence.

The side of the magnetosphere closer to the Sun does indeed resemble the giant series of concentric loops described earlier. These loops are enormous, such that the forward, or sunlit, edge of the magnetosphere is located at a distance of some 10 Earth radii (about 35,000 mi., or 65,000 km). On the other side from the Sun—the rear, or dark, side—the shape of the magnetosphere is quite different. Instead of forming relatively small loops that curve right back around into Earth's poles, the lines of magnetic force on this side shoot straight out into space a distance of some 40 Earth radii (about 140,000 mi., or 260,000 km).

The shape of the magnetosphere, then, is a bit like that of a comet moving toward the Sun. Surrounding it is the magnetopause, a sort of magnetic dead zone about 62 mi. (100 km) thick, which shields Earth from most of the solar wind. In front of it (toward the Sun) is an area of magnetic turbulence known as the magnetosheath, and still closer to the Sun is a boundary called the bow shock, a shock-wave front that slows particles of solar wind considerably. Since Earth is turning, the side of the planet away from the Sun is continually changing, of course, but the shape of the magnetosphere remains more or less intact. It is, however, highly affected by solar activity, such that an increase in solar wind can cause a depression in the magnetosphere. (See Sun, Moon, and Earth for a discussion of auroras, produced by an interaction between the solar wind and the magnetosphere.)

THE SOURCE OF EARTH'S MAGNETIC FIELD

As noted earlier, scientists at present understand little with regard to the *origins* of Earth's magnetic field—that is, the original action or actions that resulted in the creation of a geomagnetic field that has remained active for billions of years. No less a figure than Albert Einstein (1879–1955) identified this question as one of the great unsolved problems of science. On the other hand, scientists do have a good understanding of the geomagnetic field's *source,* in terms of the physical conditions that make it possible. (This distinction is rather like that between efficient cause and material cause, as discussed in Earth, Science, and Nonscience.)

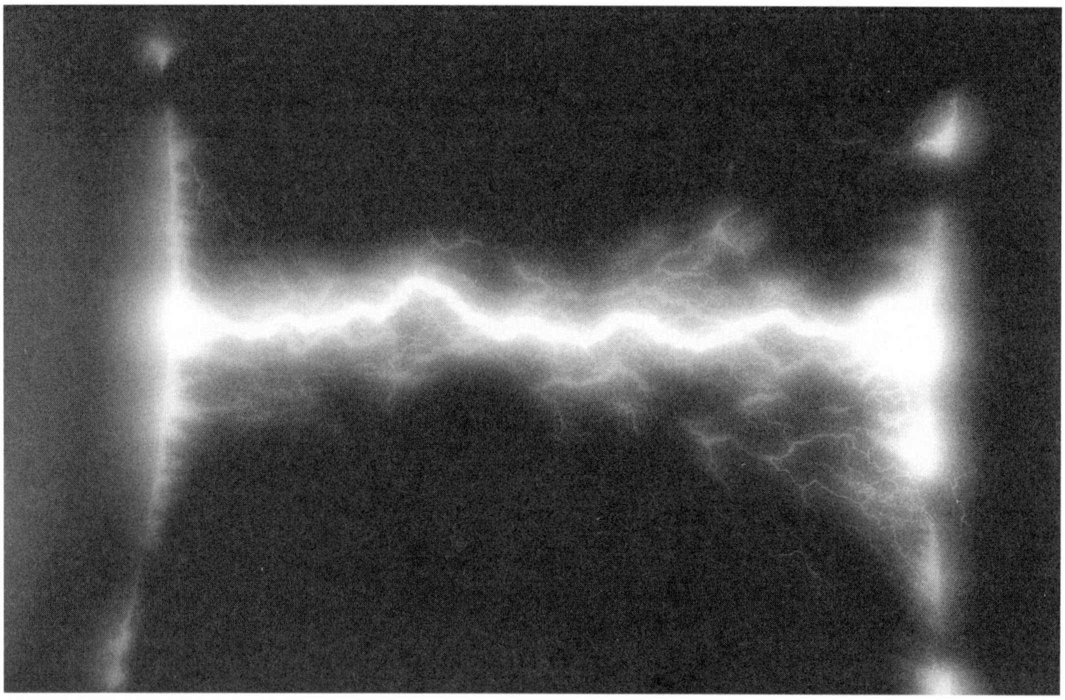

THE ELECTRIC DISCHARGE BETWEEN TWO METAL OBJECTS. *(© P. Jude/Photo Researchers. Reproduced by permission.)*

It is believed that the source of Earth's magnetism lies in a core of molten iron some 4,320 mi. (6,940 km) across, constituting half the planet's diameter. Within this core run powerful electric currents that create the geomagnetic field. Actually, the field seems to originate in the outer core, consisting of an iron-nickel alloy that is kept fluid owing to the exceedingly great temperatures. The materials of the outer core undergo convection, vertical circulation that results from variations in density brought about by differences in temperature.

This process of convection (imagine giant spirals moving vertically through the molten metal) creates the equivalent of a solenoid, described earlier. Even so, there had to be an original source for the magnetic field, and it is possible that it came from the Sun. In any case, the magnetic field could not continue to exist if the fluid of the outer core were not in constant convective motion. If this convection stopped, within about 10,000 years (which, in terms of Earth's life span, is like a few seconds to a human being), the geomagnetic field would decay and cease to operate. Likewise, if Earth's core ever cooled and solidified, Earth would become like the Moon, a body whose magnetic field has disappeared, leaving only the faintest traces of magnetism in its rocks.

CHANGES IN THE MAGNETIC FIELD

Though there is no reason to believe that anything so dramatic will happen, there are curious and perhaps troubling signs that Earth's magnetic field is changing. According to data recorded by the U.S. Geological Survey, which updates information on magnetic declination, the field is shifting—and weakening. Over the course of about a century, scientists have recorded data suggesting a reduction of about 6% in the strength of the magnetic field.

The behavior, in terms of both weakening and movement, appears to be similar to changes taking place in the magnetic field of the Sun. Indeed, as we have seen already, Earth's magnetism is heavily affected by the Sun, and it is possible that a period of strong solar-flare activity could shut down Earth's magnetic field. Even the present trend of weakening, if it were to continue for just 1,500 years, would wipe out the magnetic field. Some scientists believe that the planet is simply experiencing a fluctuation, however, and that the geomagnetic field will recover. Others

KEY TERMS

CONSERVATION: In physics and other sciences, "to conserve" something means "to result in no net loss of" that particular component. It is possible that within a given system the component may change form or position, but as long as the net value of the component remains the same, it has been conserved.

DIPOLE: A pair of equal and opposite electric charges, or an entire body having the characteristics of a dipole—for instance, a magnet with north and south poles.

ELECTROMAGNETIC ENERGY: A form of energy with electric and magnetic components that travels in waves.

ELECTROMAGNETIC FORCE: The total force on an electrically charged particle, which is a combination of forces due to electric and magnetic fields around the particle. Electromagnetic force reflects electromagnetic interaction, one of the four fundamental interactions in nature.

ELECTRON: A negatively charged particle in an atom, which spins around the nucleus.

ELEMENT: A substance made up of only one kind of atom. Unlike compounds, elements cannot be broken chemically into other substances.

FIELD: A region of space in which it is possible to define the physical properties of each point in the region at any given moment in time.

FUNDAMENTAL INTERACTION: The basic mode by which particles interact. There are four known fundamental interactions in nature: gravitational, electromagnetic, strong nuclear, and weak nuclear.

GEOMAGNETISM: A term referring to the magnetic properties of Earth as a whole rather than those possessed by a single object or place on Earth.

MAGNETIC DECLINATION: The angle between magnetic north and geographic north.

MAGNETOSPHERE: An area surrounding Earth, reaching far beyond the atmosphere, in which ionized particles (i.e., ones that have lost or gained electrons so as to acquire a net electric charge) are affected by Earth's magnetic field.

PALEOMAGNETISM: An area of historical geology devoted to studying the direction and intensity of magnetic fields in the past, as discerned from the residual magnetization of rocks.

POTENTIAL: Position in a field, such as a gravitational force field.

SOLAR WIND: A stream of particles continually emanating from the Sun and moving outward through the solar system.

SYSTEM: Any set of interactions that can be set apart mentally from the rest of the universe for the purposes of study, observation, and measurement.

maintain that the geomagnetic field is on its way to a reversal.

A reversal? Odd as it may sound, the direction of the geomagnetic field has reversed itself many, many times in the past. Furthermore, the planet has attempted unsuccessfully to reverse its geomagnetic field many more times—as recently as 30,000 to 40,000 years ago. These reversals are among the interests of paleomagnetism, the area of geology devoted to the direction and intensity of magnetic fields in the past, as discerned from the residual magnetization of rocks.

A compass works, of course, because the metal points toward Earth's magnetic north pole, which is close to its geographic north pole. Likewise, the magnetic materials in the rocks of Earth point north—or rather, they *would* point north if the direction of the magnetic field had not changed over time. Around the turn of the nineteenth century, geologists noticed that whereas some magnetic rocks pointed toward Earth's current North Pole, some were pointing in the opposite direction. This led to the realization that the magnetic field had reversed and to the development of paleomagnetism as a field of study. Studies in paleomagnetism, in turn, have provided confirmation of the powerful theory known as plate tectonics. (See Plate Tectonics for more on paleomagnetism and the shifting of plates beneath Earth's surface.)

WHERE TO LEARN MORE

Campbell, Wallace Hall. *Earth Magnetism: A Guided Tour Through Magnetic Fields.* New York: Harcourt Academic Press, 2001.

Dispezio, Michael A., and Catherine Leary. *Awesome Experiments in Electricity and Magnetism.* New York: Sterling, 1998.

Geomagnetism (Web site). <http://ub.nmh.ac.uk/>.

The Great Magnet, the Earth (Web site). <http://istp.gsfc.nasa.gov/earthmag/dmglist.htm>.

Hancock, Paul L., and Brian J. Skinner. *The Oxford Companion to the Earth.* New York: Oxford University Press, 2000.

Laboratory of Earth's Magnetism (Web site). <http://geo.phys.spbu.ru/Magnetism/>.

Lafferty, Peter. *Magnets to Generators.* Illus. Alex Pang and Galina Zolfanghari. New York: Gloucester Press, 1989.

Lefkowitz, R. J., and J. Carol Yake. *Forces in the Earth: A Book About Gravity and Magnetism.* New York: Parents' Magazine Press, 1976.

Smith, David G. *The Cambridge Encyclopedia of Earth Sciences.* New York: Cambridge University Press, 1981.

Stern, David P. *400 Years of "De magnete"* (Web site). <http://www-istp.gsfc.nasa.gov/earthmag/demagint.htm>.

CONVECTION

CONCEPT

Convection is the name for a means of heat transfer, as distinguished from conduction and radiation. It is also a term that describes processes affecting the atmosphere, waters, and solid earth. In the atmosphere, hot air rises on convection currents, circulating and creating clouds and winds. Likewise, convection in the hydrosphere circulates water, keeping the temperature gradients of the oceans stable. The term convection generally refers to the movement of fluids, meaning liquids and gases, but in the earth sciences, convection also can be used to describe processes that occur in the solid earth. This geologic convection, as it is known, drives the plate movement that is one of the key aspects of plate tectonics.

HOW IT WORKS

INTRODUCTION TO CONVECTION

Some concepts and phenomena cross disciplinary boundaries within the earth sciences, an example being the physical process of convection. It is of equal relevance to scientists working in the geologic, atmospheric, and hydrologic sciences, or the realms of study concerned with the geosphere, atmosphere, and hydrosphere, respectively. The only major component of the earth system not directly affected by convection is the biosphere, but given the high degree of interconnection between different subsystems, convection indirectly affects the biosphere in the air, waters, and solid earth.

Convection can be defined as vertical circulation that results from differences in density ultimately brought about by differences in temperature, and it involves the transfer of heat through the motion of hot fluid from one place to another. In the physical sciences, the term fluid refers to any substance that flows and therefore has no definite shape. This usually means liquids and gases, but in the earth sciences it can refer even to slow-flowing solids. Over the great expanses of time studied by earth scientists, the net flow of solids in certain circumstances (for example, ice in glaciers) can be substantial.

CONVECTION AND HEAT

As indicated in the preceding paragraph, convection is related closely to heat and temperature and indirectly related to another phenomenon, thermal energy. What people normally call *heat* is actually thermal energy, or kinetic energy (the energy associated with movement) produced by molecules in motion relative to one another.

Heat, in its scientific meaning, is internal thermal energy that flows from one body of matter to another or from a system at a higher temperature to a system at a lower temperature. Temperature thus can be defined as a measure of the average molecular kinetic energy of a system. Temperature also governs the direction of internal energy flow between two systems. Two systems at the same temperature are said to be in a state of thermal equilibrium; when this occurs, there is no exchange of heat, and therefore heat exists only in transfer between two systems.

There is no such thing as cold, only the absence of heat. If heat exists only in transit between systems, it follows that the direction of heat flow must *always* be from a system at a higher temperature to a system at a lower temperature. (This fact is embodied in the second law of thermodynamics, which is discussed, along with other topics mentioned here, in Energy and Earth.) Heat transfer occurs through three means: conduction, convection, and radiation.

CONDUCTION AND RADIATION. Conduction involves successive molecular collisions and the transfer of heat between two bodies in contact. It usually occurs in a solid. Convection requires the motion of fluid from one place to another, and, as we have noted, it can take place in a liquid, a gas, or a near solid that behaves like a slow-flowing fluid. Finally, radiation involves electromagnetic waves and requires no physical medium, such as water or air, for the transfer.

If you put one end of a metal rod in a fire and then touch the "cool" end a few minutes later, you will find that it is no longer cool. This is an example of heating by conduction, whereby kinetic energy is passed from molecule to molecule in the same way as a secret is passed from one person to another along a line of people standing shoulder to shoulder. Just as the original phrasing of the secret becomes garbled, some kinetic energy is inevitably lost in the series of transfers, which is why the end of the rod outside the fire is still much cooler than the one sitting in the flames.

As for radiation, it is distinguished from conduction and convection by virtue of the fact that it requires no medium for its transfer. This explains why space is cold yet the Sun's rays warm Earth: the rays are a form of electromagnetic energy, and they travel by means of radiation through space. Space, of course, is the virtual absence of a medium, but upon entering Earth's atmosphere, the heat from the electromagnetic rays is transferred to various media in the atmosphere, hydrosphere, geosphere, and biosphere. That heat then is transferred by means of convection and conduction.

HEAT TRANSFER THROUGH CONVECTION. Like conduction and unlike radiation, convection requires a medium. However, in conduction the heat is transferred from one molecule to another, whereas in convection the heated fluid itself is actually moving. As it does, it removes or displaces cold air in its path. The flow of heated fluid in this situation is called a convection current.

Convection is of two types: natural and forced. Heated air rising is an example of natural convection. Hot air has a lower density than that of the cooler air in the atmosphere above it and therefore is buoyant; as it rises, however, it loses energy and cools. This cooled air, now denser than the air around it, sinks again, creating a repeating cycle that generates wind.

Forced convection occurs when a pump or other mechanism moves the heated fluid. Examples of forced-convection apparatuses include some types of ovens and even refrigerators or air conditioners. As noted earlier, it is possible to transfer heat only from a high-temperature reservoir to a low-temperature one, and thus these cooling machines work by removing hot air. The refrigerator pulls heat from its compartment and expels it to the surrounding room, while an air conditioner pulls heat from a room or building and releases it to the outside.

Forced convection does not necessarily involve man-made machines: the human heart is a pump, and blood carries excess heat generated by the body to the skin. The heat passes through the skin by means of conduction, and at the surface of the skin it is removed from the body in a number of ways, primarily by the cooling evaporation of perspiration.

REAL-LIFE APPLICATIONS

CONVECTIVE CELLS

One important mechanism of convection, whether in the air, water, or even the solid earth, is the convective cell, sometimes known as the convection cell. The latter may be defined as the circular pattern created by the rising of warmed fluid and the sinking of cooled fluid. Convective cells may be only a few millimeters across, or they may be larger than Earth itself.

These cells can be observed on a number of scales. Inside a bowl of soup, heated fluid rises, and cooled fluid drops. These processes are usually hard to see unless the dish in question happens to be one such as Japanese miso soup. In this case, pieces of soybean paste, or miso, can be observed as they rise when heated and then drop down into the interior to be heated again.

On a vastly greater scale, convective cells are present in the Sun. These vast cells appear on the Sun's surface as a grainy pattern formed by the

A CUMULONIMBUS CLOUD—THUNDERHEAD—IS A DRAMATIC EXAMPLE OF A CONVECTION CELL. *(© Keith Kent/Photo Researchers. Reproduced by permission.)*

variations in temperature between the parts of the cell. The bright spots are the top of rising convection currents, while the dark areas are cooled gas on its way to the solar interior, where it will be heated and rise again.

A cumulonimbus cloud, or "thunderhead," is a particularly dramatic example of a convection cell. These are some of the most striking cloud formations one ever sees, and for this reason the director Akira Kurosawa used scenes of

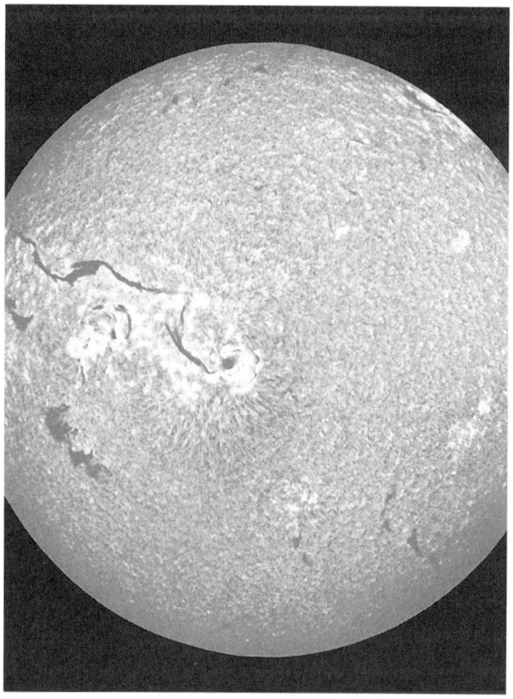

CONVECTIVE CELLS APPEAR ON THE SUN'S SURFACE AS A GRAINY PATTERN FORMED BY VARIATIONS IN TEMPERATURE. *(© Noao/Photo Researchers. Reproduced by permission.)*

rolling thunderheads to add an atmospheric quality (quite literally) to his 1985 epic *Ran.* In the course of just a few minutes, these vertical towers of cloud form as warmed, moist air rises, then cools and falls. The result is a cloud that seems to embody both power and restlessness, hence Kurosawa's use of cumulonimbus clouds in a scene that takes place on the eve of a battle.

A SEA BREEZE. Convective cells, along with convection currents, help explain why there is usually a breeze at the beach. At the seaside, of course, there is a land surface and a water surface, both exposed to the Sun's light. Under such exposure, the temperature of land rises more quickly than that of water. The reason is that water has an extraordinarily high specific heat capacity—that is, the amount of heat that must be added to or removed from a unit of mass for a given substance to change its temperature by 33.8°F (1°C). Thus a lake, stream, or ocean is always a good place to cool down on a hot summer day.

The land, then, tends to heat up more quickly, as does the air above it. This heated air rises in a convection current, but as it rises and thus overcomes the pull of gravity, it expends energy and therefore begins to cool. The cooled air then sinks. And so it goes, with the heated air rising and the cooling air sinking, forming a convective cell that continually circulates air, creating a breeze.

CONVECTIVE CELLS UNDER OUR FEET. Convective cells also can exist in the solid earth, where they cause the plates (movable segments) of the lithosphere—the upper layer of Earth's interior, including the crust and the brittle portion at the top of the mantle—to shift. They thus play a role in plate tectonics, one of the most important areas of study in the earth sciences. Plate tectonics explains a variety of phenomena, ranging from continental drift to earthquakes and volcanoes. (See Plate Tectonics for much more on this subject.)

Whereas the Sun's electromagnetic energy is the source of heat behind atmospheric convection, the energy that drives geologic convection is geothermal, rising up from Earth's core as a result of radioactive decay. (See Energy and Earth.) The convective cells form in the asthenosphere, a region of extremely high pressure at a depth of about 60–215 mi. (about 100–350 km), where rocks are deformed by enormous stresses.

In the asthenosphere, heated material rises in a convection current until it hits the bottom of the lithosphere (the upper layer of Earth's interior, comprising the crust and the top of the mantle), beyond which it cannot rise. Therefore it begins moving laterally or horizontally, and as it does so, it drags part of the lithosphere. At the same time, this heated material pushes away cooler, denser material in its path. The cooler material sinks lower into the mantle (the thick, dense layer of rock, approximately 1,429 mi. [2,300 km] thick, between Earth's crust and core) until it heats again and ultimately rises up, thus propagating the cycle.

SUBSIDENCE: FAIR WEATHER AND FOUL

As with convective cells, subsidence can occur in the atmosphere or geosphere. The term subsidence can refer either to the process of subsiding, on the part of air or solid earth, or, in the case of solid earth, to the resulting formation. It thus is defined variously as the downward movement of air, the sinking of ground, or a depression in the earth. In the present context we will discuss atmospheric subsidence, which is more closely related to convection. (For more about geologic

KEY TERMS

ASTHENOSPHERE: A region of extremely high pressure underlying the lithosphere, where rocks are deformed by enormous stresses. The asthenosphere lies at a depth of about 60–215 mi. (about 100–350 km).

ATMOSPHERE: In general, an atmosphere is a blanket of gases surrounding a planet. Unless otherwise identified, however, the term refers to the atmosphere of Earth, which consists of nitrogen (78%), oxygen (21%), argon (0.93%), and other substances that include water vapor, carbon dioxide, ozone, and noble gases such as neon, which together comprise 0.07%.

BIOSPHERE: A combination of all living things on Earth—plants, animals, birds, marine life, insects, viruses, single-cell organisms, and so on—as well as all formerly living things that have not yet decomposed.

CONDUCTION: The transfer of heat by successive molecular collisions. Conduction is the principal means of heat transfer in solids, particularly metals.

CONVECTION: Vertical circulation that results from differences in density ultimately brought about by differences in temperature. Convection involves the transfer of heat through the motion of hot fluid from one place to another and is of two types, natural and forced. (See *natural convection, forced convection.*)

CONVECTION CURRENT: The flow of material heated by means of convection.

CONVECTIVE CELL: The circular pattern created by the rising of warmed fluid and the sinking of cooled fluid. This is sometimes called a convection cell.

CORE: The center of Earth, an area constituting about 16% of the planet's volume and 32% of its mass. Made primarily of iron and another, lighter element (possibly sulfur), it is divided between a solid inner core with a radius of about 760 mi. (1,220 km) and a liquid outer core about 1,750 mi. (2,820 km) thick.

CRUST: The uppermost division of the solid earth, representing less than 1% of its volume and varying in depth from 3 to 37 mi. (5 to 60 km). Below the crust is the mantle.

FLUID: In the physical sciences, the term fluid refers to any substance that flows and therefore has no definite shape. Fluids can be both liquids and gases. In the earth sciences, occasionally substances that appear to be solid (for example, ice in glaciers) are, in fact, flowing slowly.

FORCED CONVECTION: Convection that results from the action of a pump or other mechanism (whether man-made or natural), directing heated fluid toward a particular destination.

GEOSPHERE: The upper part of Earth's continental crust, or that portion of the solid earth on which human beings live and which provides them with most of their food and natural resources.

HEAT: Internal thermal energy that flows from one body of matter to another.

HYDROSPHERE: The entirety of Earth's water, excluding water vapor in the atmosphere but including all oceans, lakes, streams, groundwater, snow, and ice.

KINETIC ENERGY: The energy that an object possesses by virtue of its motion.

KEY TERMS CONTINUED

LITHOSPHERE: The upper layer of Earth's interior, including the crust and the brittle portion at the top of the mantle.

MANTLE: The dense layer of rock, approximately 1,429 mi. (2,300 km) thick, between Earth's crust and its core.

NATURAL CONVECTION: Convection that results from the buoyancy of heated fluid, which causes it to rise.

PLATE TECTONICS: The name both of a theory and of a specialization of tectonics. As an area of study, plate tectonics deals with the large features of the lithosphere and the forces that shape them. As a theory, it explains the processes that have shaped Earth in terms of plates and their movement.

PLATES: Large, movable segments of the lithosphere.

RADIATION: The transfer of energy by means of electromagnetic waves, which require no physical medium (for example, water or air) for the transfer. Earth receives the Sun's energy via the electromagnetic spectrum by means of radiation.

SUBSIDENCE: A term that refers either to the process of subsiding, on the part of air or solid Earth, or, in the case of solid Earth, to the resulting formation. Subsidence thus is defined variously as the downward movement of air, the sinking of ground, or a depression in Earth's crust.

SYSTEM: Any set of interactions that can be set apart mentally from the rest of the universe for the purposes of study, observation, and measurement.

TECTONICS: The study of tectonism, including its causes and effects, most notably mountain building.

TECTONISM: The deformation of the lithosphere.

TEMPERATURE: The direction of internal energy flow between two systems when heat is being transferred. Temperature measures the average molecular kinetic energy in transit between those systems.

THERMAL ENERGY: Heat energy, a form of kinetic energy produced by the motion of atomic or molecular particles in relation to one another. The greater the relative motion of these particles, the greater the thermal energy.

subsidence, see the entries Geomorphology and Mass Wasting.)

In the atmosphere, subsidence results from a disturbance in the normal upward flow of convection currents. These currents may act to set up a convective cell, as we have seen, resulting in the flow of breeze. The water vapor in the air may condense as it cools, changing state to a liquid and forming clouds. Convection can create an area of low pressure, accompanied by converging winds, near Earth's surface, a phenomenon known as a cyclone. On the other hand, if subsidence occurs, it results in the creation of a high-pressure area known as an anticyclone.

Air parcels continue to rise in convective currents until the density of their upper portion is equal to that of the surrounding atmosphere, at which point the column of air stabilizes. On the other hand, subsidence may occur if air at an altitude of several thousand feet becomes denser than the surrounding air without necessarily being cooler or moister. In fact, this air is unusually dry, and it may be warm or cold. Its density then makes it sink, and, as it does, it compresses the air around it. The result is high pressure at the surface and diverging winds just above the surface.

The form of atmospheric subsidence described here produces pleasant results, explaining why high-pressure systems usually are associated with fair weather. On the other hand, if the subsiding air settles onto a cooler lay of air, it creates what is known as a subsidence inversion, and the results are much less beneficial. In this situation a warm air layer becomes trapped between cooler layers above and below it, at a height of several hundred or even several thousand feet. This means that air pollution is trapped as well, creating a potential health hazard. Subsidence inversions occur most often in the far north during the winter and in the eastern United States during the late summer.

WHEN A NON-FLUID ACTS LIKE A FLUID

Up to this point we have spoken primarily of convection in the atmosphere and the geosphere, but it is of importance also in the oceans. The miso soup example given earlier illustrates the movement of fluid, and hence of particles, that can occur when a convective cell is set up in a liquid.

Likewise, in the ocean convection—driven both by heat from the surface and, to a greater extent, by geothermal energy at the bottom—keeps the waters in constant circulation. Oceanic convection results in the transfer of heat throughout the depths and keeps the ocean stably stratified. In other words, the strata, or layers, corresponding to various temperature levels are kept stable and do not wildly fluctuate.

Ocean waters fit the most common, everyday definition of fluid, but as noted at the beginning of this essay, a fluid can be anything that flows—including a gas or, in special circumstances, a solid. Solid rocks or solid ice, in the form of glaciers, can be made to flow if the materials are deformed sufficiently. This occurs, for instance, when the weight of a glacier deforms ice at the bottom, thus causing the glacier as a whole to move. Likewise, geothermal energy can heat rock and cause it to flow, setting into motion the convective process of plate tectonics, described earlier, which literally moves the earth.

WHERE TO LEARN MORE

Educator's Guide to Convection (Web site). <http://www.solarviews.com/eng/edu/convect.htm>.

Erickson, Jon. *Plate Tectonics: Unraveling the Mysteries of the Earth.* New York: Facts on File, 1992.

Hess, Harry. "History of Ocean Basins" (Web site). <http://www-geology.ucdavis.edu/~GEL102/hess/jesse.htm>.

Jones, Helen. *Open-Ocean Deep Convection: A Field Guide* (Web site). <http://puddle.mit.edu/~helen/oodc.html>.

Ocean Oasis Teacher's Guide Activity 4 (Web site). <http://www.oceanoasis.org/teachersguide/activity4.html>.

Santrey, Laurence, and Lloyd Birmingham. *Heat.* Mahwah, NJ: Troll Associates, 1985.

Scorer, R. S., and Arjen Verkaik. *Spacious Skies.* Newton Abbot, England: David and Charles, 1989.

Sigurdsson, Haraldur. *Melting the Earth: The History of Ideas on Volcanic Eruptions.* New York: Oxford University Press, 1999.

Skinner, Brian J., Stephen C. Porter, and Daniel B. Botkin. *The Blue Planet: An Introduction to Earth System Science.* 5th ed. New York: John Wiley and Sons, 1999.

Smith, David G. *The Cambridge Encyclopedia of Earth Sciences.* New York: Cambridge University Press, 1981.

ENERGY AND EARTH

CONCEPT

Earth is a vast flow-through system for the input and output of energy. The overwhelming majority of the input to Earth's energy budget comes from the Sun in the form of solar radiation, with geothermal and tidal energy rounding out the picture. Each form of energy is converted into heat and re-radiated to space, but the radiation that leaves Earth travels in longer wavelengths than that which entered the planet. This is in accordance with the second law of thermodynamics, which shows that energy output will always be smaller than energy input and that energy which flows through a system will return to the environment in a degraded form. Yet what the Earth system does in processing that energy, particularly the portion that passes through the biosphere, is amazing. Some biological matter decays and, over the course of several hundred million years, produces fossil fuels that have given Earth a slight energy surplus. Human use of fossil fuels is rapidly depleting those sources, however, while posing new environmental problems, and this has encouraged the search for alternative forms of energy. Many of those forms, most notably geothermal energy, come from Earth itself.

HOW IT WORKS

ENERGY, WORK, AND POWER

Physicists define energy as the ability of an object (and in some cases a nonobject, such as a magnetic force field) to accomplish work. "Work" in this context does not have the same meaning as it does in everyday life; along with the closely related concept of power, it is defined very specifically in a scientific context.

Work is the exertion of force over a given distance, and therefore it is measured in units of force multiplied by units of length. In the English system used by most Americans, a pound is the unit of force, and the foot-pound (ft-lb) would be the unit of work. However, scientists worldwide use SI, or the International System, which applies metric units. The metric unit of force is the newton (N), and the metric unit of work is the joule (J), equal to 1 newton-meter ($N \times m$).

Power is the rate at which work is accomplished over time and therefore is measured in units of work divided by units of time. The metric unit of power is the watt (W), named after James Watt (1736–1819), the Scottish inventor who developed the first fully viable steam engine and thus helped inaugurate the Industrial Revolution. A watt is equal to 1 J per second, but this is such a small unit that kilowatts, or units of 1,000 W, are more frequently used. Discussing the vast energy budget of Earth itself, however, requires use of an even larger unit: the terawatt (TW), equal to 10^{12} (one trillion) W.

Ironically, Watt himself—like most people in the British Isles and America—lived in a world that used the British system, in which the unit of power is the foot-pound per second. The latter unit, too, is very small, so for measuring the power of his steam engine, Watt suggested a unit based on something quite familiar to the people of his time: the power of a horse. One horsepower (hp) is equal to 550 ft-lb per second.

In the present context, we will rely as much as possible on SI units, especially because the watt is widely used in America. Horsepower typically is applied in the United States only for measuring the power of a mechanical device, such as an automobile or even a garbage disposal. For measuring electrical power, particularly in larger quantities, the SI kilowatt (kW) is used. When an electric utility performs a meter reading on a family's power usage, for instance, it measures that usage in terms of electrical "work" performed for the family and thus bills them by the kilowatt-hour (kWh).

VARIETIES OF ENERGY

In thc most fundamental sense, there are only three kinds of energy: kinetic, potential, and mass, or rest, energy. These types are, respectively, the energy an object possesses by virtue of its motion, its position (or its ability to perform work), and its mass. The first two are understood in relation to each other: for example, a ball held over the side of a building has a certain gravitational potential energy, but once it is dropped, it begins to lose potential energy and gain kinetic energy. The faster it moves, the greater the kinetic energy; but as it covers more distance, the less its potential energy is. (See Earth Systems for more about the kinetic-potential energy system.)

As Earth moves around the Sun, the gravitational interaction between the two bodies is not unlike that of the baseball and the ground in the illustration just given. Earth makes an elliptical, or oval-shaped, path in its orbit, meaning that the distance between it and the Sun is not uniform. At its furthest distance, Earth's potential energy is maximized, but as it comes closer to the Sun in its orbital path, its kinetic energy increases, with a corresponding decrease of potential energy.

MASS AND ENERGY. Mass, or rest, energy is identified in the famous formula $E = mc^2$, derived by Albert Einstein (1879–1955). In simple terms Einstein's formula means that every object possesses an amount of energy equal to its mass multiplied by the speed of light squared. Given that light travels at 186,000 mi. (299,339 km) per second, this is an enormous figure, even for a small object. A mere baseball, which weighs about 0.333 lb. (0.15 kg), possesses enough energy to yield about 3.75 billion kWh worth of power—enough to run all the lights and appliances in a typical American home for more than 156,000 years!

To release this energy in significant quantities, it would be necessary to accelerate the baseball to a speed close to that of light. Even in ordinary experience, however, very small amounts of mass are converted to energy. For instance, when a fire burns, the mass of the ashes combined with that of the particles and gases sent into the atmosphere is smaller (by an almost imperceptible fraction) than the mass of the original wood. The "lost" mass is converted to energy. These mass-energy conversions occur on a much larger level in nuclear reactions, such as the nuclear fusion of hydrogen atoms to form helium in the solar core (see Sun, Moon, and Earth).

MANIFESTATIONS OF ENERGY. In discussing kinetic and potential energy, the example of dropping a baseball from a height illustrates these two types of energy in a gravitational field—that is, the gravitational field of Earth. Yet the concept of potential and kinetic energy translates to a situation involving an electromagnetic field as well. For instance, the positive or negative attraction between two electromagnetically charged particles is analogous to the force of gravity, and a system of two or more charges possesses a certain amount of kinetic and potential electromagnetic energy.

Electromagnetic energy, which is the form in which solar power reaches Earth, is a type of energy that (as its name suggests) combines both electrical and magnetic energy. Another important form of energy in the Earth system is thermal, or heat, energy, which is the kinetic energy of molecules, since heat is simply the result of molecular motion.

Other types of energy include sound, chemical, and nuclear energy. Sound waves, which require a physical medium such as air in which to travel, are simply pressure fluctuations that carry varying levels of energy, depending on the frequency (pitch) and amplitude (volume) of the waves. Chemical energy makes possible the forming and releasing of molecular bonds, and, for this reason, chemical reactions often are accompanied by the production of heat. Whereas chemical energy concerns the bonds *between* atoms, nuclear energy relates to the bonds *within* them. Nuclear fission reactions involve the splitting of an atomic nucleus, while nuclear fusion is the joining of nuclei.

HEAT AND THERMODYNAMICS

Thermodynamics is the study of the relationships between heat, work, and energy. As with work, energy, and power, heat and the related concept of temperature are terms that have special definitions in the physical sciences. Heat itself is not to be confused with thermal energy, which, as noted earlier, is the kinetic energy that arises from the motion of particles at the atomic or molecular level. The greater the movement of these particles relative to one another, the greater the thermal energy.

Heat is internal thermal energy that flows from one body of matter to another. It is not the same as the energy contained in a system—that is, the internal thermal energy of the system. Rather than being "energy-in-residence," heat is "energy-in-transit." This may seem a little confusing, but all it means is that heat, in its scientific sense, exists only when internal energy is being transferred. As for temperature, it is not (as is commonly believed) a measure of heat and cold. Instead, temperature indicates the direction of internal energy flow between bodies and the average molecular kinetic energy in transit between those bodies.

In any case, temperature could not be a measure of heat and cold, as though these two were equal and opposing entities, because, scientifically speaking, there is no such thing as cold—only an absence of heat. When we place an ice cube in a cup of coffee, we say that the ice is there to "cool the coffee down," but, in fact, the opposite is happening: the coffee is warming up the ice cube, and in the process of doing so, it loses heat. This may seem like a difference of semantics, but it is not. It is a physical law that the flow of heat is always from a high-temperature reservoir to a low-temperature reservoir. Even air conditioners and refrigerators work by pulling heat out of a compartment rather than by bringing cold *in*.

MEASURING TEMPERATURE AND HEAT. Temperature, of course, can be measured by either the Fahrenheit or the Centigrade scales familiar in everyday life. Scientists, however, prefer the Kelvin (K) scale, established by William Thomson, Lord Kelvin (1824–1907). Drawing on the discovery made by the French physicist and chemist J. A. C. Charles (1746–1823) that gas at 0°C (32°F) regularly contracts by about 1/273 of its volume for every Celsius degree drop in temperature, Thomson derived the value of absolute zero (the temperature at which molecular motion virtually ceases) as –273.15°C (–459.67°F). The Kelvin and Celsius scales are thus directly related: Celsius temperatures can be converted to Kelvin units (for which neither the word nor the symbol for "degree" is used) by adding 273.15.

Heat, on the other hand, is measured not by degrees but by the same units as work. Energy is the ability to perform work, so heat or work units are also units of energy. Aside from the joule, heat often is measured by the kilocalorie, or the amount of heat that must be added to or removed from 1 kg of water to change its temperature by 1°C. As its name suggests, a kilocalorie is 1,000 calories, a calorie being the amount of heat required to change the temperature in 1 g of water by 1°C. The dietary calorie with which most people are familiar, however, is the same as a kilocalorie.

THE LAWS OF THERMODYNAMICS

The three laws of thermodynamics collectively show that it is impossible for a system to produce more energy than was put into it or even to produce an equal amount of usable energy. In other words, a perfectly efficient system—whether an engine or the entire Earth—is an impossibility. Derived during a period of about 60 years beginning in the 1840s, the laws of thermodynamics helped scientists and engineers improve the machines that powered the height of the Industrial Age. They also revealed the impossibility of constructing anything approaching a perpetual-motion machine, that great quest of dreamers over the ages, which the laws of thermodynamics proved to be an impossible dream.

THE FIRST LAW OF THERMODYNAMICS. The first law of thermodynamics is related to the conservation of energy, a physical law whereby the total energy in a system remains the same, though transformations of energy from one form to another take place. Such transformations occur frequently in the Earth system, as when a plant receives electromagnetic energy from the Sun and converts it to chemical potential energy in the form of carbohydrates. Likewise humans, by building dams, can harness the gravitational potential energy of flowing water and convert it into electromagnetic energy.

The conservation of energy, in effect, states that "the glass is half full," meaning that we can obtain as much energy from a system as we put into it. While saying the same thing, the first law of thermodynamics in effect states that "the glass is half empty,": that is, that we can obtain *no more* energy from a system than we put into it. According to this law, because the amount of energy in a system remains constant, it is impossible to perform work that results in an energy output greater than the energy input.

The term *law* in the physical sciences is no empty expression; it means that a principle has been shown to be the case always and may be expected to remain the case in all situations. It is possible, of course, for a physical law to be overturned in light of later evidence. It is not likely, however, that any set of circumstances in the universe will ever disprove the core truth behind this law, which may be stated colloquially as "You can't get something for nothing."

THE SECOND LAW OF THERMODYNAMICS. In a 1959 lecture published as *The Two Cultures and the Scientific Revolution,* the British writer and scientist C. P. Snow (1905–1980) compared transfers of heat and energy to a game. The laws of thermodynamics are its rules, and, as Snow stated, the first law proves that it is impossible to win at this game, while the second law shows the impossibility of breaking even.

The second law of thermodynamics is more complicated than the first and is stated in a number of ways, though they are all interrelated. According to this law, spontaneous or unaided transfers of energy are irreversible and impossible without an increase of entropy in the universe. Entropy is the tendency of natural systems toward breakdown, specifically, the tendency for the energy in a system to be dissipated or degraded. (Later in this essay, we discuss examples of energy that has been degraded—for instance, wood that has been burned to produce ashes.) The second law means that spontaneous processes are irreversible and that it is impossible, without the additional input of energy, to transfer heat from a colder to a hotter body or to convert heat into an equal amount of work.

Whereas the first law showed engineers the impossibility of building a perpetual-motion machine, the second law proves that it is impossible to build even a perfectly efficient engine. Of all the energy we put into our automobiles in the form of gasoline (which is chemical potential energy in the form of hydrocarbons derived from the fossilized remains of dinosaurs in the earth), only about 30% of it goes into moving the car forward. The rest is dissipated in a number of ways, chiefly through heat and sound. Entropy, as it turns out, is inescapable and as inevitable as death. In fact, death itself is a result of entropy in the systems of all living things.

THE THIRD LAW OF THERMODYNAMICS. The third law of thermodynamics is not as well known as the other two and has little bearing on the discussion at hand, but it deserves at least brief mention. According to the third law, at the temperature of absolute zero entropy also approaches zero, which might sound like a way out of the restrictions imposed by the first two laws. All it really means is that absolute zero is impossible to reach—or, as Snow put it, the third law shows that "you can't escape the game."

In 1824 the French physicist and engineer Nicolas Léonard Sadi Carnot (1796–1832) had shown that an engine could achieve maximum efficiency if its lowest operating temperature were absolute zero. His work influenced that of Kelvin, who established the absolute-temperature scale mentioned earlier. Additionally, Carnot's discoveries informed the development of the third law. Whereas the second law is not derived from the first (though it is certainly consistent with it), the third law relies on the second: if it is impossible to build a perfectly efficient engine, as the second law states, it is likewise impossible to reach absolute zero.

This, of course, has not stopped scientists from attempting to achieve absolute zero, most properly defined as the temperature at which the motion of the average atom or molecule is zero. Helium atoms, in fact, never fully cease their motion, even at temperatures very close to 0K—and scientists have come very, very close. In 1993 physicists at the Helsinki University of Technology Low Temperature Laboratory in Finland used a nuclear demagnetization device to achieve a temperature of 2.8×10^{-10}K, or 0.00000000028K. This amounts to a difference of only 28 parts in 100 billion between that temperature and absolute zero.

EARTH'S ENERGY INPUT

Just as households have financial budgets, a system such as Earth (see Earth Systems) has an energy budget. The latter may be defined as the total amount of energy available to a system or, more specifically, the difference between the

THE POHUTU GEYSER AT THE ROTORUA WHAKAREWAREWA THERMAL AREA IN NORTH ISLAND, NEW ZEALAND. *(© A. N. T./Photo Researchers. Reproduced by permission.)*

energy flowing into the system and the energy lost by it. Having reviewed the laws of thermodynamics, one might suspect that a great deal of energy is dissipated in the operation of the Earth system, and, of course, that is absolutely correct.

Earth receives 174,000 TW of energy, or 174 quadrillion J per second. Human civilization, by contrast, uses only 10 TW, or about 0.00574% as much as Earth's total energy. Of that total, there are three principal sources, though one of these sources—the Sun—dwarfs the other two in importance. The breakdown of Earth's energy input, along with the percentage of the total that each portion constitutes, is as follows:

- Solar radiation: 99.985%
- Geothermal energy: 0.013%
- Tidal energy: 0.002%

SOLAR RADIATION. The Sun radiates electromagnetic energy, which, as mentioned previously, is a form of energy that produces both electric and magnetic fields. Electromagnetic energy travels in waves, and since waves follow regular patterns, it is possible to know that those waves with shorter wavelengths have a higher frequency and thus higher energy levels.

The electromagnetic spectrum contains a variety of waves, each with progressively higher energy levels, including long-wave and short-wave radio; microwaves (used for TV transmissions); infrared, visible, and ultraviolet light; x rays; and ultra-high-energy gamma rays. Visible light is only a very small portion of this spectrum, and each color has its own narrow wavelength range.

Red has the least energy and purple or violet the most; hence, the names *infrared* for light with less energy than red, and *ultraviolet* for light with more energy than violet. The order of these wavelengths of light, along with the colors between, is remembered easily by the mnemonic device ROY G. BIV (standing for red, orange, yellow, green, blue, indigo, and violet). (Actually, there are only six major color ranges, and the name really should be *ROY G. BV.*)

Although it covers the entire electromagnetic spectrum, energy from the Sun is referred to by earth scientists as short-wavelength radiation. This is because the solar energy that enters the Earth system is shorter in wavelength (and thus higher in energy level) than the energy returned to space by Earth. (We discuss the degradation of energy in the Earth system later in this essay.) Without solar radiation, the life-giving processes of the hydrosphere, biosphere, and atmosphere would be impossible. An example is photosynthesis, the biological conversion of electromagnetic energy to chemical energy in plants. (See the later discussion of photosynthesis and the food web.)

GEOTHERMAL ENERGY. A much smaller, but still significant component of Earth's energy budget is geothermal energy, the planet's internal heat energy. Much of this heat comes from Earth's core, which has temperatures as high as 8,132°F (4,500°C) and from whence thermal energy circulates throughout the planet's interior. Also significant is the heat from radioactive elements, most notably uranium and thorium, near Earth's surface.

This thermal energy heats groundwater, and thus the principal visible sources of geothermal energy include geysers, hot springs, and fumaroles—fissures, created by volcanoes, from which hot gases pour. There are several types of geothermal energy reserves, among them dry and wet steam fields. The first of these reserves occurs when groundwater boils normally, whereas in the second type of reserve, groundwater is superheated, or prevented from boiling even though its temperature is above the boiling point. In both cases the waters have a much higher concentration of gases and minerals than ordinary groundwater. Another type of reserve can be found under the ocean floors, where natural gas mixes with very hot water.

Geothermal energy powers seismic activity as well as volcanic eruptions and mountain building, which together have played a significant role in shaping Earth as we know it today. Aside from its obvious impact on the planet's terrain, geothermal energy has had an indirect influence on the transfer of vital elements from beneath Earth's surface, a benefit of volcanic activity. (See the later discussion of the human use of geothermal energy in this essay.)

TIDAL ENERGY. Whereas the principal form of energy in Earth's budget comes from the Sun and the secondary source from Earth itself, the third type of energy input to the Earth system comes chiefly from the Moon. The Sun also affects tides, but because of its close proximity to Earth, the Moon has more influence over the movements of our planet's ocean waters.

Though the Moon is much smaller than Earth, it is larger, in proportion to the planet it orbits, than any satellite in the solar system (with the possible exception of Pluto's moon Charon). Given this fact, combined with its close proximity to Earth, it is understandable that the Moon would exert a powerful pull on its host planet. The gravitational pull of the Moon (and, to a lesser extent, that of the Sun) on Earth causes the oceans to bulge outward on the side of Earth closest to the Moon. At the same time, the oceans on the opposite side of the planet bulge in response. (See Sun, Moon, and Earth for more about tides and the bulges that result from the Moon's gravitational pull.)

This gravitational pull creates a torque that acts as a brake on Earth's rotation, producing a relatively small amount of energy that is dissipated primarily within the waters of the ocean. Incidentally, the lunar-solar tidal torque, by increasing the amount of time it takes Earth to turn on its axis, is causing a gradual increase in the length of a day. Today, of course, there are 365.25 days in a year, but about 650 million years ago there were 400 days. In other words, Earth made 400 revolutions on its axis in the period of time it took it to revolve around the Sun. The change is a result of the fact that Earth's rotation is being slowed by 24 microseconds a year.

ENERGY PROFIT AND LOSS

Focusing now on solar radiation, since it is by far the greatest source of energy input to the Earth system, let us consider Earth's energy budget in terms of "profit and loss." In other words, how much useful energy output is denied to Earth owing to the laws of thermodynamics and other factors?

First of all, a good 30% of the Sun's energy input is reflected back into space unchanged, without entering Earth's atmosphere. This results from our planet's *albedo,* or reflective power. Albedo is the proportion of incoming radiation that is reflected by a body (e.g., a planet) or surface such as a cloud: the higher the proportion of incoming radiation that a planet deflects, the higher its albedo. The latter is influenced by such factors as solar angle, amount of cloud cover, particles in the atmosphere, and the character of the planetary surface.

Another 25% of solar radiation is absorbed by the atmosphere, while about 45% is absorbed at the planetary surface by living and nonliving materials. Thus, electromagnetic energy from the Sun enters the atmosphere, biosphere, and hydrosphere, where it is converted to other forms of energy, primarily thermal. Some of this thermal energy, for instance, causes the evaporation of water, which cycles through the atmosphere and then reenters the hydrosphere as precipitation. In other cases, absorbed radiation drives atmospheric and hydrologic distribution mechanisms, including winds, water currents, and waves. A very small, but extremely significant portion of incoming solar radiation goes into plant photosynthesis, discussed later.

ENERGY DEGRADATION. The energy that enters the Earth system—not only solar radiation but also geothermal and tidal energy—ultimately leaves the system. As shown by the second law of thermodynamics, however, the energy that departs the Earth system will be in a degraded form compared with the energy that entered it.

In a steam engine, water in the form of steam goes to work to power gears or levers. In the process, it cools, and the resulting cool water constitutes a degraded form of energy. Likewise, the ashes that remain after a fire or the fumes that are a by-product of an internal combustion engine's operation contain degraded forms of energy compared with that in the original wood or gasoline, respectively. In the same way, Earth receives short-wavelength energy from the Sun, but the energy it radiates to space is in a long-wavelength form.

All physical bodies with a temperature greater than absolute zero emit electromagnetic energy in accordance with their surface temperatures, and the hotter the body, the shorter the wavelength of the radiation. The sunlight that enters Earth's atmosphere is divided between the visible portion of the spectrum and the high-frequency side of the infrared portion. (Note that the Sun emits energy across the entire electromagnetic spectrum, but only a small part gets through Earth's atmospheric covering.) Earth, with an average surface temperature of 59°F (15°C), is much cooler than the Sun, with its average surface temperature of about 10,000°F (5,538°C). The radiation Earth sends back into space, then, is on the low-frequency, long-wavelength side of the infrared spectrum.

NO ENERGY LOSS. In accordance with the conservation of energy, no energy truly has been lost. In a relatively simple system, such as an automobile, chemical potential energy enters the vehicle in the form of gasoline and, after being processed by the engine, exits in a variety of forms. There is the kinetic energy that turns the wheels; the thermal energy of the engine and exhaust; electromagnetic energy from the battery for the headlights, dashboard lights, radio, air conditioning, and so on; and the sound energy dissipated in the noise of the car. If one could add all those energy components together, one would find that all the energy that entered the system left the system. Note, however, that once again the process is irreversible: one can use gasoline to power a battery and hence a car radio, but the radio or the battery cannot generate gasoline.

The Earth system is much more complex, of course, but the same principle applies: about 174,000 TW of energy enter the system, and about 174,000 TW are used in the form of heat. Of the portion that enters the atmosphere, some goes into warming the planet, some into moving the air and water, and a very small part into the all-important biological processes described later, but *all* of it is used. It should be noted that Earth has a very small energy surplus, owing to the accumulation of undecomposed biomass that ultimately becomes fossil fuels; however, the amount of energy involved is minor compared with the larger energy budget. (For more information about biomass and fossil fuels, see the following discussion.)

THE GREENHOUSE EFFECT. Not only is there no net loss of energy in the universe, but Earth itself also possesses a remarkably efficient system for making use of the energy it receives. This is the greenhouse effect, whereby the planet essentially recycles the degraded energy it is in the process or returning to space.

Water vapor and carbon dioxide, as well as methane, nitrous oxide, and ozone, all absorb long-wavelength radiated energy as the latter makes its way up through the atmosphere. When heated, these radiatively active gases (as they are called) re-radiate the energy, now at even longer wavelengths. In so doing, they slow the planet's rate of cooling. Without the greenhouse effect, surface temperatures would be about 50°F (10°C) cooler than they are—that is, around 17.6°F (–8°C). This, of course, is well below the freezing temperature of water and much too cold for Earth's biological processes. Thus, the greenhouse effect literally preserves life on the planet.

It may be surprising to learn that the greenhouse effect, a term often heard in the context of dire environmental warnings, is a natural and healthful part of the Earth system. Like many useful things, the greenhouse effect is not necessarily better in larger doses, however, and that is the problem. It is believed that human activities have resulted in an increase of radiatively active gases, which could lead to global warming. Among such gases are the chlorofluorocarbons (CFCs) used in aerosol cans, now banned by international treaty. Also of concern are the high levels of carbon dioxide produced by the burning of fossil fuels.

A GERMINATING GARDEN PEA SEEKS SUNLIGHT FOR PHOTOSYNTHESIS. *(© J. Burgess/Photo Researchers. Reproduced by permission.)*

REAL-LIFE APPLICATIONS

ENERGY AND THE LIVING WORLD

One of the most interesting components of the entire energy picture is the relationship of energy input, energy output, and the biosphere. Particularly important is the use of solar radiation for the purposes of photosynthesis, an activity that constitutes a small but vital sector of Earth's energy budget.

Though it accounts for only about 1% of the energy received from solar radiation, photosynthesis is essential to the sustenance of life. In photosynthesis, plants receive solar radiation, carbon dioxide, and water, which chemically react to produce carbohydrates and oxygen. Animals depend not only on the oxygen but also on carbohydrates, such as sugars, and thus photosynthesis makes possible the development of the food chains that constitute the world of living creatures.

FOOD WEBS. Even though *food chain* is a well-known expression, modern scientists prefer the term *food web,* because it more accurately portrays the complicated relationships involved. The term food chain, as it is commonly used, implies a strict hierarchical structure in which (to use another popular phrase) "the big fish eat the little fish." In fact, the relationship between participants is not quite so neatly defined.

It is, however, possible to describe a food web in terms of a few key players, or types of players. There are primary producers, which are green plants, and primary consumers—herbivores, or plant-eating animals. Secondary consumers (and those at further levels, such as tertiary and quaternary) are either carnivores—that is, meat-eating animals who eat the herbivores—or omnivores, which are both plant and animal eaters. For example, omnivorous humans eat herbivorous cows, who have eaten plants.

Carnivores and omnivores, however, are not really at the top of the food chain, so to speak; rather, in line with the non-hierarchical idea of the food web, they represent points in an interlocking set of relationships. Materials from plants, herbivores, carnivores, and omnivores ultimately will all be consumed by the lowliest of creatures, that is, detritivores, or decomposers, including bacteria and worms. At each stage energy is transferred, and, as always, the second law of thermodynamics comes into play. The energy is degraded in transfer; specifically, the further away an organism is from the original plant source, the less a given quantity of fuel contributes to its growth. It is interesting to note the economy of energy use at the detritivore stage.

Worms and other decomposers are exceedingly efficient feeders, working the same food particles over and over and extracting more stored energy each time. They then produce waste products that increase the vitamin content of the soil, thus enabling the growth of plants and the continuation of the biological cycle. Also, detritivores may contribute directly (and, from the larger energy-cycle perspective, less efficiently) to providing fuel for carnivores, as when a bird eats a worm.

BIOENERGY

The food web is the mechanism whereby energy is cycled through the biosphere, as fuel in the form of food. Plant matter and other biological forms also can serve as direct sources of heat, providing fuel that can be either burned without processing or converted into gas or alcohol. In such situations the plant matter is described as bioenergy, or energy derived from biological sources that are used as fuel.

Materials that are burned or processed to produce bioenergy are called biomass. Examples of the latter include wood logs burned on a fire, probably the oldest type of fuel known to humankind and still one of the principal forms of heating available in many developing countries. In fact, some of the least technologically advanced and most technologically advanced nations are alike in their use of another variety of biomass: waste.

Dried animal dung provides heating material in many a third world village where electricity is unknown, while at the other end of the technological spectrum, some Western municipalities extract burnable biomass from processed sewage. Since Western countries have at their disposal plenty of other energy sources, they typically burn off the methane gas and dried waste material from treated sewage simply as a means of removing it, rather than as an energy source. Those materials could provide usable energy, however.

The products of sewage treatment, of course, are the result of processing, which involves the conversion of biomass to either gases (for example, methane, as mentioned previously) or alcohol. Farmers in rural China, for instance, often place agricultural waste and sewage in small closed pits, from whence they extract burnable methane gas. In the United States, Brazil, and other countries with abundant farmland, some of the agricultural output is directed not toward production of food but toward production of fuel in the form of ethanol, a type of alcohol made from sugarcane, corn, or sorghum grain. Ethanol can be mixed with gasoline to run an automobile or burned alone in specially modified engines. In either case, the fuel burns much more cleanly, producing less poisonous carbon monoxide than ordinary gasoline.

FOSSIL FUELS

Biomass is potentially renewable, whereas another source of bioenergy is not. This is the bioenergy from fossil fuels—buried deposits of petroleum, coal, peat, natural gas, and other organic compounds. (Actually, fossil fuels typically are considered separate from other forms of bioenergy, because they are nonrenewable.)

Fossil fuels are the product of plants and animals that lived millions of years ago, died, decomposed, and became part of Earth's interior. Over the ages, as more sediment weighed down on these organic deposits, the weight applied more pressure, generated more heat, and led to the concentration of this decomposed material, which became a valuable source of energy.

USING UP RESOURCES. Given the vast spans of time that have passed, as well as the almost inconceivable numbers of plants and animals, the supplies of fossil-fuel energy stored under Earth's surface are enormous. But they are *not* infinite, and, as noted earlier, they are nonrenewable; once they are gone, they might as well be gone forever, because it would take hundreds of millions of years to produce additional deposits. Furthermore, humans are using up these energy sources at an alarming rate.

Up until about 1750, Earth's fossil-fuel deposits were largely intact, but after that time industrialized societies began to extract coal for heating, transportation, and industrial uses. Today it remains a leading means of generating electrical power. During the twentieth century, petroleum increasingly was directed toward transportation, and this led to the extraction of still more of Earth's fossil-fuel deposits. If civilization continues to consume these products at its current rate, reserves will be exhausted long before the end of the twenty-first century.

SEARCHING FOR ENERGY SOURCES. There are several reasons not to panic over the loss of fossil-fuel resources. First, known reserves are just that—they are the ones that energy companies, and their geologists, know about at the present time. As long as plentiful resources are available, corporations and governments do not feel a pressing need to search for more, but as those resources are used up, such searches become economically necessary. For a time at least, these searches will continue to yield new (though increasingly harder to reach) deposits.

Also, fears about Earth running out of fuel are built on the assumption that no one will develop other sources of energy, a few of which are discussed at the conclusion of this essay. Though most of the known alternative energy sources face their own challenges, it is thoroughly conceivable that scientists of the future will develop means of completely replacing fossil fuels as the source of electrical power, fuel for transport, and other uses. After all, there was once a time (during the second half of the nineteenth century) when Americans became increasingly anxious over dwindling reserves of a vital energy resource essential for powering the nation's lamps: whale oil.

ENVIRONMENTAL CONCERNS

While the loss of energy reserves may not be an immediate cause for panic, the effects of fossil-fuel burning on the environment have alarmed scientists and environmentalists alike. At the most basic level, there is the environmental impact posed by the extraction of fossil fuels. Coal mines, for instance, have been used up and now sit abandoned, their land worthless for any purpose. There is also the environmental danger created by hazards in misuse or transport of fossil fuels—for example, the vast oil spill caused by the grounding of the Exxon *Valdez* near Alaska's Prince William Sound in 1989.

By far the greatest environmental concern raised by fossil fuels, however, is the effect they produce in the atmosphere when burned. For instance, one of the impurities in coal is sulfur, and when coal burns, the sulfur reacts with oxygen in the combustion process to create sulfur dioxide and sulfur trioxide. Sulfur trioxide reacts with water in the air, creating sulfuric acid and thus acid rain, which can endanger plant and animal life as well as corrode metals and building materials.

THE GREENHOUSE EFFECT REVISITED. Even greater fears center on the release into the atmosphere of carbon monoxide and carbon dioxide, both of which are by-products in the burning of fossil fuels. The first is a poison, whereas the second is a vital part of the life cycle, yet carbon dioxide, in fact, may pose the greater threat.

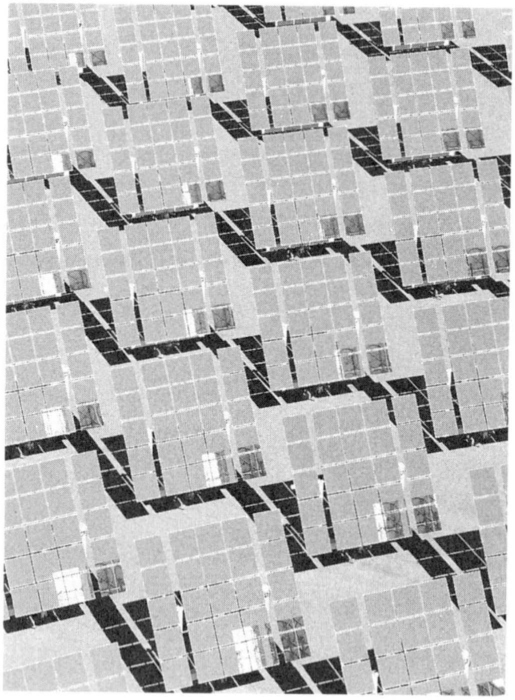

AN ARRAY OF SOLAR REFLECTORS NEAR ALBUQUERQUE, NEW MEXICO, HARNESSES SOLAR POWER. *(© J. Mead/Photo Researchers. Reproduced by permission.)*

Earlier we discussed the greenhouse effect, which, when it occurs naturally, is important to the preservation of life on the planet. The large number of internal-combustion engines in operation on Earth today produce an inordinate amount of carbon dioxide, which in turn provides the atmosphere with more radiatively active gas than it needs. According to many environmentalists, the result is, or will be, global warming. If it takes place over a long period of time, global warming could bring about serious hazards—in particular, the melting of the polar ice caps. It should be noted, however, that not all scientists are in agreement that global warming is occurring or that humans are the principal culprits inducing these environmental changes.

ALTERNATIVE ENERGY SOURCES

Whatever the merits of the various sides in the debate on global warming or the exhaustion of fuel resources, one need hardly be an environmentalist to agree that the world cannot forever rely only on existing fossil fuels and the technology that uses them. Today, even as scientists in laboratories around the world work to develop viable alternative means of powering industry, utilities, and transportation, many alternative energy sources are already in use.

Some of these energy forms are very old, for instance, burnable biomass, water power, or wind power, all of which date back to ancient times. Others are extremely new in concept, most notably, nuclear energy, which was developed in the twentieth century. Still others are new, high-tech versions of old-fashioned energy sources, the best example being solar power. In fact, all three major contributors to Earth's energy budget—solar radiation, tidal energy, and geothermal energy (discussed further later in this essay)—have been harnessed by human societies.

HIGH-TECH ENERGY SOLUTIONS. Several of the more ambitious ideas for energy creation, while they may capture people's imaginations, have significant drawbacks. There is the proposal, for instance, to extract hydrogen gas from water by means of electrolysis, potentially providing an extremely clean, virtually limitless source of energy. Hydrogen gas, however, is highly flammable, as the 1937 explosion of the airship *Hindenburg* illustrated, and, in any case, the fuel to provide the electricity necessary for electrolysis would have to come from somewhere, presumably, the burning of fossil fuels.

Nuclear energy, of course, has frightened many people in the wake of such well-known disasters as those that occurred at Three Mile Island, Pennsylvania, in 1979 and Chernobyl, in the former Soviet Union, in 1986. In fact, those two situations illustrate more about governments than they do about technology itself. No one died at Three Mile Island, and with an open society and media access to the site, the public outcry became so great that the plant was closed. By contrast, the Soviets' outmoded technology helped bring about the Chernobyl disaster, and the communist dictatorship's practice of censorship and suppression led to a massive cover-up that greatly increased the death toll. As a result, thousands died at Chernobyl, and thousands more died as the result of the indirect effects of nuclear pollution in the environment.

There is no question, however, that nuclear energy does pose an enormous potential environmental threat from its waste products. Spent fuel rods, if simply buried, eventually leak radioactive waste into the water table and could kill or harm vast populations. This all relates to nuclear *fission,* the only type of peaceful nuclear

KEY TERMS

ABSOLUTE ZERO: The temperature at which all molecular motion virtually ceases.

ALBEDO: The reflective power of a surface or body or, more specifically, the proportion of incoming radiation that the surface or body reflects.

ATMOSPHERE: In general, an atmosphere is a blanket of gases surrounding a planet. Unless otherwise identified, however, the term refers to the atmosphere of Earth, which consists of nitrogen (78%), oxygen (21%), argon (0.93%), and other substances that include water vapor, carbon dioxide, ozone, and noble gases such as neon, which together comprise 0.07%.

BIOENERGY: Energy derived from biological sources that are used directly as fuel (as opposed to food, which becomes fuel).

BIOMASS: Materials that are burned or processed to produce bioenergy.

BIOSPHERE: A combination of all living things on Earth—plants, mammals, birds, reptiles, amphibians, aquatic life, insects, viruses, single-cell organisms, and so on—as well as all formerly living things that have not yet decomposed.

CALORIE: A measure of heat or energy in the SI, or metric, system, equal to the heat that must be added to or removed from 1 g of water to change its temperature by 1°C. The dietary calorie with which most people are familiar is the same as the kilocalorie, or 1,000 calories.

CONSERVATION OF ENERGY: A law of physics that holds that within a system isolated from all outside factors, the total amount of energy remains the same, though transformations of energy from one form to another take place. The first law of thermodynamics is the same as the conservation of energy.

ELECTROMAGNETIC ENERGY: A form of energy with electric and magnetic components, which travels in waves.

ELECTROMAGNETIC SPECTRUM: The complete range of electromagnetic waves on a continuous distribution from a very low range of frequencies and energy levels, with a correspondingly long wavelength, to a very high range of frequencies and energy levels, with a correspondingly short wavelength. Included on the electromagnetic spectrum are long-wave and short-wave radio; microwaves; infrared, visible, and ultraviolet light; x rays; and gamma rays.

ENERGY: The ability of an object (or in some cases a nonobject, such as a magnetic force field) to accomplish work.

ENERGY BUDGET: The total amount of energy available to a system or, more specifically, the difference between the energy flowing into the system and the energy lost by it.

ENTROPY: The tendency of natural systems toward breakdown and, specifically, the tendency for the energy in a system to be dissipated. Entropy is related closely to the second law of thermodynamics.

ENVIRONMENT: In discussing systems, the term environment refers to the surroundings—everything external to and separate from the system.

KEY TERMS CONTINUED

FIRST LAW OF THERMODYNAMICS: A law of physics stating that the amount of energy in a system remains constant, and therefore it is impossible to perform work that results in an energy output greater than the energy input. This is the same as the conservation of energy.

FOSSIL FUELS: Nonrenewable forms of bioenergy, including petroleum, coal, peat, natural gas, and other organic compounds usable as fuel.

FREQUENCY: The number of waves, measured in Hertz, passing through a given point during the interval of one second. The higher the frequency, the shorter the wavelength.

GEOTHERMAL ENERGY: Heat, or thermal, energy from Earth's interior.

GREENHOUSE EFFECT: Warming of the lower atmosphere and surface of Earth. This occurs because of the absorption of long-wavelength radiation from the planet's surface by certain radiatively active gases, such as carbon dioxide and water vapor, in the atmosphere. These gases are heated and ultimately re-radiate energy at an even longer wavelength to space.

HEAT: Internal thermal energy that flows from one body of matter to another.

HERTZ: A unit for measuring frequency, equal to one cycle per second. High frequencies are expressed in terms of kilohertz (kHz; 10^3 or 1,000 cycles per second), megahertz (MHz; 10^6 or one million cycles per second), and gigahertz (GHz; 10^9 or one billion cycles per second.)

HYDROSPHERE: The entirety of Earth's water, excluding water vapor in the atmosphere but including all oceans, lakes, streams, groundwater, snow, and ice.

JOULE: The SI measure of work. One joule (J) is equal to the work required to accelerate 1 kg of mass by 1 m per second squared (1 m/s^2) over a distance of 1 m. Owing to the small size of the joule, however, it often is replaced by the kilowatt-hour, equal to 3.6 million (3.6×10^6) J.

KELVIN SCALE: Established by William Thompson, Lord Kelvin (1824–1907), the Kelvin scale measures temperature in relation to absolute zero, or 0K. (Note that units in the Kelvin system, known as kelvins, do not include the word or symbol for "degree.") The Kelvin scale, which is the system usually favored by scientists, is related directly to the Celsius scale; hence, Celsius temperatures can be converted to kelvins by adding 273.15.

KINETIC ENERGY: The energy that an object possesses by virtue of its motion.

LAW: A scientific principle that is shown always to be the case and for which no exceptions are deemed possible.

MASS ENERGY: The energy an object possesses by virtue of its mass. Sometimes called *rest energy.*

NUCLEAR FISSION: A nuclear reaction that involves the splitting of an atomic nucleus.

NUCLEAR FUSION: A nuclear reaction that involves the joining of atomic nuclei.

NUCLEUS: The center of an atom, a region where protons and neutrons are located and around which electrons spin.

KEY TERMS CONTINUED

PHOTOSYNTHESIS: The biological conversion of light energy (that is, electromagnetic energy) to chemical energy in plants.

POTENTIAL ENERGY: The energy that an object possesses by virtue of its position or its ability to perform work.

POWER: The rate at which work is accomplished over time, a figure rendered mathematically as work divided by time. The SI unit of power is the watt, while the British unit is the foot-pound per second.

RADIATION: The transfer of energy by means of electromagnetic waves, which require no physical medium (for example, water or air) for the transfer. Earth receives the Sun's energy via the electromagnetic spectrum by means of radiation.

RADIOACTIVITY: A term describing a phenomenon whereby certain materials are subject to a form of decay brought about by the emission of high-energy particles or radiation. Forms of particles or energy include alpha particles (positively charged helium nuclei), beta particles (either electrons or subatomic particles called positrons), or gamma rays, which occupy the highest energy level in the electromagnetic spectrum.

SECOND LAW OF THERMODYNAMICS: A law of physics stating that spontaneous or unaided transfers of energy are irreversible and impossible without an increase of entropy in the universe. It is therefore impossible, without the additional input of energy, to transfer heat from a colder to a hotter body or to convert heat into an equal amount of work.

SI: An abbreviation of the French term *Système International d'Unités,* or International System of Units. Based on the metric system, SI is the system of measurement units in use by scientists worldwide.

SYSTEM: Any set of interactions that can be set apart mentally from the rest of the universe for the purposes of study, observation, and measurement.

TEMPERATURE: The direction of internal energy flow between two systems when heat is being transferred. Temperature measures the average molecular kinetic energy in transit between those systems.

TERAWATT: See *watt.*

THERMAL ENERGY: Heat energy, a form of kinetic energy produced by the motion of atomic or molecular particles in relation to one another. The greater the relative motion of these particles, the greater the thermal energy.

WATT: The metric unit of power, equal to 1 J per second. Because this is such a small unit, scientists and engineers typically speak in terms of kilowatts, or units of 1,000 W. Very large figures, such as those relating to Earth's energy budget, usually are given in terawatts, or 10^{12} (one trillion) W.

WAVELENGTH: The distance between a crest and the adjacent crest or the trough and an adjacent trough of a wave. Wavelength is related inversely to frequency, meaning that the shorter the wavelength, the higher the frequency.

WORK: The exertion of force over a given distance. In the metric, or SI, system, work is measured by the joule (J) and in the British system by the foot-pound (ft-lb).

power in use today. In building the hydrogen bomb in the mid-twentieth century, physicists and chemists used the much greater power of nuclear *fusion,* or the bonding of atomic nuclei. If nuclear fusion could be produced in controlled reactions, for peaceful use, it would provide safe, cheap, limitless power to the planet.

GEOTHERMAL ENERGY. Unless or until nuclear fusion or some more advanced alternative energy source is developed, societies will continue to rely on fossil fuels for the bulk of their energy needs. At the same time, alternative sources will continue to supply energy in certain situations. An excellent example of such a source is geothermal energy, harnessed by peoples in areas as widespread as New Zealand and Iceland centuries ago.

Long before the first Europeans arrived in New Zealand, the native Maori people used geothermal energy from geysers to cook food. Modern applications of geothermal energy began with the creation of the first geothermal well, by workers who accidentally drilled into one in Hungary in 1867. Today Hungary is the world's leading producer and consumer of geothermal energy, followed closely by Italy and Iceland. The word *fumarole,* used earlier in this essay, is Italian in origin, and it is said that the fumaroles near the town of Lardarello, used for the production of electricity since 1904, once inspired Dante Alighieri's (1265–1321) vision of the Inferno, as captured in his celebrated work by that name. As for Iceland, more than 99% of the buildings in the capital city of Reykjavik use geothermal energy for heat.

Heat is not the primary human application for geothermal energy. As noted earlier, in the context of the first law of thermodynamics, energy can be converted from one form to another. Thus, geothermal energy is applied for the creation of electromagnetic energy: steam or heated water from the ground runs turbines, which produce electricity.

Geothermal energy has enormous advantages, including the fact that its raw materials (heated water and steam) are free and relatively inexpensive to extract. It is also inexpensive environmentally, causing virtually no air pollution. Will geothermal energy ever significantly compete with fossil fuels as a significant source of energy for humans? It is conceivable, but at present a number of barriers exist. Geothermal resources exist only in very specific parts of the world, and the extraction of the raw materials may release noxious gases, such as hydrogen sulfide (the same compound that gives intestinal gas its smell). Also, ironically, there are environmental concerns, not because of true damage but because geothermal mines often pose a threat of sight pollution in the midst of otherwise gorgeous natural settings.

WHERE TO LEARN MORE

Altman, Linda Jacobs. *Letting Off Steam: The Story of Geothermal Energy.* Minneapolis, MN: Carolrhoda Books, 1989.

Brown, Warren. *Alternative Sources of Energy.* New York: Chelsea House, 1994.

Cox, Reg, and Neil Morris. *The Natural World.* Philadelphia: Chelsea House, 2000.

Earth's Energy Budget (Web site). <http://radar.metr.ou.edu/OK1/meteorology/EnergyBudget.html>.

Earth's Energy Budget, or Can You Spare a Sun? (Web site). <http://www.strategies.org/LESSON8.html>.

Fowler, Allan. *Energy from the Sun.* New York: Children's Press, 1997.

Gutnik, Martin J., and Natalie Browne-Gutnik. *The Energy Question: Thinking About Tomorrow.* Hillside, NJ: Enslow Publishers, 1993.

Hancock, Paul L., and Brian J. Skinner. *The Oxford Companion to the Earth.* New York: Oxford University Press, 2000.

Knapp, Brian J. *Earth Science: Discovering the Secrets of the Earth.* Illus. David Woodroffe and Julian Baker. Danbury, CT: Grolier Educational, 2000.

Skinner, Brian J., Stephen C. Porter, and Daniel B. Botkin. *The Blue Planet: An Introduction to Earth System Science.* 2nd ed. New York: John Wiley and Sons, 1999.

EARTH'S INTERIOR

EARTH'S INTERIOR

CONCEPT

For the most part, this book is concerned with geologic, geophysical, and geochemical processes that take place on or near Earth's surface. Even the essay Plate Tectonics, which takes up one of the central ideas in modern earth sciences, discusses only the lithosphere and crust but not the depths of the mantle or the core. Yet there are several good reasons to study Earth's interior, even if it is not immediately apparent why this should be the case. At first glance it would seem that activities in Earth's interior could hardly be removed further from day-to-day experience. By contrast, even the Moon seems more related to daily life. At least it is something we can see and a place to which humans have traveled; on the other hand, no human has ever seen the interior of our planet, nor is anyone likely to do so. What could Earth's interior possibly have to do with everyday life? The answer may be a bit surprising. As it turns out, many factors that sustain life itself are the result of phenomena that take place far below our feet.

HOW IT WORKS

The Core: Gravity and Density

In the essay on Planetary Science, there is a discussion of an age-old question: "Why is the earth round?" or rather "Why is Earth a sphere?" The answer, explained in more detail within the context of that essay, is that gravitational force dictates a spherical shape. As far as we know, there is no such thing as a planet or sun in the shape of a cube, because for every large object, the gravitational pull from the interior forces it to assume a more or less uniform shape. Since there is no shape more perfectly uniform than that of a sphere, this is the typical form of bodies that possess large mass.

In fact, the greater the mass, the greater the tendency toward roundness. Although they are less dense than Earth, Jupiter and Saturn are certainly more massive, and therefore they are more perfectly round. Mars and the Moon, on the other hand, are less so. Earth is not perfectly round, owing to the fact that its mass bulges at the equator because it is moving; if it were still, it would be quite round indeed, a result of the great mass at its core.

A MASSIVE CORE. As we shall see, nearly a third of Earth's mass is at its core, even though the core accounts for only about a fifth of its total volume. In other words, Earth's core is exceptionally dense, and this has several implications. First of all, the planet has a powerful gravitational pull, which not only serves to keep people and other objects rooted on the surface of the solid earth but also holds our atmosphere in place.

The gravitational attraction between any two objects is related directly to mass and inversely to the distance between them. Everything in the universe exerts some degree of gravitational pull on everything else, but unless at least one of the objects is of significant mass, the total gravitational force is negligible. The reason for this—as determined by the English mathematician and physicist Sir Isaac Newton (1642–1727)—is that gravitational force between two objects is the product of their mass divided by the distance between them and multiplied by

SIR ISAAC NEWTON *(Library of Congress.)*

an extremely small quantity known as the *gravitational constant.*

In the case of Earth, there is an extremely large amount of mass at the interior. Moreover, that mass is at a relatively short distance from objects on the planet's surface—or, to put it another way, Earth has a relatively small radius. Hence its powerful gravitational pull—one of many ways that the interior of Earth affects the overall conditions of the planet.

DENSITY OF TERRESTRIAL AND JOVIAN PLANETS. Saying that a large amount of mass is concentrated in a small area on Earth is another way of saying that the planet's interior is extremely dense. As it turns out, Earth is, in fact, the densest planet in the solar system; indeed the only other planets that come close are Mercury and Venus.

Mercury, Venus, Earth, and Mars together are designated as the terrestrial planets: bodies that are small, rocky, and dense; have relatively small amounts of gaseous elements; and are composed primarily of metals and silicates. (See the essay Minerals for more on metals as well as the extremely abundant silicates.) By contrast, the Jovian planets—Jupiter, Saturn, Uranus, and Neptune—are large, low in density, and composed primarily of gases. (Scientists know little about Pluto, which was discovered in the early twentieth century. It has a density higher than any Jovian planet, but there is little basis for classifying it as a terrestrial planet.)

Saturn, which is the least dense among the planets, has a mass only about 100 times as great as that of Earth, while its volume is almost 800 times greater. Thus its density is only about 12% of Earth's. And whereas Jovian planets, such as Saturn, are mostly gaseous and solid only in their small, dense cores, Earth is extremely solid. For a Jovian planet, there is little distinction between the "atmosphere" and the surface of the planet itself, whereas anyone who has ever jumped from a great height on Earth can attest to the sharp difference between thin air and solid ground.

Beneath that solid ground is a planetary interior composed of iron, nickel, and traces of other elements. The vast mass of the interior not only gives Earth a strong gravitational pull but also, in combination with the comparatively high speed of the planet's rotation, causes Earth to have a powerful magnetic field. Furthermore, Earth is distinguished even from most terrestrial planets (among which the Moon sometimes is counted) owing to the high degree of tectonic activity beneath its surface.

PLATE TECTONICS AND THE INTERIOR

Of all the terrestrial planets, Earth is the only one on which the processes of plate tectonics take place. Tectonism is the deformation of the lithosphere, the brittle area of Earth's interior that includes the crust and upper mantle. (We take a closer look at these regions later in this essay.) The lithosphere is characterized by large, movable segments called plates, and plate tectonics is the name both of a theory and of a specialization of tectonics, or the study of tectonism.

As a realm of study, plate tectonics deals with the large features of the lithosphere and the forces that shape them. As a theory, it explains the processes that have shaped Earth in terms of plates and their movement. This theory, discussed in detail within the Plate Tectonics essay, brings together aspects of seismic (earthquake) and volcanic activity, the structures of Earth's crust, and other phenomena to provide a unifying model of Earth's evolution. It is one of the dominant concepts in the modern earth sciences.

THE FOUR PLANETS OF THE INNER SOLAR SYSTEM, WITH THE SUN. MERCURY, VENUS, EARTH, AND MARS ARE DESIGNATED THE TERRESTRIAL PLANETS—SMALL, ROCKY, DENSE, AND COMPOSED OF METALS AND SILICATES WITH FEW GASEOUS ELEMENTS. *(© Photo Researchers. Reproduced by permission.)*

THE IMPORTANCE OF TECTONIC ACTIVITY. As discussed in Plate Tectonics, there is a difference in thickness between continental and oceanic plates on Earth. By contrast, the other terrestrial planets have crusts of fairly uniform thickness, suggesting that they have experienced little in the way of tectonic activity. Several other factors indicate that Earth is by far the most prone to tectonic activity.

Earth's core is enormous, larger than the entire planet Mercury. This means that there is a large area of high pressure and high heat driving tectonic processes, as we discuss later in this essay. In addition, Earth has a relatively thin lithosphere, meaning that the effects of heat below the lithosphere are manifested dramatically above it in the form of shifting plates and the results of such shifts—for instance, mountain building.

REAL-LIFE APPLICATIONS

"DIGGING TO CHINA"

As children, many people growing up in the West heard something along these lines: "If you could dig a hole straight through the earth, you would end up in China." This might be more or less literally true, since eastern China is on the opposite side of the planet from eastern North America. (Southeast Asia, however, is farther away, because it is more exactly opposite the eastern seaboard.) Even with the most sophisticated equipment imaginable, however, it is unlikely that anyone will put a hole straight through Earth.

The idea of "digging to China" may have raised a new question in many a child's mind. Suppose a person were to dig a hole through the Earth and jump down into it. What would happen? Gravity would carry the person to the center of the earth, but after that, would he or she just go on flying past the gravitational center of the planet? It is a good question, and the likelihood is that the powerful gravitational force at the center of the earth would hold the person there. Again, however, the likelihood of ever conducting such an experiment—for instance, with a steel ball that emitted a radio signal—is slim.

HOW DEEP? The reason for this slim likelihood can be illustrated by visiting some of the world's deepest mines. There is, for instance, the Homestake Gold Mine in South Dakota, one of the deepest mines in the United States, which extends to about 8,500 ft. (2,591 m) below the surface. This is about 1.6 mi. (2.6 km), almost six times as deep as the height of the world's tallest building, the Petronas Towers in Malaysia.

Impressive as the Homestake is, it is almost insignificant when compared with the Western Deeps Gold Mine, near Carletonville, South Africa, which reaches down about 13,000 ft. (3,962 m)! In a mine such as the Western Deeps, or even the Homestake, temperatures can reach 140°F (60°C), which makes working in such an environment extremely hazardous. Mines are air-conditioned to make them bearable, but even so, there are other dangers associated with the great depth. For instance, the pressure caused by the rocks lying above the mine may become so great that rocks in the wall shatter spontaneously.

It is no wonder, then, that workers in extremely deep gold mines and diamond mines are well paid or that their insurance premiums are very expensive. Yet even the Western Deeps is not the deepest spot where humans have drilled holes on Earth. Scientists in Sweden and Russia have overseen the drilling of deep holes purely for research purposes, while in Louisiana and Oklahoma, a few such holes have been drilled in the process of exploring for petroleum. The deepest of these holes are at Andarko Basin, Oklahoma, and the Kola Peninsula, Russia, where artificial holes extend to a staggering 7.5 mi. (12 km). This is more than three times as far down as the Western Deeps, and it is hard to imagine how any human could survive at such depths.

HOW DO WE KNOW? Even these deepest excavations represent only 0.2% of the distance from Earth's surface to its core, which is about 3,950 mi. (6,370 km) below our feet. Given that fact, one might wonder exactly how it is that earth scientists—particularly geophysicists—claim to know so much about what lies beneath the crust. In fact, they have a number of fascinating tools and methods at their disposal.

Among these tools are such rocks as kimberlite and ophiolite, which originate deep in the crust and mantle but move upward to the surface. In addition, meteorites that have landed on Earth are believed to be similar to the rocks at the mantle and core, since the planet was originally a cloud of gas around which solid materials began to form as a result of bombardment from outer space (see the essays Sun, Moon, and Earth and Planetary Science). Most important of all are seismic, or earthquake, waves, whose speed, motion, and direction tell us a great deal about the materials through which they have passed and the distances over which they have traveled.

AN IMAGINARY JOURNEY

Having established just how far humans would have to go to penetrate even just below Earth's crust with existing technology, let us now pretend that such obstacles have been overcome. In this imaginary situation, through a miracle of science let us say that there really is a hole straight through the earth and an elevator that passes through it.

This, of course, raises still more complications, aside from the gravitational problem mentioned earlier. Among other things, our elevator would have to be made of a heat-resistant material, given the temperatures we are likely to encounter in our descent. It may have sounded hot in the gold mines of South Africa and South Dakota, but that will seem cool by the time we reach Earth's core, which is as hot as the Sun's surface.

STARTING OUT. For now, however, we will throw all those logistical problems out the window and begin our journey to the center of the earth. In so doing, we will pass through three major regions—crust, mantle, and core—as well as several subsidiary realms within. By far the smallest of these is the crust, which is also the only part about which we know anything from direct experience.

Very quickly we find ourselves passing through the A, B, and C horizons of soil discussed in the Soil essay, and soon we are passing through bedrock into the main part of the crust. Bedrock might be only 5-10 ft. deep (1.5–3 m), or it might be half a mile deep (0.8 km) or perhaps even deeper. Although this is a long way for a person to dig, we still have barely scratched the surface.

As noted earlier, there is a difference between continental crust and oceanic crust. We will ignore the details here, except to say that the continental crust is thicker but the oceanic crust is denser. Thus, the continents are at a higher elevation than the oceans around them. Depending on whether the crust is oceanic or continental, we have between 3 mi. and 40 mi. (5–70 km) to travel before we begin to pass out of the crust and into the mantle.

THE LITHOSPHERE, SEISMOLOGY, AND REMOTE SENSING. The transition from crust to mantle is an abrupt one, marked by the boundary zone known as the Mohorovicic discontinuity. Sometimes called the M-discontinuity or, more commonly, the Moho, it was the discovery of the Croatian geologist Andrija Mohorovicic (1857–1936). On October 8, 1909, while studying seismic waves from an earthquake in southeastern Europe, Mohorovicic noticed that the speed of the waves increased dramatically at a depth of about 30 mi. (50 km).

Since waves travel faster through denser materials, Mohorovicic reasoned that there must be an abrupt transition from the rocky material in the Earth's crust to denser rocks below. His discovery is an excellent example of remote sensing (see Remote Sensing), whereby earth scientists are able to study places and phenomena that are impossible to observe directly.

After the Moho, which is only about 0.1-1.9 mi. (0.2–3 km) thick, we enter the mantle—or, more specifically, the lithospheric mantle. This subregion may extend to depths between 30 mi. and 60 mi. (50–100 km) and is much more dense than the crust. Like the crust, it is brittle, solid, and relatively cool compared with the regions below; hence, the crust and lithospheric mantle are lumped together as the lithosphere.

THE ASTHENOSPHERE AND ITS IMPACT

At the base of the lithosphere, we pass through another transition zone, known as the Gutenberg low-velocity zone (named after the German-born American seismologist Beno Gutenberg [1889–1960]), where the speed of seismic waves again increases dramatically. After that, we enter a layer of much softer material, known as the asthenosphere. The material in the asthenosphere is soft not because it is weak—on the contrary, it is made of rock—but because it is under extraordinarily high pressure.

What happens in the asthenosphere plays a powerful role in life on the surface. The plates of the lithosphere float, as it were, atop the molten rock of the asthenosphere, which forces these plates against one another as though they were ice cubes floating in a bowl of water in constant motion. This motion is the phenomenon of plate tectonics, which, as we have discussed, quite literally shapes the world we know.

VOLCANISM AND THE ATMOSPHERE. Plate tectonics is responsible not only for such phenomena as the creation of mountains but also, by influencing the development of volcanoes, indirectly for Earth's atmosphere. In the first few billion years of the planet's existence, the action of volcanoes brought water vapor, carbon dioxide, nitrogen, sulfur, and sulfur compounds from the planet's interior to its surface. This was critical to the formation of the air we breathe today. Additionally, volcanic activity plays a significant role in the carbon cycle, whereby that vital element is circulated through various earth systems (see Biogeochemical Cycles and Carbon Cycle).

Earth and Venus stand alone among terrestrial planets as the only two still prone to volcanic activity. (By contrast, Mercury and the Moon have long been dead volcanically, and volcanism on Mars seems to have ended at some point during the past billion years.) This is significant, because even though all the planets possess more or less the same chemical elements, volcanoes are critical to distributing those elements.

In addition, volcanic activity, as well as the heat from Earth's interior that drives it and other tectonic phenomena, is an important influence on the separation of chemical compounds. When Earth formed some 4.5 billion years ago, heavier compounds—among them, those containing iron—sank toward the planet's core. At the same time, lighter ones began to rise into the atmosphere. Among these compounds was oxygen, which is clearly essential to the life of humans and other animals. This separation of compounds continues on Earth, owing to the large amount of heat that emanates from the interior.

GEOTHERMAL ENERGY. One would hardly guess that our atmosphere—or the circulation of carbon, a key component in all lifeforms—could be the indirect product of activity that takes place at least 60 mi.(100 km) below our feet. Nor is this the only illustration of the impact that Earth's interior exerts on our world. The interior of Earth is also responsible for the action that produces geothermal energy, discussed in detail within Energy and Earth.

Geothermal energy provides heating and electricity for several countries and is responsible for the dramatic effect of such phenomena as "Old Faithful" at Yellowstone Park in Wyoming. It is also the source behind the soothing natural springs found in such well-known resorts as Warm Springs, Georgia (a favorite getaway for President Franklin D. Roosevelt, who died there in 1945), and Hot Springs, Arkansas, the hometown of another president, Bill Clinton.

Mesosphere to Inner Core: Geomagnetism and Gravity

After we pass through the base of the asthenosphere, we are still only 155 mi. (250 km) deep. Now we are in the mesosphere, which extends to a depth of 1,800 mi. (2,900 km) and includes several other discontinuities, or thresholds of change. We will not discuss the details of these discontinuities here, except to note that they indicate changes in geochemical composition: for example, at 400 mi. (650 km) there appears to be a marked increase in the ratio of iron to magnesium.

The Gutenberg discontinuity, or the core-mantle boundary (CMB), marks our entrance to the core. By now it has become very, very hot. Whereas the lithospheric mantle is about 1,600°F (870°C), the bottom of the lithosphere is about 4,000–6,700°F (2,200–3,700°C). By the time we get to the inner core, we may be confronted with temperatures as high as 13,000°F (7,200°C)—which, if this is true, would make the center of Earth about 50% hotter than the surface of the Sun!

EARTH'S FIERY HEART. At such temperatures, one would expect the rock of the outer core to be entirely molten, and indeed it is. In fact, it is this difference in phase or state of matter that marks the change from mantle (which is partially solid) to the liquid outer core. The region between the mantle and the outer core is one of undulating boundaries due to convection (see Convection), which may be the driving force behind the plate tectonic activity that occurs at a much higher level. In addition, the eddies and currents of molten iron in the core are ultimately responsible for the planet's magnetic field (see Geomagnetism).

The boundary between the mantle and core is lower today than it once was, a sign that our planet is slowly aging. If there ever comes a point when the heat is entirely dissipated, as may perhaps happen many billions of years from now, that could well be the end of Earth as a "living" planet. If we did not have a mantle and core, with all their heat, pressure, and resulting tectonic activity, Earth would be as dead as the Moon, whose interior is relatively cool.

The distinction between the outer and inner cores, which starts at a depth of about 3,150 mi. (5,100 km), comes from the fact that here, too, there is a phase change—in this case from liquid, molten material back to solid. This has nothing to do with cooling, since, as we have noted, the inner core is almost unimaginably hot; rather, it is a result of the immense pressures apparent at this depth.

GRAVITY ALWAYS WINS. It is interesting to note that the core constitutes only about 16% of the planet's volume but 32% of its mass. As we discussed near the beginning of this essay, enormous gravitational force exists between two objects when at least one of them has a relatively large amount of mass and the distance between them is great. Thus, Earth's mass, concentrated deep in its interior, helps hold our world—people, animals, plants, buildings, and so forth—in place. It also keeps our atmosphere firmly rooted as well. Without an extensive gravitational field of the kind that Earth possesses, significantly less massive bodies, such as the Moon or Mercury, have no atmosphere. In this and many another way, it turns out that life on

THE ASTRONAUT BRUCE MCCANDLESS FLOATS FREELY IN SPACE, OUTSIDE EARTH'S ATMOSPHERE, DURING A SHUTTLE MISSION. EARTH'S VAST INTERIOR MASS GIVES IT A STRONG GRAVITATIONAL PULL, HELPING ROOT THE PEOPLE AND MATERIALS OF OUR WORLD AND HOLDING OUR ATMOSPHERE IN PLACE. *(NASA. Reproduced by permission.)*

Earth's surface depends heavily on what goes on its ultra-hot, extremely pressurized interior.

A BIZARRE POSTSCRIPT

Given the vast amount of power in Earth's interior, it is no wonder that it has long fascinated humans—even before science possessed any sort of intelligent understanding with regard to the contents of that interior. The ancients offered all manner of fascinating speculation regarding the contents of Earth: it was hollow, some said, while others claimed that it contained one substance or another—perhaps even a heart of gold.

Such imaginative musings continued well into the Middle Ages, when the Italian poet Dante Alighieri (1265–1321) described an allegorical journey through Earth's interior in his epochal *Divine Comedy.* This epic poem depicts the inside of the planet as concentric circles of hell, descending toward the fiery core, where Satan himself resides. Beyond this lies Purgatory and further still—on the other side of Earth—Heaven, the New Jerusalem.

By the time the French writer Jules Verne (1828–1905) wrote *Journey to the Center of the Earth* almost six centuries later, scientific knowledge regarding Earth's interior had increased dramatically, though many of the significant discoveries we have examined here—for example, the Moho—still lay in the future. In any case, the

KEY TERMS

ASTHENOSPHERE: A region of extremely high pressure underlying the lithosphere, where rocks are deformed by enormous stresses. The asthenosphere lies at a depth of about 60 mi. to 215 mi. (about 100–350 km).

ATMOSPHERE: In general, an atmosphere is a blanket of gases surrounding a planet. Unless otherwise identified, however, the term refers to the atmosphere of Earth, which consists of nitrogen (78%), oxygen (21%), argon (0.93%), and other substances that include water vapor, carbon dioxide, ozone, and noble gases such as neon (0.07%).

CORE: The center of Earth, an area constituting about 16% of the planet's volume and 32% of its mass. Made primarily of iron and another, lighter element (possibly sulfur), it is divided between a solid inner core with a radius of about 760 mi. (1,220 km) and a liquid outer core about 1,750 mi. (2,820 km) thick. For terrestrial planets, in general, core refers to the center, which in most cases is probably molten metal of some kind.

CRUST: The uppermost division of the solid Earth, representing less than 1% of its volume and varying in depth from 3 mi. to 37 mi. (5–60 km). Below the crust is the mantle.

GEOCHEMISTRY: A branch of the earth sciences, combining aspects of geology and chemistry, that is concerned with the chemical properties and processes of Earth—in particular, the abundance and interaction of chemical elements and their isotopes.

GEOLOGY: The study of the solid earth, in particular, its rocks, minerals, fossils, and land formations.

GEOPHYSICS: A branch of the earth sciences that combines aspects of geology and physics. Geophysics addresses the planet's physical processes as well as its gravitational, magnetic, and electric properties and the means by which energy is transmitted through its interior.

GEOSPHERE: The upper part of Earth's continental crust, or that portion of the solid earth on which human beings live and which provides them with most of their food and natural resources.

tone of Verne's work was that of a new literary style, science fiction. Pioneered by Verne and the British writer H. G. Wells, science fiction could not have been less like Dante's poetry, infused as it was with spirituality and mystery.

SCREAMS OF THE DAMNED? In the early 1990s, an urban legend of sorts brought together the science fiction of Verne, the religious vision of Dante, and a number of other, less pleasant strains—including ignorance and, on the part of its originators, the willingness to deceive. This "urban legend" did not involve crocodiles in sewers, ghostly hitchhikers, or the other usual fodder; instead, it concerned the center of the earth—where, it was claimed, hell had been discovered.

The full account appeared on Ship of Fools (see "Where to Learn More"), a Web site operated by Rich Buhler—himself a Christian minister and a debunker of what he has called "Christian urban legends." As Buhler reported, the story gained so much support that it appeared on Trinity Broadcasting Network (TBN), a major evangelical television outlet. According to the TBN

KEY TERMS CONTINUED

JOVIAN PLANETS: The planets between Mars (the last terrestrial planet) and Pluto, all of which are large, low in density, and composed primarily of gases.

LITHOSPHERE: The upper layer of Earth's interior, including the crust and the brittle portion at the top of the mantle.

MANTLE: The thick, dense layer of rock, approximately 1,429 mi. (2,300 km) thick, between Earth's crust and its core. In reference to the other terrestrial planets, mantle simply means the area of dense rock between the crust and core.

ORGANIC: At one time chemists used the term *organic* only in reference to living things. Now the word is applied to most compounds containing carbon, with the exception of carbonates (which are minerals) and oxides, such as carbon dioxide.

PLATE TECTONICS: The name both of a theory and of a specialization of tectonics. As an area of study, plate tectonics deals with the large features of the lithosphere and the forces that shape them. As a theory, it explains the processes that have shaped Earth in terms of plates and their movement. Plate tectonics theory brings together aspects of continental drift, seafloor spreading, seismic and volcanic activity, and the structures of Earth's crust to provide a unifying model of Earth's evolution. It is one of the dominant concepts in the modern earth sciences.

PLATES: Large, movable segments of the lithosphere.

SEISMIC WAVE: A packet of energy resulting from the disturbance that accompanies a strain on rocks in the lithosphere.

SEISMOLOGY: The study of seismic waves as well as the movements and vibrations that produce them.

TECTONICS: The study of tectonism, including its causes and effects, most notably mountain building.

TECTONISM: The deformation of the lithosphere.

TERRESTRIAL PLANETS: The four inner planets of the solar system: Mercury, Venus, Earth, and Mars. These are all small, rocky, and dense; have relatively small amounts of gaseous elements; and are composed primarily of metals and silicates. Compare with Jovian planets.

report, Russian geologists had drilled a hole some 8.95 mi. (14.4 km) into Earth's crust and heard screams, which supposedly came from condemned souls in the nether regions.

As the embellished details of the story began to unfold, it turned out that the Russian geologists had found the temperatures to be much higher than expected: 2,000°F (1,093°C). Also, their drilling had unleashed a bat that flew out of hell with the words "I have conquered" inscribed in Russian on its wings. Buhler and his team traced this bizarre tale to Finland and then back to southern California. As to how the story originated, Buhler noted the drilling at the Kola Peninsula, which we mentioned earlier in this essay. The depth cited in the rumor, however, was greater than that which the drilling at Kola reached, and the temperatures claimed were much higher than what one actually would encounter at that depth.

"It is possible that somewhere in the world there has been a spooky experience during deep drilling operations," Buhler concluded. Nonetheless, "characteristic of many urban legends, this story was alleged to have occurred in an obscure part of the world where it would be virtually impossible to track down the facts. And once the story got started, people began quoting one another's newsletters to validate their own. This is the stuff of which tabloid newspapers are made." In the end, the "screams of hell" offered nothing of value in terms of either science or religion, but it proved to be an excellent example of human beings' fascination with, and latent terror of, Earth's interior.

WHERE TO LEARN MORE

Buhler, Rich. "Drilling for Hell," *Ship of Fools* (Web site). <http://ship-of-fools.com/Myths/03Myth.html#Top>.

Darling, David J. *Could You Ever Dig a Hole to China?* Minneapolis, MN: Dillon Press, 1990.

De Bremaecker, Jean-Claude. *Geophysics: The Earth's Interior.* New York: John Wiley and Sons, 1985.

Earth's Interior (Web site). <http://www.seismo.unr.edu/ftp/pub/louie/class/100/interior.html>.

Earth's Interior and Plate Tectonics (Web site). <http://www.star.le.ac.uk/edu/solar/earthint.html>.

Earth's Interior and Plate Tectonics (Web site). <http://www.solarviews.com/eng/earthint.htm>.

Hancock, Paul L., and Brian J. Skinner. *The Oxford Companion to the Earth.* New York: Oxford University Press, 2000.

Physical Geology: Interior of the Earth (Web site). <http://www.uh.edu/~jbutler/physical/chapter19.html>.

Rockdoctors Guide: Earth's Interior (Web site). <http://www.cobweb.net/~bug2/rock7.htm>.

Skinner, Brian J., Stephen C. Porter, and Daniel B. Botkin. *The Blue Planet: An Introduction to Earth System Science.* 2d ed. New York: John Wiley and Sons, 1999.

PLATE TECTONICS

CONCEPT

The earth beneath our feet is not dead; it is constantly moving, driven by forces deep in its core. Nor is the planet's crust all of one piece; it is composed of numerous plates, which are moving steadily in relation to one another. This movement is responsible for all manner of phenomena, including earthquakes, volcanoes, and the formation of mountains. All these ideas, and many more, are encompassed in the concept of plate tectonics, which is the name for a branch of geologic and geophysical study and for a powerful theory that unites a vast array of ideas. Plate tectonics works hand in hand with several other striking concepts and discoveries, including continental drift and the many changes in Earth's magnetic field that have taken place over its history. No wonder, then, that this idea, developed in the 1960s but based on years of research that preceded that era, is described as "the unifying theory of geology."

HOW IT WORKS

TECTONICS AND TECTONISM

The lithosphere is the upper layer of Earth's interior, including the crust and the brittle portion at the top of the mantle. Tectonism is the deformation of the lithosphere, and the term tectonics refers to the study of this deformation, including its causes and effects, most notably mountain building. This deformation is the result of the release and redistribution of energy from Earth's core.

The interior of Earth itself is divided into three major sections: the crust, mantle, and core. The first is the uppermost division of the solid earth, representing less than 1% of its volume and varying in depth from 3 mi. to 37 mi. (5–60 km). Below the crust is the mantle, a thick, dense layer of rock approximately 1,429 mi. (2,300 km) thick. The core itself is even more dense, as illustrated by the fact that it constitutes about 16% of the planet's volume and 32% of its mass. Composed primarily of iron and another, lighter element (possibly sulfur), it is divided between a solid inner core with a radius of about 760 mi. (1,220 km) and a liquid outer core about 1,750 mi. (2,820 km) thick.

Tectonism results from the release and redistribution of energy from Earth's interior. There are two components of this energy: gravity, a function of the enormous mass at the core, and heat from radioactive decay. (For more about gravity, see Gravity and Geodesy. The heat from Earth's core, the source of geothermal energy, is discussed in Energy and Earth.) Differences in mass and heat within the planet's interior, known as pressure gradients, result in the deformation of rocks.

DEFORMATION OF ROCKS. Any attempt to deform an object is referred to as stress, and stress takes many forms, including tension, compression, and shear. Tension acts to stretch a material, whereas compression—a type of stress produced by the action of equal and opposite forces, whose effect is to reduce the length of a material—has the opposite result. (Compression is a form of pressure.) As for shear, this is a kind of stress resulting from equal and opposite forces that do not act along the same line. If a thick, hardbound book is lying flat and one pushes the front cover from the side so that the covers and pages are no longer perfectly aligned, this is an example of shear.

Under the effects of these stresses, rocks may bend, warp, slide, or break. They may even flow, as though they were liquids, or melt and thus truly become liquid. As a result, Earth's interior may manifest faults, or fractures in rocks, as well as folds, or bends in the rock structure. The effects of this activity can be seen on the surface in the form of subsidence, which is a depression in the crust, or uplift, which is the raising of crustal materials. Earthquakes and volcanic eruptions also may result.

There are two basic types of tectonism: orogenesis and epeirogenesis. *Orogenesis* is taken from the Greek words *oros* ("mountain") and *genesis* ("origin") and involves the formation of mountain ranges by means of folding, faulting, and volcanic activity. The Greek word *epeiros* means "mainland," and epeirogenesis takes the form of either uplift or subsidence. Of principal concern in the theory of plate tectonics, as we shall see, is orogenesis, which involves more lateral, as opposed to vertical, movement.

CONTINENTAL DRIFT

If one studies a world map for a period of time, one may notice something interesting about the shape of Africa's west coast and that of South America's east coast: they seem to fit together like pieces of a jigsaw puzzle. Early in the twentieth century, two American geologists, Frank Bursley Taylor (1860–1938) and Howard Baker, were among the first scientists to point out this fact. According to Taylor and Baker, Europe, the Americas, and Africa all had been joined at one time. This was an early version of continental drift, a theory concerning the movement of Earth's continents.

Continental drift is based on the idea that the configuration of continents was once different than it is today, that some of the individual landmasses of today once were joined in other continental forms, and that the landmasses later moved to their present locations. Though Taylor and Baker were early proponents, the theory is associated most closely with the German geophysicist and meteorologist Alfred Wegener (1880–1930), who made the case for continental drift in *The Origin of Continents and Oceans* (1915).

PANGAEA, LAURASIA, AND GONDWANALAND. According to Wegener, the continents of today once formed a single supercontinent called Pangaea, from the Greek words *pan* ("all") and *gaea* ("Earth"). Eventually, Pangaea split into two halves, with the northern continent of Laurasia and the southern continent of Gondwanaland, sometimes called Gondwana, separated by the Tethys Sea. In time, Laurasia split to form North America, the Eurasian landmass with the exception of the Indian subcontinent, and Greenland. Gondwanaland also split, forming the major southern landmasses of the world: Africa, South America, Antarctica, Australia, and India.

The Austrian geologist Eduard Suess (1831–1914) and the South African geologist Alexander du Toit (1878–1948), each of whom contributed significantly to continental drift theory, were responsible for the naming of Gondwanaland and Laurasia, respectively. Suess preceded Wegener by many years with his theory of Gondwanaland, named after the Gondwana region of southern India. There he found examples of a fern that, in fossilized form, had been found in all the modern-day constituents of the proposed former continent. Du Toit, Wegener's contemporary, was influenced by continental drift theory and improved on it greatly.

FORMATION OF THE CONTINENTS. Today continental drift theory is accepted widely, in large part owing to the development of plate tectonics, "the unifying theory in geology." We examine the evidence for continental drift, the arguments against it, and the eventual triumph of plate tectonics in the course of this essay. Before going on, however, let us consider briefly the now-accepted timeline of events described by Wegener and others.

About 1,100 million years ago (earth scientists typically abbreviate this by using the notation 1,100 Ma), there was a supercontinent named Rodinia, which predated Pangaea. It split into Laurasia and Gondwanaland, which moved to the northern and southern extremes of the planet, respectively. Starting at about 514 Ma, Laurasia drifted southward until it crashed into Gondwanaland about 425 Ma. Pangaea, surrounded by a vast ocean called Panthalassa ("All Ocean"), formed approximately 356 Ma.

In the course of Pangaea's formation, what is now North America smashed into northwestern

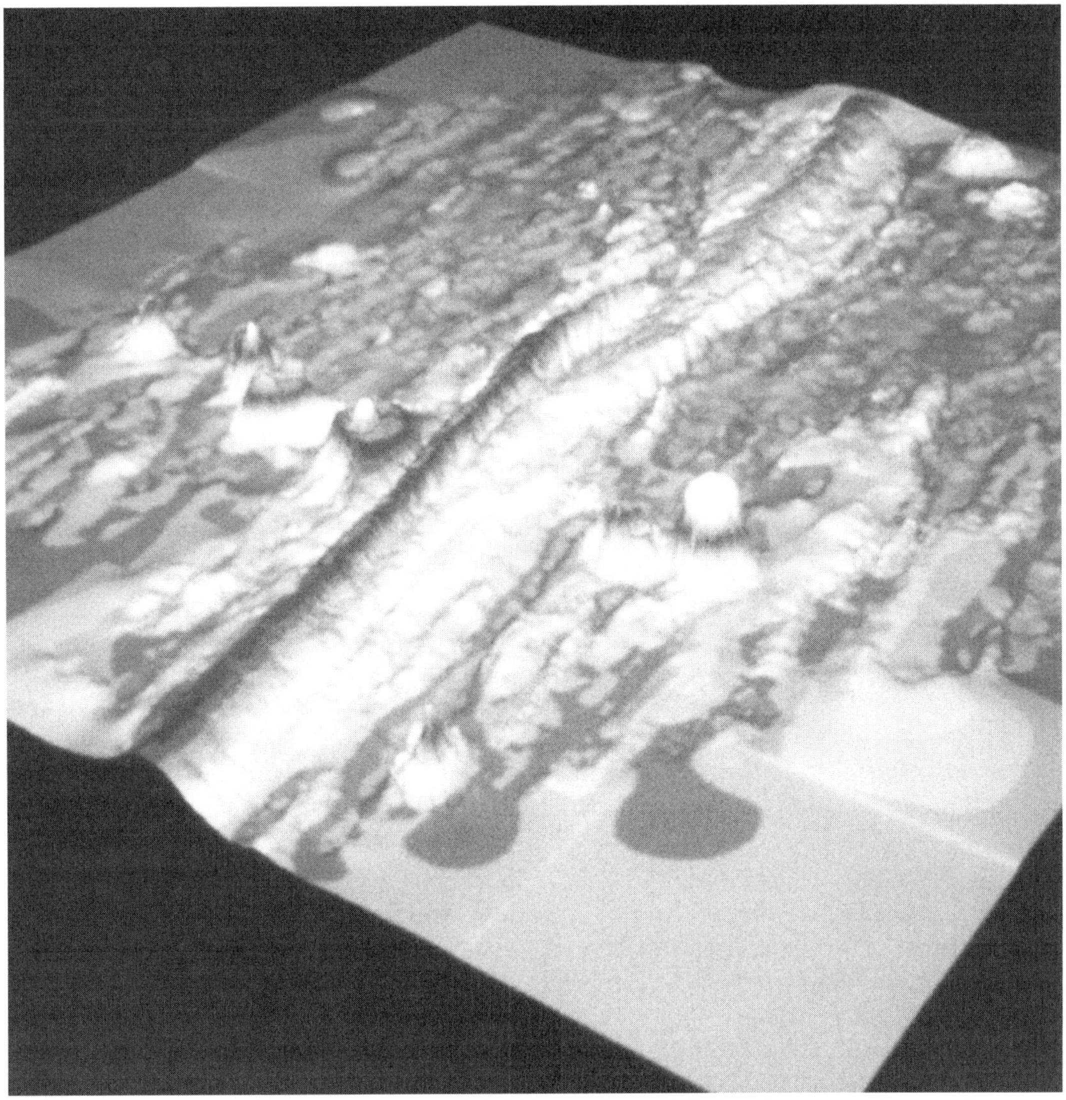

A MAP OF PART OF THE EAST PACIFIC RISE, A MID-OCEAN RIDGE TO THE WEST OF CENTRAL AMERICA THAT MARKS THE BOUNDARY BETWEEN THE PACIFIC AND COCOS TECTONIC PLATES. *(© Dr. Ken MacDonald/Photo Researchers. Reproduced by permission.)*

Africa, forming a vast mountain range. Traces of these mountains still can be found on a belt stretching from the southern United States to northern Europe, including the Appalachians. As Pangaea drifted northward and smashed into the ocean floor of Panthalassa, it formed a series of mountain ranges from Alaska to southern South America, including the Rockies and Andes. By about 200 Ma, Pangaea began to break apart, forming a valley that became the Atlantic Ocean. But the separation of the continents was not a "neat" process: today a piece of Gondwanaland lies sunken beneath the eastern United States, far from the other landmasses to which it once was joined.

By about 152 Ma, in the late Jurassic period, the continents as we know them today began to take shape. By about 65 Ma, all the present continents and oceans had been formed for the most part, and India was drifting north, eventually smashing into southern Asia to shape the world's tallest mountains, the Himalayas, the Karakoram Range, and the Hindu Kush. This process is not finished, however, and geologists believe that some 250–300 million years from now, Pangaea will re-form.

EVIDENCE AND ARGUMENTS. As proof of his theory, Wegener cited a wide variety of examples, including the apparent fit between the coastlines of South America and western Africa as well as that of North America and northwestern Africa. He also noted the existence of rocks apparently gouged by glaciers in southern Africa, South America, and India, far from modern-day glacial activity. Fossils in South America matched those in Africa and Australia, as Suess had observed. There were also signs that mountain ranges continued between continents—not only those apparently linking North America and Europe but also ranges that seemed to extend from Argentina to South Africa and Australia.

By measurements conducted over a period of years, Wegener even showed that Greenland was drifting slowly away from Europe, yet his theory met with scorn from the geoscience community of his day. If continents could plow through oceanic rock, some geologists maintained, then they would force up mountains so high that Earth would become imbalanced. As for his claim that matching fossils in widely separated regions confirmed his theory of continental drift, geologists claimed that this could be explained by the existence of land bridges, now sunken, that once had linked those areas. The apparent fit between present-day landmasses could be explained away as coincidence or perhaps as evidence that Earth simply was expanding, with the continents moving away from one another as the planet grew.

INTRODUCTION TO PLATE TECTONICS

Though Wegener was right, as it turned out, his theory had one major shortcoming: it provided no explanation of exactly *how* continental drift had occurred. Even if geologists had accepted his claim that the continents are moving, it raised more questions than answers. A continent is a very large thing simply to float away; even an aircraft carrier, which is many millions of times lighter, has to weigh less than the water it displaces, or it would sink like a stone. In any case, Wegener never claimed that continents floated. How, then, did they move?

The answer is plate tectonics, the name both of a theory and of a specialization of tectonics. As an area of study, plate tectonics deals with the large features of the lithosphere and the forces that fashion them. As a theory, it explains the processes that have shaped Earth in terms of plates (large movable segments of the lithosphere) and their movement. Plate tectonics theory brings together aspects of continental drift, seafloor spreading (discussed later), seismic and volcanic activity, and the structures of Earth's crust to provide a unifying model of Earth's evolution.

It is hard to overemphasize the importance of plate tectonics in the modern earth sciences; hence, its characterization as the "unifying theory." Its significance is demonstrated by its inclusion in the book *The Five Biggest Ideas of Science,* cited in the bibliography for this essay. Alongside plate tectonics theory in that volume are four towering concepts of extraordinary intellectual power: the atomic model, or the concept that matter is made up of atoms; the periodic law, which explains the chemical elements; big bang theory, astronomers' explanation of the origins of the universe (see Planetary Science); and the theory of evolution in the biological sciences.

THE PIECES COME TOGETHER. In 1962 the United States geologist Harry Hammond Hess (1906–1969) introduced a new concept that would prove pivotal to the theory of plate tectonics: seafloor spreading, the idea that seafloors crack open along the crest of mid-ocean ridges and that new seafloor forms in those areas. (Another American geologist, Robert S. Dietz [1914–1995], had published his own theory of seafloor spreading a year before Hess's, but Hess apparently developed his ideas first.) According to Hess, a new floor forms when molten rock called magma rises up from the asthenosphere, a region of extremely high pressure underlying the lithosphere, where rocks are deformed by enormous stresses. The magma wells up through a crack in a ridge, runs down the sides, and solidifies to form a new floor.

Three years later, the Canadian geologist John Tuzo Wilson (1908–1993) coined the term plates to describe the pieces that make up Earth's rigid surface. Separated either by the mid-ocean rifts identified earlier by Heezen or by mountain chains, the plates move with respect to one another. Wilson presented a model for their behavior and established a global pattern of faults, a sort of map depicting the movable plates. The pieces of a new theory were forming (an apt metaphor in this instance!), but as yet it had no name.

That name appeared in 1967, when D. P. Mackenzie of England and R. L. Parker of the

United States introduced the term plate tectonics. They maintained that the surface of Earth is divided into six major as well as seven minor movable plates and compared the continents to enormous icebergs—much as Wegener had described them half a century earlier. Subsequent geologic research has indicated that there may be as many as nine major plates and as many as 12 minor ones.

To test these emerging ideas, the U.S. National Science Foundation authorized a research voyage by the vessel *Glomar Challenger* in 1968. On their first cruise, through the Gulf of Mexico and the Atlantic, the *Challenger*'s scientific team collected sediment, fossil, and crust samples that confirmed the basics of seafloor spreading theory. These results led to new questions regarding the reactions between rocks and the heated water surrounding them, spawning new research and necessitating additional voyages. In the years that followed, the *Challenger* made more and more cruises, its scientific teams collecting a wealth of evidence for the emerging theory of plate tectonics.

REAL-LIFE APPLICATIONS

EARLY EVIDENCE OF PLATE TECTONICS

No single person has been as central to plate tectonics as Wegener was to continental drift or as the English naturalist Charles Darwin (1809–1882) was to evolution. The roots of plate tectonics lie partly in the observations of Wegener and other proponents of continental drift as well as in several discoveries and observations that began to gather force in the third quarter of the twentieth century.

During World War II, submarine warfare necessitated the development of new navigational technology known as sonar (*SO*und *N*avigation *A*nd *R*anging). Sonar functions much like radar (see Remote Sensing), but instead of using electromagnetic waves, it utilizes ultrasonic, or high-frequency, sound waves projected through water. Sonar made it feasible for geologists to study deep ocean basins after the war, making it possible for the first time in history to map and take samples from large areas beneath the seas. These findings raised many questions, particularly concerning the vast elevation differences beneath the seas.

EWING AND THE MOUNTAINS UNDER THE OCEAN. One of the first earth scientists to notice the curious aspects of underwater geology was the American geologist William Maurice Ewing (1906–1974), who began his work long before the war. He had gained his first experience in a very practical way during the 1920s, as a doctoral student putting himself through school. Working summers with oil exploration teams in the Gulf of Mexico off the coast of Texas had given him a basic understanding of the subject, and in the following decade he went to work exploring the structure of the Atlantic continental shelf and ocean basins.

His work there revealed extremely thick sediments covering what appeared to be high mountainous regions. These findings sharply contradicted earlier ideas about the ocean floor, which depicted it as a flat, featureless plain rather like the sandy-bottomed beaches found in resort areas. Instead, the topography at the bottom of the ocean turned out to be at least as diverse as that of the land above sea level.

HEEZEN AND THE RIFT VALLEY. During the 1950s, a team led by another American geologist, Bruce Charles Heezen (1924–1977), worked on developing an overall picture of the ocean basin's topography. Earlier work had identified a mountain range running the length of the Atlantic, but Heezen's team discovered a deep valley down the middle of the chain, running parallel to it. They described it as a rift valley, a long trough bounded by two or more faults, and compared it to a similar valley in eastern Africa.

Around the same time, a group of transatlantic telephone companies asked Heezen to locate areas of possible seismic or earthquake activity in the Atlantic. Phone company officials reasoned that if they could find the areas most likely to experience seismic activity, they could avoid placing their cables in those areas. As it turned out, earthquakes tended to occur in exactly the same region that Heezen and his team had identified as the rift valley.

THE PLATES AND THEIR INTERACTIONS

The most significant plates that make up Earth's surface are as follows:

Selected Major Plates

- North American (almost all of North America and Mexico, along with Greenland and the northwestern quadrant of the Atlantic)
- South American (all of South America and the southwestern quadrant of the Atlantic)
- African (Africa, the southeastern Atlantic, and part of the Indian Ocean)
- Eurasian (Europe and Asia, excluding the Indian subcontinent, along with surrounding ocean areas)
- Indo-Australian (India, much of the Indian Ocean, Australia, and parts of the Indonesian archipelago and New Zealand)
- Antarctic (Antarctica and the Antarctic Ocean)

In addition to these plates, there are several plates that while they are designated as "major" are much smaller: the Philippine, Arabian, Caribbean, Nazca (off the west coast of South America), Cocos (off the west coast of Mexico), and Juan de Fuca (extreme western North America). Japan, one of the most earthquake-prone nations in the world, lies at the nexus of the Philippine, Eurasian, and Pacific plates.

MOVEMENT OF THE PLATES. One of the key principles of geology, discussed elsewhere in this book, is uniformitarianism: the idea that processes occurring now also occurred in the past. The reverse usually is also true; thus, as we have noted, the plates are still moving, just as they have done for millions of years. Thanks to satellite remote sensing, geologists are able to measure this rate of movement. (See Remote Sensing for more on this subject.) Not surprisingly, its pace befits the timescale of geologic, as opposed to human, processes: the fastest-moving plates are careening forward at a breathtaking speed of 4 in. (10 cm) per year. The ground beneath Americans' feet (assuming they live in the continental United States, east of the Juan de Fuca) is drifting at the rate of 1.2 in. (3 cm) every year, which means that in a hundred years it will have shifted 10 ft. (3 m).

WHEN PLATES INTERACT. Plates interact by moving toward each other (convergence), away from each other (divergence), or past each other (transform motion). Convergence usually is associated with subduction, meaning that one plate is forced down into the mantle and eventually undergoes partial melting. This typically occurs in the ocean, creating a depression known as an oceanic trench. Divergence results in the separation of plates and most often is associated either with seafloor spreading or the formation of rift valleys.

There are three types of plate margins, or boundaries between plates, depending on the two types of crusts that are interacting: oceanic with oceanic, continental with continental, or continental with oceanic. The rift valleys of the Atlantic are an example of an oceanic margin where divergence has occurred, while oceanic convergence is illustrated by a striking example in the Pacific. There, subduction of the Philippine Plate by the Pacific Plate has created the Mariana Trench, which at 36,198 ft. (10,911 m) is the deepest depression on Earth.

When continental plates converge, neither plate subducts; rather, they struggle against each other like two warriors in a fight to the death, buckling, folding, and faulting to create huge mountain ranges. The convergence of the Indo-Australian and Eurasian plates has created the *highest* spots on Earth, in the Himalayas, where Mount Everest (on the Nepal-Indian border) rises to 29,028 ft. (8,848 m). Continental plates also may experience divergence, resulting in the formation of seas. An example is the Red Sea, formed by the divergence of the African and Arabian plates.

Given these facts about the interactions of oceanic and continental plates with each other, what occurs when continental plates meet oceanic ones is no surprise. In this situation, the oceanic plate meeting the continental plate is like a high-school football player squaring off against a National Football League pro tackle. It is no match: the oceanic plate easily subducts. This leads to the formation of a chain of volcanoes along the continental crust, examples being the Cascade Range in the U.S. Pacific Northwest (Juan de Fuca and Pacific plates) or the Andes (South American and Nazca plates).

Transform margins may occur with any combination of oceanic or continental plates and result in the formation of faults and earthquake zones. Where the North American Plate slides against the Pacific Plate along the California coast, it has formed the San Andreas Fault, the source of numerous earthquakes, such as the dramatic San Francisco quakes of 1906 and 1989 and the Los Angeles quake of 1994.

SHUTTLE PHOTOGRAPH OF EASTERN EGYPT SHOWS THE RED SEA AT THE TOP, WITH THE GULF OF SUEZ CONNECTING IT TO THE MEDITERRANEAN SEA. THE RED SEA WAS FORMED BY THE DIVERGENCE OF THE AFRICAN AND ARABIAN PLATES. *(© NASA/Photo Researchers. Reproduced by permission.)*

PALEOMAGNETISM

As noted earlier, plate tectonics brings together numerous areas of study in the geologic sciences that developed independently but which came to be seen as having similar roots and explanations. Among these disciplines is paleomagnetism, an area of historical geology devoted to studying the direction and intensity of magnetic fields in the past, as discerned from the residual magnetization of rocks.

Earth has a complex magnetic field whose principal source appears to be the molten iron of the outer core. In fact, the entire planet is like a giant bar magnet, with a north pole and a south pole. It is for this reason that the magnetized material in a compass points north; however, Earth's magnetic north pole is not the same as its geographic north pole. It so happens that magnetic north lies in more or less the same direction as geographic north, but as geologists in the mid–nineteenth century discovered, this has not always been the case. (For more about magnetic north and other specifics of Earth's magnetic field, see Geomagnetism.)

In 1849 the French physicist Achilles Delesse (1817–1881) observed that magnetic minerals tend to line up with the planet's magnetic field, pointing north as though they were compass needles. Nearly 60 years later, however, another French physicist, Bernard Brunhes (1867–1910),

KEY TERMS

ASTHENOSPHERE: A region of extremely high pressure underlying the lithosphere, where rocks are deformed by enormous stresses. The asthenosphere lies at a depth of about 60-215 mi. (about 100–350 km).

COMPRESSION: A form of stress produced by the action of equal and opposite forces, the effect of which is to reduce the length of a material. Compression is a form of pressure.

CONTINENTAL DRIFT: The theory that the configuration of Earth's continents was once different than it is today; that some of the individual landmasses of today once were joined in other continental forms; and that these landmasses later separated and moved to their present locations.

CONVERGENCE: A tectonic process whereby plates move toward each other. Usually associated with subduction, convergence typically occurs in the ocean, creating an oceanic trench. It is one of the three ways, along with divergence and transform motion, that plates interact.

CORE: The center of Earth, an area constituting about 16% of the planet's volume and 32% of its mass. Made primarily of iron and another, lighter element (possibly sulfur), it is divided between a solid inner core with a radius of about 760 mi. (1,220 km) and a liquid outer core about 1,750 mi. (2,820 km) thick.

CRUST: The uppermost division of the solid earth, representing less than 1% of its volume and varying in depth from 3-37 mi. (5–60 km). Below the crust is the mantle.

DIVERGENCE: A tectonic process whereby plates move away from each other. Divergence results in the separation of plates and is associated most often either with seafloor spreading or with the formation of rift valleys. It is one of the three ways, along with convergence and transform motion, that plates interact.

EPEIROGENESIS: One of two principal forms of tectonism, the other being orogenesis. Derived from the Greek words *epeiros* ("mainland") and *genesis* ("origins"), epeirogenesis takes the form of either uplift or subsidence.

FAULT: An area of fracturing between rocks resulting from stress.

FOLD: An area of rock that has been bent by stress.

GEOPHYSICS: A branch of the earth sciences that combines aspects of geology and physics. Geophysics addresses the planet's physical processes as well as its gravitational, magnetic, and electric properties, and the means by which energy is transmitted through its interior.

HISTORICAL GEOLOGY: The study of Earth's physical history. Historical geology is one of two principal branches of geology, the other being physical geology.

LITHOSPHERE: The upper layer of Earth's interior, including the crust and the brittle portion at the top of the mantle.

MA: An abbreviation used by earth scientists, meaning "million years." When an event is designated as, for instance, 160 Ma, it means that it happened 160 million years ago.

MANTLE: The thick, dense layer of rock, approximately 1,429 mi. (2,300 km) thick, between Earth's crust and its core.

KEY TERMS CONTINUED

MID-OCEAN RIDGES: Submarine mountain ridges where new seafloor is created by seafloor spreading.

OCEANIC TRENCH: A deep depression in the ocean floor caused by the convergence of plates and the resulting subduction of one plate.

OROGENESIS: One of two principal forms of tectonism, the other being epeirogenesis. Derived from the Greek words *oros* ("mountain") and *genesis* ("origin"), orogenesis involves the formation of mountain ranges by means of folding, faulting, and volcanic activity. The processes of orogenesis play a major role in plate tectonics.

PALEOMAGNETISM: An area of historical geology devoted to studying the direction and intensity of magnetic fields in the past, as discerned from the residual magnetization of rocks.

PLATE MARGINS: Boundaries between plates.

PLATE TECTONICS: The name both of a theory and of a specialization of tectonics. As an area of study, plate tectonics deals with the large features of the lithosphere and the forces that fashion them. As a theory, it explains the processes that have shaped Earth in terms of plates and their movement. Plate tectonics theory brings together aspects of continental drift, seafloor spreading, seismic and volcanic activity, and the structures of Earth's crust to provide a unifying model of Earth's evolution. It is one of the dominant concepts in the modern earth sciences.

PLATES: Large movable segments of the lithosphere.

RADIOACTIVITY: A term describing a phenomenon whereby certain materials are subject to a form of decay brought about by the emission of high-energy particles or radiation. Forms of particles or energy include alpha particles (positively charged helium nuclei), beta particles (either electrons or subatomic particles called *positrons*), or gamma rays, which occupy the highest energy level in the electromagnetic spectrum.

REMOTE SENSING: The gathering of data without actual contact with the materials or objects being studied.

RIFT: A split between two bodies (for example, two plates) that once were joined.

RIFT VALLEY: A long trough bounded by two or more faults.

SEAFLOOR SPREADING: The theory that seafloors crack open along the crests of mid-ocean ridges and that new seafloor forms in those areas.

SHEAR: A form of stress resulting from equal and opposite forces that do not act along the same line. If a thick, hardbound book is lying flat and one pushes the front cover from the side so that the covers and pages are no longer perfectly aligned, this is an example of shear.

STRESS: In general terms, any attempt to deform a solid. Types of stress include tension, compression, and shear. More specifically, stress is the ratio of force to unit area F/A, where F is force and A area.

SUBDUCTION: A tectonic process that results when plates converge and one plate forces the other down into Earth's mantle. As a result, the subducted plate eventually undergoes partial melting.

KEY TERMS CONTINUED

SUBSIDENCE: A depression in Earth's crust.

TECTONICS: The study of tectonism, including its causes and effects, most notably mountain building.

TECTONISM: The deformation of the lithosphere.

TENSION: A form of stress produced by a force that acts to stretch a material.

THEORY: A general statement derived from a hypothesis that has withstood sufficient testing.

TRANSFORM MOTION: A tectonic process whereby plates slide past each other. It is one of the three ways, along with convergence and divergence, that plates interact.

noted that in some rocks magnetic materials point *south.* This suggested one of two possibilities: either the planet's magnetic field had reversed itself over time, or the ground containing the magnetized rocks had moved. Both explanations must have seemed far-fetched at the time, but as it turned out, both are correct.

Earth's magnetic field has shifted, meaning that the magnetic north and south poles have changed places many times over the eons. In addition, the magnetic poles have wandered around the southern and northern portions of the globe: for instance, whereas magnetic north today lies in the frozen islands to the north of Canada, at about 300 Ma it was located in eastern Siberia. The movement of magnetic rocks on Earth's surface, however, has turned out to be too great to be explained either by magnetic shifts or by regional wandering of the poles. This is where plate tectonics and paleomagnetism come together.

CONFIRMATION OF PLATE TECTONIC THEORY. Rocks in Alaska have magnetic materials aligned in such a way that they once must have been at or near the equator. In addition, the orientation of magnetic materials on South America's east coast shows an affinity with that of similar materials on the west coast of Africa. In both cases, continental drift, with its driving mechanism of plate tectonics, seems the only reasonable explanation.

Thus, paleomagnetic studies have served to confirm the ideas of continental drift and plate tectonics, while research conducted at sea bolsters seafloor spreading theory. Using devices called magnetometers, geologists have found that the orientation of magnetic minerals on one side of a rift mirrors that of materials on the other side. This suggests that the new rock on either side of the rift was formed simultaneously, as seafloor spreading theory indicates.

EARTHQUAKES AND VOLCANOES

Several findings relating to earthquakes and volcanic activity also can be explained by plate tectonics. If one follows news stories of earthquakes, one may begin to wonder why such places as California or Japan have so many quakes, whereas the northeastern United States or western Europe have so few. The fact is that earthquakes occur along belts, and the vast majority of these belts coincide with the boundaries between Earth's major tectonic plates.

The same is true of volcanoes, and it is no mistake that places famous for earthquakes—the Philippines, say, or Italy—often also are known for their volcanoes. Although they are located near the center of the Pacific Plate, the islands of Hawaii are subject to plate movement, which has helped generate the volcanoes that gave those islands their origin. At the southern end of the island chain, many volcanoes are still active, while those at the northern end tend to be dormant. The reason is that the Pacific Plate as a whole is moving northward over a stationary lava source in the mantle below Hawaii. The southern islands remain poised above that source, while the northern islands have moved away from it.

The Oceanic and Continental Crusts

Given what we have seen about continental drift and seafloor spreading, it should come as no surprise to learn that, generally speaking, the deeper one goes in the ocean, the newer the crust. Specifically, the crust is youngest near the center of ocean basins and particularly along mid-ocean ridges, or submarine mountain ridges where new seafloor is created by seafloor spreading.

It also should not be surprising to learn that oceanic and continental crusts differ both in thickness and in composition. Basalt, an igneous rock (rock formed from the cooling of magma), makes up the preponderance of ocean crust, whereas much of the continental crust is made up of granite, another variety of igneous rock. Whereas the ocean crust is thin, generally 3–6 mi. (5–10 km) in depth, the continental crust ranges in thickness from 12.5–55 mi. (20–90 km). This results in a difference in thickness for the lithosphere, which is only about 60 mi. (100 km) thick beneath the oceans but about 2.5 times as thick—150 mi. (250 km)—under the continents.

Where to Learn More

Erickson, Jon. *Plate Tectonics: Unraveling the Mysteries of the Earth.* New York: Facts on File, 1992.

Gallant, Roy A. *Dance of the Continents.* New York: Benchmark Books, 2000.

Geology: Plate Tectonics (Web site). <http://www.ucmp.berkeley.edu/geology/tectonics.html>.

Kious, W. Jacquelyne, and Robert I. Tilling. *This Dynamic Earth: The Story of Plate Tectonics. U.S. Geological Survey* (Web site). <http://pubs.usgs.gov/publications/text/dynamic.html>.

Miller, Russell. *Continents in Collision.* Alexandria, VA: Time-Life Books, 1987.

Plate Tectonics (Web site). <http://www.platetectonics.com/>.

Plate Tectonics (Web site). <http://observe.ivv.nasa.gov/nasa/earth/tectonics/Tectonics1.html>.

Plate Tectonics, the Cause of Earthquakes (Web site). <http://www.seismo.unr.edu/ftp/pub/louie/class/100/plate-tectonics.html>.

Silverstein, Alvin, Virginia B. Silverstein, and Laura Silverstein Nunn. *Plate Tectonics.* Brookfield, CT: Twenty-First Century Books, 1998.

Wynn, Charles M., Arthur W. Wiggins, and Sidney Harris. *The Five Biggest Ideas in Science.* New York: John Wiley and Sons, 1997.

SEISMOLOGY

CONCEPT

Disturbances within Earth's interior, which is in a constant state of movement, result in the release of energy in packets known as seismic waves. An area of geophysics known as seismology is the study of these waves and their effects, which often can be devastating when experienced in the form of earthquakes. The latter do not only take lives and destroy buildings, but they also produce secondary effects, most often in the form of a tsunami, or tidal wave. Using seismographs and seismometers, seismologists study earthquakes and other seismic phenomena, including volcanoes and even explosions resulting from nuclear testing. They measure earthquakes according to their magnitude or energy as well as their intensity or human impact. Seismology also is used to study Earth's interior, about which it has revealed a great deal.

HOW IT WORKS

STRESS AND STRAIN IN EARTH'S INTERIOR

Modern earth scientists' studies in seismology, as in many other areas, are informed by plate tectonics, and to understand the causes of earthquakes and volcanoes, it is necessary to understand the basics of tectonics as well as plate tectonics theory. The latter subject is discussed in depth within a separate essay, which the reader is encouraged to consult for a more detailed explanation of concepts covered briefly here.

The term tectonism refers to deformation of the lithosphere, the upper layer of Earth's interior. Tectonics is the study of this deformation, which results from the release and redistribution of energy from Earth's core. The core is an extremely dense region, composed primarily of iron and another, lighter element (possibly sulfur), and is divided between a solid inner core with a radius of about 760 mi. (1,220 km) and a liquid outer core about 1,750 mi. (2,820 km) thick.

Earth's core possesses enormous energy, both gravitational and thermal. Gravitational energy is a result of the core's great mass (see Gravity and Geodesy for more about the role of mass in gravity), while thermal energy results from the radioactive decay of elements. In the context of radioactivity, decay does not mean "rot" rather, it refers to the release of high-energy particles. The release of these particles results in the generation of thermal energy, commonly referred to as heat. (See Energy and Earth for more about the scientific definition of heat as well as a discussion of geothermal energy.)

Differences in mass and temperature within the planet's interior, known as pressure gradients, result in the deformation of rocks in the lithosphere. The lithosphere includes the brittle upper portion of the mantle, a dense layer of rock approximately 1,429 mi. (2,300 km) thick, as well as the crust, which varies in depth from 3 mi. to 37 mi. (5–60 km). Deformation is the result of stress—that is, tension (stretching), compression, or shear. (The last of these stresses results from equal and opposite forces that do not act along the same line. To visualize shear, one need only imagine a thick hardbound book with its front cover pushed from the side so that the covers and pages are no longer perfectly aligned.)

A CHASM ALONG A FAULT SCARP IN SAN BERNARDINO COUNTY, CALIFORNIA. *(© Ken M. Johns/Photo Researchers. Reproduced by permission.)*

Under the effects of these stresses, rocks experience strain, or a change in dimension as they bend, warp, slide, break, flow as though they were liquids, or melt. This strain, in turn, leads to a release of energy in the form of seismic waves. These waves may cause faults, or fractures, as well as folds, or bends in the rock structure, which manifest on the surface in the form of earthquakes, volcanoes, and other varieties of seismic activity. Seismology is the study of these waves as well as the movements and vibrations that produce them.

CONTINENTAL DRIFT AND PLATE TECTONICS

The theory of continental drift, discussed in Plate Tectonics, is based on the idea that the configuration of Earth's continents was once different than it is today. Integral to this theory is the accompanying idea that some of the individual land masses of today once were joined in other continental forms and that the land masses later moved to their present locations.

Continental drift theory was introduced in 1915 by the German geophysicist and meteorologist Alfred Wegener (1880–1930), but it failed to gain acceptance for half a century, in large part because it offered no explanation as to how the continents drifted. That explanation came in the 1960s with the development of plate tectonics, the name both of a theory and of a specialization of tectonics. As an area of study, plate tectonics deals with the large features of the lithosphere and the forces that shape them. As a theory, it explains the processes that have shaped Earth in terms of plates (large movable segments of the lithosphere) and their movement.

THE PLATES AND SEISMIC ACTIVITY. There are several major plates, some of which are listed in Plate Tectonics. That essay also discusses modern theories regarding the means by which continents broke apart many millions of years ago and then drifted back together, slamming into one another to form a number of notable features, such as the high mountains between the Indian subcontinent and the Eurasian landmass. Nor have the continents stopped moving; they continue to do so, though at a rate too slow to be noticed in a lifetime or even over the course of several generations. Based on its current rate of movement, in another 6,000 years—approximately the span of time since human civilization began—North America will have drifted about 600 ft. (183 m).

For the most part, the continents we know today are composed of single plates. For instance, South America sits on its own plate, which includes the southwestern quadrant of the Atlantic. But there are exceptions, an example being India itself, which is part of the Indo-Australian plate. Also notable is the Juan de Fuca Plate, a small portion of land attached to the North American continent and comprising the region from northern California to southern British Columbia.

It so happens that this area is home to an unusual amount of volcanic activity. Southern California, where the North American and Pacific plates meet on the San Andreas fault, also is extremely prone to earthquakes, as is Japan, whose islands straddle the Philippine, Eurasian, and Pacific plates. Hawaii is another site of seismic activity in the form of volcanoes, but it does not lie at the nexus of any major plates. Instead, it is situated squarely atop the Pacific Plate, which is moving northward over a hot spot, a region of high volcanic activity. The hot spot remains more or less stationary, while the Pacific Plate moves across it; this explains why the volcanoes of northern Hawaii are generally dormant, whereas many volcanoes in the southern part of the island chain are still active.

Plates interact by moving toward each other (convergence), away from each other (divergence), or past each other (transform motion). When a continental plate converges with an oceanic plate (the differences between these types are discussed in Plate Tectonics), the much sturdier continental plate plows over the oceanic one. This is called subduction. The subducted plate undergoes partial melting, leading to the formation of volcanic chains, as in the nexus of the Juan de Fuca and Pacific or the South American and Nazca plates. The subduction of the Nazca Plate, which lies to the west of South America, helped form the Andes. Transform margins result in the formation of faults and earthquake zones, an example being the volatile San Andreas Fault.

Seismic Waves

The first scientific description of seismic waves was that of John William Strutt, Baron Rayleigh (1842–1919), who in 1885 characterized them as having aspects both of longitudinal and of transverse waves. These are, respectively, waves in which the movement of vibration is in the same direction as the wave itself and those in which the vibration or motion is perpendicular to the direction in which the wave is moving. (Ocean waves, for example, are longitudinal, whereas sound waves are transverse.)

Rayleigh waves later would be distinguished from Love waves, named after the English mathematician and geophysicist Augustus Edward Hough Love (1863–1940). The motion of Love waves is entirely horizontal, or longitudinal. Both are examples of surface waves, or seismic waves whose line of propagation is along the surface of a medium, such as the solid earth. These waves tend to be slower and more destructive than body waves, defined as waves whose line of propagation is through the body of a medium. Body waves include P-waves (primary waves), which are extremely fast moving and longitudinal, and S-waves (secondary waves), which move somewhat less fast and are transverse. The respective

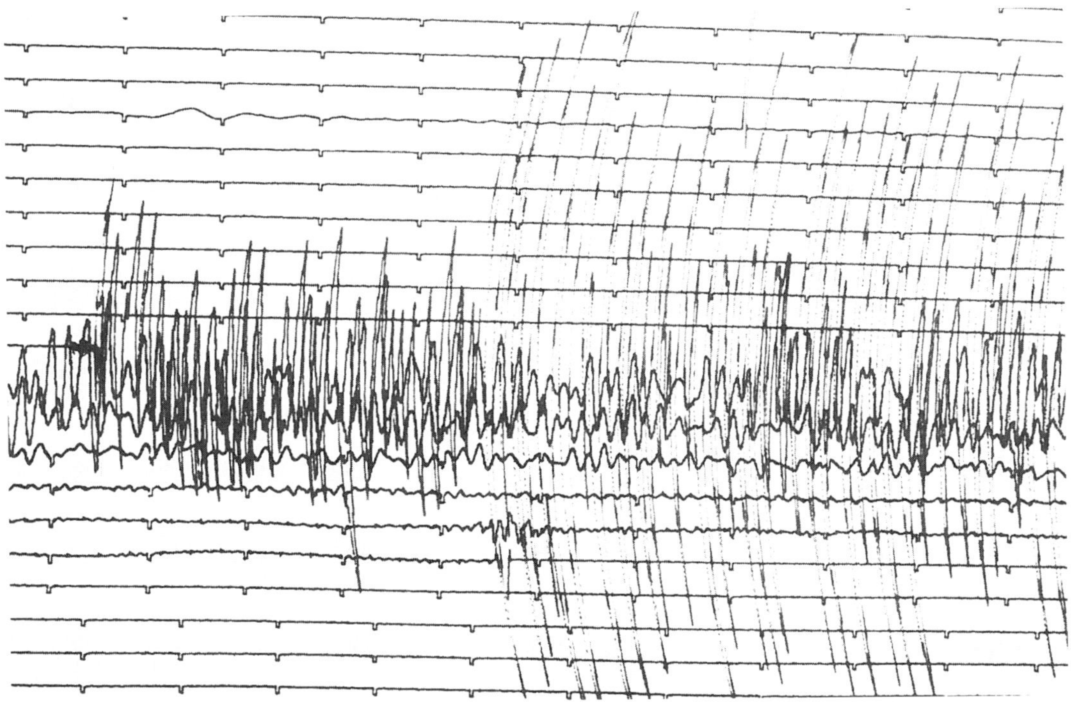

A SEISMOGRAPH READING FROM THE 1989 LOMA PRIETA, CALIFORNIA, EARTHQUAKE. *(© Russell D. Curtis/Photo Researchers. Reproduced by permission.)*

waves' rates of propagation through the solid earth are as follows:

- P-waves: about 4 mi. (6.4 km) per second
- S-waves: about 2 mi. (3.2 km) per second
- Rayleigh and Love waves: less than 2 mi. per second

REAL-LIFE APPLICATIONS

THE LISBON QUAKE AND ITS EFFECTS

On November 1, 1755, the Portuguese capital of Lisbon became the site of one of the worst earthquakes in European history. The event had a number of aftereffects, natural and immediate as well as human and longer term. The results from nature were devastating; the earthquake caused a tsunami, or tidal wave, that flooded the Tagus River even as a fire, also caused by the earthquake, raged through the city.

Estimates of the deaths related directly or indirectly to the Lisbon quake range from 10,000 to as many as 60,000, making it the worst European earthquake since 1531. That earlier quake, incidentally, also had occurred in Lisbon—another example of the fact that certain areas are more prone to seismic activity. It so happens that Portugal lies near the boundary between the Eurasian and African plates.

As for the human response to the quake, it is best represented by the French writer Voltaire (1694–1778). Always a critic of religious faith, Voltaire saw in the incident evidence that called into question Christians' belief in a loving God. He made this case both in the philosophical poem *Le désastre de Lisbonne* (The disaster of Lisbon, 1756) and, more memorably, in the satirical novel *Candide* (1759).

MICHELL AND THE BIRTH OF SEISMOLOGY. Another, much less famous, thinker responded to the Lisbon earthquake in quite a different fashion. This was English geologist and astronomer John Michell (*ca.* 1724–1793), who studied the event and concluded that quakes are accompanied by shock waves. In an article published in 1760, he noted that earthquakes are found to occur near volcanoes and suggested that they are caused by pressure produced by water that boils from volcanic heat. He also indicated that one can calculate the center of an earthquake by making note of the time at which the motions are felt.

Today Michell is regarded as the father of seismology, a discipline that began to mature in the nineteenth century. The name itself was coined by the Irish engineer Robert Mallet (1810–1881), who in 1846 compiled the first modern catalogue of earthquakes. Eleven years after publishing the book, which listed all known quakes of any significance since 1606 B.C., Mallet conducted experiments with shock waves by exploding gunpowder and measuring the rate at which the waves travel through various types of material.

DETECTING AND MEASURING SEISMIC ACTIVITY

As noted earlier, seismology is concerned with seismic waves, which generally are caused by movements within the solid earth. These waves also may be produced by man-made sources. Seismologic studies assist miners in knowing how much dynamite to use for a quarry blast so as to be effective without destroying the mine itself or the resources being sought. In addition, seismology can be used to reveal the location of such materials as coal and oil.

Thanks to seismometers (instruments for detecting seismic waves) and seismographs, which record information regarding those waves, seismologists are able to detect not only natural seismic activity but also the effects of underground nuclear testing. Underground testing is banned by international treaty, and if a "rogue nation" were to conduct such testing, it would come to the attention of the World-Wide Standardized Seismograph Network (WWSSN), which consists of 120 seismic stations in some 60 countries.

Most of the remainder of this essay is devoted to a single type of seismic phenomenon: earthquakes. As noted, they are far from the only effect of seismic activity; however, they are the most prevalent and well documented. A close second would be volcanoes, which are discussed in the essay Mountains.

EARLY SEISMOGRAPHIC INSTRUMENTS. In A.D. 132, the Chinese scientist Chang Heng (78–139) constructed what may have been the first seismographic instrument, which was designed to detect not only the presence of seismic activity but also the direction from which it came. His invention ultimately was discarded, however, and understanding of earthquakes progressed little for more than 1,600 years.

The first crude seismograph was invented in 1703 by the French physicist Jean de Hautefeuille (1647–1724), long before Michell formally established a connection between shock waves and earthquakes. Historians date the starting point of modern seismographic monitoring, however, to an 1880 invention by the English geologist John Milne (1850–1913). Milne's creation, the first precise seismograph, measured motion with a horizontal pendulum attached to a pen that recorded movement on a revolving drum. Milne used his device to record earthquakes from as far away as Japan and helped establish seismologic stations around the world. The first modern seismograph in the United States was installed at the University of California at Berkeley and proved its accuracy in recording the 1906 San Francisco quake, discussed later in this essay.

MAGNITUDE: THE RICHTER SCALE. An earthquake can be measured according to either its magnitude or its intensity. The first refers to the amount of energy released by the earthquake, and its best-known scale of measurement is the Richter scale. Developed in 1935 by the American geophysicist Charles Richter (1900–1985), the Richter scale is logarithmic rather than arithmetic, meaning that increases in value involve multiplication rather than addition.

The numbers on the Richter scale, from 1.0 to 10.0, should be thought of as exponents rather than integers. Each whole-number increase represents a tenfold increase in the amplitude (size from crest to trough) of the seismic wave. Therefore 2.0 is not twice as much as 1.0; it is 10 times as much. To go from 1.0 to 3.0 is an increase by a factor of 100, and to go from 1.0 to 4.0 indicates an increase by a factor of 1,000. The scales of magnitude thus become ever greater, and while a whole-number increase on the Richter scale indicates an increase of amplitude by a factor of 10, it represents an increase of energy by a factor of about 31.

INTENSITY: THE MERCALLI SCALE. The amplitude and energy measured by the Richter scale are objective and quantitative, whereas intensity is more subjective and qualitative. Intensity, an indication of the earthquake's effect on human beings and structures, is measured by the Mercalli scale, named after the Italian seismologist Giuseppe Mercalli (1850–1914). The 12 levels on the Mercalli scale range from I, which means that few people felt the quake, to XII, which indicates total damage. A few comparisons serve to illustrate the scales' relationship to each other.

A score of I on the Mercalli scale equates to a value between 1.0 and a 3.0 on the Richter scale and indicates a tremor felt only by a very few people under very specific circumstances. At 5.0 to 5.9 on the Richter scale (VI to VII on the Mercalli scale), everyone feels the earthquake, and many people are frightened, but only the most poorly built structures are damaged significantly. Above 7.0 on the Richter scale and VIII on the Mercalli scale, wooden and then masonry structures collapse, as do bridges, while railways are bent completely out of shape. In populated areas, as we shall see, the death toll can be enormous.

FAMOUS QUAKES

The great San Francisco earthquake, which struck on April 18, 1906, spawned a massive fire, and these events resulted in the deaths of some 700 people, including 270 inmates of a mental institution. Another 300,000 people were left homeless, and 490 city blocks were destroyed. Ultimately, the financial impact of the San Francisco quake proved to be one of the contributing factors in the March 13, 1907, stock market crash that played a key role in the panic of 1907.

At 5:04 P.M. on October 17, 1989, another quake struck San Francisco. It lasted just 15 seconds, long enough to kill some 90 people and cause $6 billion in property damage. Though it was the biggest quake since the 1906 tremor, it was much smaller: 7.19 on the Richter scale, or about one-fifth of the 7.7 measured for the 1906 quake. The 1989 Loma Prieta quake cost much more than the earlier tragedy, which had caused $500 million in damage, but, of course, half a billion dollars in 1906 was worth a great deal more than $6 billion 83 years later.

Neither earthquake, however, was the greatest in American history; in fact, the 1989 quake does not rank among the top 15, even for the continental United States. The eight worst earthquakes in U.S. history all occurred in one state: Alaska. Greatest of all was the March 27, 1964, quake at Prince William Sound, which registered a staggering 9.2 on the Richter scale and took 125 lives. Of that number, 110 were killed in a tsunami resulting from the quake.

The high incidence of earthquakes in Alaska is understandable enough, given the fact that its southern edge abuts a subduction zone and, along with the panhandle, sits astride the boundary between the North American and Pacific plates. Although this may not be much comfort to people in Alaska, it is fortunate that the most earthquake-prone state is also the most sparsely populated. Had the epicenter (the point on Earth's surface directly above the hypocenter, or focal point from which a quake originates) of the 1964 earthquake been in New York City, the death toll would have been closer to 125,000 than 125.

GREATEST QUAKES IN THE CONTINENTAL UNITED STATES. Similarly, it is fortunate that the greatest quakes to strike the continental United States outside California have been in low-population centers. Of the 15 worst earthquakes in U.S. history, only one was outside Alaska, California, or Hawaii. In fact, it was the site of both the worst and the fifth-worst earthquakes in the continental United States: New Madrid, Missouri, site of a 7.9 quake on February 7, 1812, and a 7.7 quake just two months earlier, on December 16, 1811.

New Madrid lies at the extreme southeastern tip of Missouri, near the Mississippi River and within a few hundred miles of several major cities: St. Louis, Missouri; Memphis and Nashville, Tennessee; and Louisville, Kentucky. Had the 1811 and 1812 quakes occurred today, they undoubtedly would have taken a vast human toll owing to the resulting floods. As it was, some lakes rose by as much as 15 ft. (4.6 m), streams changed direction, and the Mississippi and Ohio rivers flowed backward. Fortunately, however, they occurred at a time when the Missouri Territory—it was not even a state yet—and surrounding areas were sparsely populated. The combined death toll was in the single digits.

Of the top 15 earthquakes in the continental United States, all but the 1906 San Francisco quake (which ranks sixth) took place in areas with small populations. Ten were in California but generally in less populous areas or at times when there were fewer people there (e.g., no. 2: Fort Tejon, 1857; no. 3: Owens Valley, 1872; and no. 4: Imperial Valley, 1892). Other than the two New Madrid quakes, the remainder took place in Nevada (no. 12: Dixie Valley, 1954), Montana (no. 13: Hebgen Lake, 1959), and Idaho (no. 14:

Earthquake damage in California. *(© David Weintraub/Photo Researchers. Reproduced by permission.)*

Borah Peak, 1983). As of late 2001, the Idaho quake was the second most recent, after no. 9, at Landers, California, in 1992. (The 1994 Northridge quake, in the Los Angeles area, ranked 6.7 on the Richter scale, well below the 7.3 registered by no. 15, west of Eureka, California, in 1922.)

THE WORLD'S MOST DESTRUCTIVE QUAKES. None of these U.S. quakes, however, compares with the July 27, 1976, earthquake in T'ang-shan, China. The worst earthquake in modern history, it shattered some 20 sq. mi. (32 km sq.) near the capital city of Beijing and killed about 242,000 people while injuring an estimated 600,000 more. There are several interesting aspects to this quake, aside from its sheer scale.

One is sociological, involving the human response to the quake. As in Portugal in 1755, people saw events in a cosmic light; in this case, though, they did not interpret the quake as evidence of divine unconcern but quite the opposite. Mao Tse-tung (1893–1976), by far the most influential Chinese leader of modern times, had just died, and the Chinese saw the natural disaster as fitting into a larger historical pattern. In the traditional Chinese view, earthquakes, floods, and other signs from the gods attend the change of dynasties.

Also interesting is the fact that the T'ang-shan quake was merely the most destructive in a worldwide series of quakes that took place between February and November 1976. In the course of these events, 23,000 people died in Guatemala after a February 4 quake; 3,000 people were reported dead, and 3,000 more were missing in Indonesia, as a result of a series of quakes and landslides on June 26 (later, the U.S. Federal Emergency Management Agency, or FEMA, placed the number of dead from the Indonesia quake at just 443); as many as 8,000 people died in an earthquake and tsunami that hit the southern Philippines on August 16; and 4,000 more perished in a November 24 quake in eastern Turkey.

Similarly, a few months before the 1755 Lisbon earthquake, a quake hit northern Iran. This is an aspect of seismology that cannot be explained readily by plate tectonics: Iran and Portugal are not on the same plate margins; in fact, northern Iran is not on a plate margin at all. Likewise, the areas hit in the 1976 quakes were not on the same plate margins, and T'ang-shan (unlike the other places affected) is not on a major plate margin at all. Nor is Shansi in north-central China, site of history's most destructive earthquake on January 24, 1556, which killed more than 830,000 people.

Note that the 1556 and 1976 Chinese quakes were the worst, respectively, of all history and of modern times—but *worst* in terms of intensity, not magnitude. One might say that they were the most destructive but not the worst in pure terms. The 1976 quake is not even on the list of the 10 worst earthquakes—those of the greatest magnitude—in the twentieth century. Whereas the T'ang-shan quake registered 8.0, a quake in Chile on May 22, 1960, had a magnitude of 9.5, or about 50 times greater, yet the death toll was much smaller—2,000 people killed. Three thousand more were injured in the Chilean quake, and two million were rendered homeless. The last statistic perhaps best signifies the magnitude of the 1960 quake, which caused tsunamis that brought death and destruction as far away as Hawaii, Japan, the Philippines, and the west coast of the United States.

Learning from Seismology

As noted, plate tectonics does not explain every earthquake, but it does explain most, probably about 90%. Not that it is much help in predicting earthquakes, because the processes of plate tectonics take place on an entirely different time scale than the ones to which humans are accustomed. These processes happen over millions of years, so it is hard to say, for any particular year, just what will happen to a particular plate.

Plate tectonics, then, tells us only areas of likelihood for earthquakes—specifically, plate boundaries of the types discussed near the end of Plate Tectonics. And even though the processes that create the conditions for an earthquake are extremely slow, usually the discernible indications that an earthquake is coming appear only seconds before the quake itself. Thus, as sophisticated as modern seismometers are, they generally do not provide enough advance notice of earthquakes to offer any lifesaving value.

There are not just a few earthquakes each year but many thousands of tremors, most of them too small to register. Sometimes these tremors may be foreshocks, or indicators that a quake is coming to a particular area. In addition, studies of other phenomena, from tidal behavior to that of animals (probably a result of some creatures' extremely acute hearing), may offer suggestions as to the locations of future quakes.

Earth's Core and the Moho. Seismology is useful for learning about more than just earthquakes or volcanoes. During the early years of the twentieth century, the Irish geologist Richard Dixon Oldham (1858–1936) studied data from a number of recent earthquakes and noticed a difference in the behavior of compression waves and shear waves. (These terms merely express the differ-

KEY TERMS

AMPLITUDE: The maximum displacement of a vibrating material, or the "size" of a wave from crest to trough.

BODY WAVES: Waves whose line of propagation is through the body of a medium. These include P-waves (primary waves), which move extremely fast and are longitudinal, and S-waves (secondary waves), which are move somewhat less fast and are transverse. Compare with *surface waves.*

COMPRESSION: A form of stress produced by the action of equal and opposite forces, the effect of which is to reduce the length of a material. Compression is a form of pressure.

CONTINENTAL DRIFT: The theory that the configuration of Earth's continents was once different than it is today, that some of the individual landmasses of today once were joined in other continental forms, and that these landmasses later separated and moved to their present locations.

CONVERGENCE: A tectonic process whereby plates move toward each other. Usually associated with subduction, convergence typically occurs in the ocean, creating an oceanic trench. It is one of the three ways, along with divergence and transform motion, that plates interact.

CORE: The center of Earth, an area constituting about 16% of the planet's volume and 32% of its mass. Made primarily of iron and another, lighter element (possibly sulfur), it is divided between a solid inner core with a radius of about 760 mi. (1,220 km) and a liquid outer core about 1,750 mi. (2,820 km) thick.

CRUST: The uppermost division of the solid earth, representing less than 1% of its volume and varying in depth from 3–37 mi. (5–60 km). Below the crust is the mantle.

DIVERGENCE: A tectonic process whereby plates move away from each other. Divergence results in the separation of plates and most often is associated either with seafloor spreading or the formation of rift valleys. It is one of the three ways, along with convergence and transform motion, that plates interact.

ELASTICITY: The response of solids to stress.

EPICENTER: The point on Earth's surface directly above the hypocenter, or the focal point from which an earthquake originates.

FAULT: An area of fracturing, as a result of stress, between rocks.

FOLD: An area of rock that has been bent by stress.

GEOPHYSICS: A branch of the earth sciences that combines aspects of geology and physics. Geophysics addresses the planet's physical processes as well as its gravitational, magnetic, and electric properties and the means by which energy is transmitted through its interior.

HEAT: Internal thermal energy that flows from one body of matter to another.

HOT SPOT: A region of high volcanic activity.

KEY TERMS CONTINUED

INTENSITY: Where earthquakes are concerned, intensity refers to the amount of damage to humans and buildings. Subjective and qualitative (as opposed to magnitude, which is objective and quantitative), intensity is measured by the Mercalli scale.

KINETIC ENERGY: The energy that an object possesses by virtue of its motion.

LITHOSPHERE: The upper layer of Earth's interior, including the crust and the brittle portion at the top of the mantle.

LONGITUDINAL WAVE: A wave in which the movement of vibration is in the same direction as the wave itself. This is contrasted with a *transverse wave.*

LOVE WAVES: See *surface waves.*

MAGNITUDE: Where earthquakes are concerned, magnitude refers to the amount of energy released by the quake as well as the amplitude of the seismic waves. Objective and quantitative (as opposed to intensity, which is subjective and qualitative), magnitude is measured by the Richter scale.

MANTLE: The thick, dense layer of rock, approximately 1,429 mi. (2,300 km) thick, between Earth's crust and its core. In reference to the other terrestrial planets, *mantle* simply means the area of dense rock between the crust and core.

MERCALLI SCALE: See *intensity.*

PLATE MARGINS: Boundaries between plates.

PLATE TECTONICS: The name both of a theory and of a specialization of tectonics. As an area of study, plate tectonics deals with the large features of the lithosphere and the forces that shape them. As a theory, it explains the processes that have shaped Earth in terms of plates and their movement. Plate tectonics theory brings together aspects of continental drift, seafloor spreading, seismic and volcanic activity, and the structures of Earth's crust to provide a unifying model of Earth's evolution. It is one of the dominant concepts in the modern earth sciences.

PLATES: Large, movable segments of the lithosphere.

PROPAGATION: The act or state of traveling from one place to another.

P-WAVES: See *body waves.*

RADIOACTIVITY: A term describing a phenomenon whereby certain materials are subject to a form of decay brought about by the emission of high-energy particles or radiation. Forms of particles or energy include alpha particles (positively charged helium nuclei); beta particles (either electrons or subatomic particles called *positrons*); or gamma rays, which occupy the highest energy level in the electromagnetic spectrum.

RAYLEIGH WAVES: See *surface waves.*

RICHTER SCALE: See *magnitude.*

SEISMIC WAVE: A packet of energy resulting from the disturbance that accompanies a strain on rocks in the lithosphere.

SEISMOGRAPH: An instrument designed to record information regarding seismic waves.

KEY TERMS CONTINUED

SEISMOLOGY: The study of seismic waves as well as the movements and vibrations that produce them.

SEISMOMETER: An instrument for detecting seismic waves.

SHEAR: A form of stress resulting from equal and opposite forces that do not act along the same line. If a thick, hardbound book is lying flat and one pushes the front cover from the side so that the covers and pages no longer constitute parallel planes, this is an example of shear.

STRAIN: The ratio between the change in dimension experienced by an object that has been subjected to stress and the original dimensions of the object.

STRESS: In general terms, any attempt to deform a solid. Types of stress include tension, compression, and shear.

SUBDUCTION: A tectonic process that results when plates converge, and one plate forces the other down into Earth's mantle. As a result, the subducted plate eventually undergoes partial melting.

SURFACE WAVES: Seismic waves whose line of propagation is along the surface of a medium such as the solid earth. These waves tend to be slower and more destructive than body waves. Examples include Rayleigh waves (waves with both transverse and longitudinal characteristics) and Love waves (purely longitudinal). Compare with *body waves.*

S-WAVES: See *body waves.*

TECTONICS: The study of tectonism, including its causes and effects, most notably mountain building.

TECTONISM: The deformation of the lithosphere.

TENSION: A form of stress produced by a force that acts to stretch a material.

THERMAL ENERGY: Heat energy, a form of kinetic energy produced by the motion of atomic or molecular particles in relation to one another. The greater the relative motion of these particles, the greater the thermal energy.

TRANSFORM MOTION: A tectonic process whereby plates slide past each other. It is one of the three ways, along with convergence and divergence, that plates interact.

TRANSVERSE WAVE: A wave in which the vibration or motion is perpendicular to the direction in which the wave is moving. Compare with *longitudinal wave.*

TSUNAMI: A tidal wave produced by an earthquake or volcanic eruption. The term comes from the Japanese words for "harbor" and "wave."

ences in stress produced by seismic waves.) As it turns out, shear waves are deflected as they pass through the center of Earth. Since liquid cannot experience shear, this finding told him that the planet's core must be made of molten material.

Oldham's findings, published in 1906—the same year as the great San Francisco quake—made him a pioneer in the application of seismology to the study of Earth's interior. Three years later, studies of earthquake waves by the Croatian geologist Andrija Mohorovicic (1857–1936) revealed still more about the interior of the planet. Based on his analysis of wave speeds and arrival times, Mohorovicic was able to calculate the depth at which the crust becomes the mantle. This change is abrupt rather than gradual, and the boundary on which it occurs is today known as the Mohorovicic discontinuity, or simply the Moho.

WHERE TO LEARN MORE

Abbott, Patrick. *Natural Disasters.* Dubuque, IA: William C. Brown Publishers, 1996.

National Earthquake Information Center—World Data Center for Seismology, Denver (Web site). <http://neic.usgs.gov/>.

Prager, Ellen J., Kate Hutton, Costas Synolakis, et al. *Furious Earth: The Science and Nature of Earthquakes, Volcanoes, and Tsunamis.* New York: McGraw-Hill, 2000.

Seismology Info Page—Netherlands (Web site). <http://home.wish.net/~riknl/>.

Seismology Research Centre—Australia (Web site). <http://www.seis.com.au/>.

Sigurdsson, Haraldur. *Encyclopedia of Volcanoes.* San Diego: Academic Press, 2000.

Sleh, Keery E., and Simon LeVay. *The Earth in Turmoil: Earthquakes, Volcanoes, and Their Impact on Humankind.* New York: W. H. Freeman, 1998.

University of Alaska Fairbanks Seismology (Web site). <http://www.aeic.alaska.edu/>.

University of Washington Seismology and Earthquake Information (Web site). <http://www.geophys.washington.edu/SEIS/>.

"Earthquake Hazards Program," USGS (United States Geological Survey) (Web site). <http://earthquake.usgs.gov/>.

Wade, Nicholas. *The Science Times Book of Natural Disasters.* New York: Lyons Press, 2000.

GEOMORPHOLOGY

GEOMORPHOLOGY

CONCEPT

The surface of Earth is covered with various landforms, a number of which are discussed in various entries throughout this book. This essay is devoted to the study of landforms themselves, a subdiscipline of the geologic sciences known as geomorphology. The latter, as it has evolved since the end of the nineteenth century, has become an interdisciplinary study that draws on areas as diverse as plate tectonics, ecology, and meteorology. Geomorphology is concerned with the shaping of landforms, through such processes as subsidence and uplift, and with the classification and study of such landforms as mountains, volcanoes, and islands.

HOW IT WORKS

An Evolving Area of Study

Geomorphology is an area of geology concerned with the study of landforms, with the forces and processes that have shaped them, and with the description and classification of various physical features on Earth. The term, which comes from the Greek words *geo,* or "Earth," and *morph,* meaning "form," was coined in 1893 by the American geologist William Morris Davis (1850–1934), who is considered the father of geomorphology.

During Davis's time, geomorphology was concerned primarily with classifying different structures on Earth's surface, examples of which include mountains and islands, discussed later in this essay. This view of geomorphology as an essentially descriptive, past-oriented area of study closely aligned with historical geology prevailed throughout the late nineteenth and early twentieth centuries.

By the mid-twentieth century, however, the concept of geomorphology inherited from Davis had fallen into disfavor, to be replaced by a paradigm, or model, oriented toward physical rather than historical geology. (These two principal branches of geology are concerned, in the first instance, with Earth's past and the processes that shaped it and, in the second instance, with Earth's current physical features and the processes that continue to shape it.)

RETHINKING GEOMORPHOLOGY. As reconceived in the 1950s and thereafter, geomorphology became an increasingly exact science. As has been typical of many sciences in their infancy, early geomorphology focused on description rather than prediction and tended to approach its subject matter in a qualitative fashion. The term qualitative suggests a comparison between qualities that are not defined precisely, such as "fast" and "slow" or "warm" and "cold." On the other hand, a quantitative approach, as has been implemented for geomorphology from the mid–twentieth century onward, centers on a comparison between precise quantities—for instance, 10 lb. (4.5 kg) versus 100 lb. (45 kg) or 50 MPH (80.5 km/h) versus 120 MPH (193 km/h).

As part of its shift in focus, geomorphology began to treat Earth's physical features as systems made up of complex and ongoing interactions. This view fell into line with a general emphasis on the systems concept in the study of Earth. (See Earth Systems for more about the systems concept.) As geomorphology evolved, it became more interdisciplinary, as we shall see. This, too,

MOUNT MACHHAPUCHHARE IN THE HIMALAYAS. THE HIMALAYAS WERE FORMED THROUGH THE PROCESS OF UPLIFT, OR RISING OF EARTH'S SURFACE. *(© George Turner/Photo Researchers. Reproduced by permission.)*

was part of an overall trend in the earth sciences toward an approach that viewed subjects in broad, cross-disciplinary terms as opposed to a narrow focus on specific areas of study.

LANDFORMS AND PROCESSES

Two concerns are foremost within the realm of geomorphology, and these concerns reflect the stages of its history. First, in line with Davis's original conception of geomorphology as an area of science devoted to classifying and describing natural features, there is its concern with topography. The latter may be defined as the configuration of Earth's surface, including its relief (elevation and other inequalities) as well as the position of physical features.

These physical features are called landforms, examples of which include mountains, plateaus, and valleys. Geomorphology always has involved classification, and early scientists working in this subdiscipline addressed the classification of landforms. Other systems of classification, however, are not so concerned with cataloging topographical features themselves as with differentiating the processes that shaped them. This brings us to the other area of interest in geomorphology: the study of how landforms came into being.

SHAPING THE EARTH. Among the processes that drive the shaping of landforms is plate tectonics, or the shifting of large, movable segments of lithosphere (the crust and upper layer of Earth's mantle). Plate tectonics is discussed in detail within its own essay and more briefly in other areas throughout this book, as befits its status as one of the key areas of study in the earth sciences.

Other processes also shape landforms. Included among these processes are weathering, the breakdown of rocks and minerals at or near the surface of Earth due to physical or chemical processes; erosion, the movement of soil and rock due to forces produced by water, wind, glaciers, gravity, and other influences; and mass wasting or mass movement, the transfer of earth material, by processes that include flow, slide, fall, and creep, down slopes. Also of interest are fluvial and eolian processes (those that result from water flow and wind, respectively) as well as others related to glaciers and coastal formations.

Human activity also can play a significant role in shaping Earth. This effect may be direct, as when the construction of cities, the building of dams, or the excavation of mines alters the landscape. On the other hand, it can be indirect. In the latter instance, human activity in the biosphere exerts an impact, as when the clearing of forest land or the misuse of crop land results in the formation of a dust bowl.

INTERDISCIPLINARY STUDIES

As noted earlier, geomorphology is characteristic of the earth sciences as a whole in its emphasis on an interdisciplinary approach. As is true of earth scientists in general, those studying landforms and the processes that shape them do not work simply in one specialty. Among the areas of interest in geomorphology are, for example, deep-sea geomorphology, which draws on oceanography, and planetary geomorphology, the study of landscapes on other planets.

When studying coastal geomorphology, a geologist may draw on realms as diverse as fluid mechanics (an area of physics that studies the behavior of gases and liquids at rest and in motion) and sedimentology. The investigation of such processes as erosion and mass wasting calls on knowledge in the atmospheric sciences as well as the physics and chemistry of soil. It is almost inevitable that a geomorphologic researcher will draw on geophysics as well as on such subspecialties as volcanology. These studies may go beyond the "hard sciences," bringing in such social sciences as geography.

REAL-LIFE APPLICATIONS

SUBSIDENCE

Subsidence refers to the process of subsiding (settling or descending), on the part of either an air column or the solid earth, or, in the case of solid earth, to the resulting formation or depression. Subsidence in the atmosphere is discussed briefly in the entry Convection. Subsidence that occurs in the solid earth, known as geologic subsidence, is the settling or sinking by a body of rock or sediment. (The latter can be defined as material deposited at or near Earth's surface from a number of sources, most notably preexisting rock.)

As noted earlier, many geomorphologic processes can be caused either by nature or by human beings. An example of natural subsidence takes place in the aftermath of an earthquake, during which large areas of solid earth may simply drop by several feet. Another example can be observed at the top of a volcano some time after it has erupted, when it has expelled much of its material (i.e., magma) and, as a result, has collapsed.

Natural subsidence also may result from cave formation in places where underground water has worn away limestone. If the water erodes too much limestone, the ceiling of the cave will subside, usually forming a sinkhole at the surface. The sinkhole may fill with water, making a lake; the formation of such sinkholes in many spots throughout an area (whether the sinkholes become lakes or not), is known as karst topography.

In places where the bedrock is limestone—particularly in the sedimentary basins of rivers—karst topography is likely to develop. The United States contains the most extensive karst region in the world, including the Mammoth cave system in Kentucky. Karst topography is very pronounced in the hills of southern China, and karst landscapes have been a prominent feature of Chinese art for centuries. Other extensive karst regions can be found in southern France, Central America, Turkey, Ireland, and England.

MAN-MADE SUBSIDENCE. Man-made subsidence often ensues from the removal of groundwater or fossil fuels, such as petroleum or coal. Groundwater removal can be perfectly safe, assuming the area experiences sufficient rainfall to replace, or recharge, the lost water. If recharging does not occur in the necessary proportions, however, the result will be the eventual collapse of the aquifer, a layer of rock that holds groundwater.

In so-called room-and-pillar coal mining, pillars, or vertical columns, of coal are left standing, while the areas around them are extracted. This method maintains the ceiling of the "room" that has been mined of its coal. After the mine is abandoned, however, the pillar eventually may experience so much stress that it breaks, leading to the collapse of the mined room. As when the ceiling of a cave collapses, the subsidence of a coal mine leaves a visible depression above ground.

UPLIFT

As its name implies, uplift describes a process and results opposite to those of subsidence. In uplift the surface of Earth rises, owing either to a decrease in downward force or to an increase in upward force. One of the most prominent examples of uplift is seen when plates collide, as when India careened into the southern edge of the Eurasian landmass some 55 million years ago. The result has been a string of mountain ranges, including the Himalayas, Karakoram Range, and Hindu Kush, that contain most of the world's tallest peaks.

Plates move at exceedingly slow speeds, but their mass is enormous. This means that their inertia (the tendency of a moving object to keep moving unless acted upon by an outside force) is likewise gargantuan in scale. Therefore, when plates collide, though they are moving at a rate equal to only a few inches a year, they will keep pushing into each other like two automobiles crumpling in a head-on collision. Whereas a car crash is over in a matter of seconds, however, the crumpling of continental masses takes place over hundreds of thousands of years.

When sea floor collides with sea floor, one of the plates likely will be pushed under by the other one, and, likewise, when sea floor collides with continental crust, the latter will push the sea floor under. (See Plate Tectonics for more about oceanic-oceanic and continental-oceanic collisions.) This results in the formation of volcanic mountains, such as the Andes of South America or the Cascades of the Pacific Northwest, or volcanic islands, such as those of Japan, Indonesia, or Alaska's Aleutian chain.

ISOSTATIC COMPENSATION. In many other instances, collision, compression, and extension cause uplift. On the other hand, as noted, uplift may result from the removal of a weight. This occurs at the end of an ice age, when glaciers as thick as 1.9 mi. (3 km) melt, gradually removing a vast weight pressing down on the surface below.

This movement leads to what is called isostatic compensation, or isostatic rebound, as the crust pushes upward like a seat cushion rising after a person is longer sitting on it. Scandinavia is still experiencing uplift at a rate of about 0.5 in. (1 cm) per year as the after-effect of glacial melting from the last ice age. The latter ended some 10,000 years ago, but in geologic terms this is equivalent to a few minutes' time on the human scale.

ISLANDS

Geomorphology, as noted earlier, is concerned with landforms, such as mountains and volcanoes as well as larger ones, including islands and even continents. Islands present a particularly interesting area of geomorphologic study. In general, islands have certain specific characteristics in terms of their land structure and can be analyzed from the standpoint of the geosphere, but particular islands also have unique ecosystems, requiring an interdisciplinary study that draws on botany, zoology, and other subjects.

In addition, there is something about an island that has always appealed to the human imagination, as evidenced by the many myths, legends, and stories about islands. Some examples include Homer's *Odyssey,* in which the hero Odysseus visits various islands in his long wanderings; Thomas More's *Utopia,* describing an idealized island republic; *Robinson Crusoe,* by Daniel Defoe, in which the eponymous hero lives for many years on an island with no companion but the trusty native Friday; *Treasure Island,* by Robert Louis Stevenson, in which the island is the focus of a treasure hunt; and Mark Twain's *Adventures of Huckleberry Finn,* depicting Jackson Island in the Mississippi River, to which Huckleberry Finn flees to escape "civilization."

One of the favorite subjects of cartoonists is that of a castaway stranded on a desert island, a mound of sand with no more than a single tree.

Movies, too, have long portrayed scenarios, from the idyllic to the brutal, that take place on islands, particularly deserted ones, a notable example being *Cast Away* (2000). A famous line by the English poet John Donne (1572–1631) warns that "no man is an island," implying that many wish they could enjoy the independence suggested by the concept of an island. Within the Earth system, however, nothing is fully independent, and, as we shall see, this is certainly the case where islands are concerned.

THE ISLANDS OF EARTH. Earth has literally tens of thousands of islands. Just two archipelagos (island chains), those that make up the Philippines and Indonesia, include thousands of islands each. While there are just a few dozen notable islands on Earth, many more dot the planet's seas and oceans. The largest are these:

- Greenland (Danish, northern Atlantic): 839,999 sq. mi.(2,175,597 sq km)
- New Guinea (divided between Indonesia and Papua New Guinea, western Pacific): 316,615 sq. mi. (820,033 sq km)
- Borneo (divided between Indonesia and Malaysia, western Pacific): 286,914 sq. mi. (743,107 sq km)
- Madagascar (Malagasy Republic, western Indian Ocean): 226,657 sq. mi. (587,042 sq km)
- Baffin (Canadian, northern Atlantic): 183,810 sq. mi. (476,068 sq km)
- Sumatra (Indonesian, northeastern Indian Ocean): 182,859 sq. mi. (473,605 sq km)

The list could go on and on, but it stops at Sumatra because the next-largest island, Honshu (part of Japan), is less than half as large, at 88,925 sq. mi. (230,316 sq km). Clearly, not all islands are created equal, and though some are heavily populated or enjoy the status of independent nations (e.g., Great Britain at number eight or Cuba at number 15), they are not necessarily the largest. On the other hand, some of the largest are among the most sparsely populated.

Of the 32 largest islands in the world, more than a third are in the icy northern Atlantic and Arctic, with populations that are small or practically nonexistent. Greenland's population, for instance, was just over 59,000 in 1998, while that of Baffin Island was about 13,200. On both islands, then, each person has about 14 frozen sq. mi. (22 sq km) to himself or herself, making them among the most sparsely populated places on Earth.

A TINY ISLAND IN THE TRUK LAGOON, MICRONESIA. (© *Stuart Westmorland/Photo Researchers. Reproduced by permission.)*

CONTINENTS, OCEANS, AND ISLANDS. Australia, of course, is not an island but a continent, a difference that is not related directly to size. If Australia *were* an island, it would be by far the largest. Australia is regarded as a continent, however, because it is one of the principal landmasses of the Indo-Australian plate, which is among a handful of major continental plates on Earth. Whereas continents are more or less permanent (though they have experienced considerable rearrangement over the eons), islands come and go, seldom lasting more than 10 million years. Erosion or rising sea levels remove islands, while volcanic explosions can create new ones, as when an eruption off the coast of Iceland resulted in the formation of an island, Surtsey, in 1963.

Islands are of two types, continental and oceanic. Continental islands are part of continental shelves (the submerged, sloping ledges of continents) and may be formed in one of two ways. Rising ocean waters either cover a coastal region, leaving only the tallest mountains exposed as islands or cut off part of a peninsula,

KEY TERMS

BIOSPHERE: A combination of all living things on Earth—plants, animals, birds, marine life, insects, viruses, single-cell organisms, and so on—as well as all formerly living things that have not yet decomposed.

CONVERGENCE: A tectonic process whereby plates move toward each other. Usually associated with subduction, convergence typically occurs in the ocean, creating an oceanic trench. It is one of the three ways, along with divergence and transform motion, in which plates interact.

CRUST: The uppermost division of the solid earth, representing less than 1% of its volume and varying in depth from 3 to 37 mi. (5 to 60 km). Below the crust is the mantle.

EROSION: The movement of soil and rock due to forces produced by water, wind, glaciers, gravity, and other influences.

GEOLOGY: The study of the solid earth, in particular its rocks, minerals, fossils, and land formations.

GEOMORPHOLOGY: An area of geology concerned with the study of landforms, with the forces and processes that have shaped them, and with the description and classification of various physical features on Earth.

GEOPHYSICS: A branch of the earth sciences that combines aspects of geology and physics. Geophysics addresses the planet's physical processes as well as its gravitational, magnetic, and electric properties and the means by which energy is transmitted through its interior.

GEOSPHERE: The upper part of Earth's continental crust, or that portion of the solid earth on which human beings live and which provides them with most of their food and natural resources.

HISTORICAL GEOLOGY: The study of Earth's physical history. Historical geology is one of two principal branches of geology, the other being physical geology.

LANDFORM: A notable topographical feature, such as a mountain, plateau, or valley.

LITHOSPHERE: The upper layer of Earth's interior, including the crust and the brittle portion at the top of the mantle.

MASS WASTING: The transfer of earth material, by processes that include flow, slide, fall, and creep, down slopes. Also known as mass movement.

which then becomes an island. Most of Earth's significant islands are continental and are easily spotted as such, because they lie at close proximity to continental landmasses. Many other continental islands are very small, however; examples include the barrier islands that line the East Coast of the United States. Formed from mainland sand brought to the coast by rivers, these are technically not continental islands, but they more clearly fit into that category than into the grouping of oceanic islands.

Oceanic islands, of which the Hawaiian-Emperor island chain and the Aleutians off the Alaskan coast are examples, form as a result of volcanic activity on the ocean floor. In most cases, there is a region of high volcanic activity,

KEY TERMS CONTINUED

PHYSICAL GEOLOGY: The study of the material components of Earth and of the forces that have shaped the planet. Physical geology is one of two principal branches of geology, the other being historical geology.

PLATE TECTONICS: The name both of a theory and of a specialization of tectonics. As an area of study, plate tectonics deals with the large features of the lithosphere and the forces that shape them. As a theory, it explains the processes that have shaped Earth in terms of plates and their movement.

PLATES: Large, movable segments of the lithosphere.

QUALITATIVE: Involving a comparison between qualities that are not defined precisely, such as "fast" and "slow" or "warm" and "cold."

QUANTITATIVE: Involving a comparison between precise quantities—for instance, 10 lb. versus 100 lb. or 50 mi. per hour versus 120 mi. per hour.

RELIEF: Elevation and other inequalities on a land surface.

SEDIMENT: Material deposited at or near Earth's surface from a number of sources, most notably preexisting rock.

SEDIMENTOLOGY: The study and interpretation of sediments, including sedimentary processes and formations.

SUBSIDENCE: A term that refers either to the process of subsiding (settling or descending), on the part of either air or solid earth or, in the case of solid earth, to the resulting formation. Subsidence thus is defined variously as the downward movement of air, the sinking of ground, or a depression in Earth's crust.

SYSTEM: Any set of interactions that can be set apart mentally from the rest of the universe for the purposes of study, observation, and measurement.

TECTONICS: The study of tectonism, including its causes and effects, most notably mountain building.

TECTONISM: The deformation of the lithosphere.

TOPOGRAPHY: The configuration of Earth's surface, including its relief, as well as the position of physical features.

UPLIFT: A process whereby the surface of Earth rises, due to either a decrease in downward force or an increase in upward force.

WEATHERING: The breakdown of rocks and minerals at or near the surface of Earth due to physical or chemical processes.

called a hot spot, beneath the plates, which move across the hot spot. This is the situation in Hawaii, and it explains why the volcanoes on the southern islands are still active while those to the north are not: the islands themselves are moving north across the hot spot. If two plates converge and one subducts (see Plate Tectonics for an explanation of this process), a deep trench with a parallel chain of volcanic islands may develop. Exemplified by the Aleutians, these chains are called island arcs.

ISLAND ECOSYSTEMS. The ecosystem, or community of all living organisms, on islands can be unique owing to their separation from continents. The number of life-forms on an island is relatively small and can encompass some unusual circumstances compared with the larger ecosystems of continents. Ireland, for instance, has no native snakes, a fact "explained" by the legend that Saint Patrick drove them away. Hawaii and Iceland are also blessedly free of serpents.

Oceanic islands, of course, tend to have more unique ecosystems than do continental islands. The number of land-based animal lifeforms is necessarily small, whereas the varieties of birds, flying insects, and surrounding marine life will be greater owing to those creatures' mobility across water. Vegetation is relatively varied, given the fact that winds, water currents, and birds may carry seeds.

Nonetheless, ecosystems of islands tend to be fairly delicate and can be upset by the human introduction of new predators (e.g., dogs) or new creatures to consume plant life (e.g., sheep). These changes sometimes can have disastrous effects on the overall balance of life on islands. Overgrazing may even open up the possibility of erosion, which has the potential of bringing an end to an island's life.

WHERE TO LEARN MORE

Color Landform Atlas of the United States (Web site). <http://fermi.jhuapl.edu/states/states.html>.

Erickson, Jon. *Rock Formations and Unusual Geologic Structures: Exploring the Earth's Surface.* New York: Facts on File, 1993.

Erickson, Jon. *Making the Earth: Geologic Forces That Shape Our Planet.* New York: Facts on File, 2000.

Gerrard, John. *Rocks and Landforms.* Boston: Unwin Hyman, 1988.

Image Gallery of Landforms (Web site). <http://www.athena.ivv.nasa.gov/curric/land/landform/landform.html>.

Selby, Michael John. *Earth's Changing Surface: An Introduction to Geomorphology.* New York: Clarendon Press, 1985.

Tinkler, K. J. *A Short History of Geomorphology.* Totowa, NJ: Barnes and Noble, 1985.

The Virtual Geomorphology (Web site). <http://main.amu.edu.pl/~sgp/gw/gw.htm>.

Wells, Lisa. "Images Illustrating Principles of Geomorphology" (Web site). <http://geoimages.berkeley.edu/GeoImages/Wells/wells.html>.

Zoehfeld, Kathleen Weidner, and James Graham Hale. *How Mountains Are Made.* New York: HarperCollins, 1995.

MOUNTAINS

CONCEPT

Among the most striking of geologic features are mountains, created by several types of tectonic forces, including collisions between continental masses. Mountains have long had an impact on the human psyche, for instance by virtue of their association with the divine in the Greek myths, the Bible, and other religious or cultural traditions. One does not need to be a geologist to know what a mountain is; indeed there is no precise definition of mountain, though in most cases the distinction between a mountain and a hill is fairly obvious. On the other hand, the defining characteristics of a volcano are more apparent. Created by violent tectonic forces, a volcano usually is considered a mountain, and almost certainly is one after it erupts, pouring out molten rock and other substances from deep in the earth.

HOW IT WORKS

PLATE TECTONICS

Earth is constantly moving, driven by forces beneath its surface. The interior of Earth itself is divided into three major sections: the crust, mantle, and core. The lithosphere is the upper layer of Earth's interior, including the crust and the brittle portion at the top of the mantle. Tectonism is the deformation of the lithosphere, and the term tectonics refers to the study of this deformation. Most notable among examples of tectonic deformation is mountain building, or orogenesis, discussed later in this essay.

The planet's crust is not all of one piece: it is composed of numerous plates, which are steadily moving in relation to one another. This movement is responsible for all manner of phenomena, including earthquakes, volcanoes, and mountain building. All these ideas and many more are encompassed in the concept of plate tectonics, which is the name for a branch of geologic and geophysical study and of a dominant principle often described as "the unifying theory of geology" (see Plate Tectonics).

CONTENTS UNDER PRESSURE. Tectonism results from the release and redistribution of energy from Earth's interior. This energy is either gravitational, and thus a function of the enormous mass at the planet's core, or thermal, resulting from the heat generated by radioactive decay. Differences in mass and heat within the planet's interior, known as pressure gradients, result in the deformation of rocks, placing many forms of stress and strain on them.

In scientific terms, *stress* is any attempt to deform an object, and strain is a change in dimension resulting from stress. Rocks experience stress in the form of tension, compression, and shear. Tension acts to stretch a material, whereas compression is a form of stress produced by the action of equal and opposite forces, whose effect is to reduce the length of a material. (Compression is a form of pressure.) Shear results from equal and opposite forces that do not act along the same plane. If a thick, hardbound book is lying flat, and one pushes the front cover from the side so that the covers and pages are no longer in alignment, is an example of shear.

Rocks manifest the strain resulting from these stresses by warping, sliding, or breaking. They may even flow, as though they were liquids, or melt and thus truly become liquid. As a result, Earth's interior may manifest faults, or fractures in rocks, as

MOUNTAINS SOMETIMES ARISE IN ISOLATION, AS WAS THE CASE WITH MOUNT KILIMANJARO IN TANZANIA. *(© T. Davis/Photo Researchers. Reproduced by permission.)*

well as folds, or bends in the rock structure. The effects can be seen on the surface in the form of subsidence, which is a depression in the crust; or uplift, the raising of crustal materials. Earthquakes and volcanic eruptions also may result.

OROGENESIS

There are two basic types of tectonism: epeirogenesis and orogenesis. The first takes its name from the Greek words *epeiros,* meaning "mainland," and *genesis,* or "origin." Epeirogenesis, which takes the form of either uplift or subsidence, is a chiefly vertical form of movement and plays little role in either plate tectonics or mountain building.

Orogenesis, on the other hand, *is* mountain building, as the prefix *oros* ("mountain") shows. Orogenesis involves the formation of mountain ranges by means of folding, faulting, and volcanic activity—lateral movements as opposed to vertical ones. Geologists typically use the term *orogenesis,* instead of just "mountain building," when discussing the formation of large belts of mountains from tectonic processes.

PLATE MARGINS. Plates may converge (move toward one another), diverge (move away from one another), or experience transform motion, meaning that they slide against one another. Convergence usually is associated with subduction, in which one plate is forced down into the mantle and eventually undergoes partial melting. This typically occurs in the ocean, creating a depression known as an oceanic trench.

There are three types of plate margins, or boundaries between plates, depending on the two types of crusts that interact: oceanic with oceanic, continental with continental, or continental with oceanic. Any of these margins may be involved in mountain formation. Orogenic belts, or mountain belts, typically are situated in subduction zones at convergent plate boundaries and consist of two types.

The first type occurs when igneous material (i.e., rock from volcanoes) forms on the upper plate of a subduction zone, causing the surface to rise. This can take place either in the oceanic crust, in which case the mountains formed are called island arcs, or along continental-oceanic margins. The Aleutian Islands are an example of an island arc, while the Andes range represent mountains formed by the subduction of an oceanic plate under a continental one.

The second type of mountain belt occurs when continental plates converge or collide. When continental plates converge, one plate may "try" to subduct the other, but ultimately the buoyancy of the lower plate (which floats, as it were, on the lithosphere) pushes it upward. The result is the creation of a wide, unusually thick or "tall" belt. An example is the Himalayas, the world's tallest mountain range, which is still being pushed upward as the result of a collision between India and Asia that happened some 30 million years ago. (See Plate Tectonics for more about continental drift and collisions between plates.)

REAL-LIFE APPLICATIONS

WHAT IS A MOUNTAIN?

In the 1995 film *The Englishman Who Went Up a Hill But Came Down a Mountain,* the British actor Hugh Grant plays an English cartographer, or mapmaker, sent in 1917 by his government to measure what is purportedly "the first mountain inside Wales." He quickly determines that according to standards approved by His Majesty, the "mountain" in question is, in fact, a hill. Much of the film's plot thereafter revolves around attempts on the part of the villagers to rescue their beloved mountain from denigration as a "hill," a fate they prevent by piling enough rocks and dirt onto the top to make it meet specifications.

This comedy aptly illustrates the somewhat arbitrary standards by which people define mountains. The British naturalist Roderick Peattie (1891–1955), in his 1936 book *Mountain Geography,* maintained that mountains are distinguished by their impressive appearance, their individuality, and their impact on the human imagination. This sort of qualitative definition, while it is certainly intriguing, is of little value to science; fortunately, however, more quantitative standards exist.

In Britain and the United States, a mountain typically is defined as a landform with an elevation of 985 ft. (300 m) above sea level. This was the standard applied in *The Englishman,* but the Welsh villagers would have had a hard time raising their "hill" to meet the standards used in continental Europe: 2,950 ft. (900 m) above sea level. This seems to be a more useful standard, because the British and American one would take in high plains and other nonmountainous regions of relatively great altitude. On the other hand, there are landforms in Scotland that rise only a few hundred meters above sea level, but their morphologic characteristics or shape seem to qualify them as mountains. Not only are their slopes steep, but the presence of glaciers and snow-capped peaks, with their attendant severe weather and rocky, inhospitable soil, also seem to indicate the topography associated with mountains.

MOUNTAIN GEOMORPHOLOGY

One area of the geologic sciences especially concerned with the study of mountains is geomorphology, devoted to the investigation of landforms. Geomorphologists studying mountains must draw on a wide variety of disciplines, including geology, climatology, biology, hydrology, and even anthropology, because, as discussed at the conclusion of this essay, mountains have played a significant role in the shaping of human social groups.

From the standpoint of geology and plate tectonics, mountain geomorphology embraces a complex of characteristic formations, not all of which are necessarily present in a given orogen, or mountain. These include forelands and foredeeps along the plains; foreland fold-and-thrust belts, which more or less correspond to "foothills" in layperson's terminology; and a crystalline core zone, composed of several types of rock, that is the mountain itself.

ENVIRONMENTAL ZONES. Mountain geomorphology classifies various environmental zones, from lowest to highest altitude. Near the bottom are flood plains, river terraces, and alluvial fans, all areas heavily affected by rivers flowing from higher elevations. (In fact, many of the world's greatest rivers flow from mountains, examples being the Himalayan Ganges and Indus rivers in Asia and the Andean Amazon in South America.) Farming villages may be found as high as the 9,845-ft. to 13,125-ft. range (3,000–4,000 m), an area known as a submontane, or forested region.

The tree line typically lies at an altitude of 14,765 ft. (4,500 m). Above this point, there is little human activity but plenty of geologic activity, including rock slides, glacial flow, and, at very high altitudes, avalanches. From the tree line upward, the altitude levels that mark a particular region are differentiated for the Arctic and tropical zones, with much lower altitudes in the Arctic mountains. For instance, the tree line lies at about 330 ft. (100 m) in the much colder Arctic zone.

Above the tree line is the subalpine, or montane, region. The mean slope angle of the mountain is less steep here than it is at lower or higher elevations: in the submontane, or forested region, below the tree line, the slope is about 30°, and above the subalpine, in the high alpine, the slope can become as sharp at 65°. In the subalpine, however, it is only about 20°, and because grass (if not trees) grows in this region, it is suited for grazing.

It may seem surprising to hear of shepherds bringing sheep to graze at altitudes of 16,400 ft. (5,000 m), as occurs in tropical zones. This does not necessarily mean that people live at such altitudes; more often than not, mountain dwellers have their settlements at lower elevations, and shepherds simply take their flocks up into the heights for grazing. Yet the ancient Bolivian city of Tiahuanaco, which flourished in about A.D. 600—some four centuries before the rise of the Inca—lay at an almost inconceivable altitude of 13,125 ft. (4,000 m), or about 2.5 times the elevation of Denver, Colorado, America's Mile-High City.

CLASSIFYING MOUNTAINS

There are several ways to classify mountains and groups of mountains. Mountain belts, as described earlier, typically are grouped according to formation process and types of plates: island arcs, continental arcs (formed with the subduction of an oceanic plate by a continental plate), and collisional mountain belts. Sometimes a mountain arises in isolation, an example being Kilimanjaro in Tanzania, Africa. Another example is Stone Mountain outside Atlanta, an exposed pluton, or a mass of crystalline igneous rock that forms deep in Earth's crust and rises. Many volcanoes, which we discuss later, arise individually, but mountains are most likely to appear in conjunction with other mountains. One such grouping, though far from the only one, is a mountain range, which can be defined as a relatively localized series of peaks and ridges.

RANGES, CHAINS, AND MASSES. Some of the world's most famous mountain ranges include the Himalayas, Karakoram Range, and Pamirs in central Asia; the Alps and Urals in Europe; the Atlas Mountains in Africa; the Andes in South America; and the Cascade Range, Sierra Nevada, Rocky Mountains, and Appalachians as well as their associated ranges in North America. Ranges affiliated with the Appalachians, for instance, include the Great Smokies in the south and the Adirondacks, Alleghenies, and Poconos in the north.

Several of the examples given here illustrate the fact that ranges are not the largest groupings of mountains. Sometimes series of ranges stretch across a continent for great distances in what are called mountain chains, an example of which is the Mediterranean chain of Balkans, Apennines, and Pyrenees that stretches across southern Europe.

There also may be irregular groupings of mountains, which lack the broad linear sweep of mountain ranges or chains and which are known as mountain masses. The mountains surrounding the Tibetan plateau represent an example of a mountain mass. Finally, ranges, chains, and masses of mountains may be combined to form vast mountain systems. An impressive example is the Alpine-Himalayan system, which unites parts of the Eurasian, Arabian, African, and Indo-Australian continental plates.

OTHER TYPES OF MOUNTAIN. There are certain special types of orogeny, as when ocean crust subducts continental crust—something that is not supposed to happen but occasionally does. This rare variety of subduction is called obduction, and the mountains produced are called ophiolites. Examples include the uplands near Troodos in Cyprus and the Taconic Mountains in upstate New York.

Fault-block mountains appear when two continental masses push against each other and the upper portion of a continental plate splits from the deeper rocks. A portion of the upper crust, usually several miles thick, begins to move slowly across the continent. Ultimately it runs into another mass, creating a ramp. This can result in unusually singular mountains, such as Chief Mountain in Montana, which slid across open prairie on a thrust sheet.

Under the ocean is the longest mountain chain on Earth, the mid-ocean ridge system, which runs down the center of the Atlantic Ocean and continues through the Indian and Pacific oceans. Lava continuously erupts along this ridge, releasing geothermal energy and opening up new strips of ocean floor. This brings us to a special kind of mountain, typically resulting from the sort of dramatic plate tectonic processes that also produce earthquakes: volcanoes.

VOLCANOES

Most volcanoes are mountains, and for this reason, it is appropriate to discuss them together; however, a volcano is not necessarily a mountain. A volcano may be defined as a natural opening in Earth's surface through which molten (liquid), solid, and gaseous material erupts. The word *volcano* also is used to describe the cone of erupted material that builds up around the opening or fissure. Because these cones are often quite impressive in height, they frequently are associated with mountains.

THE POPOCATEPETL VOLCANO ERUPTS, SPEWING ASH, ROCKS, AND GASES. *(© Wesley Bocxe/Photo Researchers. Reproduced by permission.)*

Though volcanic activity has been the case of death and destruction, it is essential to the planet's survival. Volcanic activity is the principal process through which chemical elements, minerals, and other compounds from Earth's interior reach its surface. These substances, such as carbon dioxide, have played a major role in the development of the planet's atmosphere, waters, and soils. Even today, soil in volcanic areas is among the richest on Earth. Volcanoes provide additional benefits in their release of geothermal energy, used for heating and other purposes in such countries as Iceland, Italy, Hungary, and New Zealand (see Energy and Earth). In addition, volcanic activity beneath the oceans promises to supply almost limitless geothermal energy, once the technology for its extraction becomes available.

FORMATION OF VOLCANOES. As noted earlier, land volcanoes are formed in coastal areas where continental and oceanic plates converge. As the oceanic plate is subducted and pushed farther and farther beneath the continental surface, the buildup of heat and pressure results in the melting of rock. This molten rock, or magma, tends to rise toward the surface and collect in magma reservoirs. Pressure buildup in the magma reservoir ultimately pushes the magma upward through cracks in Earth's crust, creating a volcano.

Volcanoes also form underwater, in which case they are called seamounts. Convergence of oceanic plates causes one plate to sink beneath the other, creating an oceanic trench; as a result, magma rises from the subducted plate to fashion volcanoes. If the plates diverge, magma seeps upward at the ridge or margin between plates, producing more seafloor. This process, known as seafloor spreading, leads to the creation of volcanoes on either side of the ridge.

In some places a plate slides over a stationary area of volcanic activity, known as a hot spot. These are extremely hot plumes of magma that well up from the crust, though not on the edge or margin of a plate. A tectonic plate simply drifts across the hot spot, and as it does, the area just above the hot spot experiences volcanic activity. Hot spots exist in Hawaii, Iceland, Samoa, Bermuda, and America's Yellowstone National Park.

CLASSIFYING VOLCANOES. Volcanoes can be classified in terms of their volcanic activity, in which case they are labeled as active (currently erupting), dormant (not currently erupting but likely to do so in the future), or extinct. In the case of an extinct volcano, no eruption has been noted in recorded history, and it is likely that the volcano has ceased to erupt permanently.

In terms of shape, volcanoes fall into four categories: cinder cones, composite cones, shield volcanoes, and lava domes. These types are distinguished not only by morphologic characteristics but also by typical sizes and even angles of slope. For instance, cinder cones, built of lava fragments, have slopes of 30° to 40°, and are seldom more than 1,640 ft. (500 m) in height.

Composite cones, or stratovolcanoes, are made up of alternating layers of lava (cooled magma), ash, and rock. (The prefix *strato* refers to these layers.) They may slope as little as 5° at the base and as much as 30° at the summit. Stratovolcanoes may grow to be as tall as 2–3 mi. (3.2–4.8 km) before collapsing and are characterized by a sharp, dramatic shape. Examples include Fuji, a revered mountain that often serves as a symbol of Japan, and Washington state's Mount Saint Helens.

A shield volcano, which may be a solitary formation and often is located over a hot spot, is built from lava flows that pile one on top of another. With a slope as little as 2° at the base and no more than 10° at the summit, shield volcanoes are much wider than stratovolcanoes, but sometimes they can be impressively tall. Such is the case with Mauna Loa in Hawaii, which at 13,680 ft. (4,170 m) above sea level is the world's largest active volcano. Likewise, Mount Kilimanjaro, though long ago gone dormant, is the tallest mountain in Africa.

Finally, there are lava domes, which are made of solid lava that has been pushed upward. Closely related is a volcanic neck, which often forms from a cinder cone. In the case of a volcanic neck, lava rises and erupts, leaving a mountain that looks like a giant gravel heap. Once it has become extinct, the lava inside the volcano begins to solidify. Over time the rock on the exterior wears away, leaving only a vent filled with solidified lava, usually in a funnel shape. A dramatic example of this appears at Shiprock, New Mexico.

VOLCANIC ERUPTIONS. Volcanoes frequently are classified by the different ways in which they erupt. These types of eruption, in turn, result from differences in the material being disgorged from the volcano. When the magma is low in gas and silica (silicon dioxide, found in sand and rocks), the volcano erupts in a relatively gentle way. Its lava is thin and spreads quickly. Gas and silica–rich magma, on the other hand, brings about a violent explosion that yields tarlike magma.

There are four basic forms or phases of volcanic eruption: Hawaiian, Strombolian, Vulcanian, and Peleean. The Hawaiian phase is simply a fountain-like gush of runny lava, without any explosions. The Strombolian phase (named after a volcano on a small island off the Italian penin-

CRATER LAKE IN OREGON IS THE RESULT OF THE COLLAPSE OF A MAGMA CHAMBER AFTER A VOLCANIC ERUPTION. THIS COLLAPSE FORMS A BOWL-LIKE CRATER CALLED A CALDERA, WHICH FILLS WITH WATER. *(© Francois Gohier/Photo Researchers. Reproduced by permission.)*

sula) involves thick lava and mild explosions. In a Vulcanian phase, magma has blocked the volcanic vent, and only after an explosion is the magma released, with the result that tons of solid material and gases are hurled into the sky. Most violent of all is the Peleean, named after Mount Pelée on Martinique in the Caribbean (discussed later). In the Peleean phase, the volcano disgorges thick lava, clouds of gas, and fine ash, all at formidable velocities.

Accompanying a volcanic eruption in many cases are fierce rains, the result of the expulsion of steam from the volcano, after which the steam condenses in the atmosphere to form clouds. Gases thrown into the atmosphere are often volatile and may include hydrogen sulfide, fluorine, carbon dioxide, and radon. All are detrimental to human beings when present in sufficient quantity, and radon is radioactive.

Not surprisingly, the eruption of a volcano completely changes the morphologic characteristics of the landform. During the eruption a crater is formed, and out of this flows magma and ash, which cool to form the cone. In some cases, the magma chamber collapses just after the eruption, forming a caldera, or a large, bowl-shaped crater. These caldera (the plural as well as singular form) may fill with water, as was the case at Oregon's Crater Lake.

INFAMOUS VOLCANIC DISASTERS. Volcanoes result from some of the same tectonic forces as earthquakes (see Seismology), and, not surprisingly, they often have resulted in enormous death and destruction. Some remarkable examples include:

- *Vesuvius, Italy, A.D. 79 and 1631:* Situated along the Bay of Naples in southern Italy, Vesuvius has erupted more than 50 times during the past two millennia. Its most famous eruption occurred in A.D. 79, when the Roman Empire was near the height of its power. The first-century eruption buried the nearby towns of Pompeii and Herculaneum, where bodies and buildings were preserved virtually intact until excavation of the area in 1748. Another eruption, in 1631, killed some 4,000 people.
- *Krakatau, Indonesia, A.D. 535 (?) and 1883:* The most famous eruption of Krakatau occurred in 1883, resulting in the loss of some 36,000 lives. The explosion, which was heard 3,000 mi. away, threw 70-lb. (32-kg) boulders as far as 50 mi. (80 km). It also produced a tsunami, or tidal wave, 130 ft.

(40 m) high, which swept away whole villages. In addition, the blast hurled so much dust into the atmosphere that the Moon appeared blue or green for two years. It is also possible that Krakatau erupted in about A.D. 535, causing such a change in the atmosphere that wide areas of the world experienced years without summer. (See Earth Systems for more on this subject.)

- *Tambora, Indonesia, 1815:* Another Indonesian volcano, Tambora, killed 12,000 people when it erupted in 1815. As with Krakatau in 535, this eruption was responsible for a year without summer in 1816 (see Earth Systems).
- *Pelée, Martinique, 1902:* When Mount Pelée erupted on the Caribbean island of Martinique, it sent tons of poisonous gas and hot ash spilling over the town of Saint-Pierre, killing all but four of its 29,937 residents.
- *Saint Helens, Washington, 1980:* Relatively small compared with earlier volcanoes, the Mount Saint Helens blast is still significant because it was so recent and took place in the United States. The eruption sent debris flying upward 1,300 ft. (396 m) and caused darkness over towns as far as 85 mi. (137 km) away. Fifty-seven people died in the eruption and its aftermath.
- *Pinatubo, Philippines, 1991:* Dormant for 600 years, Mount Pinatubo began to rumble one day in 1991 and, after a few days, erupted in a cloud that spread ash 6 ft. (1.83 m) deep along a radius of 2 mi. (3.2 km). A U.S. air base 15 mi. (24 km) away was buried. The blast threw 20 million tons (18,144,000 metric tons) of sulfuric acid 12 mi. (19 km) into the stratosphere, and the cloud ultimately covered the entire planet, resulting in moderate cooling for a few weeks.

THE IMPACT OF MOUNTAINS

Volcanic eruptions are among the most dramatic effects produced by mountains, but they are far from the only ones. Every bit as fascinating are the effects mountains produce on the weather, on the evolution of species, and on human society. In each case, mountains serve as a barrier or separator—between masses of air, clouds, and populations.

Wind pushes air and moisture-filled clouds up mountain slopes, and as the altitude increases, the pressure decreases. As a result, masses of warm, moist air become larger, cooler, and less dense. This phenomenon is known as *adiabatic expansion,* and it is the same thing that happens when an aerosol can is shaken, reducing the pressure of gases inside and cooling the surface of the can. Under the relatively high-pressure and high-temperature conditions of the flatlands, water exists as a gas, but in the heights of the mountaintops, it cools and condenses, forming clouds.

RAIN SHADOWS. As the clouds rise along the side of the mountain, they begin to release heavy droplets in the form of rain and, at higher altitudes, snow. By the time the cloud crosses the top of the mountain, however, it will have released most of its moisture, and hence the other side of the mountain may be arid. The leeward side, or the side opposite the wind, becomes what is called a rain shadow.

Although they are only 282 mi. (454 km) apart, the cities of Seattle and Spokane, Washington, have radically different weather patterns. Famous for its almost constant rain, Seattle lies on the windward, or wind-facing, side of the Cascade Range, toward the Pacific Ocean. On the leeward side of the Cascades is Spokane, where the weather is typically warm and dry. Though it is only on the other side of the state, Spokane might as well be on the other side of the continent. Indeed, it is associated more closely with the arid expanses of Idaho, whereas Seattle belongs to a stretch of cold, wet Pacific terrain that includes San Francisco and Portland, Oregon.

Much of the western United States consists of deserts formed by rain shadows or, in some cases, double rain shadows. Much of New Mexico, for instance, lies in a double rain shadow created by the Rockies in the west and Mexico's Sierra Madres to the south. In southern California, tall redwoods line the lush windward side of the Sierra Nevadas, while Death Valley and the rest of the Mojave Desert lies in the rain shadow on the eastern side. The Great Basin that covers eastern Oregon, southern Idaho, much of Utah, and almost all of Nevada, likewise is created by the rain shadow of the Sierra Nevada-Cascade chain.

MOUNTAINS AND SPECIES. One of the most intriguing subjects involved in the study of mountains is their effects on large

KEY TERMS

ACTIVE: A term to describe a volcano that is currently erupting.

COMPRESSION: A form of stress produced by the action of equal and opposite forces, the effect of which is to reduce the length of a material. Compression is a form of pressure.

CONVERGENCE: A tectonic process whereby plates move toward each other.

CRUST: The uppermost division of the solid earth, representing less than 1% of its volume and varying in depth from 3 mi. to 37 mi. (5–60 km). Below the crust is the mantle.

DIVERGENCE: A tectonic process whereby plates move away from each other.

DORMANT: A term to describe a volcano that is not currently erupting but is likely to do so in the future.

EPEIROGENESIS: One of two principal forms of tectonism, the other being orogenesis. Derived from the Greek words *epeiros* ("mainland") and *genesis* ("origins"), epeirogenesis takes the form of either uplift or subsidence.

EXTINCT: A term to describe a volcano for which no eruption has been known in recorded history. In this case, it is likely that the volcano has ceased to erupt permanently.

GEOMORPHOLOGY: An area of physical geology concerned with the study of landforms, with the forces and processes that have shaped them, and with the description and classification of various physical features on Earth.

HOT SPOT: A region of high volcanic activity.

LANDFORM: A notable topographical feature, such as a mountain, plateau, or valley.

LITHOSPHERE: The upper layer of Earth's interior, including the crust and the brittle portion at the top of the mantle.

MANTLE: The thick, dense layer of rock, approximately 1,429 mi. (2,300 km) thick, between Earth's crust and its core.

MORPHOLOGY: Structure or form or the study thereof.

MOUNTAIN CHAIN: A series of ranges stretching across a continent for a great distance.

MOUNTAIN MASS: An irregular grouping of mountains, which lacks the broad linear sweep of a range or chain.

MOUNTAIN RANGE: A relatively localized series of peaks and ridges.

MOUNTAIN SYSTEM: A combination of ranges, chains, and masses of mountains that stretches across vast distances, usually encompassing more than one continent.

OROS: A Greek word meaning "mountain," which appears in such words as *orogeny,* a variant of *orogenesis; orogen,* another term for "mountain" and *orogenic,* as in "orogenic belt."

OROGENESIS: One of two principal forms of tectonism, the other being epeirogenesis. Derived from the Greek words *oros* ("mountain") and *genesis* ("origin"), orogenesis involves the formation of mountain ranges by means of folding, faulting, and volcanic activity. The processes of orogenesis play a major role in plate tectonics.

PLATE MARGINS: Boundaries between plates.

KEY TERMS CONTINUED

PLATE TECTONICS: The name both of a theory and of a specialization of tectonics. As an area of study, plate tectonics deals with the large features of the lithosphere and the forces that shape them. As a theory, it explains the processes that have shaped Earth in terms of plates and their movement.

PLATES: Large, movable segments of the lithosphere.

SHEAR: A form of stress resulting from equal and opposite forces that do not act along the same line. If a thick, hardbound book is lying flat, and one pushes the front cover from the side so that the covers and pages are no longer aligned, this is an example of shear.

STRAIN: The ratio between the change in dimension experienced by an object that has been subjected to stress and the original dimensions of the object.

STRESS: In general terms, any attempt to deform a solid. Types of stress include tension, compression, and shear.

SUBSIDENCE: A term that refers either to the process of subsiding, on the part of air or solid earth, or, in the case of solid earth, to the resulting formation. Subsidence thus is defined variously as the downward movement of air, the sinking of ground, or a depression in Earth's crust.

TECTONICS: The study of tectonism, including its causes and effects, most notably mountain building.

TECTONISM: The deformation of the lithosphere.

TENSION: A form of stress produced by a force that acts to stretch a material.

TOPOGRAPHY: The configuration of Earth's surface, including its relief as well as the position of physical features.

UPLIFT: A process whereby the surface of Earth rises, as the result of either a decrease in downward force or an increase in upward force.

VOLCANO: A natural opening in Earth's surface through which molten (liquid), solid, and gaseous material erupts. The word volcano is also used to describe the cone of erupted material that builds up around the opening or fissure.

groups of plants, animals, and humans. Mountains may separate entire species, creating pockets of flora and fauna virtually unknown to the rest of the world. Thus, during the 1990s, huge numbers of species that had never been catalogued were discovered in the mountains of southeast Asia.

The formation of mountains and other landforms may even lead to speciation, a phenomenon in which members of a species become incapable of reproducing with other members, thus creating a new species. When the Colorado River cut open the Grand Canyon, it separated groups of squirrels that lived in the high-altitude pine forest. Over time these populations ceased to interbreed, and today the Kaibab squirrel of the north rim and the Abert squirrel of the south are separate species, no more capable of interbreeding than humans and apes.

HUMAN SOCIETIES AND MOUNTAINS. Although the Appalachians of the eastern United States are hundreds of millions of years old, most ranges are much younger. Most will erode or otherwise cease to exist in a relatively short time (short, that is, by geologic standards), yet to humans throughout the ages, mountains have seemed a symbol of permanence. This is just one aspect of mountains' impact on the human psyche.

In his 1975 study of symbolism in political movements, *Utopia and Revolution,* Melvin J. Lasky devoted considerable space to the mountain and its association with divinity through figures such as the Greek Olympians and Noah and Moses in the Bible. Clearly, mountains have proved enormously influential on human attitudes, and nowhere is this more obvious than in relation to the people who live *in* the mountains. Whether the person is a coal miner from Appalachia or a rancher from the Rockies, a Scottish highlander or a Quechua-speaking Peruvian, the mentality is similar, characterized by a combination of hardiness, fierce independence, and disdain for lowland ways.

These characteristics, combined with the harsh weather of the mountains, have made mountain warfare a challenge to lowland invaders. This explains the fact that Switzerland has kept itself free from involvement in European wars since Napoleon's time, and why the independent Scottish Highlands were long a thorn in England's side. It also explains why neither the British nor the Russian empires could manage to control Afghanistan fully during their struggle over that mountainous nation in the late nineteenth and early twentieth centuries.

Britain eventually pulled out of the "Great Game," as this struggle was called, but Russia never really did. Many years later, the Soviets became bogged down in a war in Afghanistan that they could not win. The war, which lasted from 1979 to 1989, helped bring about the end of the Soviet Union and its system of satellite dictatorships. More than a decade later, as the United States launched strikes against Afghanistan in 2001, a superpower once again faced the challenge posed by one of the poorest, most inhospitable nations on Earth.

But the independence of the mountaineer is deceptive; in fact, mountains have little to offer, economically, other than their beauty and the resources deep beneath their surfaces. In other words, they are really of value only to flatland tourists and mining companies. Since few mountain environments offer much promise agriculturally, the people of the mountains are dependent on the flatlands for sustenance. Gorgeous and rugged as they are, such mountainous states as Colorado or Wyoming might be as poor as Afghanistan were it not for the fact that they belong to a larger political unit, the United States.

WHERE TO LEARN MORE

A Geological History of Rib Mountain, Wisconsin (Web site). <http://www.uwmc.uwc.edu/geography/ribmtn/ribmtn.htm>.

Gore, Pamela. "Physical Geology at Georgia Perimeter College" (Web site). <http://www.gpc.peachnet.edu/~pgore/geology/geo101.htm>.

Kraulis, J. A., and John Gault. *The Rocky Mountains: Crest of a Continent.* New York: Facts on File, 1987.

Michigan Technological University Volcanoes Page (Web site). <http://www.geo.mtu.edu/volcanoes/>.

Oregon Geology—Cascade Mountains (Web site). <http://sarvis.dogami.state.or.us/learnmore/Cascades.HTM>.

Prager, Ellen J., Kate Hutton, Costas Synolakis, et al. *Furious Earth: The Science and Nature of Earthquakes, Volcanoes, and Tsunamis.* New York: McGraw-Hill, 2000.

Schaer, Jean-Paul, and John Rodgers. *The Anatomy of Mountain Ranges.* Princeton, NJ: Princeton University Press, 1987.

Sigurdsson, Haraldur. *Encyclopedia of Volcanoes.* San Diego: Academic Press, 2000.

Silver, Donald M., and Patricia Wynne. *Earth: The Ever-Changing Planet.* New York: Random House, 1989.

Volcanoes, Glaciers, and Plate Tectonics: The Geology of the Mono Basin (Web site). <http://www.r5.fs.fed.us/inyo/vvc/mono/vg&26pt.htm>.

EROSION

CONCEPT

Erosion is a broadly defined group of processes involving the movement of soil and rock. This movement is often the result of flowing agents, whether wind, water, or ice, which sometimes behaves like a fluid in the large mass of a glacier. Gravitational pull may also influence erosion. Thus, erosion, as a concept in the earth sciences, overlaps with mass wasting or mass movement, the transfer of earth material down slopes as a result of gravitational force. Even more closely related to erosion is weathering, the breakdown of rocks and minerals at or near the surface of Earth owing to physical, chemical, or biological processes. Some definitions of erosion even include weathering as an erosive process. Though most widely known as a by-product of irresponsible land use by humans and for its negative effect on landforms, erosion is neither unnatural nor without benefit. Far more erosion occurs naturally than as a result of land development, and a combination of weathering and erosion is responsible for producing the soil from which Earth's plants grow.

HOW IT WORKS

WEATHERING

The first step in the process of erosion is weathering. Weathering, in a general sense, occurs everywhere: paint peels; metal oxidizes, resulting in its tarnishing or rusting; and any number of products, from shoes to houses, begin to show the effects of physical wear and tear. The scuffing of a shoe, cracks in a sidewalk, or the chipping of glass in a gravel-spattered windshield are all examples of physical weathering. On the other hand, the peeling of paint is usually the result of chemical changes, which have reduced the adhesive quality of the paint. Certainly oxidation is a chemical change, meaning that it has not simply altered the external properties of the item but also has brought about a change in the way that the atoms are bonded.

Weathering, as the term is used in the geologic sciences, refers to these and other types of physical and chemical changes in rocks and minerals at or near the surface of Earth. A mineral is a substance that occurs naturally and is usually inorganic, meaning that it contains carbon in a form other than that of an oxide or a carbonate, neither of which is considered organic. It typically has a crystalline structure, or one in which the constituent parts have a simple and definite geometric arrangement repeated in all directions. Rocks are simply aggregates or combinations of minerals or organic material or both.

TWO AND ONE-HALF KINDS OF WEATHERING. There are three kinds of weathering (or perhaps two and one-half, since the third incorporates aspects of the first two): physical or mechanical, chemical, and biological. Physical or mechanical weathering takes place as a result of such factors as gravity, friction, temperature, and moisture. Gravity may cause a rock to drop from a height, such that it falls to the ground and breaks into pieces, while the friction of wind-borne sand may wear down a rock surface. Changes in temperature and moisture cause expansion and contraction of materials, as when water seeps into a crack in a rock and then freezes, expanding and splitting the rock.

NICOLA RIVER CANYON IN BRITISH COLUMBIA SHOWS THE EFFECTS OF FREEZE-THAW AND EROSION BY WIND AND RAIN. *(© K. Svensson/Photo Researchers. Reproduced by permission.)*

Minerals are chemical compounds; thus, whereas physical weathering attacks the rock as a whole, chemical weathering effects the breakdown of the minerals that make up the rock. This breakdown may lead to the dissolution of the minerals, which then are washed away by water or wind or both, or it may be merely a matter of breaking the minerals down into simpler compounds. Reactions that play a part in this breakdown may include oxidation, mentioned earlier, as well as carbonation, hydrolysis (a reaction with water that results in the separation of a compound to form a new substance or substances), and acid reactions. For instance, if coal has been burned in an area, sulfur impurities in the air react with water vapor (an example of hydrolysis) to produce acid rain, which can eat away at rocks. Rainwater itself is a weak acid, and over the years it slowly dissolves the marble of headstones in old cemeteries.

As noted earlier, there are either three or two and one-half kinds of weathering, depending on whether one considers biological weathering a third variety or merely a subset of physical and chemical weathering. The weathering exerted by organisms (usually plants rather than animals) on rocks and minerals is indeed chemical and physical, but because of the special circumstances, it is useful to consider it individually. There is likely to be a long-term interaction between the organism and the geologic item, an obvious example being a piece of moss that grows on a rock. Over time, the moss will influence both physical and chemical weathering through its attendant moisture as well as its specific chemical properties, which induce decomposition of the rock's minerals.

UNCONSOLIDATED MATERIAL

The product of weathering in rocks or minerals is unconsolidated, meaning that it is in pieces, like gravel, though much less uniform in size. This is called regolith, a general term that describes a layer of weathered material that rests atop bedrock. Sand and soil, including soil mixed with loose rocks, are examples of regolith. Regolith is, in turn, a type of sediment, material deposited at or near Earth's surface from a number of sources, most notably preexisting rock.

Every variety of unconsolidated material has its own angle of repose, or the maximum angle at which it can remain standing. Piles of rocks may have an angle of repose as high as 45°, whereas dry sand has an angle of only 34°. The addition of water can increase the angle of repose, as anyone who has ever strengthened a sand castle by adding water to it knows. Suppose one builds a sand castle in the morning, sloping the sand at angles that would be impossible if it were dry. By afternoon, as wind and sunlight dry out the sand, the sand castle begins to fall apart, because its angle of repose is too high for the dry sand.

Water gives sand surface tension, the same property that causes water that has been spilled on a table to bead up rather than lie flat. If too much water is added to the sand, however, the sand becomes saturated and will flow, a process called lateral spreading. On the other hand, with too little moisture, the material is susceptible to erosion. Unconsolidated material in nature generally has a slope less than its angle of repose, owing to the influence of wind and other erosive forces.

Introduction to Mass Wasting

There are three general processes whereby a piece of earth material can be moved from a high outcropping to the sea: weathering, mass wasting, and erosion. In the present context, we are concerned primarily with the last of these processes, of course, and secondarily with weathering, inasmuch as it contributes to erosion. A few words should be said about mass wasting, however, which, in its slower forms (most notably, creep), is related closely to erosion.

Mechanical or chemical processes, or a combination of the two, acting on a rock to dislodge it from a larger sample (e.g., separating a rock from a boulder) is an example of weathering, as we have seen. If the pieces of rock are swept away by a river in a valley below the outcropping, or if small pieces of rock are worn away by high winds, the process is erosion. Between the outcropping and the river below, if a rock has been broken apart by weathering, it may be moved farther along by mass-wasting processes, such as creep or fall.

Real-Life Applications

Mass Wasting in Action

One of the principal sources of erosion is gravity, which is also the force behind creep, the slow downward movement of regolith along a hill slope. The regolith begins in a condition of unstable equilibrium, like a soda can lying on its side rather than perpendicular to a table's surface: in both cases, the object remains in place, yet a relatively small disturbance would be enough to dislodge it.

Changes in temperature or moisture are among the leading factors that result in creep. A variation in either can cause material to expand or contract, and freezing or thawing may be enough to shake regolith from its position of unstable equilibrium. Water also can provide lubrication, or additional weight, that assists the material in moving. Though it is slow, over time creep can produce some of the most dramatic results of any mass-wasting process. It can curve tree trunks at the base, break or dislodge retaining walls, and overturn objects ranging from fence posts to utility poles to tombstones.

Other Varieties of Flow. Creep is related to another slow mass-wasting process, known as solifluction, that occurs in the active layer of permafrost—that is, the layer that thaws in the summertime. The principal difference between creep and solifluction is not the speed at which they take place (neither moves any faster than about 0.5 in. [1 cm] per year) but the materials involved. Both are examples of flow, a chaotic form of mass wasting in which masses of material that are not uniform move downslope. With the exception of creep and solifluction, most forms of flow are comparatively rapid, and some are extremely so.

Because it involves mostly dry material, creep is an example of granular flow, which is composed of 0% to 20% water; on the other hand, solifluction, because of the ice component, is an instance of slurry flow, consisting of 20% to 40% water. If the water content is more than 40%, a slurry flow is considered a stream. Types of granular flow that move faster than creep range from earth flow to debris avalanche. Both earth flow and debris flow, its equivalent in slurry form, move at a broad range of speeds, anywhere from about 4 in. (10 cm) per year to 0.6 mi. (1 km) per *hour.* Grain flow can be as fast as 60 mi. (100 km) per hour, and mud flow is even faster. Fastest of all is debris avalanche, which may achieve speeds of 250 mi. (400 km) per hour.

Other Types of Mass Wasting. Other varieties of mass wasting include slump, slide, and fall. Slump occurs when a mass of regolith slides over or creates a concave surface

ROCK ARCHES FORMED BY EROSION. *(© N. R. Rowan/Photo Researchers. Reproduced by permission.)*

(one shaped like the inside of a bowl.) The result is the formation of a small, crescent-shaped cliff, known as a scarp, at the upper end—rather like the crest of a wave. Slump often is classified as a variety of slide, in which material moves downhill in a fairly coherent mass (i.e., more or less in a section or group) along a flat or planar surface. These movements are sometimes called rock slides, debris slides, or, in common parlance, landslides.

In contrast to most other forms of mass wasting, in which there is movement along slopes that are considerably less than 90°, fall occurs at angles almost perpendicular to the ground. The "Watch for Falling Rock" signs on mountain roads may be frightening, and rock or debris fall is certainly one of the more dramatic forms of mass wasting. Yet the variety of mass wasting that has the most widespread effects on the morphology or shape of landforms is the slowest one—creep. (For more about the varieties of mass wasting, see Mass Wasting.)

What Causes Erosion?

As noted earlier, the influences behind erosion are typically either gravity or flowing media: water, wind, and even ice in glaciers. Liquid water is the substance perhaps most readily associated with erosion. Given enough time, water can wear away just about anything, as proved by the carving of the Grand Canyon by the Colorado River.

Dubbed the *universal solvent* for its ability to dissolve other materials, water almost never appears in its pure form, because it is so likely to contain other substances. Even "pure" mountain water contains minerals and pieces of the rocks over which it has flowed, a testament to the power of water in etching out landforms bit by bit. Nor does it take a rushing mountain stream or crashing waves to bring about erosion; even a steady drip of water is enough to wear away granite over time.

Moving water. Along coasts, pounding waves continually alter the shoreline. The sheer force of those walls of water, a result of the Moon's gravitational pull (and, to a lesser extent, the Sun's), is enough to wear away cliffs, let alone beaches. In addition, waves carry pieces of pebble, stone, and sand that cause weathering in rocks. Waves even can bring about small explosions in pockmarked rock surfaces by trapping air in small cracks; eventually the pressure becomes great enough that the air escapes, loosening pieces of the rock.

In addition to the erosive power of saltwater waves on the shore, there is the force exerted by running water in creeks, streams, and rivers. As the river moves, pushing along sediment and other materials eroded from the streambed or riverbed, it carves out deep chasms in the bedrock beneath. These moving bodies of water continually reshape the land, carrying soil and debris downslope, or from the source of the river to its mouth or delta. A delta is a region of sediment formed when a river enters a larger body of water, at which point the reduction in velocity on the part of the river current leads to the widespread deposition (depositing) of sediment. It is so named because its triangular shape resembles that of the Greek letter delta, Δ.

Water at the bottom of a large body, such as a pond or lake, also exerts erosive power. Then there is the influence of falling rain. Assuming ground is not protected by vegetation, raindrops can loosen particles of soil, sending them scattering in all directions. A rain that is heavy enough may dislodge whole layers of topsoil and send them rushing away in a swiftly moving current. The land left behind may be rutted and scarred, much of its best soil lost for good.

Just as erosion gives to the soil, it also can take away. Whereas erosion on the Nile delta acted to move rich, black soil into the region (hence, the ancient Egyptians' nickname for their country, the "black land"), erosion also can remove soil layers. As is often the case, it is much easier to destroy than to create: 1 in. (2.5 cm) of soil may take as long as 500 years to form, yet a single powerful rainstorm or windstorm can sweep it away.

Glaciers

Ice, of course, is simply another form of water, but since it is solid, its physical (*not* its chemical) properties are quite different. Generally, physical sciences, such as physics or chemistry, treat as fluid all forms of matter that flow, whether they are liquid or gas. Normally, no solids are grouped under the heading of "fluid," but in the earth sciences there is at least one type of solid object that behaves as though it were fluid: a glacier.

A glacier is a large, typically moving mass of ice either on or adjacent to a land surface. It does not flow in the same way that water does; rather, it is moved by gravity, as a consequence of its extraordinary weight. Under certain conditions, a glacier may have a layer of melted water surrounding it, which greatly enhances it mobility. Regardless of whether it has this lubricant, however, a glacier steadily moves forward, carrying pieces of rock, soil, and vegetation with it.

These great rivers of ice gouge out pieces of bedrock from mountain slopes, fashioning deep valleys. Ice along the bottom of the glacier pulls away rocks and soil, which assist it in wearing

A "DUST DEVIL," OR A SMALL WHIRLWIND, CARRIES WITH IT DEBRIS AND SAND. *(© Clem Haagner/Photo Researchers. Reproduced by permission.)*

away bedrock. The fjords of Norway, where high cliffs surround narrow inlets whose depths extend many thousands of feet below sea level, are a testament to the power of glaciers in shaping the Earth. The fact that the fjords came into existence only in the past two million years, a product of glacial activity associated with the last ice age, is evidence of something else remarkable about glaciers: their speed.

"Speed," of course, is a relative term when speaking about processes involved in the shaping of the planet. A "fast" glacier, one whose movement is assisted by a wet and warm (again, *relatively* warm!) maritime climate, moves at the rate of about 980 ft. (300 m) per year. Examples include not only the glaciers that shaped the fjords, but also the active Franz Josef glacier in southern New Zealand. By contrast, in the dry, exceptionally cold, inland climate of Antarctica, the Meserve glacier moves at the rate of just 9.8 ft. (3 m) per year.

WIND

The erosion produced by wind often is referred to as an eolian process, the name being a reference to Aeolus, the Greek god of the winds encountered in Homer's *Odyssey* and elsewhere. Eolian processes include the erosion, transport, and deposition of earth material owing to the action of wind. It is most pronounced in areas that lack effective ground cover in the form of solidly rooted, prevalent vegetation.

Eolian erosion in some ways is less forceful than the erosive influence of water. Water, after all, can lift heavier and larger particles than can the winds. Wind, however, has a much greater frictional component in certain situations. This is particularly true when the wind carries sand, every grain of which is like a cutting tool. In some desert regions the bases of rocks or cliffs have been sandblasted, leaving a mushroom-shaped formation. The wind could not lift the fine grains of sand very high, but in places where it has been able to do its work, it has left an indelible mark.

THE DUST BOWL AND HUMAN CONTRIBUTION TO EROSION

Though human actions are not a direct cause of erosion, human negligence or mismanagement often has prepared the way for erosive action by wind, water, or other agents. Interesting, soil itself, formed primarily by chemical weathering and enhanced by biological activity in the sediment, is a product of nature's erosive powers. Erosion transports materials from one place to another, robbing the soil in one place and greatly enhancing it in another.

This is particularly the case where river deltas are concerned. By transporting sediment and depositing it in the delta, the river creates an area of extremely fertile soil that, in some cases, has become literally the basis for civilizations. The earliest civilizations of the Western world, in Egypt and Sumer, arose in the deltas of the Nile and the Tigris-Euphrates river systems, respectively.

EROSION ON THE GREAT PLAINS. An extreme example of the negative effects on the soil that can come from erosion (and, ultimately, from human mismanagement) took place in Texas, Oklahoma, Colorado, and Kansas during the 1930s. In the preceding years, farmers unwittingly had prepared the way for vast erosion by overcultivating the land and not taking proper steps to preserve its moisture against drought. In some places farmers alternated between wheat cultivation and livestock grazing on particular plots of land.

The soil, already weakened by raising wheat, was damaged further by the hooves of livestock, and thus when a period of high winds began at the height of the Great Depression (1929–41), the land was particularly vulnerable. The winds carried dust to places as far away as the eastern seaboard, in some cases removing topsoil to a depth of 3–4 in. (7–10 cm). Dunes of dust as tall as 15–20 ft. (4.6–6.1 m) formed, and the economic blight of the Depression was compounded for the farmers of the plains states, many of whom lost everything.

Out of the Dust Bowl era came some of the greatest American works of art: the 1939 film *Wizard of Oz,* John Steinbeck's book *The Grapes of Wrath* and the acclaimed motion picture (1939 and 1940, respectively), as well as Dorothea Lange's haunting photographs of Dust Bowl victims. The Dust Bowl years also taught farmers and agricultural officials a lesson about land use, and in later years farming practices changed. Instead of alternating one year of wheat growing with one year in which a field lay fallow, or unused, farmers discovered that a wheat-sorghum-fallow cycle worked better. They also enacted other measures, such as the planting of trees to serve as windbreaks around croplands.

THE STRIKING LANDSCAPE OF EROSION

Among the by-products of erosion are some of the most dramatic landscapes in the world, many of which are to be found in the United States. A particularly striking example appears in Colorado, where the Arkansas River carved out the Royal Gorge. Though it is not nearly as deep as the Grand Canyon, this one has something the more famous gorge does not: a bridge. Motorists with the stomach for it can cross a span 1,053 ft. (0.32 km) above the river, one of the most harrowing drives in America.

Another, perhaps equally taxing, drive is that down California 1, a gorgeous scenic highway whose most dramatic stretches lie between Carmel and San Simeon. Drivers headed south find themselves pressed up against the edge of the cliffs, such that the slightest deviation from the narrow road would send an automobile and its passengers plummeting to the rocks many hundreds of feet below. These magnificent, terrifying landforms are yet another product of erosion, in this case, the result of the pounding Pacific waves.

Also striking is the topography produced by the erosion of material left over from a volcanic eruption. As discussed in the Mountains essay, Devils Tower National Monument in Wyoming is

KEY TERMS

CREEP: A form of mass wasting involving the slow downward movement of regolith as a result of gravitational force.

DELTA: A region of sediment formed when a river enters a larger body of water, at which point the reduction in velocity on the part of the river current leads to the widespread deposition of sediment.

DEPOSITION: The process whereby sediment is laid down on the Earth's surface.

EROSION: The movement of soil and rock due to forces produced by water, wind, glaciers, gravity, and other influences. In most cases, a fluid medium, such as air or water, is involved.

FLOW: A form of mass wasting in which a body of material that is not uniform moves rapidly downslope.

GEOMORPHOLOGY: An area of physical geology concerned with the study of landforms, with the forces and processes that have shaped them, and with the description and classification of various physical features on Earth.

GLACIER: A large, typically moving mass of ice either on or adjacent to a land surface.

LANDFORM: A notable topographical feature, such as a mountain, plateau, or valley.

MASS WASTING: The transfer of earth material, by processes that include creep, slump, slide, flow, and fall, down slopes. Also known as *mass movement.*

MORPHOLOGY: Structure or form or the study thereof.

REGOLITH: A general term describing a layer of weathered material that rests atop bedrock.

SEDIMENT: Material deposited at or near Earth's surface from a number of sources, most notably preexisting rock.

SLIDE: A variety of mass wasting in which material moves downhill in a fairly coherent mass (i.e., more or less in a section or group) along a flat or planar surface.

SLUMP: A form of mass wasting that occurs when a mass of regolith slides over or creates a concave surface (one shaped like the inside of a bowl).

TOPOGRAPHY: The configuration of Earth's surface, including its relief as well as the position of physical features.

WEATHERING: The breakdown of rocks and minerals at or near the surface of Earth due to physical, chemical, or biological processes.

the remains of an extinct volcano whose outer surface long ago eroded, leaving just the hard lava of the volcanic "neck." Erosion of lava also can produce mesas. Lava that has settled in a river valley may be harder than the rocks of the valley walls, such that the river eventually erodes the rocks, leaving only the lava platform. What was once the floor of the valley thus becomes the top of a mesa.

CONTROLLING EROSION

The force that shapes valleys and coastlines is certainly enough to destroy hill slopes, often with disastrous consequences for nearby residents. Such has been the case in California, where, during the 1990s, areas were dealt a powerful one-two punch of drought followed by rain. The drought killed off much of the vegetation that might have held the hillsides, and when rains came, they brought about mass wasting in the form of mudflows and landslides.

Over the surface of the planet, the average rate of erosion is about 1 in. (2.2 cm) in a thousand years. This is the *average,* however, meaning that in some places the rate is much, much higher, and in others it is greatly lower. The rate of erosion depends on several factors, including climate, the nature of the materials, the slope and angle of repose, and the role of plant and animal life in the local environment.

Whereas many types of plants help prevent erosion, the wrong types of planting can be detrimental. The dangers of improper land usage for crops and livestock are illustrated by the Dust Bowl experience, which highlights the fact that the organism most responsible for erosion is humanity itself. On the other hand, people also can protect against erosion by planting vegetation that holds the soil, by carefully managing and controlling land usage, and by lessening slope angle in places where gravity tends to erode the soil.

WHERE TO LEARN MORE

Cherrington, Mark. *Degradation of the Land.* New York: Chelsea House, 1991.

"Coastal and Nearshore Erosion." United States Geological Survey (USGS) (Web site). <http://walrus.wr.usgs.gov/hazards/erosion.html>.

Dean, Cornelia. *Against the Tide: The Battle for America's Beaches.* New York: Columbia University Press, 1999.

Hecht, Jeff. *Shifting Shores: Rising Seas, Retreating Coastlines.* New York: Scribners, 1990.

Middleton, Nick. *Atlas of Environmental Issues.* Illus. Steve Weston and John Downes. New York: Facts on File, 1989.

Protecting Your Property from Erosion (Web site). <http://www.abag.ca.gov/bayarea/enviro/erosion/erosion.html>.

Rybolt, Thomas R., and Robert C. Mebane. *Environmental Experiments About Land.* Hillside, NJ: Enslow Publishers, 1993.

"Soil Erosion on Farmland." New Zealand Ministry of Agriculture and Forestry (Web site). <http://www.maf.govt.nz/MAFnet/publications/erosion-risks/httoc.htm>.

Weathering and Erosion (Web site). <http://vishnu.glg.nau.edu/people/jhw/GLG101/Weathering.html>.

Wind Erosion Research Unit. United States Department of Agriculture/Kansas State University (Web site). <http://www.weru.ksu.edu/>.

MASS WASTING

CONCEPT

The term mass wasting (sometimes called mass movement) encompasses a broad array of processes whereby earth material is transported down a slope by the force of gravity. It is related closely to weathering, which is the breakdown of minerals or rocks at or near Earth's surface through physical, chemical, or biological processes, and to erosion, the transport of material through a variety of agents, most of them flowing media, such as air or water. Varieties of mass wasting are classified according to the speed and force of the process, from extremely slow creep to very rapid, dramatic slide or fall. Examples of rapid mass wasting include landslides and avalanches, which can be the cause of widespread death and destruction when they occur in populated areas.

HOW IT WORKS

Moving Earth and Rocks

In discussing mass wasting, the area of principal concern is Earth's surface rather than its interior. Thus, mass wasting is related most closely to the realm of geomorphology, a branch of physical geology concerned with the study of landforms, with the forces and processes that have shaped them, and with the description and classification of various physical features on Earth. Though plate tectonics (which involves the movement of giant plates beneath the earth's surface) can influence mass wasting, plate tectonics entails interior processes that humans usually witness only indirectly, by seeing their effects. Mass wasting, on the other hand, often can be observed directly, particularly in its more rapid forms, such as rock fall.

There are three general processes whereby a piece of earth material can be moved from a high outcropping to the sea: weathering, mass wasting, or erosion. If mechanical, biological, or chemical processes act on the material, dislodging it from a larger sample of material (e.g., separating a rock from a boulder), it is an example of weathering, which is discussed later in this essay. Supposing that a rock has been broken apart by weathering, it may be moved further by mass-wasting processes, such as creep or fall. Pieces of rock swept away by a river in a valley below the outcropping and small bits of rock worn away by high winds are examples of erosion. Erosion and weathering are examined in separate essays within this book.

As for the relationships between erosion, weathering, and mass wasting, the lines are not clearly drawn. Some authors treat weathering and mass wasting as varieties of erosion, and some apply a strict definition of erosion as resulting only from flowing media. (In the physical sciences, fluid means anything that flows, not just liquids.) Weathering, mass wasting, and erosion also can be viewed as stages in a process, as described in the preceding paragraph. This broad array of approaches, while perhaps confusing, only serves to illustrate the fact that the earth sciences are relatively young compared with such ancient disciplines as astronomy and biology. Not all definitions in the earth sciences are, as it were, "written in stone."

Weathering

A mineral is a substance that occurs naturally, is usually inorganic, and typically has a crystalline structure. The term *organic* does not necessarily mean "living" rather, it refers to all carbon-containing compounds other than oxides, such as carbon dioxide, and carbonates, which are often found in Earth's rocks. A crystalline solid is one in which the constituent parts have a simple and definite geometric arrangement repeated in all directions.

Rocks, scientifically speaking, are simply aggregates or combinations of minerals or organic material or both, and weathering is the process whereby rocks and minerals are broken down into simpler materials. Weathering is the mechanism through which soil is formed, and therefore it is a geomorphologic process essential to the sustenance of life on Earth. There are three varieties of weathering: physical or mechanical, chemical, and biological.

THE THREE TYPES OF WEATHERING. Physical or mechanical weathering involves such factors as gravity, friction, temperature, and moisture. Gravity, for instance, may cause a rock to drop from a height, such that it falls to the ground and breaks into pieces. If wind-borne sand blows constantly across a rock surface, the friction will have the effect of sandpaper, producing mechanical weathering. In addition, changes in temperature and moisture will cause expansion and contraction of materials, bringing about sometimes dramatic changes in their physical structure.

Chemical weathering not only is a separate variety of weathering but also is regarded as a second stage, one that follows physical weathering. Whereas physical changes are typically external, chemical changes affect the molecular structure of a substance, bringing about a rearrangement in the ways that atoms are bonded. Important processes that play a part in chemical weathering include acid reactions, hydrolysis (a reaction with water that results in the separation of a compound to form a new substance or substances), and oxidation. The latter can be defined as any chemical reaction in which oxygen is added to or hydrogen is removed from a substance.

An example of biological weathering occurs when a plant grows from a crevice in a rock. As the plant grows, it gradually forces the sides of the crevice apart even further, and it ultimately may tear the rock apart. Among the most notable agents of biological weathering are algae and fungi, which may be combined in a mutually beneficial organism called a lichen. (Reindeer moss is an example of a lichen.) Through a combination of physical and chemical processes, organisms ranging from lichen to large animals can wear away rock gradually.

PROPERTIES OF UNCONSOLIDATED MATERIAL

Regolith is a general term that describes a layer of weathered material that rests atop bedrock. It is unconsolidated, meaning that it is in pieces, like gravel, though much less uniform in size. Sand and soil, including soil mixed with loose rocks, are examples of regolith.

Every variety of unconsolidated material has its own angle of repose, or the maximum angle at which it can remain standing. Everyone who has ever attempted to build a sand castle at the beach has experienced angle of repose firsthand, perhaps without knowing it. Imagine that you are trying to build a sand castle with a steep roof. Dry sand would not be good for this purpose, because it is loose and has a tendency to flow easily. Much better would be moist sand, which can be shaped into a sharper angle, meaning that it has a higher angle of repose.

A certain amount of water gives sand surface tension, the same property that causes water to bead up on a table rather than lying flat. If too much water is added to the sand, however, the sand becomes saturated and will flow, a process called lateral spreading. Thus, to a point, the addition of water increases the angle of repose for sand, which is only about 34° when the sand is dry. (This is the angle of repose for sand in an hourglass.) On the other hand, piles of rocks may have an angle of repose as high as 45°. In practice, most aggregates of materials in nature have slopes less than their angle of repose, owing to the influence of wind and other erosive forces.

TYPES OF MASS WASTING

As noted earlier, there is some disagreement among writers in the geologic sciences regarding the types of mass wasting. Indeed, even the term mass wasting is not universal, since some writers refer to it as mass movement. Others do not even treat the subject as a category unto itself, preferring instead to address related concepts, such as weathering and erosion, as well as instances of mass wasting, such as avalanches and landslides.

An avalanche on Mount McKinley in Alaska. *(© W. Bacon/Photo Researchers. Reproduced by permission.)*

For this reason, the classification of mass-wasting processes presented here is by no means universal and instead represents a composite of several schools of thought. Generally speaking, geologists and geomorphologists classify processes of mass wasting according to the rapidity with which they occur. Most sources recognize at least three types of mass wasting: flow, slide, and fall. Some sources include slump among the categories of relatively rapid mass-wasting process, as opposed to the slower, less dramatic (but ultimately more important) process known as creep. Some writers classify uplift and subsidence with mass wasting; however, in this book, uplift and subsidence are treated separately, in the Geomorphology essay.

REAL-LIFE APPLICATIONS

Creep

Creep is the slow downward movement of regolith as a result of gravitational force. Before the initiation of the creeping process, the regolith is in what physicists call a condition of unstable equilibrium: it remains in place, yet a relatively small disturbance would be enough to dislodge

A mudflow caused by heavy winter rains brings down the hillside under homes in Millbrae, California. *(AP/Wide World Photos. Reproduced by permission.)*

it. Though it is slow, creep can produce some of the most dramatic results over time. It can curve tree trunks at the base, break or overturn retaining walls, and cause objects from fence posts to utility poles to tombstones to be overturned.

Changes in temperature or moisture are among the leading factors that result in the disturbance of regolith. A change in either can cause material to expand or contract, and freezing or thawing may be enough to shake regolith from its position of unstable equilibrium. In fact, some geomorphologists cite a distinct mass-wasting process, known as solifluction, that occurs in the active layer of permafrost, which thaws in the summertime. Water also can provide lubrication or additional weight that assists the material in moving. One of the only causes of creep not associated with changes in temperature or moisture is the burrowing of small animals.

Slump and Slide

Slump occurs when a mass of regolith slides over or creates a concave surface (one shaped like the inside of a bowl). The result is the formation of a small, crescent-shaped cliff, known as a scarp, at the upper end—rather like the crest of a wave. Soil flow takes place at the bottom end of the slump. One is likely to see slumps in any place where forces, whether man-made or natural, have graded material to a slope too steep for its angle of repose. This may happen along an interstate highway, where a road crew has cut the slope too sharply, or on a riverbank, where natural erosion has done its work.

Often, slump is classified as a variety of slide, in which material moves downhill in a fairly coherent mass (i.e., more or less in a section or group) along a flat or planar surface. These movements sometimes are called rock slides, debris slides, or, in common parlance, landslides. Among the most destructive types of mass wasting, they may be set in motion by earthquakes, which are caused by plate tectonic processes, or by hydrologic agents (i.e., excessive rain or melting snow and ice).

Flow

When a less uniform, or more chaotic, mass of material moves rapidly downslope, it is called flow. Flow is divided into categories, depending on the amounts of water involved: granular flow (0-20% water) and slurry flow (20-40% water). Creep and solifluction often are classified as very slow forms of granular and slurry flow, respectively. In order of relative speed, these categories are as follows:

Granular Flow (0-20% Water)

- Slowest: Creep
- Slower: Earth flow
- Faster: Grain flow
- Fastest: Debris avalanche

Slurry Flow (20-40% Water)

- Slow: Solifluction
- Medium: Debris flow
- Fast: Mudflow

Earth flow moves at a rate anywhere from 3.3 ft. (1 m) per year to 330 ft. (100 m) per hour. Grain flow can be nearly 60 mi. (100 km) per hour, and debris avalanche may achieve speeds of 250 mi. (400 km) per hour, making it extremely dangerous. Among types of slurry flow, debris flow is roughly analogous to earth flow, falling into a range from about 4 in. (10 cm) per year to 0.6 mi. (1 km) per hour. Mudflow is slightly faster than grain flow. If the water content is more than 40%, a slurry flow is considered a stream.

Earth flows involve fine-grained materials, such as clay or silt, and typically occur in humid areas after heavy rains or the melting of snow. Debris flows usually result from heavy rains as well and may start with slumps before flowing downhill, forming lobes with a surface broken by ridges and furrows. Grain flows can be caused by a small disturbance, which forces the dry, unconsolidated material rapidly downslope. Debris avalanches are commonly the result of earthquakes or volcanic eruptions.

Seismic disturbances or volcanic activity may cause the collapse of a mountain slope, sending debris avalanches moving swiftly even along the gentler slopes of the mountainside. Likewise, mudflows may be the result of volcanic activity, in which case they are known as lahars. In some situations, the material in a lahar is extremely hot. Mudflows tend to be highly fluid mixtures of sediment (material deposited at or near Earth's surface from a number of sources, most notably preexisting rock) and water and typically flow along valley floors.

Fall

Most other forms of mass wasting entail movement along slopes that are considerably less than 90°, whereas fall takes place at angles almost perpendicular to the ground. Anyone who has driven through a wide mountain area, with steep cliffs on either side, has seen signs that say "Watch for Falling Rock." These warnings, which appear regularly on the drive through the Rockies in Colorado or on highways across the Blue Ridge and Great Smoky mountains in the southern United States, indicate the threat of rock fall.

The mechanism behind rock fall is simple enough. When a rock at the top of a slope is in unstable equilibrium, it can be dislodged such that it either falls directly downward or bounces and rolls. Usually, the bottom of the slope or cliff contains accumulated talus, or fallen rock material. Freezing and thawing as well as the growth of plant roots may cause fall. The latter is not limited to rock fall: debris fall, which is closely related, includes soil, vegetation, and regolith as well as rocks.

Mass Wasting and Natural Disasters

Among the most dramatic and well-known varieties of mass wasting are avalanches, a variety of flow, and landslides, which (as their name suggests) are a type of slide. These can result, and have resulted, in enormous loss of life and property. Some notable modern occurrences of mass wasting, and the type of movement involved, are listed below. With each incident, the approximate number of fatalities is shown in parentheses.

- *China, 1920:* Landslide caused by an earthquake (200,000)
- *Peru, 1970:* Debris avalanche related to an earthquake (70,000)
- *Colombia, 1985:* Mudflow related to a volcanic eruption (23,000)
- *Soviet Union, 1949:* Landslide caused by an earthquake (12,000–20,000)
- *Italy and Austria, 1916:* Landslide (10,000)
- *Peru, 1962:* Landslide (4,000–5,000)
- *Italy, 1963:* Landslide (2,000)
- *Japan, 1945:* Landslide caused by a flood (1,200)
- *Ecuador, 1987:* Landslide related to an earthquake (1,000)
- *Austria, 1954:* Landslide (200)

The Role of Plate Tectonics

Note how many times an instance of mass wasting was either caused by or "related to" (meaning that geologists could not establish a full causal relationship) volcanic or seismic activity. Both, in

KEY TERMS

ANGLE OF REPOSE: The maximum slope at which a relatively large sample of unconsolidated material can remain standing. Often, the addition of water increases the angle of repose, up to the point at which the material becomes saturated.

AVALANCHE: See *flow.*

CREEP: A form of mass wasting involving the slow downward movement of regolith as a result of gravitational force.

EROSION: The movement of soil and rock due to forces produced by water, wind, glaciers, gravity, and other influences. In most cases, a fluid medium, such as air or water, plays a part.

FALL: A form of mass wasting in which rock or debris moves downward along extremely steep angles.

FLOW: A form of mass wasting in which a body of material that is not uniform moves rapidly downslope. Flow is divided into categories, depending on the amounts of water involved: granular flow (0-20% water) and slurry flow (20-40% water). An avalanche is an example of flow and may involve either rock (granular) or snow (slurry).

FLUID: In the physical sciences, the term fluid refers to any substance that flows and therefore has no definite shape—that is, both liquids and gases. In the earth sciences, occasionally substances that appear to be solid, for example, ice in glaciers, are, in fact, flowing slowly.

GEOMORPHOLOGY: An area of physical geology concerned with the study of landforms, with the forces and processes that have shaped them, and with the description and classification of various physical features on Earth.

LANDSLIDE: See *slide.*

LITHOSPHERE: The upper layer of Earth's interior, including the crust and the brittle portion at the top of the mantle.

turn, are the result of plate movement in most instances, and thus it is not surprising that several of the locales noted here are either at plate margins or in mountainous regions where plate tectonic and other processes are at work. (For more on this subject, see the entries Plate Tectonics and Mountains.)

To set mass wasting into motion, it is necessary to have a steep slope and some type of force to remove material from its position of unstable equilibrium. Plate tectonic processes provide both. Not only does an earthquake, for instance, jar rocks loose from the upper portion of a slope, but the movement of plates also helps create steep slopes, for example, the collision of the Indo-Australian and Eurasian belts that produced the Himalayas.

Some of the most vigorous plate tectonic activity occurs underwater, and, likewise, there are remarkable manifestations of mass wasting beneath the seas. Off Moss Landing, a research facility that serves a consortium of state universities in northern California, is an underwater canyon more than 0.6 mi. (1 km) deep. At one time, Monterey Canyon was thought to be the result of erosion by a river flowing into the ocean; however, today it is believed to be the result of underwater mass wasting.

DETECTING AND PREVENTING MASS WASTING

The dramatic instances of mass wasting discussed here hardly require any effort at detection. Their effect is obvious and, to those unfortunate enough

KEY TERMS

MASS WASTING: The transfer of earth material down slopes by processes that include creep, slump, slide, flow, and fall. Also known as *mass movement.*

PLATE MARGINS: Boundaries between plates.

PLATE TECTONICS: The name both of a theory and of a specialization of tectonics. As an area of study, plate tectonics deals with the large features of the lithosphere and the forces that shape them. As a theory, it explains the processes that have shaped Earth in terms of plates and their movement.

PLATES: Large, movable segments of the lithosphere.

REGOLITH: A general term describing a layer of weathered material that rests atop bedrock.

SEDIMENT: Material deposited at or near Earth's surface from a number of sources, most notably preexisting rock.

SLIDE: A variety of mass wasting in which material moves downhill in a fairly coherent mass (i.e., more or less in a section or group) along a flat or planar surface.

SLUMP: A form of mass wasting that occurs when a mass of regolith slides over or creates a concave surface (one shaped like the inside of a bowl).

SURFACE TENSION: An attractive force exerted by molecules in the interior of a liquid on molecules at the exterior. This force draws the material inward such that it occupies less than its maximum horizontal area. The surface tension of water is high, causing it to bead on most surfaces.

UNSTABLE EQUILIBRIUM: A situation in which an object remains in place, yet a relatively small disturbance would be enough to dislodge it.

WEATHERING: The breakdown of rocks and minerals at or near the surface of Earth due to physical, chemical, or biological processes.

to be nearby, inescapable. Other types of mass wasting occur so slowly that they do not invite immediate detection. This can be unfortunate, because in some cases slow mass wasting is a harbinger of much more rapid movements to follow.

A dwelling atop a hill is subject to enormous gravitational force, and the more massive the dwelling, the greater the pull of gravity. (Weight is, after all, nothing but gravitational force.) If a homeowner adds a swimming pool or other items that contribute to the weight of the dwelling, it only increases the chances that it may experience mass wasting. Heavy rains can bring so much water that it saturates the soil, reducing its surface tension and causing it to slide—as occurred, for instance, in the area around Malibu, California, during the late 1990s.

The California mud slides and landslides are a dramatic example of mass wasting, but more often than not mass wasting takes the form of creep, which is detectable only over a matter of years. When creep occurs, the upper layer of soil moves, while the layer below remains stationary. One way to keep the upper layer in place is to plant vegetation that will put down roots deep enough to hold the soil.

This may create unintended consequences. During the 1930s, New Deal officials imported kudzu plants from China, intending to protect the hillsides of the American South from creep and erosion. The kudzu protected the slopes, but as it turned out, this voracious plant had a tendency to creep as well. Before communities began taking steps to eradicate it, or at least push it back, in the 1970s, kudzu seemingly threatened to cover the entire southern United States.

To prevent some of the more dramatic varieties of mass wasting, such as landslides in a residential area, a homeowner or group of homeowners may commission an engineer's study. The engineer can test the material of the slope, measure the stresses acting on it, and perform other calculations to predict the likelihood that a slope will succumb to a given amount of force. For this reason, zoning laws in areas with steep slopes are typically strict. These laws are geared toward preventing homeowners and builders from erecting structures likely to create a threat of mass wasting in a period of heavy rains.

WHERE TO LEARN MORE

Abbott, Patrick. *Natural Disasters.* Dubuque, IA: William C. Brown Publishers, 1996.

Allen, Missy, and Michel Peissel. *Dangerous Natural Phenomena.* New York: Chelsea House, 1993.

Armstrong, Betsy R., and Knox Williams. *The Avalanche Book.* Golden, CO: Fulcrum, 1986.

Goodwin, Peter. *Landslides, Slumps, and Creeps.* New York: Franklin Watts, 1997.

Gore, Pamela. "Mass Wasting" (Web site). <http://www.gpc.peachnet.edu/~pgore/geology/geo101/masswasting.html>.

Mass Wasting (Web site). <http://www.es.ucsc.edu/~jsr/EART10/Lectures/HTML/lecture.07.html>.

"Mass Wasting Features of North Dakota." North Dakota State University (Web site). <http://www.ndsu.nodak.edu/nd_geology/nd_mwast/index_mw.htm>.

Murck, Barbara Winifred, Brian J. Skinner, and Stephen C. Porter. *Dangerous Earth: An Introduction to Geologic Hazards.* New York: John Wiley, 1996.

Nelson, Stephen A. *Mass-Wasting and Mass-Wasting Processes. Tulane University* (Web site). <http://www.tulane.edu/~sanelson/geol204/masswastproc.htm>.

Weathering and Mass Wasting Learning Module (Web site). <http://home.aol.com/rhaberlin/mwmod.htm>.

SEDIMENTOLOGY AND SOIL SCIENCE

SEDIMENT AND SEDIMENTATION

SOIL

SOIL CONSERVATION

SEDIMENT AND SEDIMENTATION

CONCEPT

The materials that make up Earth are each products of complex cycles and interactions, as a study of sediment and sedimentation shows. Sediment is unconsolidated material deposited at or near Earth's surface from a number of sources, most notably preexisting rock. There are three kinds of sediment: chemical, organic, and rock, or clastic sediment. Weathering removes this material from its source, while erosion and mass wasting push it along to a place where it is deposited. After deposition, the material may become a permanent part of its environment, or it may continue to undergo a series of cycles in which it experiences ongoing transformation.

HOW IT WORKS

TRANSPORTING SEDIMENT

There are three types of sediment: rocks, or clastic sediment; mineral deposits, or chemical sediment; and organic sediment, composed primarily of organic material. (In this context, the term *organic* refers to formerly living things and parts or products of living things; however, as discussed in Minerals, the term actually has a much broader meaning.) There are also three general processes involved in the transport of sediment from higher altitudes to lower ones, where they eventually are deposited: weathering, mass wasting, and erosion.

The lines between these three processes are not always clearly drawn, but, in general, the following guidelines apply. When various processes act on the material, causing it to be dislodged from a larger sample (for example, separating a rock from a boulder), this is an example of weathering. Assuming that a rock has been broken apart by weathering, it may be moved farther by mass-wasting processes, such as creep or fall, for which gravity is the driving factor. If the pieces of rock are swept away by a river, high winds, or a glacier (all of which are flowing media), this is an example of erosion.

WEATHERING. Weathering is divided further into three different types: physical, chemical, and biological. Physical or mechanical weathering takes place as a result of such factors as gravity, friction, temperature, and moisture. Gravity may cause a rock to roll down a hillside, breaking to pieces at the bottom; friction from particles of matter borne by the wind may wear down a rock surface; and changes in temperature and moisture can cause expansion and contraction of materials.

Whereas physical weathering attacks the rock as a whole, chemical weathering involves the breakdown of the minerals or organic materials that make up the rock. Chemical breakdown may lead to the dissolution of the materials in the rock, which then are washed away by water or wind or both, or it may be merely a matter of breaking the materials down into simpler compounds.

Biological weathering is not so much a third type of weathering as it is a manifestation of chemical and physical breakdown caused by living organisms. Suppose, for instance, that a plant grows within a crack in a rock. Over time, the plant will influence physical weathering through its moisture and the steady force of its growth pushing at the walls of the fissure in which it is rooted. At the same time, its specific chemical

SEDIMENT CAN BE TRANSPORTED BY WEATHERING, EROSION, AND MASS WASTING. HERE A WINTER STORM HAS CAUSED COASTAL EROSION OF A BEACH. *(© Rafael Macia/Photo Researchers. Reproduced by permission.)*

properties likely will induce decomposition of the rock.

MASS WASTING. Mass wasting, sometimes known as mass movement, comprises a number of types of movement of earth material, all of them driven by gravity. Creep is the slow downward drift of regolith (unconsolidated material produced by weathering), while slump occurs when a mass of regolith slides over, or creates, a concave surface—that is, one shaped like the inside of a bowl. Slump sometimes is classified as a variety of slide, in which material moves downhill in a fairly coherent mass along a flat or planar surface. Such movements, sometimes called rock slides, debris slides, or landslides, are among the most destructive types of mass wasting.

When a less uniform, or more chaotic, mass of material moves rapidly down a slope, it is called flow. Flow is divided into categories, depending on the specific amounts of water: granular flows (0–20% water) and slurry flows (20–40% water), the fastest varieties of which are debris avalanche and mudflow, respectively. Mudflows can be more than 60 mi. (100 km) per hour, while debris avalanches may achieve speeds of 250 mi. (400 km) per hour.

Even these high-speed varieties of mass wasting entail movement along slopes that are considerably less than 90°, whereas a final variety of mass wasting, that is, fall, takes place at angles almost perpendicular to the ground. Typically the bottom of a slope or cliff contains accumulated talus, or fallen rock material. Nor is fall limited to rock fall: debris fall, which is closely related, includes soil, vegetation, and regolith as well as rocks. (For more on these subjects, see Mass Wasting.)

EROSION

Erosion typically is caused either by gravity (in which case it is generally known as mass wasting, discussed earlier) or by flowing media, such as water, wind, and even ice in glaciers. It removes sediments in one of three ways: by the direct impact of the agent (i.e., the flowing media that is discussed in the following sections); by abrasion, another physical process; or by corrosion, a chemical process.

In the case of direct impact, the wind, water, or ice removes sediment, which may or may not be loose when it is hit. On the other hand, abrasion involves the impact of solid earth materials carried by the flowing agent rather than the impact of the flowing agent itself. For example, sand borne by the wind, as discussed later, or pebbles carried by water may cause abrasion.

Corrosion is chemical and is primarily a factor only in water-driven, as opposed to wind-driven or ice-driven, erosion. Streams slowly dissolve rock, removing minerals that are carried downstream by the water.

WATER. Of the fluid substances driving erosion, liquid water is perhaps the one most readily associated in most people's minds with erosion. In addition to the erosive power of waves on the seashore, there is the force exerted by running water in creeks, streams, and rivers. As a river moves, pushing along sediment eroded from the streambeds or riverbeds, it carves out deep chasms in the bedrock beneath.

Moving bodies of water continually reshape the land, carrying soil and debris down slopes, from the source of the river to its mouth, or delta. A delta is a region formed when a river enters a larger body of water, at which point the reduction in velocity on the part of the river current leads to the widespread deposition of sediment. It is so named because its triangular shape resembles that of the Greek letter delta, Δ.

Water at the bottom of a large body, such as a pond or lake, also exerts erosive power; then there is the influence of falling rain. Assuming that the ground is not protected by vegetation, raindrops can loosen particles of soil, sending them scattering in all directions. A rain that is heavy enough may dislodge whole layers of topsoil and send them rushing away in a swiftly moving current. The land left behind may be rutted and scarred, much of its best soil lost for good.

ICE. Ice, of course, is simply another form of water, but since it is solid, its physical properties are quite different. It is a solid rather than a fluid, such as liquid water or air (the physical sciences treat gases and liquids collectively as "fluids"), yet owing to the enormous volume of ice in glaciers, these great masses are capable of flowing. Glaciers do not flow in the same way as a fluid does; instead, they are moved by gravity, and like giant bulldozers made of ice, they plow through rock, soil, and plants.

Under certain conditions a glacier may have a layer of melted water surrounding it, which greatly enhances its mobility. Even without such lubricant, however, these immense rivers of ice move steadily forward, gouging out pieces of bedrock from mountain slopes, fashioning deep valleys, removing sediment from some regions and adding it to others. In unglaciated areas, or places that have never experienced any glacial activity, sediment is formed by the weathering and decomposition of rock. On the other hand, formerly glaciated areas are distinguished by layers of till, or glacial sediment, from 200 to 1,200 ft. (61–366 m) thick.

WIND. The processes of wind erosion sometimes are called *eolian* processes, after Aeolus, the Greek god of the winds. Eolian erosion is in some ways less forceful than the erosive influence of water. Water, after all, can lift heavier and larger particles than can the winds. Wind, however, has a much greater frictional component in certain situations. This is particularly true when the wind carries sand, every grain of which is like a cutting tool.

Wind erosion, in fact, is most pronounced in precisely those places where sand abounds, in deserts and other areas that lack effective ground cover in the form of solidly rooted, prevalent vegetation. In some desert regions the bases of rocks or cliffs have been sandblasted, leaving a mushroom-shaped formation owing to the fact that the wind could not lift the fine grains of sand very high.

SEDIMENT LOAD

Eroded particles become part of what is called the sediment load transported by the fluid medium. Sediment load falls into three categories: dissolved load, suspended load, and bed load. The amount of each type of load that a fluid medium is capable of carrying depends on the density of the fluid medium itself: in other words, wind can carry the least of each and ice the most.

The wind does not carry any dissolved load, since solid particles (unlike gases) cannot be dissolved in air. Ice or water, on the other hand, is able to dissolve materials, which become invisible within them. Typically, about 90% of the dissolved load in a river is accounted for by five different ions, or atoms that carry a net electric charge: the anions (negative ions) chloride, sulfate, and bicarbonate and the cations (positive ions) of sodium and calcium.

Suspended load is sediment that is suspended, or floating, in the erosive medium. In this instance, wind is just as capable as water or ice of suspending particles of the sediment load, which are likely to color the medium that carries them. Hence, water or wind carrying suspended particles is usually murky. The thicker the medium, the larger the particles it is capable of suspending. In other words, ice can suspend extremely large pieces of sediment, whereas water can suspend much more modest ones. Wind can suspend only tiny particles.

Then there is bed load, large sediment that never becomes suspended but rather is almost always in contact with the substrate or bottom, whether "the bottom" is a streambed or the ground itself. Instead of being lifted up by the medium, bed load is nudged along, rolling, skipping, and sliding as it makes its way over the substrate. Once again, the density of the medium itself has a direct relationship to the size of the bed load it is capable of carrying. Wind rarely transports bed load thicker than fine sand, and water usually moves only pebbles, though under flood conditions it can transport boulders. As with suspended load, glaciers can transport virtually any size of bed load.

SEDIMENT SIZES AND SHAPES

Geologists and sedimentologists use certain terms to indicate sizes of the individual particles in sediment. Many of these terms are familiar to us from daily life, but whereas people typically use them in a rather vague way, within the realm of sedimentology they have very specific meanings. Listed below are the various sizes of rock, each with measurements or measurement ranges for the rock's diameter:

- Clay: Smaller than 0.00015 in. (0.004 mm)
- Silt: 0.00015 in. (0.004 mm) to 0.0025 in. (0.0625 mm)
- Sand: 0.0025 in. (0.0625 mm) to 0.08 in. (2 mm)
- Pebble: 0.08 in. (2 mm) to 2.5 in. (64 mm)
- Cobble: 2.5 in. (64 mm) to 10 in. (256 mm)
- Boulder: Larger than 10 in. (256 mm).

This listing is known as the Udden–Wentworth scale, which was developed in 1898 by J.A. Udden (1859–1932), an American sedimentary petrologist (a scientist who studies rocks). In 1922 the British sedimentary petrologist C. K. Wentworth) expanded Udden's scale, adapting the definitions of various particle sizes to fit more closely with the actual usage and experience of researchers in the field. The scale uses modifiers to pinpoint the relative sizes of particles. In ascending order of size, these sizes are very fine, fine, medium, coarse, and very coarse.

REAL-LIFE APPLICATIONS

SEDIMENTS AND DUST BOWLS

Sediment makes possible the formation of soil, which of course is essential for growing crops. Therefore it is a serious matter indeed when wind and other forces of erosion remove sediment, creating dust-bowl conditions. The term "Dust Bowl," with capital letters, refers to the situation that struck the United States Great Plains states during the 1930s, devastating farms and leaving thousands of families without home or livelihood. (See Erosion for much more about the Dust Bowl.)

During the late 1990s, some environmentalists became concerned that farming practices in the western United States were eroding sediment, putting in place the possibility of a return to the conditions that created the Dust Bowl. However, in August 1999, the respected journal *Science* reported studies showing that sediment in farmlands was not eroding at anything like the rate that had been feared. Soil scientist Stanley Trimble at the University of California, Los Angeles, studied Coon Creek, Wisconsin, and its tributaries, a watershed for which 140 years' worth of erosion data were available. As Trimble discovered, the rate of sediment erosion in the area had dramatically decreased since the 1930s, and was now at 6% of the rate during the Dust Bowl years.

Some studies from the 1970s onward had indicated that farming techniques, designed to improve the crop output from the soil, had created a situation in which sediment was being washed away at alarming rates. However, if such sediment removal were actually taking place, there would have to be some evidence—if nothing more, the sediment that had been washed away would have had to go somewhere. Instead, as Trimble reported, "We found that much of the sediment in Coon Creek doesn't move very far, and that it moves in complex ways." The sediment, as he went on to explain, was moving within the Coon Creek basin, but the amount that actually made it to the Mississippi River (which could be counted as true erosion, since it was removing sediment from the area) had stayed essentially the same for the past 140 years.

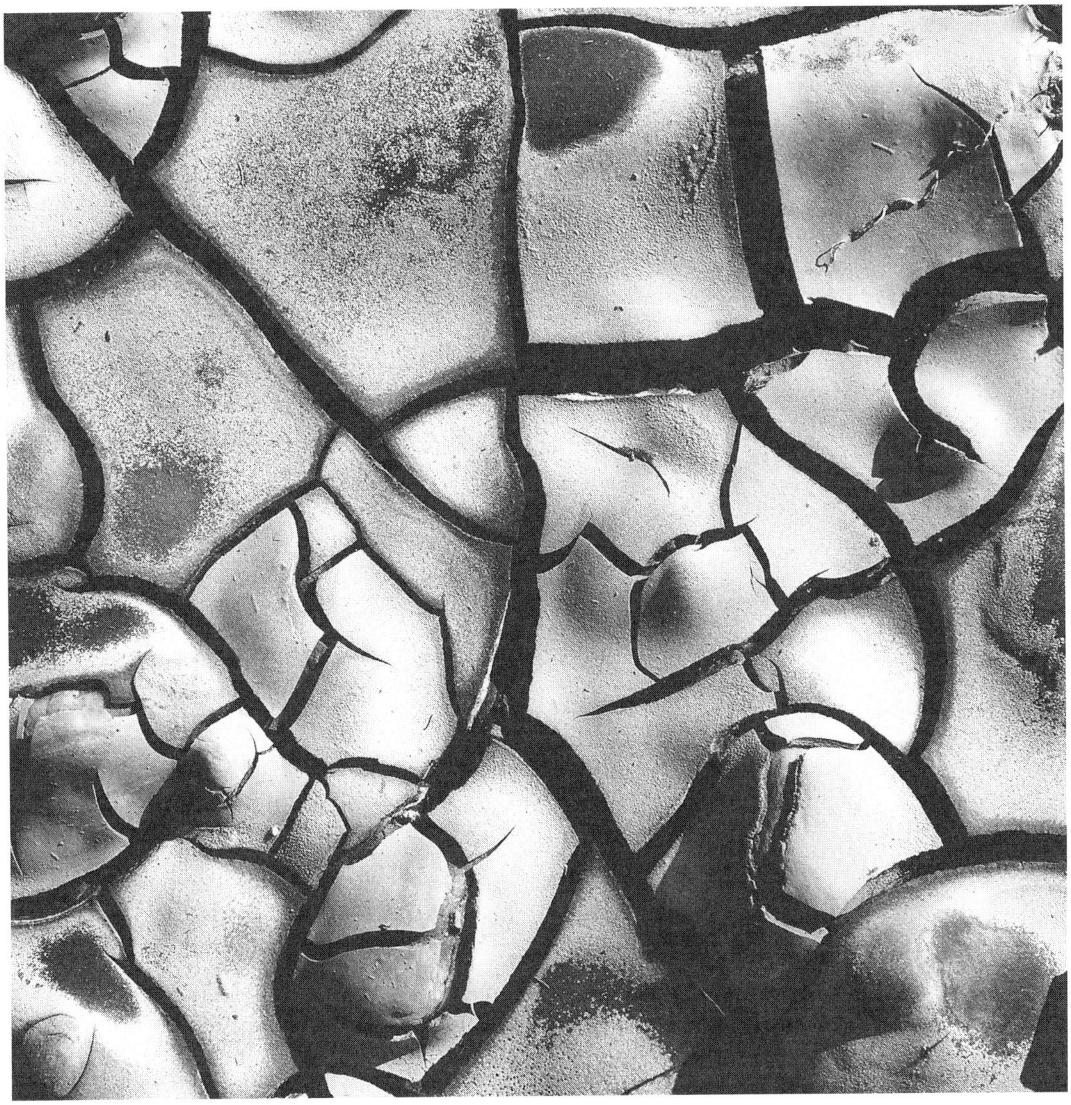

SEDIMENTARY STRUCTURES REMAINING IN A DRIED RIVER BED. CLAY SOILS CRACK AS THEY LOSE MOISTURE AND CONTRACT WHEN TRAPPED WATER EVAPORATES AS THE RESULT OF DROUGHT. *(© B. Edmaier/Photo Researchers. Reproduced by permission.)*

DEPOSITION AND DEPOSITIONAL ENVIRONMENTS

Eventually everything in motion—including sediment—comes to rest somewhere. A piece of sediment traveling on a stream of water may stop hundreds of times, but there comes a point when it comes to a complete stop. This process of coming to rest is known as deposition, which may be of two types, mechanical or chemical. The first of these affects clastic and organic sediment, while the second applies (fittingly enough) to chemical sediment.

In mechanical deposition, particles are deposited in order of their relative size, the largest pieces of bed load coming to a stop first. These large pieces are followed by medium-size pieces and so on until both bed load and suspended load have been deposited. If the sediment has come to a full stop, as, for instance, in a stagnant pool of water, even the finest clay suspended in the water eventually will be deposited as well.

Unlike mechanical deposition, chemical deposition is not the result of a decrease in the velocity of the flow; rather, it comes about as a result of chemical precipitation, when a solid particle crystallizes from a fluid medium. This often happens in a saltwater environment, where waters may become overloaded with salt and other minerals. In such a situation, the water is unable to maintain the minerals in a dissolved state (i.e., in solution) and precipitates part of its content in the form of solids.

DEPOSITIONAL ENVIRONMENTS. The matter of sediment deposition in water is particularly important where reservoirs are concerned, since in that case the water is to be used for drinking, cooking, bathing, and other purposes by humans. One of the biggest problems for the maintenance of clean reservoirs is the transport of sediment from agricultural areas, in which the soil is likely to contain pesticides and other chemicals, including the phosphorus found in fertilizer. A number of factors, including precipitation, topography, and land use, affect the rate at which sediment is deposited in reservoirs.

The area in which sediment is deposited is known as its depositional environment, of which there are three basic varieties: terrestrial, marginal marine, and marine. These are, respectively, environments on land (and in landlocked waterways, such as creeks or lakes), along coasts, and in the open ocean. A depositional environment may be a large-scale one, known as a regional environment, or it may be a smaller subenvironment, of which there may be hundreds within a given regional environment.

SEDIMENTARY STRUCTURES. There are many characteristic physical formations, called sedimentary structures, that sediment forms after it has reached a particular depositional environment. These formations include bedding planes and beds, channels, cross-beds, ripples, and mud cracks. A bed is a layer, or stratum, of sediment, and bedding planes are surfaces that separate beds. The bedding plane indicates an interruption in the regular order of deposition. (These are concepts that also apply to the field of stratigraphy. For more on that subject, see the essay Stratigraphy.)

Channels are simply depressions in a bed that reflect the larger elongated depression made by a river as it flows along its course. Cross-beds are portions of sediment that are at an angle to the beds above and below them, as a result of the action of wind and water currents—for example, in a flowing stream. As for ripples, they are small sandbar-like protuberances that form perpendicular to the direction of water flow. At the beach, if you wade out into the water and look down at your feet, you are likely to see ripples perpendicular to the direction of the waves. Finally, mud cracks are the sedimentary structures that remain when water trapped in a muddy pool evaporates. The clay, formerly at the bottom of the pool, begins to lose its moisture, and as it does, it cracks.

THE IMPACT OF SEDIMENT

It is estimated that the world's rivers carry as much as 24 million tons (21,772,800 metric tons) of sediment to the oceans each year. There is also the sediment carried by wind, glaciers, and gravity. Where is it all going? The answer depends on the type of sediment. Clastic and organic sediment may wind up in a depositional environment and experience compaction and cementation in the process of becoming sedimentary rock. (For more on sedimentary rock, see Rocks.)

On the other hand, clastic and organic particles may be buried, but before becoming lithified (turned to rock), they once again may be exposed to wind and other forces of nature, in which case they go through the entire cycle again: weathering, erosion, transport, deposition, and burial. This cycle may repeat many times before the sediment finally winds up in a permanent depositional environment. In the latter case, particles of clastic and organic sediment ultimately may become part of the soil, which is discussed elsewhere in this book (See Soil).

A chemical sediment also may become part of the soil, or it may take part in one or more biogeochemical cycles (also discussed elsewhere; see Biogeochemical Cycles). These chemicals may wind up as water in underground reservoirs, as ice at Earth's poles, as gases in the atmosphere, as elements or compounds in living organisms, or as parts of rocks. Indeed, all three types of sediment—clastic, chemical, and organic—are part of what is known as the rock cycle, whereby rocks experience endlessly repeating phases of destruction and renewal. (See Rocks for more details.)

SEDIMENTARY MINERAL DEPOSITS

Among the most interesting aspects of sediment are the mineral deposits it contains—deposits that may, in the case of placer gold, be of significant value. A placer deposit is a concentration of heavy minerals left behind by the effect of gravity on moving particles, and since gold is the densest of all metals other than uranium (which

KEY TERMS

BED LOAD: Sediment that is capable of being transported by an erosive medium (wind, water, or air) but only under conditions in which it remains in nearly constant contact with the substrate or bottom (e.g., a streambed or the ground). Bed load, along with dissolved load and suspended load, is one of three types of sediment load.

COMPOUND: A substance made up of atoms of more than one element chemically bonded to one another.

CONSOLIDATION: A process whereby materials become compacted, or experience an increase in density. This takes place through a number of processes, including recrystallization and cementation.

DEPOSITION: The process whereby sediment is laid down on the Earth's surface.

DIAGENESIS: A term referring to all the changes experienced by a sediment sample under conditions of low temperature and low pressure following deposition.

DISSOLVED LOAD: Sediment load that is absorbed completely by the erosive medium (either water or ice) that carries it. Dissolved load is one of three types of sediment load, the others being suspended load and bed load.

EROSION: The movement of soil and rock due to forces produced by water, wind, glaciers, gravity, and other influences. In most cases, a fluid medium, such as air or water, is involved.

FLUID: In the physical sciences, the term fluid refers to any substance that flows and therefore has no definite shape—that is, both liquids and gases. Occasionally, substances that appear to be solid (for example, ice in glaciers), in fact, are flowing slowly; therefore, within the earth sciences, ice often is treated as another fluid medium.

ION: An atom or group of atoms that has lost or gained one or more electrons and thus has a net electric charge. Positively charged ions are called *cations,* and negatively charged ones are called *anions.*

MASS WASTING: The transfer of earth material down slopes by processes that include creep, slump, slide, flow, and fall. Also known as *mass movement.*

MINERAL: A naturally occurring, typically inorganic substance with a specific chemical composition and a crystalline structure.

ORGANIC: At one time, chemists used the term *organic* only in reference to living things. Now the word is applied to most compounds containing carbon, with the exception of carbonates (which are minerals) and oxides, such as carbon dioxide.

PRECIPITATION: In the context of chemistry, precipitation refers to the formation of a solid from a liquid.

REGOLITH: A general term describing a layer of weathered material that rests atop bedrock.

ROCK: An aggregate of minerals or organic matter, which may be consolidated or unconsolidated.

KEY TERMS CONTINUED

SEDIMENT: Material deposited at or near Earth's surface from a number of sources, most notably preexisting rock. There are three types of sediment: rocks, or clastic sediment; mineral deposits, or chemical sediment; and organic sediment, composed primarily of organic material.

SEDIMENTARY ROCK: One of the three major types of rock, along with igneous and metamorphic rock. Sedimentary rock usually is formed by the deposition, compaction, and cementation of rock that has experienced weathering. It also may be formed as a result of chemical precipitation.

SEDIMENTATION: The process of erosion, transport, and deposition undergone by sediment.

SEDIMENT LOAD: A term for the particles transported by a flowing medium of erosion (wind, water, or ice). The types of sediment load are dissolved load, suspended load, and bed load.

SEDIMENTOLOGY: The study and interpretation of sediments, including sedimentary processes and formations.

SUSPENDED LOAD: Sediment that is suspended, or floating, in the erosive medium (wind, water, or ice). Suspended load is one of three types of sediment load, along with dissolved load and bed load.

TILL: A general term for the sediments left by glaciers that lack any intervening layer of melted ice.

UNCONSOLIDATED ROCK: Rock that appears in the form of loose particles, such as sand.

WEATHERING: The breakdown of rocks and minerals at or near the surface of Earth due to physical, chemical, or biological processes.

is even more rare), it is among the most notable of placer deposits.

Of course, the fact that gold is valuable has done little to hurt, and a great deal to help, human fascination with placer gold deposits. Placer gold played a major role from the beginning of the famous California Gold Rush (1848–49), which commenced with discovery of a placer deposit by prospector James Marshall on January 24, 1848, along the American River near the town of Coloma. This discovery not only triggered a vast gold rush, as prospectors came from all over the United States in search of gold, but it also proved a major factor in the settlement of the West. Most of the miners who went to the West failed to make a fortune, of course, but instead they found something much better than gold: a gorgeous, fertile land like few places in the United States—California, a place that today holds every bit as much allure for many Americans as it did in 1848.

Despite the attention it naturally attracts, gold is far from the only placer mineral. Other placer minerals, all with a high specific gravity (density in comparison to that of water), include platinum, magnetite, chromite, native copper, zircon, and various gemstones. Nor are placer minerals found only in streams and other flowing bodies of water; wave action and shore currents can leave behind what are called beach placers. Among the notable beach placers in the world are gold deposits near Nome, Alaska, as well as zircon in Brazil and Australia, and marine gravel near Namaqualand, South Africa, which contains diamond particles.

An entirely different process can result in the formation of evaporites, minerals that include carbonates, gypsum, halites, and magnesium and potassium salts. (These specific mineral types are discussed in Minerals.) Formed when the evaporation of water leaves behind ionic, or electrically charged, chemical compounds, evaporites sometimes undergo physical processes similar to those of clastic sediment. They may even have graded bedding, meaning that the heavier materials fall to the bottom. In addition to their usefulness in industry and commerce (e.g., the use of gypsum in sheetrock for building), physical and chemical aspects of evaporites also provide scientists with considerable information regarding the past climate of an area.

WHERE TO LEARN MORE

Cherrington, Mark. *Degradation of the Land.* New York: Chelsea House, 1991.

"Erosion—Fast or Slow or None?" (Web site). <http://www.crcwater.org/issues9/19990828erosion.html>.

Middleton, Nick. *Atlas of Environmental Issues.* Illus. Steve Weston and John Downes. New York: Facts on File, 1989.

Schneiderman, Jill S. *The Earth Around Us: Maintaining a Livable Planet.* New York: W. H. Freeman, 2000.

Snedden, Robert. *Rocks and Soil.* Illus. Chris Fairclough. Austin, TX: Raintree Steck–Vaughn, 1999.

Soil Erosion and Sedimentation in the Great Lakes Region (Web site). <http://www.great-lakes.net/envt/pollution/erosion.html>.

Sediment/Soil Quality Related Links (Web site). <http://response.restoration.noaa.gov/cpr/sediment/sedlinks.html>.

State of the Land—Sedimentation and Water Quality (Web site). <http://www.nhq.nrcs.usda.gov/land/env/wq4.html>.

USDA–ARS–National Sedimentation Laboratory. United States Department of Agriculture, Agriculture Research Laboratory National Sedimentation Laboratory (Web site). <http://www.sedlab.olemiss.edu/>.

USGS Sediment Database. United States Geological Survey (Web site). <http://webserver.cr.usgs.gov/sediment/>.

Wyler, Rose. *Science Fun with Mud and Dirt.* Illus. Pat Ronson Stewart. New York: Julian Messner, 1986.

SOIL

CONCEPT

If there is anything on Earth that seems simple and ordinary, it is the soil beneath our feet. Other than farmers, people hardly think of it except when tending to their lawns, and even when we do turn our attention to the soil, we tend to view it as little more than a place where grass grows and earthworms crawl. Yet the soil is a complex mixture of minerals and organic material, built up over billions of years, and without it, life on this planet would be impossible. It is home to a vast array of species that continually process it, enriching it as they do. Nor are all soils the same; in fact, there are a great variety of soil environments and a great deal of difference between the soil at the surface and that which lies further down, closer to the bedrock.

HOW IT WORKS

The Beginnings of Soil Formation

It has taken billions of years to yield the soil as we know it now. Over the course of these mind-boggling stretches of time, the chemical elements on Earth came into existence, and the uniformly rocky surface of the planet gradually gave way to deposits of softer material. This softer matter, the earliest ancestor of soil, became enriched by the presence of minerals from the rocks and, over a longer period, by decaying organic matter.

After its formation from a cloud of hot gas some 4.5 billion years ago, Earth was pelted by meteorites. These meteorites brought with them solid matter along with water, forming the basis for the oceans. There was no atmosphere as such, but by about four billion years ago, volcanic activity had ejected enough carbon dioxide and other substances into the air to form the beginnings of one. The oceans began to cool, making possible the earliest forms of life—that is, molecules of carbon-based matter that were capable of replicating themselves. (For more on these subjects, see Sun, Moon, and Earth and Geologic Time. On the relationship between carbon and life-forms, see Carbon Cycle.)

All of these conditions—Earth itself, an atmosphere, waters, and life-forms—went into the creation of soil. Soil has its origins in the rocks that now lie below Earth's surface, from which the rain washed minerals. For rain to exist, of course, it was necessary to have water on the planet, along with some form of atmosphere into which it could evaporate. Once these conditions had been established (as they were, over hundreds of millions of years) and the rains came down to cool the formerly molten rock of Earth's surface, a process of leaching began.

Leaching is the removal of soil particles that have become dissolved in water, but at that time, of course, there was no soil. There were only rocks and minerals, but these features of the geosphere, along with the chemical elements in the atmosphere and hydrosphere, were enough to set in motion the development of soil. While the atmosphere and hydrosphere supplied the falling rain, with its vital activity of leaching minerals from the rocks, the minerals themselves supplied additional chemical elements necessary to the formation of soil. (The chemical elements are discussed in several places, most notably Biogeochemical Cycles. See also Minerals and Rocks.)

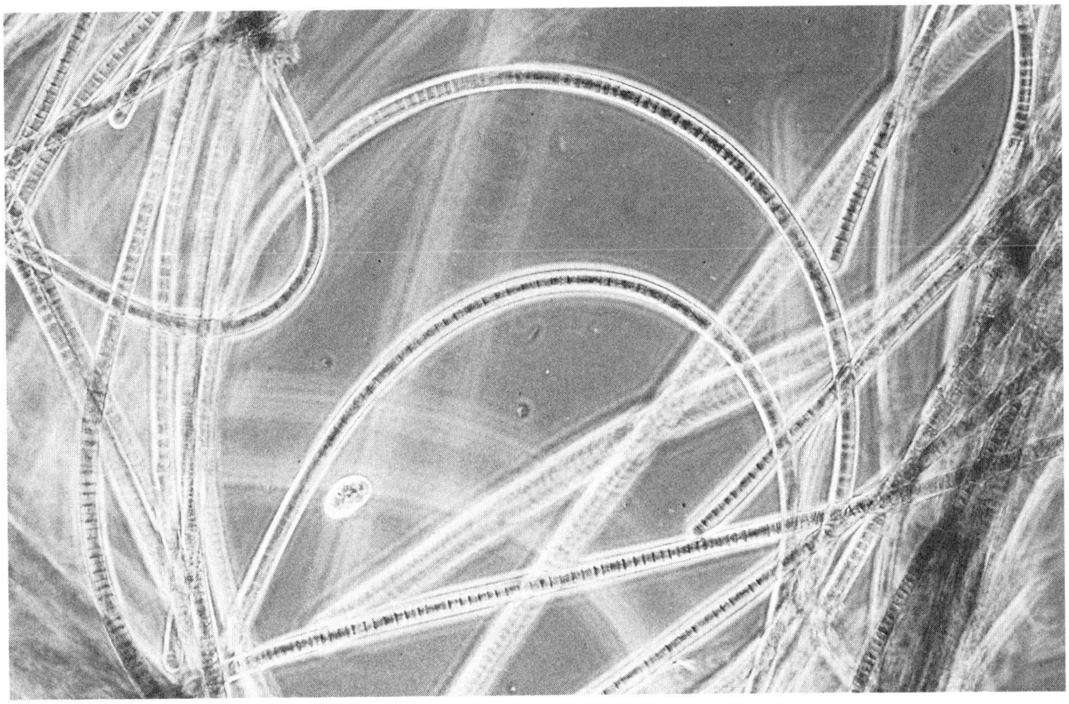

LIGHT MICROGRAPH OF BLUE-GREEN ALGAE, AN EXAMPLE OF THE SIMPLEST PLANT ORGANISMS THAT WERE THE FORERUNNERS OF LIFE ON THE EARTH. PLANT LIFE WAS MADE POSSIBLE BY THE LEACHING OF POTASSIUM, CALCIUM, AND MAGNESIUM FROM ROCK, AND, IN TURN, PLANT DEATH LENT ORGANIC MATTER TO THE GROUND TO HELP FORM THE BASIS FOR SOIL. *(© S. Stammers/Photo Researchers. Reproduced by permission.)*

THE FIRST PLANTS. Among the elements leached from the rock by the falling rains were potassium, calcium, and magnesium, all of which are essential for the growth of plant life. Thus, the foundation was laid for the first botanical forms, a fact that had several important consequences. First and most obviously, it helped set in motion the formation of the complex biosphere we have around us today. Not only did the simplest algae-like plants serve as forerunners for more complex varieties of plant and animal life to follow, but they also played a major role in the beginnings of an atmosphere breathable by animal life. As the plants absorbed carbon dioxide from their surroundings, there gradually evolved a process whereby the plant received carbon dioxide and, as a result of a chemical reaction, released oxygen.

In addition, plant life meant plant death, and as each plant died, it added just a bit more organic material—and with it nutrients and energy—to the ground. Notice the word ground as opposed to soil, which took a long, long time to form from the original rock and mineral material. Indeed, the processes we are describing here did not take shape over the course of centuries or millennia but over whole eons—the longest phases of geologic time, stretching for half a billion years or more (see Geologic Time). Only around the beginning of the present eon, the Phanerozoic, more than 500 million years ago, did soil as such begin to take shape.

WHAT IS SOIL?

As the soil began to form, processes of weathering, erosion, and sedimentation (see the entries Erosion and Sediment and Sedimentation) slowly added to the soil buildup. Today the soil forms a sheath over much of the solid earth; just inches deep or nonexistent in some places, it is many feet deep in others. It separates the planet's surface from its rocky interior and brings together a number of materials that contribute to and preserve life.

Though its origins lie in pulverized rock and decayed organic material, soil looks and feels like neither. Whether brown, red, or black, moist or dry, sandy or claylike, it is usually fairly uniform within a given area, a fact for which the organisms living in it can be thanked. Under the surface of the soil live bacteria, fungi, worms, insects, and other creatures that continually churn through it and process its chemical contents.

A filter for water and a reservoir for air, soil provides a sort of stage on which the drama of an ecosystem (a community of mutually interdependent organisms) is played out. It receives rain and other forms of precipitation, which it filters through its layers, replenishing the groundwater supplies. This natural filtration system, sometimes augmented by a little human ingenuity, is amazingly efficient for leaching out harmful microorganisms and toxins at relatively low levels. (Thus, for instance, septic tank drainage systems process wastewater, with the help of soil, before returning it to the water table.)

By collecting rainwater, soil also gives the rain a place to go and thus helps prevent flooding. Water is not the only substance it stores; soil also collects air, which accounts for a large percentage of its volume. Thus, oxygen is made available to the roots of plants and to the large populations of organisms living underground. The creatures that live in the soil also die there, providing organic material that decays along with a vast collection of dead organisms from aboveground: trees and other plants as well as dead animals—including humans, whose decomposed bodies eventually become part of the soil as well.

FACTORS THAT INFLUENCE SOIL

The processes that formed soil over the eons and that continue to contribute to the soil under our feet today are similar to those by which sedimentary rock is formed. Sedimentary rocks, such as shale and sandstone, have their origins in the deposition, compaction, and cementation of rock that has experienced weathering. Added to this is organic material derived from its ecosystem—for example, fossilized remains of animals.

Both sedimentary rock and soil are made up of sediment, which originates from the weathering, or breakdown, of rock. Weathered remains of rocks ultimately are transported by forces of erosion to what is known as a depositional environment, a location where they are sedimented. (See Sediment and Sedimentation for more about these processes.) The nature of the "parent material," or the rock from which the soil is derived, ranks among five key factors influencing the characteristics of soil in a given environment. The others are climate, living organisms, topography, and time.

PARENT MATERIAL, CLIMATE, AND ORGANISMS. Minerals, such as feldspars and micas, react strongly to natural acids carried by rain and other forms of water; therefore, when these minerals are present in the rock that makes up the parent material, they break apart quite easily into small fragments. On the other hand, a mineral that is harder—for example, quartz—will break into larger pieces of clastic, or rock, sediment. Thus, the parent material itself has a great deal to do with the initial grain of the sediment that will become soil, and this in turn influences such factors as the rate at which water leaches through it.

The release of chemical compounds and elements from minerals in weathering provides plants with the nutrients they need to grow, setting in motion the first of several steps whereby living organisms take root in, and ultimately contribute to, the soil. As the plant dies, it leaves behind material to feed decomposers, such as bacteria and fungi. The latter organisms play a highly significant role in the biogeochemical cycles whereby certain life-sustaining elements are circulated through the various earth systems.

In addition, still-living plants provide food to animals, which, when they die, likewise will become one with the soil. This is achieved through the process of decomposition, aided not only by decomposers but by detritivores as well. The latter, of which earthworms are a great example, are much more complex organisms than the typically single-cell decomposers. Detritivores consume the remains of plant and animal life, which usually contains enzymes and proteins far too complex to benefit the soil in their original state. By feeding on organic remains, detritivores cycle these complex chemicals through their systems, causing them to undergo chemical reactions that result in the breakdown of their components. As a result, simple and usable nutrients are made available to the soil.

TOPOGRAPHY AND TIME. Then there is the matter of topography, or what one might call landscape—the configuration of Earth's surface, including its relief or elevation. Soil at the top of a hill, for instance, is liable to experience considerable leaching and loss of nutrients. On the other hand, if soil is located in a basin area, it is likely to benefit from the vitamins and minerals lost to soils at higher elevations, which lose these nutrients through leaching and erosion.

In addition, topography influences the presence or absence of organic material, which is vital if the soil is to sustain plant life. Organic matter in mountainous areas accounts for only 1% to 6% of the soil composition, while in wet lowland regions it may constitute as much as 90% of soil content. Because erosion tends to bring soil, water, and organic material from the highlands to the lowlands, it is no wonder that lowlands are almost always more fertile than the mountains that surround them.

Finally, time is a factor in determining the quality of soil. As with everything else that either is living or contains living things, soil goes through a progression from immaturity to a peak to old age. In the earth sciences, age often is measured not in years, which is an absolute dating method, but by the relative dating technique of judging layers, beds, or strata of earth materials. (For more about studying rock strata as well as relative dating techniques, see Stratigraphy.)

LEAVES ON THE FOREST FLOOR ARE AN EXAMPLE OF HUMUS, A COMPONENT OF THE A HORIZON, OR TOPSOIL. *(© Michael Hubrich/Photo Researchers. Reproduced by permission.)*

REAL-LIFE APPLICATIONS

LAYERS IN THE SOIL

If you dig down into the dirt of your backyard, you will see a miniature record of your regions's geologic history over the past few million years. Actually, most homes in urban areas and suburbs today have yards made of what is called fill dirt—loose earth that has been moved into place by a backhoe or some other earthmoving mechanism. Even though the mixed quality of fill dirt makes it difficult to discern the individual strata, the soil itself tells a tale of the long ages of time that it took to shape it.

Better than a modern fill-dirt yard, of course, would be a sample taken from an older community. Here, too, however, human activities have intervened: people have dug in their yards and holes have been filled back up, for instance, thus altering the layers of soil from what they would have been in a natural state. To find a sample of soil layers that exists in a fully natural state, it might be necessary to dig in a woodland environment.

In any case, anyone with a shovel and a piece of ground that is reasonably untouched—that is, that has not been plowed up recently—can become an amateur soil scientist. Soil scientists study soil horizons, or layers of soil that lie parallel to the surface of Earth and which have built up over time. These layers are distinguished from one another by color, consistency, and composition. A cross-section combining all or most of the horizons that lie between the surface and bedrock is called a soil profile. The most basic division of layers is between the A, B, and C horizons, which differ in depth, physical and chemical characteristics, and age.

TOPSOIL. At the top is the A horizon, or topsoil, in which humus—unincorporated, often partially decomposed plant residue—is mixed with mineral particles. Technically, humus actually constitutes something called the O horizon, the topmost layer. Examples of humus would be leaves piled on a forest floor, pine straw that covers a bare-dirt area in a yard, or grass residue that has fallen between the blades of grass on a lawn. In each case, the passage of time will make the plant materials one with the soil.

Owing to its high organic content, the soil of the A horizon may be black, or at least much darker than the soil below it. Between the A and B horizons is a noticeable layer called the *E horizon,* the depth of which is a function of the particulars in its environment, as discussed earlier. In rough terms, topsoil could be less than a foot (0.3 m) deep, or it could extend to a depth of 5 ft. (1.5 m) or more.

In any case, the E horizon, known also as the eluviation or leaching layer, is composed primarily of sand and silt, built up as water has leached down through the soil. The sediment of the E horizon is nutrient-poor, because its valuable mineral content has drained through it to the B horizon. (The E horizon is just one of several layers aside from the principal A, B, and C layers. We will mention only a few of these here, but soil scientists include several other horizons in their classification system.)

SUBSOIL, REGOLITH, BEDROCK. The appearance and consistency of the soil change dramatically again as we reach the B horizon. No longer is the earth black, even in the most organically rich environments; by this point it is more likely to exhibit shades of brown, since organic material has not reached this far below the surface. Yet subsoil, which is the consistency of clay, is certainly not poor in nutrients; on the contrary, it contains abundant deposits of iron, aluminum oxides, calcium carbonate, and other minerals, leached from the layers above it.

The rock on the C horizon is called regolith, a general term for a layer of weathered material that rests atop bedrock. Neither plant roots nor any other organic material penetrate this deeply, and the deeper one goes, the more rocky the soil. At a certain depth, it makes more sense to say that there is soil among the rocks rather than rocks in the soil.

Beneath the C horizon lies the R horizon, or bedrock. As noted earlier, depths can vary. Bedrock might be only 5–10 ft. deep (1.5–3 m), or it might be half a mile deep (0.8 km) or perhaps even deeper. Whatever the depth, it is here that the solid earth truly becomes solid, and for this reason builders of skyscrapers usually dig down to the bedrock to establish foundations there.

LIFE BENEATH THE SURFACE

The ground beneath our feet—that is, the topmost layer, the A horizon—is full of living things. In fact, there are more creatures below Earth's surface than there are above it. The term creatures in this context includes microorganisms, of which there might be several billion in a sample as small as an acorn. These include decomposers, such as bacteria and fungi, which feed on organic matter, turning fresh leaves and other material into humus. In addition, both bacteria and algae convert nitrogen into forms usable by plants in the surrounding environment (see Nitrogen Cycle).

WORMS. We cannot see bacteria, of course, but almost anyone who has ever dug in the dirt has discovered another type of organism: worms. These slimy creatures might at first seem disgusting, but without them our world could not exist as it does. As they burrow through soils, earthworms mix organic and mineral material, which they make available to plants around them. They also may draw leaves deep into their middens, or burrows, thus furnishing the soil with nutrients from the surface. In addition, earthworms provide the extraordinarily valuable service of aerating the soil, or supplying it with air: by churning up the soil continuously, they expose it to oxygen from the surface and allow air to make its way down below as well.

Nor are these visible, relatively large worms the only ones at work in the soil. Colorless worms called nematodes, which are only slightly larger than microorganisms, also live in the soil, performing the vital function of processing organic material by feeding on dead plants. Some, however, are parasites that live off the roots of such crops as corn or cotton.

ANTS AND LARGER CREATURES. Likewise there are "bad" and "good" ants. The former build giant, teeming mounds and hills that rise up like sores on the surface of the ground, and some species have the capacity to sting, causing welts on human victims. But a great number of ant species perform a positive function for the environment: like earthworms, they aerate soil and help bring oxygen and organic material from the surface while circulating soils from below.

In some areas, much larger creatures call the soil home. Among these creatures are moles, who live off earthworms and other morsels to be found beneath the surface, including grubs (insect larvae) and the roots of plants. As with ants and earthworms, by burrowing under the ground, they help loosen the soil, making it more porous and thus receptive both to moisture and air. Other large burrowing creatures include mice, ground squirrels, and prairie dogs. They typically live in dry areas, where they perform the valuable function of aerating sandy, gravelly soil.

SOILS AND ENVIRONMENTS

In discussing our imaginary journey through the depths of the soil, it has been necessary to use vague terms concerning depths: "less than a foot," for instance. The reason is that no solid figures can be given for the depth of the soil in any particular area, unless those figures are obtained by a soil scientist who has studied and measured the soil.

Depth is just one of the ways that the soil may vary from one place to another. Earlier we mentioned five factors that affect the character of the soil: parent material, climate, living organisms, topography, and time. These factors determine all sorts of things about the soil—most of all, its ability to support varied life-forms. Collectively, these five factors constitute the environment in which a soil sample exists.

POOR SOILS. A desert environment might be one of immature soil, defined as a sample that has only A and C horizons, with no B horizon between them. On the other hand, the soil in rainforests suffers from just the opposite condition: it has gone beyond maturity and reached old age, when plant growth and water percolation have removed most of its nutrients.

Whether in the desert or in the rainforest, soils near the equator tend to be the "oldest," and this helps explain why few equatorial regions are noted for their agricultural productivity, even though they enjoy otherwise favorable weather for growing crops. Soils there have been leached of nutrients and contain high levels of iron oxides that give them a reddish color. Moreover, red soil is never good for growing crops: the ancient Egyptians referred to the deserts beyond their realm as "the red land," while their own fertile Nile valley was "the black land."

RAINFORESTS. If soil is so poor at the equator, why do equatorial regions such as the Congo or the Amazon River valley in Brazil support the dense, lush rain-forest ecosystems for which they are noted? The answer is that the abundance of organic material at the surface of the soil continually replenishes its nutrient content. The rapid rate of decay common in warm, moist regions further supports the process of renewing minerals in the ground.

THE BURROWING PRAIRIE DOG HELPS AERATE SANDY, GRAVELLY SOIL IN DRY AREAS. *(© Rich Kirchner/Photo Researchers. Reproduced by permission.)*

This also explains why the clearing of tropical rainforests, an issue that environmentalists called to the world's attention in the 1990s, is a serious problem. When the heavy jungle canopy of tall trees is removed, the heat of the sun and the pounding intensity of monsoon rains fall directly on ground that the canopy would normally protect. With the clearing of trees and other vegetation, the animal life that these plants support also disappears, thus removing organisms whose waste products and bodies would have decayed eventually and enriched the soil. Pounded by heat and water and without vegetation to resupply it, the soil in an exposed rainforest becomes hard and dry.

DESERTS. In deserts the soil typically comes from sandstone or shale parent material, and the lack of abundant rainfall, vegetation, or

KEY TERMS

A HORIZON: Topsoil, the uppermost of the three major soil horizons.

AERATE: To make air available to soil.

ATMOSPHERE: In general, an atmosphere is a blanket of gases surrounding a planet. Unless otherwise identified, however, the term refers to the atmosphere of Earth, which consists of nitrogen (78%), oxygen (21%), argon (0.93%), and other substances that include water vapor, carbon dioxide, ozone, and noble gases such as neon (0.07%).

B HORIZON: Subsoil, beneath topsoil and above regolith.

BEDROCK: The solid rock that lies below the C horizon, the deepest layer of soil.

BIOGEOCHEMICAL CYCLES: The changes that particular elements undergo as they pass back and forth through the various earth systems and particularly between living and nonliving matter. The elements involved in biogeochemical cycles are hydrogen, oxygen, carbon, nitrogen, phosphorus, and sulfur.

BIOSPHERE: A combination of all living things on Earth—plants, mammals, birds, reptiles, amphibians, aquatic life, insects, viruses, single-cell organisms, and so on—as well as all formerly living things that have not yet decomposed.

C HORIZON: Regolith, which lies between subsoil and bedrock and constitutes the bottommost of the soil horizons.

DECOMPOSERS: Organisms that obtain their energy from the chemical breakdown of dead organisms as well as from animal and plant waste products. The principal forms of decomposer are bacteria and fungi.

DECOMPOSITION REACTION: A chemical reaction in which a compound is broken down into simpler compounds or into its constituent elements. In the earth system, this often is achieved through the help of detritivores and decomposers.

DETRITIVORES: Organisms that feed on waste matter, breaking organic material down into inorganic substances that then can become available to the biosphere in the form of nutrients for plants. Their function is similar to that of decomposers, but unlike decomposers—which tend to be bacteria or fungi—detritivores are relatively complex organisms, such as earthworms or maggots.

ECOSYSTEM: A community of interdependent organisms along with the inorganic components of their environment.

EROSION: The movement of soil and rock due to forces produced by water, wind, glaciers, gravity, and other influences. In most cases, a fluid medium, such as air or water, is involved.

FILL DIRT: Loose earth that has been moved into place by a backhoe or some other earthmoving machine, usually as part of a large construction project.

animal life gives the soil little in the way of organic sustenance. For this reason, the A horizon level is very thin and composed of light-colored earth. Then, of course, there are desert areas made up of sand dunes, where conditions are much worse, but even the best that deserts have to offer is not very good for sustaining abundant plant life.

KEY TERMS CONTINUED

GEOSPHERE: The upper part of Earth's continental crust, or that portion of the solid earth on which human beings live and which provides them with most of their food and natural resources.

HUMUS: Unincorporated, often partially decomposed plant residue that lies at the top of soil and eventually will decay fully to become part of it.

HYDROSPHERE: The entirety of Earth's water, excluding water vapor in the atmosphere but including all oceans, lakes, streams, groundwater, snow, and ice.

LEACHING: The removal of soil materials that are in solution, or dissolved in water.

ORGANIC: At one time chemists used the term *organic* only in reference to living things. Now the word is applied to most compounds containing carbon, with the exception of carbonates (which are minerals), and oxides, such as carbon dioxide.

PARENT MATERIAL: Mineral fragments removed from rocks by means of weathering. Along with organic deposits, these form the basis for soil.

REGOLITH: A general term describing a layer of weathered material that rests atop bedrock.

SEDIMENT: Material deposited at or near Earth's surface from a number of sources, most notably preexisting rock. There are three types of sediment: rocks, or clastic sediment; mineral deposits, or chemical sediment; and organic sediment, composed primarily of organic material.

SEDIMENTARY ROCK: One of the three major types of rock, along with igneous and metamorphic rock. Sedimentary rock typically has its basis in the deposition, compaction, and cementation of rock that has experienced weathering, though it also may be formed as a result of chemical precipitation. Organic sediment also may be a part of sedimentary rock.

SEDIMENTATION: The process of erosion, transport, and deposition undergone by sediment.

SOIL HORIZONS: Layers of soil, parallel to the surface of Earth, that have built up over time. They are distinguished from one another by color, consistency, and composition.

SOIL PROFILE: A cross-section combining all or most of the soil horizons that lie between Earth's surface and the bedrock below it.

TOPOGRAPHY: The configuration of Earth's surface, including its relief as well as the position of physical features.

WEATHERING: The breakdown of rocks and minerals at or near the surface of Earth due to physical, chemical, or biological processes.

Only those species that can endure a limited water supply—for example, the varieties of cactus that grow in the American Southwest—are able to survive. But lack of water is not the only problem. Desert subsoils often contain heavy deposits of salts, and when rain or irrigation adds water to the topsoil, these salts rise. Thus, watering desert topsoil can make it a worse environment for growth.

RICH SOILS. In striking contrast to the barren soil of the deserts and the potentially barren soil of the rainforest is the rich earth that lies beneath some of the world's most fertile crop-producing regions. On the plains of the midwestern United States, Canada, and Russia, the soil is black—always a good sign for growth. Below this rich topsoil is a thick subsoil that helps hold in moisture and nutrients.

The richest variety of soil on Earth is alluvial soil, a youngish sediment of sand, silt, and clay transported by rivers. Large flowing bodies of water, such as the Nile or Mississippi, pull soil along with them as they flow, and with it they bring nutrients from the regions through which they have passed. These nutrients are deposited by the river in the alluvial soil at its delta, the place where it enters a larger body of water—the Mediterranean Sea and the Gulf of Mexico, respectively. Hence the delta regions of both rivers are extremely fertile.

WHERE TO LEARN MORE

Bial, Raymond. *A Handful of Dirt.* New York: Walker, 2000.

Bocknek, Jonathan. *The Science of Soil.* Milwaukee, WI: Gareth Stevens, 1999.

Canadian Soil Information System (Web site). <http://sis.agr.gc.ca/cansis/>.

Gardner, Robert. *Science Projects About the Environment and Ecology.* Springfield, NJ: Enslow Publishers, 1999.

Scheiderman, Jill S. *The Earth Around Us: Maintaining a Livable Planet.* New York: W. H. Freeman, 2000.

Snedden, Robert. *Rocks and Soil.* Illus. Chris Fairclough. Austin, TX: Raintree Steck–Vaughn, 1999.

Soil Association (Web site). <http://www.soilassociation.org>.

Soil Science Society of America (Web site). <http://www.soils.org/>.

USDA-NRCS National Soil Survey Center (Web site). <http://www.statlab.iastate.edu/soils/nssc/>.

World Soil Resources (Web site). <http://www.nhq.nrcs.usda.gov/WSR/Welcome.html>.

SOIL CONSERVATION

CONCEPT

With the rise of the environmentalist movement in the 1960s and afterward, it has become common to speak of conserving natural resources such as trees or fossil fuels. Yet long before humans recognized the need to make responsible use of things taken from the ground, they learned to conserve the ground itself—that is, the soil. This was a hard-won lesson: failure to conserve soil has turned many a fertile farmland into temporary dust bowl or even permanent desert. Techniques such as crop rotation aid in conservation efforts, but communities continue to face hazards associated with the soil. There is, for instance, the matter of leaching, the movement of dissolved substances through the soil, which, on the one hand, can benefit it but, on the other hand, can rob it of valuable nutrients. Issues of soil contamination also raise concerns that affect not just farmers but the population as a whole.

HOW IT WORKS

Billions of Years in the Making

Earth's present wealth of soil is the result of hundreds of millions of years' worth of weathering, erosion, and sedimentation. Once, long ago, there was no soil, only rock, and it took eons' worth of weathering to dislodge particles of those rocks. These rocks, when combined with organic materials, became the basis for soil, but before the soil could even begin to take shape, a number of things had to fall into place. Chief among these was the formation of something that, at first glance, at least, does not seem to have a great deal of bearing on the soil: the atmosphere.

In combination with water in the hydrosphere (e.g., streams and rivers) as well as water in the form of evaporated moisture and precipitation in the air itself, the blanket of gases we call our atmosphere has been essential to the formation and sustenance of Earth's soil. This importance goes beyond the obvious point that rain transports water to the soil, thus making possible the abundance of plant life that grows in it. Rain, of course, is of unquestionable importance, but it is only one of several factors associated with the atmosphere (including the water vapor it contains) that have a role in shaping soil as we know it.

To move weathered rocks from highlands to lowlands, where they can become sediment and eventually begin to form soil, it is necessary to subject the rocks themselves to a process of erosion. And erosion—aside from erosion caused by gravity, which usually is considered weathering—can take place only when an atmosphere exists, along with water in the air and on the land. The chief agents of erosion are wind, water (both flowing and in the form of precipitation), and frozen water in the form of icy glaciers, all of which depend on an atmosphere or water or both (see Glaciology).

Erosion transports not only rock sediment but organic material as well. Together, these two ingredients are as essential to making soil as tea bags and water are to making tea. Obviously, the greater the organic content, the richer the soil, and here again the air plays a part. It is important that deeper layers of soil receive a supply of air from the surface to sustain the life of subterranean organisms, who not only process nutrients through the soil but (by their burrowing activities) also aerate it, or make air available to it.

A Product of Its Environment

Soil, like most people, is a product of the environment in which it was formed. That environment has five major influencing factors: the nature of the "parent material," or the rock from which the soil was derived; the local climate; the presence of living organisms; local topography; and the passage of time.

Specific classes of mineral break apart in characteristic ways, and the size of the pieces into which the original weathered rock is broken has a great deal to do with the character of the soil that it forms. This does not mean, however, that relatively large rock pieces necessarily will yield the worst soils, since erosive forces will continue to work on the rock, pulling out its nutrient-rich mineral wealth and gradually acting to break it apart.

As for climate, it is clear that rain and sun are essential for the growth of plant matter, but, of course, too much of either or both is harmful. (See Soil for a discussion of soils in rainforests.) Plants aid the soil by dying and feeding it with more organic material, but they are not the only types of organism in the soil. Indeed, the soil constitutes an ecosystem in and of itself, a realm rich in biodiversity, in which various biogeochemical cycles are played out, and through which energy flows as part of the operation of the larger Earth system.

The underground world teems with creatures ranging from bacteria to moles and prairie dogs (in some regions), each of which fulfills a function. These functions include aerating the soil by burrowing; processing material though ingestion and elimination of waste, thus converting compounds into nutrients that the soil can use; and mixing organic material with minerals. Organisms' final contribution to the soil comes when they die, as their bodies become material that feeds the earth through decomposition.

Topography, or elevation, plays a major role in making possible erosion, itself a process that can be either beneficial or detrimental. The question of whether it is one or the other may be a matter of perspective, or rather elevation. From the standpoint of lowland areas, which receive the wealth of the upland areas in the form of nutrient-rich runoff carried by gravity or flowing media, such as wind or water, erosion is a good thing. Matters do not look as good from the viewpoint of the mountains, which lose much of their best soil to low-lying areas.

The influence of time in shaping soils—as well as much else about the soil itself—can be appreciated by studying soil horizons, the various strata, or layers, of soil that lie beneath the surface. The most basic division of layers is between the A, B, and C horizons, which differ in depth and physical and chemical characteristics as well as age.

SOIL HORIZONS. Above the A horizon, or topsoil, lies humus, decomposing organic material that eventually will become soil. The A horizon itself contains a large amount of organic matter, and thus it may be black, or at least much darker than the soil below it. Between the A and B horizons is a sandy, silty later called the E horizon. Then comes the B horizon, or subsoil, which starts at a depth as shallow as 1 ft. (0.3 m) or deeper than 5 ft. (1.5 m).

Lacking a great deal of organic material but still rich in nutrients, the B horizon has a sizable impact on the A horizon. Minerals—both healthful and harmful—may rise up from the B to the A horizon, and the ability of the B horizon to hold in moisture from above greatly affects the moisture of the A horizon soil. (See Soil for a discussion of how salt deposits in the B horizon affect topsoil in deserts.) Together, A and B horizons constitute what is called the *solum,* or true soil.

The C horizon is called *regolith.* It is the home for the rocks of the parent material, which has given up much of its nutrient riches in fortifying the soil that lies above it. This far below the surface, there is no sign of plant or animal life, and below the C horizon is the R horizon, or bedrock—the top of the layers of rock and metal that descend all the way to the planet's core. Once again depths vary, with bedrock as shallow as 5–10 ft. (1.5–3 m) or as deep as 0.5 mi. (0.8 km) or more.

Differences Between Soils

The depth of the soil is a measure of wealth—wealth, that is, in terms of natural resources. A sheath over much of the solid earth, soil separates the planet's surface from its rocky interior and preserves the lives of the plants and animals

A CLOUD OF TOPSOIL IS PICKED UP BY THE WIND NEAR BOISE CITY, OKLAHOMA, DURING THE DUST BOWL OF THE 1930S. IN SOME CASES, WIND REMOVED 3-4 IN. (7.6-10.6 CM) OF TOPSOIL, TURNING ACREAGE THAT ONCE RIPPLED WITH WHEAT INTO A DESERTLIKE WASTELAND. *(AP/Wide World Photos. Reproduced by permission.)*

that live on and in it. It receives rain and other forms of precipitation, which it filters through its layers, as we discuss later, in the context of leaching. Thus, it not only provides water to organisms above and below its surface but also helps prevent flooding by acting as a reservoir.

A great deal of soil's volume is air, for which it also acts as a reservoir. Underground creatures depend on this air and also help circulate it by burrowing. This circulation, in turn, provides oxygen to the roots of plants and makes the soil more hospitable to growth. Even though soil performs these and other life-preserving functions, it would be a mistake to assume that all soils are the same. In fact, the U.S. Department of Agriculture has identified 11 major soil orders, each of which is divided into suborders, groups, subgroups, families, and series.

The specificity of soil types, as reflected in the identification and naming of soil series, illustrates the complexity of what at first seems a very simple thing. In fact, soils can be extremely specific, with names that reflect local landmarks. If soils share enough similarities, they are grouped together in a soil series, but it is safe to say that there are thousands of individual soil types on Earth.

CONSERVING SOIL

On a broad level, there are certain types of environment more or less favorable to the formation of rich soil. Some of these types are discussed in the essay Soil, and specific examples of environmental problems are provided later in this essay. Yet almost any environment can become unfavorable to plant growth if proper soil-conservation procedures are not observed.

The phrase soil conservation refers to the application of principles for maintaining the productivity and health of agricultural land by control of wind- and water-induced soil erosion. For the remainder of this essay, we examine the dangers involved in such erosion and the use of measures to prevent it. In so doing, we give the matter of soil conservation a somewhat larger scope than the preceding definition might suggest. Since soil affects the world far beyond farms, it seems only fitting to approach it not as a concern merely of agriculture but of the environment in general.

Erosion is spoken of here in a general sense, but for a more in-depth discussion of erosive processes, see Erosion. Mass Wasting examines dramatic erosion-related phenomena, such as landslides. Biogeochemical Cycles contains some discussion of erosion, inasmuch as it helps circulate life-sustaining chemical elements throughout the various earth systems. Indeed, it is important to remember that erosion is not always negative in its results; on the contrary, it is a valuable process by which landforms are shaped. The erosive processes we explore here, however, generally contribute to the loss of soil health and productivity.

REAL-LIFE APPLICATIONS

THE DUST BOWL

When people mismanage agricultural lands or when natural forces otherwise conspire to destroy soil, the results can be devastating. One of the most dramatic examples occurred in what came to be known as the dust bowl. This was the name given to a wide area covering Texas, Oklahoma, Kansas, and even agricultural parts of Colorado during the years 1934 and 1935. Over the course of a few months, once-productive farmlands turned into worthless fields of stubble and dust, good for almost nothing and highly vulnerable to violent wind erosion.

And wind erosion came, scattering vast quantities of soil from the Great Plains of the Midwest to the Atlantic Seaboard. The classic 1939 film *The Wizard of Oz* sets its fantastic, otherworldly story against this backdrop, and to viewers in the late 1930s the tornado that swept Dorothy from her Kansas farmland into the world of Oz was all too real. The only difference was that no magical adventure awaited victims of the real-life tornadoes and other windstorms.

The fate of the dust bowl farmers, many of whom lost everything, was dramatized in the novel *The Grapes of Wrath* by John Steinbeck in 1939 as well as in the acclaimed motion picture that followed a year later. A perhaps equally eloquent tribute appeared in the form of the American photographer Dorothea Lange's photographs of dust bowl refugees. The images etched by Lange are unforgettable: in one a woman stares into the distance, her face a landscape of despair, as her children huddle next to her, their eyes hidden from the camera. In another a man, obviously exhausted from months or years of overwork, hardship, and fear, sits behind the wheel of a truck, gazing somewhere beyond the camera lens. Like the woman, he seems to be looking into a future that offers scant hope.

CAUSES OF THE DUST BOWL. What happened? The sad fact is that in the years leading up to the early 1930s, the future dust bowl farmlands had seemed remarkably productive, and farmers continued to be pleasantly surprised, year after year, at the abundant yields they could draw from each field. In fact, farmers were unwittingly preparing the way for vast erosion by overcultivating the land and not taking proper steps to preserve its moisture against drought. This was particularly unfortunate because farmers in the 1930s had long known about the principle of crop rotation as a means of giving the soil a rest in order to restore its nutrients. Yet the farmers of the plains tried to push their crops to yield more and more, and for a time it worked, though at great future expense to the land.

One is tempted to see in the agricultural world of the U.S. Midwest parallels to the foolhardy attitude that, just a few years earlier, created a boom on Wall Street, followed by the devastating stock market crash of October 29, 1929, that ushered in the Great Depression. Certainly the ravages of the dust bowl, when they came, were particularly unwelcome in a land already reeling from several years of widespread unemployment and a sagging economy. And though there was no cause-effect relationship between the Wall Street crash and the dust bowl, there is no question that both were brought about in large part by a lack of planning for the future and by a naive belief that it is possible to get "something for nothing"—that is, to get more out of the world (whether the world of finances or the natural world) than one puts into it.

In some places farmers alternated between wheat cultivation and livestock grazing on particular plots of land. Thus, the hooves of the cattle damaged the soil, which had been weakened by raising wheat. The land was therefore ready to become the site of a full-fledged natural disaster, and, at the height of the depression, natural disaster came in the form of high winds. The winds in some cases removed topsoil as much as 3–4 in. (7–10 cm) thick. Dunes of dust as tall as 15–20 ft. (4.6–6.1 m) formed, turning acreage that once had rippled with wheat into desertlike wastelands.

EROSION CONTROL IN ACTION

Today the farmlands of the plains states long since have recovered, and American farmers have benefited from the lessons learned in the dust bowl. Out of the dust bowl years came the establishment, in 1935, of the Soil Conservation Service, a federal agency charged with implementing erosion-control practices. (The Soil Conservation Service was the predecessor of the modern-day Natural Resources Conservation Service.) In the wake of the legislation creating the agency, signed into law by President Franklin D. Roosevelt (1882–1945), states passed laws creating nearly 3,000 local soil conservation districts.

If one passes through agricultural lands today, one is likely to see signs identifying the local conservation district. Even more important, the lands themselves are a testament to principles put into practice as an outgrowth of the dust bowl years. For instance, instead of alternating one year of wheat with one year in which a field lies fallow, or unused, farmers in the dust bowl region discovered that a three-year cycle of wheat, sorghum, and fallow land worked much better. They also planted trees to serve as barriers against wind.

EROSION CONTROL LEGISLATION. Concerns over soil conservation in America did not end with the dust bowl. As United States farm production soared in the 1970s, American farms enjoyed such a great surplus that U.S. farmers increasingly began to sell their crops overseas—most notably, to the Soviet Union. While some Americans were upset to see the farmers of the Midwest selling wheat to the Communists in Moscow, others saw in this act a testament to the failure of the Soviet agricultural system and to the strength of U.S. farming. In the wake of these increased exports, farmers were encouraged to cultivate even marginal croplands to increase profits, thus heightening the vulnerability of their lands to erosion.

What followed was not another dust bowl, however; instead, the experience of the 1970s and 1980s shows just how much American farmers, legislators, and others had learned from the 1930s. Environmental activists in the 1970s, concerned over water quality, helped return public interest to the problem of soil erosion. They called attention to the flow of nutrients from croplands into water resources, most notably leaching of nitrogen and phosphorus that choked lakes with eutrophication (see Biogeochemical Cycles). As a result of public concerns over these and related issues, Congress in 1977 passed the Soil and Water Resources Conservation Act, mandating the conservation of soil, water, and other resources on private farmlands and other properties.

In 1985 the Food Security Act further served to encourage steps toward the reduction of soil erosion. Some 45 million acres (18 million hectares) of land vulnerable to erosion were removed from intensive cultivation by the act. The legislation also forbade the conversion of rangelands into agricultural fields, which would have raised great potential for erosion and depletion of already vulnerable soil. In addition, the act required farmers to develop and maintain practices for the control of erosion on lands susceptible to that threat.

BARRIER AND COVER. Soil-conservation practices fall under two headings: barrier and cover. Under the barrier approach, various structures act as a wall against water runoff, wind, and the movement of soil. Among such structures are banks, hedgerows, walls of earth or other materials, and silt fences such as one sees at construction sites. The cover approach is devoted to the idea of maintaining a heavy soil cover of living and dead plant material. This is achieved through the use of mulch, cover crops, and other techniques.

Local governments and property owners in nonagricultural lands often apply both the cover and barrier approaches, planting trees as well as grass not simply to beautify the land but also to hold the soil in place. Land has to have some sort of vegetative protection to stand between it and the forces of wind and water erosion, and the two approaches together serve to protect soil against nature's onslaught.

LEACHING

Like erosion, leaching—the movement of dissolved substances with water percolating through soil—can be both positive and negative. For any plot of land, assuming the rate of water input is greater than the rate of water loss through evaporation, water has to go somewhere, so it leaves the site by moving downward. Eventually it either reaches the deep groundwater or passes through subterranean springs to flow into the surface waters of streams, rivers, and lakes.

Along the way, the leached water carries all sorts of dissolved substances, ranging from nutrients to contaminants. The threat of the latter has led to widespread concern in the United States over the leaching of toxins into water supplies, and in 1980 this concern spurred a massive piece of legislation called CERLA (Comprehensive Environmental Response, Compensation, and Liability Act), better known as Superfund. Six years later, in 1986, Congress updated CERLA with the Superfund Amendments and Reauthorization Act. These laws provided for a vast array of measures directed toward environmental cleanup, including the removal of chemicals and other toxins in soil.

Drastic measures such as those outlined in CERLA and other legislation may be required for the cleanup of artificial materials introduced into soils and groundwater. But for human waste and other more natural forms of toxin, nature itself is able to achieve a certain amount of cleanup on its own. In a septic-tank system, used by people who are not connected to a municipal sewage system, bacteria process wastes, removing a great deal of their toxic content in the tank itself. The wastewater leaves the tank and passes through a filtration system, in which the water leaches through layers of gravel and other filters that help remove more of its harmful content. As the wastewater percolates from the filtration system through the soil (usually well below the A horizon by this point), it is purified further before it enters the groundwater supply.

Not only does leaching help purify the water that passes through the soil, it also is an important part of the soil-formation process, inasmuch as it passes nutrients to the depths of the A horizon and into the B horizon. Its ability to pass along nutrients is not always beneficial, and in some ecosystems, large amounts of dissolved nitrogen are lost to soil as a result of leaching. In such a situation, soil typically is fertilized with nitrate, a form of the element with which soil often has difficulty binding (see Nitrogen Cycle). For this reason, nitrate tends to leach easily, leading to an overabundance of nitrogen in the lower levels of the soil and in the groundwater. This condition, known as nitrogen saturation, can influence the eutrophication of waters (see Biogeochemical Cycles for an explanation of eutrophication) and can cause the decline and death of trees on the surface.

DESERTIFICATION

Much of North Africa lies under the cover of a vast desert, the Sahara. By far the world's largest desert, the Sahara today spreads across some 3.5 million sq. mi. (9.06 million sq km), an area larger than the continental United States. Only about 780 acres (316 hectares) of it, or little more than 1 sq. mi. (2.6 sq km), is fertile. The rest is mostly stone and dry earth with scattered shrubs—and, here and there, the rolling sand dunes typically used to depict the Sahara in movies.

Given the forbidding moonscape of the Sahara today, it might be surprising to learn that just 8,000 years ago—the blink of an eye in terms of geologic time—it was a region of flowing rivers and lush valleys. For thousands of years it served as a home to many cultures, some of them quite advanced, to judge from their artwork. Though they left behind an extraordinary record in the form of their rock-art paintings and carvings, which show an understanding of realistic representation that would not be matched until the time of the Greeks, the identity of the early Saharan peoples themselves remains largely a mystery.

Instead of identifying them by the name of a nationality or empire, archaeologists divide the phases of the early Saharan culture according to a set of four names that collectively tell the story of the region's progressive transformation into a desert. First was the Hunter period, from about 6000 to about 4000 B.C., when a Paleolithic, or Old Stone Age, people survived by hunting the many wild animals then available in the region. Next came the Herder period, from about 4000 to 1500 B.C. As their name suggests, these people maintained herds of animals and also practiced basic agriculture.

As the Sahara became drier and drier, however, there were no more herds. Egyptians began bringing in domesticated horses to cross the desert: hence the name of the Horse period (*ca.* 1500–*ca.* 600 B.C.) By about 600 B.C., not even horses could survive in the forbidding climate. There was only one creature that could survive: the hardy, seemingly inexhaustible camel. Thus began the Camel era, which continues to the present day.

ATTEMPTS TO CONTROL DESERTIFICATION. As with the dust bowl, the first question one wants to ask when confronted

A CAMEL CARAVAN IN THE SAHARA. THE WORLD'S LARGEST DESERT, IT COVERS 3.5 MILLION SQ. MI. (9 MILLION SQ KM), BUT 8,000 YEARS AGO THIS WAS A REGION OF LUSH VALLEYS AND FLOWING RIVERS. *(© Tom Hollyman/Photo Researchers. Reproduced by permission.)*

with a story such as that of the Sahara, is "What happened?" The answer is much more complex, just as the effects of desertification—the slow transformation of ordinary lands to desert—are much more permanent than those of the erosion associated with the dust bowl. Desertification does not always result in what people normally think of as a desert. It is rather a process that contributes toward making a region more dry and arid, and because it is usually gradual, it can be reversed in some cases. But doing so represents a vast challenge.

In 1977 the United Nations (UN), in the form of the UN Conference on Desertification in Nairobi, Kenya, set out to address the spread of the Sahara into the Sahel, an arid region that stretches south of the desert. Some 700 delegates from almost 100 countries adopted a number of measures designed to halt the spread of desertification in that region and others by the year 2000.

Even though there have been some successes, the Sahel region today remains a blighted area where famine is common, and this state of affairs is not entirely the result of the natural causes addressed in the conference's resolutions. Poor government management and a near-constant state of civil war in such countries as Ethiopia have played at least as important a role in spreading famine as nature itself. During the 1980s, in fact, the government of Ethiopia (at that time a Marxist-Leninist state) deliberately withheld food supplies, shipped to it from the West, as a way of exerting pressure on rebel factions and other groups it wished to subdue.

THE EXAMPLE OF IRAQ. The arid regions of Iraq provide another example of how human influences can result in desertification. Once that country, known in ancient times as Mesopotamia, was among the greenest and most lush places in the known world. For this reason, historians today use the name Fertile Crescent to describe an arc from the deltas of the Tigris and Euphrates rivers in Mesopotamia to the mouth of the Nile in Egypt. Today, of course, Iraq is mostly a dust-colored land of bare trees and brush.

What happened? Agricultural mismanagement certainly played a role, as did the simple exhaustion of the soil by some 6,000 years of human civilization. Indeed, since the Fertile Crescent was perhaps the first area settled by agricultural societies long before the beginning of full-fledged civilization as such in about 3500

KEY TERMS

A HORIZON: Topsoil, the uppermost of the three major soil horizons.

AERATE: To make air available to soil.

B HORIZON: Subsoil, beneath topsoil and above regolith.

BEDROCK: The solid rock that lies below the C horizon, the deepest layer of soil.

BIOGEOCHEMICAL CYCLES: The changes that particular elements undergo as they pass back and forth through the various earth systems and particularly between living and nonliving matter. The elements involved in biogeochemical cycles are hydrogen, oxygen, carbon, nitrogen, phosphorus, and sulfur.

C HORIZON: Regolith, which lies between subsoil and bedrock and constitutes the bottommost of the soil horizons.

DECOMPOSERS: Organisms that obtain their energy from the chemical breakdown of dead organisms as well as from animal and plant waste products. The principal forms of decomposer are bacteria and fungi.

DECOMPOSITION REACTION: A chemical reaction in which a compound is broken down into simpler compounds or into its constituent elements. In the earth system, this often is achieved through the help of detritivores and decomposers.

DETRITIVORES: Organisms that feed on waste matter, breaking organic material down into inorganic substances that then can become available to the biosphere in the form of nutrients for plants. Their function is similar to that of decomposers; however, unlike decomposers—which tend to be bacteria or fungi—detritivores are relatively complex organisms, such as earthworms or maggots.

ECOSYSTEM: A community of interdependent organisms along with the inorganic components of their environment.

EROSION: The movement of soil and rock due to forces produced by water, wind, glaciers, gravity, and other influences. In most cases, a fluid medium, such as air or water, is involved.

EUTROPHICATION: A state of heightened biological productivity in a body of water, which is typically detrimental to the ecosystem in which it takes place. Eutrophication can be caused by an excess of nitrogen or phosphorus in the form of nitrates and phosphates, respectively.

HUMUS: Unincorporated, often partially decomposed plant residue that lies at the top of soil and eventually will decay fully to become part of it.

B.C., it is safe to say that the region has been under cultivation for several thousand years longer—perhaps 8,000 or even 10,000 years. Direct human action and malice also may have played a role: some historians believe that the Mongols, during their brutal invasion in the 1250s, so badly devastated the farmlands and irrigation channels of Iraq that the land never recovered.

SOME CAUSES OF DESERTIFICATION. With regard to human involvement in the desertification process, it is not necessary for a society to be advanced agriculturally to do long-term damage to the soil. The Pueblan

KEY TERMS CONTINUED

LANDFORM: A notable topographical feature, such as a mountain, plateau, or valley.

LEACHING: The removal of soil materials that are in solution, or dissolved in water.

MASS WASTING: The transfer of earth material down slopes by processes that include creep, slump, slide, flow, and fall. Also known as *mass movement.*

MINERAL: A naturally occurring, typically inorganic substance with a specific chemical composition and a crystalline structure.

ORGANIC: At one time chemists used the term *organic* only in reference to living things. Now the word is applied to most compounds containing carbon, with the exception of carbonates (which are minerals) and oxides, such as carbon dioxide.

PARENT MATERIAL: Mineral fragments removed from rocks by means of weathering. Along with organic deposits, these fragments form the basis for soil.

REGOLITH: A general term describing a layer of weathered material that rests atop bedrock.

SEDIMENT: Material deposited at or near Earth's surface from a number of sources, most notably preexisting rock. There are three types of sediment: rocks, or clastic sediment; mineral deposits, or chemical sediment; and organic sediment, composed primarily of organic material.

SEDIMENTATION: The process of erosion, transport, and deposition undergone by sediment.

SOIL CONSERVATION: The application of principles for maintaining the productivity and health of agricultural land by control of wind- and water-induced soil erosion. The term also may be applied more broadly to encompass the maintenance and protection of nonagricultural soils.

SOIL HORIZONS: Layers of soil, parallel to the surface of the earth, which have built up over time. These layers are distinguished from one another by color, consistency, and composition.

SOIL PROFILE: A cross-section combining all or most of the soil horizons that lie between Earth's surface and the bedrock below it.

WEATHERING: The breakdown of rocks and minerals at or near the surface of Earth due to physical, chemical, or biological processes.

culture of what is now the southwestern United States depleted an already dry and vulnerable region after about A.D. 800 by removing its meager stands of mesquite trees. And though human causes, in the form of either mismanagement or deliberate damage, have contributed toward desertification, sometimes nature itself is the driving force.

Long-term changes in rainfall or general climate as well as water erosion and wind erosion such as caused the dust bowl can turn a region into a permanent desert. An ecosystem may survive short-term drought, but if soil is forced to go too long without proper moisture, it sets in motion a chain reaction in which plant life dwindles and, with it, animal life as well. Thus, the soil is denied the fresh organic material necessary to its continued sustenance, and a slow, steady process of decline begins.

WHERE TO LEARN MORE

Bear, Firman E., H. Wayne Pritchard, and Wallace E. Akin. *Earth: The Stuff of Life.* Norman: University of Oklahoma Press, 1986.

Bocknek, Jonathan. *The Science of Soil.* Milwaukee, WI: Gareth Stevens, 1999.

Bright Edges of the World: The Earth's Evolving Drylands (Web site). <http://www.nasm.edu/ceps/drylands/drylands.html>.

Cherrington, Mark. *Degradation of the Land.* New York: Chelsea House, 1991.

"Desertification." United States Geological Survey (Web site). <http://pubs.usgs.gov/gip/deserts/desertification/>.

Gardner, Robert. Science *Projects About the Environment and Ecology.* Springfield, NJ: Enslow Publishers, 1999.

Natural Resources Conservation Service (Web site). <http://www.nrcs.usda.gov/>.

Pittman, Nancy P. *From the Land.* Washington, DC: Island Press, 1988.

Soil and Water Conservation Society (Web site). <http://www.swcs.org/>.

"Voices from the Dust Bowl: The Charles L. Todd and Robert Sonkin Migrant Worker Collection, 1940–1941." Library of Congress (Web site). <http://memory.loc.gov/ammem/afctshtml/tshome.html>.

GEOCHEMISTRY

BIOGEOCHEMICAL CYCLES

THE CARBON CYCLE

THE NITROGEN CYCLE

BIOGEOCHEMICAL CYCLES

CONCEPT

Of the 92 elements produced in nature, only six are critical to the life of organisms: hydrogen, carbon, nitrogen, oxygen, phosphorus, and sulfur. Though these elements account for 95% of the mass of all living things, their importance extends far beyond the biosphere. Hydrogen and oxygen, chemically bonded in the form of water, are the focal point of the hydrosphere, while oxygen and nitrogen form the bulk of the atmosphere. All six are part of complex biogeochemical cycles in which they pass through the biosphere, atmosphere, hydrosphere, and geosphere. These cycles circulate nutrients through the soil into plants, microbes, and animals, which return the elements to the earth system through chemical processes that range from respiration to decomposition.

HOW IT WORKS

THE ELEMENTS

An element is a substance composed of a single type of atom, meaning that it cannot be broken down chemically to make a simpler substance. They are listed on the periodic table of elements, a chart that renders them in order of their atomic numbers, or the number of protons in the nucleus of the atom. The elements we want to discuss in the context of biogeochemical cycles are all low in atomic number, starting with hydrogen, which has just one proton in its nucleus. In the following list, the elements are cited by atomic number along with their chemical symbols, or the abbreviation by which they are known to chemists.

Elements Involved in Biogeochemical Cycles

- 1. Hydrogen (B)
- 6. Carbon (C)
- 7. Nitrogen (N)
- 8. Oxygen (O)
- 15. Phosphorus (P)
- 16. Sulfur (S)

Given the fact, as noted earlier, that 92 elements appear in nature, it should come as no surprise that the highest atomic number for any naturally occurring element is 92, for uranium. Beyond uranium there are about two dozen artificially created elements, but they are of little interest outside the realm of certain specialties in chemistry and physics. The naturally occurring elements are the ones that matter to the earth sciences, and of these elements, only a handful play a significant role.

In the essays Minerals and Economic Geology, other elements—most notably silicon—are discussed with regard to their importance in forming minerals, rocks, and ores. Though they are critical to Earth's systems, elements other than the six discussed here play no role in biogeochemical cycles. Indeed, it is a fact of the physical sciences that not all elements are created equal: certainly, the universe is not divided evenly 92 ways, with equal amounts of all elements. In fact, hydrogen and helium account for 99% of the mass of the entire universe.

ABUNDANCE. On Earth the ratios are quite different, however. Oxygen and silicon constitute the preponderance of the known mass of Earth's crust, while nitrogen and oxygen form the overwhelming majority of the atmosphere. Hydrogen is proportionally much, much less abundant on Earth than in the universe as a whole, but owing to its role in forming water, a substance essential to the sustenance of life, it is unquestionably of great significance.

The two following lists provide rankings for the abundance of the six elements discussed in this essay. The first table shows their ranking and share in the entire known mass of the planet, including the crust, living matter, the oceans, and atmosphere. The second shows their relative abundance and ranking in the human body.

Abundance of Selected Elements on Earth (Ranking and Percentage)

- 1. Oxygen (49.2%)
- 9. Hydrogen (0.87%)
- 12. Phosphorus (0.11%)
- 14. Carbon (0.08%)
- 15. Sulfur (0.06%)
- 16. Nitrogen (0.03%)

Abundance of Selected Elements in the Human Body (Ranking and Percentage)

- 1. Oxygen (65%)
- 2. Carbon (18%)
- 3. Hydrogen (10%)
- 4. Nitrogen (3%)
- 6. Phosphorus (1%)
- 9. Sulfur (0.26%)

Several things are interesting about these figures. First and most obviously, there is the fact that the ranking of all these elements (with the exception of oxygen) is relatively low in the total known elemental mass of Earth, whereas their ranking is much, much higher within the human body. This is significant, given the fact that these elements are all essential to the lives of organisms.

Furthermore, note that it does not take a great percentage to constitute an "abundant" element: even nitrogen, with its 0.03% share of Earth's total known mass, still is considered abundant. The presence of the vast majority of elements on Earth is measured in parts per million (ppm) or even parts per billion (ppb).

CHEMISTRY AND GEOCHEMISTRY

In a general sense, chemistry can be defined as an area of the physical sciences concerned with the composition, structure, properties, and changes of substances, including elements, compounds, and mixtures. This definition unites the phases in the history of the development of the discipline, from early modern times—when it arose from alchemy, a set of mystical beliefs based on the idea that ordinary matter can be perfected—to modern times. Our modern understanding of chemistry, however, is quite different from the model of chemistry that prevailed until about 1800, a difference that relates to a key discovery: the atom.

This change, in fact, should be described not in terms of a discovery so much as the development of a model. Long before chemists and physicists comprehended the structure of the atom, they developed an understanding of all chemical substances as composed of atomic units, each representing one and only one element. Until the development of this model—thanks to a number of chemists, most notably, John Dalton (1766–1844) of England, Antoine Lavoisier (1743–1794) of France, and Amedeo Avogadro (1776–1856) of Italy—chemistry was concerned primarily with mixing potions and observing their effects. Thanks to the atomic model, chemists never again would confuse mixtures with chemical compounds.

CHEMICAL REACTIONS. The difference between a mixture and a compound goes to the heart of the distinction between physics and chemistry. A mixture, such as coffee, is the result of a physical process—in this case, the heating of water and coffee beans—and the result does not have a uniform chemical structure. On the other hand, a compound results from chemical reactions between atoms, which form enormously powerful bonds in the process of joining to create a molecule. A molecule is the basic particle of a compound, just as an atom is to an element. It should be noted that some elements, such as nitrogen, typically appear in diatomic form, that is, two atoms bond to form a molecule of nitrogen.

A substance may undergo physical changes without experiencing any alteration in its underlying structure; on the other hand, a chemical reaction makes a fundamental change to the substance. In a chemical reaction, a substance may experience a change of state (i.e., from solid to liquid or gas) without undergoing any physical process of being heated or cooled by an outside source. Chemical reactions involve the breaking of bonds between atoms in a molecule and the formation of new bonds. As a result, an entirely new

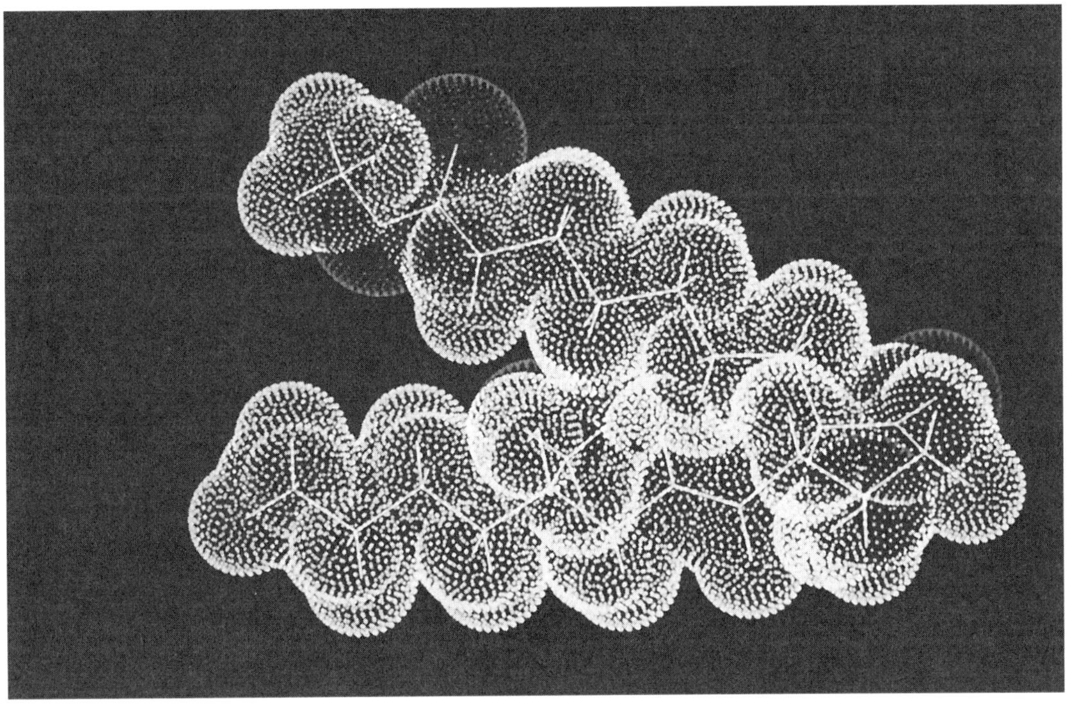

Molecular structure. A compound is formed by chemical reactions between atoms, which join together to make molecules. *(© Blair Seitz/Photo Researchers. Reproduced by permission.)*

substance is created—something that could never be achieved through mere physical processes.

REAL-LIFE APPLICATIONS

Geochemistry

Just as geochemistry is a branch of the geologic sciences that weds physics and geology, so there is a geologic subdiscipline, geochemistry, in which chemistry and the geologic sciences come together. Geochemistry is concerned with the chemical properties and processes of Earth—in particular, the abundance and interaction of chemical elements and their isotopes. (Isotopes are atoms that have an equal number of protons, and hence are of the same element, but differ in their number of neutrons. Isotopes may be either stable or unstable, in which case they are subject to the emission of high-energy particles. Some elements have numerous stable isotopes, others have only one or two, and some have none.)

Before the mid–twentieth century, geochemistry had a relatively limited scope, confined primarily to the identification of elements in rocks and minerals and the determination of the relative abundance of those elements. Since that time, however, this subdiscipline has come to encompass many other concerns, particularly those discussed in the present context. Geochemistry today focuses on such issues as the recycling of elements between the various sectors of the earth system, especially between living and nonliving things.

Biogeochemical Cycles and Earth Systems

The changes that a particular element undergoes as it passes back and forth through the various earth systems, and particularly between living and nonliving matter, are known as biogeochemical cycles. The four earth systems involved in these cycles are the atmosphere, the biosphere (the sum of all living things as well as formerly living things that have not yet decomposed), the hydrosphere (the entirety of Earth's water except for vapor in the atmosphere), and the geosphere. The last of these spheres is defined as the upper part of Earth's continental crust, or that portion of the solid earth on which human beings live and which provides them with most of their food and natural resources.

Carbon, for instance, is present in all living things on Earth. Hence, the phrase *carbon-based life-form,* a cliché found in many an old science-

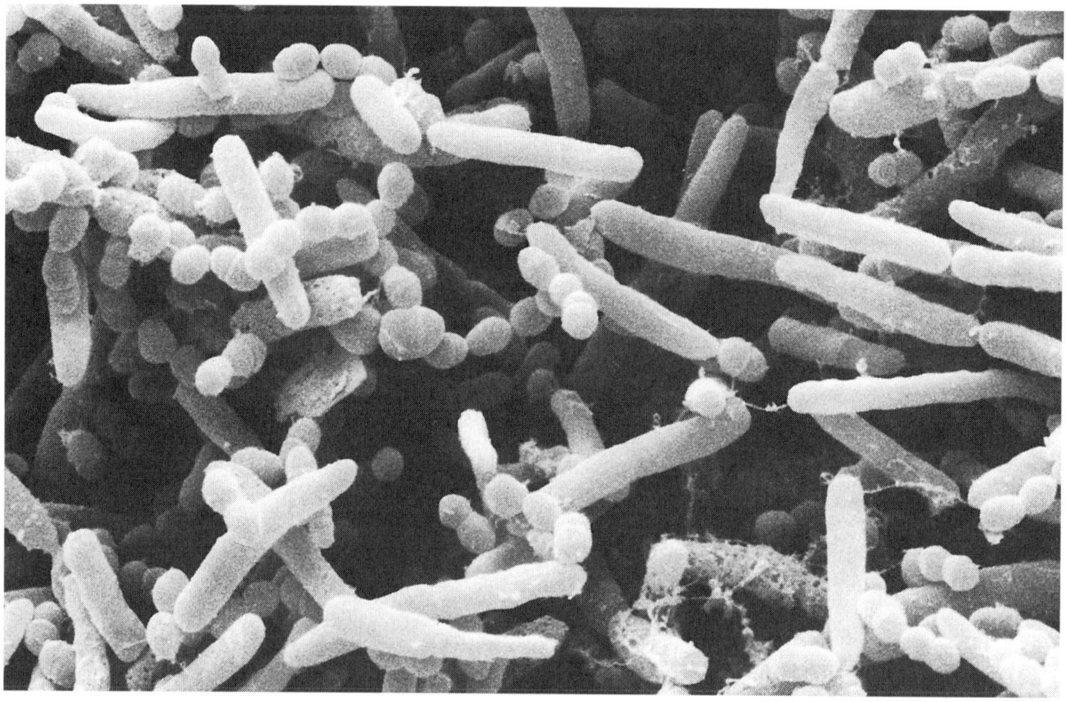

BACTERIA BEGIN THE PROCESS OF DECOMPOSITION OF ORGANIC WASTE, BREAKING DOWN PLANT MATTER AND CONVERTING IT INTO COMPOST. *(© Scimat/Photo Researchers. Reproduced by permission.)*

fiction movie, is actually a redundancy: *all* lifeforms contain carbon. In the context of the physical sciences, *organic* refers to all substances that contain carbon, with the exception of oxides, such as carbon dioxide and carbon monoxide, and carbonates, which are found in minerals. Still, carbon circulates between the organic world and the inorganic world, as when an animal exhales carbon dioxide.

The carbon cycle is of such importance to the functioning of Earth that it is discussed separately (see Carbon Cycle). So, too, is the nitrogen cycle (see Nitrogen Cycle), whereby nitrogen passes between the soil, air, and biosphere as well as the hydrosphere. The hydrosphere, as noted earlier, is based on a single substance, water, created by the chemical bonding of hydrogen and oxygen, and it is likewise discussed in detail elsewhere (see Hydrologic Cycle).

Despite the emphasis here on carbon in the biosphere, nitrogen in the geosphere, and hydrogen and oxygen in the hydrosphere, it should be noted that biogeochemical cycles involving these four elements take them through all four "spheres." The same is true of sulfur, whose biogeochemical cycle is discussed later in this essay. On the other hand, phosphorus, also discussed later, is present in only three of Earth's systems; it plays little role in the atmosphere.

DECOMPOSERS AND DETRITIVORES

Most biogeochemical cycles involve a special type of chemical reaction known as decomposition, and for this to take place, agents of decomposition—known as decomposers and detritivores—are essential. Decomposition occurs when a compound is broken down into simpler compounds or into its constituent elements. This is achieved primarily by decomposers, organisms that obtain their energy from the chemical breakdown of dead organisms as well as from animal and plant waste products.

The principal forms of decomposer are bacteria and fungi. These creatures carry enzymes, which they secrete into the materials they consume, breaking them down chemically before taking in the products of this chemical breakdown. They thus take organic matter and render it in inorganic form, such that later it can be taken in again by plants and returned to the biosphere.

Detritivores are much more complex organisms, but their role is similar to that of decomposers. They, too, feed on waste matter, breaking this organic material down into inorganic substances that then can become available to the biosphere in the form of nutrients for plants. Examples of detritivores are earthworms and maggots. As discussed in Energy and Earth, detritivores are key players in the food web, the set of nutritional interactions—sometimes called a food chain—between living organisms.

PHOSPHORUS AND THE PHOSPHORUS CYCLE

There are a few elements that were known in ancient or even prehistoric times, examples being gold, iron, lead, and tin. The vast majority, on the other hand, have been discovered since the beginning of the modern era, and the first of them was phosphorus, which is also the first element whose discoverer is known.

In 1674 the German alchemist Hennig Brand (*ca.* 1630–*ca.* 1692) was searching for the philosopher's stone, a mythical substance that allegedly would turn common or base metals into gold. Convinced that he would find this substance in the human body, Brand evaporated water from a urine sample and burned the precipitate (the solid that remained) along with sand. The result was a waxy, whitish substance that glowed in the dark and reacted violently with oxygen. Brand named it phosphorus, a name derived from a Greek term meaning "lightbearer."

Owing to its high reactivity with oxygen, phosphorus is used in the production of safety matches, smoke bombs, and other incendiary devices. It is also important in various industrial applications and in fertilizers. In fact, ancient humans used phosphorus without knowing it when they fertilized their crops with animal bones.

PHOSPHATES. In the early 1800s, chemists recognized that the critical component in bones was phosphorus, which plants use in photosynthesis—the biological conversion of energy from the Sun into chemical energy (see Energy and Earth). With this discovery came the realization that phosphorus would make an even more effective fertilizer when treated with sulfuric acid, which makes it soluble, or capable of being dissolved, in water. This compound, known as superphosphate, can be produced from phosphates, a type of mineral.

Phosphates represent one of the eight major classes of mineral (see Minerals). All phosphates contain a characteristic formation, PO_4, which is bonded to other elements or compounds—for example, with aluminum in aluminum phosphate, or $AlPO_4$. Phosphorus fertilizer is typically calcium phosphate, known as bone ash, the most important industrial mineral (see Economic Geology) produced from phosphorus. Another significant phosphate is sodium phosphate, used in dishwashing detergents. In fact, phosphates once played a much larger role in the detergent industry—with disastrous consequences, as we shall see.

THE PHOSPHORUS CYCLE. The majority of phosphorus in the earth system is located in rocks and deposits of sediment, from which it can be removed by one of three processes: weathering, the breakdown of rocks and minerals at or near the surface of Earth as the result of physical, chemical, or biological processes (see Erosion, Sediment and Sedimentation); leaching, the removal of soil materials that are in solution, or dissolved in water; and mining.

Phosphorus is highly reactive, meaning that it is likely to bond with other elements, and for this reason it often is found in compounds. Microorganisms absorb insoluble phosphorus compounds (ones that are incapable of being dissolved) and, through the action of acids within the microorganisms, turn them into soluble phosphates. Algae and other green plants absorb these phosphates and, in turn, are eaten by animals. When they die, the animals release the phosphates back into the soil.

As with all elements, the total amount of phosphorus on Earth stays constant, but the distribution of it does not. Some of the phosphorus passes from the geosphere into the biosphere, but the majority of it winds up in the ocean. It may find its way into sediments in shallow waters, in which case it continues to circulate, or it may be taken to the deep parts of the seas, in which case it is likely to be deposited for the long term.

Fish absorb particles of phosphorus, and thus some of the element returns to dry land through the catching and consumption of seafood. In addition, guano, or dung, from birds that live in an ocean environment (e.g., seagulls) also returns portions of phosphorus to the terrestrial environment. Nonetheless, geochemists

A STAND OF FIR TREES SHOWS THE DEVASTATING EFFECTS OF ACID RAIN, WHICH IS CREATED WHEN SULFURIC ACID MIXES WITH MOISTURE IN THE ATMOSPHERE. *(© Will and Demi McIntyre/Photo Researchers. Reproduced by permission.)*

believe that phosphorus is being transferred steadily to the ocean, from whence it is not likely to return. It is for this reason that phosphorus-based fertilizers are important, because they feed the soil with nutrients that otherwise would be continually lost.

EUTROPHICATION. To be sure, phosphorus, in the proper quantities, is good for the environment. But as with medicine or any other beneficial substance, if a little is good, that does not mean that a lot is necessarily better. In the case of phosphorus, an overabundance of the element in the environment can lead to a phenomenon called eutrophication, a state of heightened biological productivity in a body of water. One of the leading causes of eutrophication (from a Greek term meaning "well nourished") is a high rate of nutrient input, in the form of phosphates or nitrates, a nitrogen-oxygen compound (see Nitrogen Cycle).

As a result of soil erosion, fertilizers make their way into bodies of water, as does detergent runoff in wastewater. Excessive phosphates and nitrates stimulate growth in algae and other green plants, and when these plants die, they drift to the bottom of the lake or other body of water. There, decomposers consume the remains of the plants and, in the process, also use oxygen that otherwise would be available to fish, mollusks, and other forms of life. As a result, those species die off, to be replaced by others that are more tolerant of lowered oxygen levels—for example, worms. Needless to say, the outcome of eutrophication is devastating to the lake's ecosystem.

During the 1960s, Lake Erie—one of the Great Lakes on the U.S.-Canadian border—became an example of eutrophication gone mad. As a result of high phosphate concentrations, Erie's waters were choked with plant and algae growth. Fish were unable to live in the water, the beaches reeked with the smell of decaying algae, and Erie became widely known as a "dead" body of water. This situation led to the passage of new environmental standards and pollution controls by both the United States and Canada, whose governments acted to reduce drastically the phosphate content in fertilizers and detergents. Lake Erie proved to be an environmental success story: within a few decades the lake once again teemed with life.

SULFUR AND THE SULFUR CYCLE

If there is any element that can be said to have a bad image—and a falsely bad one at that—it is sulfur. As everyone "knows," sulfur has a foul smell, and this smell, combined with its com-

KEY TERMS

ALCHEMY: A set of mystical beliefs based on the idea that ordinary matter can be perfected. Though it was a pseudoscience, alchemy, which flourished in the late Middle Ages, was a forerunner of scientific chemistry.

ATMOSPHERE: In general, an atmosphere is a blanket of gases surrounding a planet. Unless otherwise identified, however, the term refers to the atmosphere of Earth, which consists of nitrogen (78%), oxygen (21%), argon (0.93%), and other substances that include water vapor, carbon dioxide, ozone, and noble gases such as neon (0.07%).

ATOMIC NUMBER: The number of protons in the nucleus of an atom. Since this number is different for each element, elements are listed on the periodic table in order of atomic number.

BIOGEOCHEMICAL CYCLES: The changes that particular elements undergo as they pass back and forth through the various earth systems and particularly between living and nonliving matter. The elements involved in biogeochemical cycles are hydrogen, oxygen, carbon, nitrogen, phosphorus, and sulfur.

BIOSPHERE: A combination of all living things on Earth—plants, mammals, birds, amphibians, reptiles, aquatic life, insects, viruses, single-cell organisms, and so on—as well as all formerly living things that have not yet decomposed.

CARNIVORE: A meat-eating organism.

CHEMICAL BONDING: The joining through electromagnetic force of atoms that sometimes, but not always, represent more than one chemical element. The result is the formation of a molecule.

CHEMICAL SYMBOL: A one-letter or two-letter abbreviation for the name of an element.

COMPOUND: A substance made up of atoms of more than one element chemically bonded to one another.

DECOMPOSERS: Organisms that obtain their energy from the chemical breakdown of dead organisms as well as from animal and plant waste products. The principal forms of decomposers are bacteria and fungi.

DECOMPOSITION REACTION: A chemical reaction in which a compound is broken down into simpler compounds or into its constituent elements. In the earth system, this often is achieved through the help of detritivores and decomposers.

DETRITIVORES: Organisms that feed on waste matter, breaking organic material down into inorganic substances that then can become available to the biosphere in the form of nutrients for plants. Their function is similar to that of decomposers; however, unlike decomposers—which tend to be bacteria or fungi—detritivores are relatively complex organisms, such as earthworms or maggots.

ECOSYSTEM: A term referring to a community of interdependent organisms along with the inorganic components of their environment.

ELEMENT: A substance made up of only one kind of atom. Unlike compounds, elements cannot be broken chemically into other substances.

KEY TERMS CONTINUED

EUTROPHICATION: A state of heightened biological productivity in a body of water, which is typically detrimental to the ecosystem in which it takes place. Eutrophication can be caused by an excess of nitrogen or phosphorus in the form of nitrates and phosphates, respectively.

FOOD WEB: A term describing the interaction of plants, herbivores, carnivores, omnivores, decomposers, and detritivores, each of which consumes nutrients and passes it along to other organisms.

GEOCHEMISTRY: A branch of the earth sciences, combining aspects of geology and chemistry, that is concerned with the chemical properties and processes of Earth—in particular, the abundance and interaction of chemical elements and their isotopes.

GEOSPHERE: The upper part of Earth's continental crust, or that portion of the solid earth on which human beings live and which provides them with most of their food and natural resources.

HERBIVORE: A plant-eating organism.

HYDROSPHERE: The entirety of Earth's water, excluding water vapor in the atmosphere but including all oceans, lakes, streams, groundwater, snow, and ice.

LEACHING: The removal of soil materials that are in solution, or dissolved in water.

MINERAL: A naturally occurring, typically inorganic substance with a specific chemical composition and a crystalline structure.

MIXTURE: A substance with a variable composition, meaning that it is composed of molecules or atoms of differing types. Compare with *compound.*

MOLECULE: A group of atoms, usually but not always representing more than one element, joined in a structure. Compounds typically are made up of molecules.

bustibility, led to the biblical association of *brimstone*—the ancient name for the element—with the fires of hell. It may come as a surprise, then, to learn that sulfur has no smell of its own. Only in combination with other elements does it acquire the offensive odor that has led to its unpleasant reputation.

An example of such a compound is hydrogen sulfide, a poisonous substance present in intestinal gas. The May 2001 *National Geographic* included two stories relating to the presence of natural hydrogen sulfide deposits on opposite sides of the earth, and in both cases the presence of these toxic fumes created unusual ecosystems.

A system of caves known as Villa Luz in southern Mexico contains some 20 underground springs that carry large quantities of hydrogen sulfide. Among the strange creatures that have made Villa Luz their home are species of fish colored bright red; the pigmentation is a result of the fact that they have to produce high quantities of hemoglobin (a component in red blood cells) to survive on the scant oxygen. The waters of the cave also are populated by microorganisms that oxidize the hydrogen sulfide and turn it into sulfuric acid, which dissolves the rock walls and continually enlarges the cave.

Thousands of miles away, in the Black Sea, explorers examining evidence of a great ancient flood like the one depicted in the Bible (see Earth, Science, and Nonscience) found an unexpected ally in the form of hydrogen sulfide. Because the Black Sea lacks the temperature differences that cause water to circulate from the bottom upward, hydrogen sulfide had gathered at the bottom and stayed there, covered by dense layers of saltwater. Oxygen could not reach the

KEY TERMS CONTINUED

OMNIVORE: An organism that eats both plants and other animals.

ORGANIC: At one time chemists used the term *organic* only in reference to living things. Now the word is applied to most compounds containing carbon, with the exception of carbonates (which are minerals) and oxides, such as carbon dioxide.

PERIODIC TABLE OF ELEMENTS: A chart that shows the elements arranged in order of atomic number along with their chemical symbols and the average atomic mass for each particular element.

PHOTOSYNTHESIS: The biological conversion of light energy (that is, electromagnetic energy) from the Sun to chemical energy in plants.

REACTIVITY: A term referring to the ability of one element to bond with others. The higher the reactivity, the greater the tendency to bond.

SEDIMENT: Material deposited at or near Earth's surface from a number of sources, most notably preexisting rock. There are three types of sediment: rocks, or clastic sediment; mineral deposits, or chemical sediment; and organic sediment, composed primarily of organic material.

SOLUBLE: Capable of being dissolved.

SOLUTION: A homogeneous mixture (i.e., one that is the same throughout) in which one or more substances is dissolved in another substance—for example, sugar dissolved in water.

WEATHERING: The breakdown of rocks and minerals at or near the surface of Earth due to physical, chemical, or biological processes.

bottom of the Black Sea, and thus wood-boring worms could not live in the toxic environment. As a result, a 1,500-year-old shipwreck had been virtually undisturbed.

THE SULFUR CYCLE. Sulfur is removed from rock by weathering, at which point it reacts with oxygen in the air to form sulfate, or SO_4. This sulfate is taken in by plants and microorganisms, which convert it to organic materials and pass it on to animals in the food web. Later, when these organisms die, decomposers absorb the sulfur from their bodies and return it to the environment. As with phosphorus, however, sulfur is being lost continually to the oceans as it drains through lakes and streams (and through the atmosphere) on its way to the sea.

In the ocean ecosystem, sulfur can take one of three routes. Some of it circulates through food webs, and some drifts to the bottom to bond with iron in the form of ferrous sulfide, or FeS. Ferrous sulfide contributes to the dark color of sediments at the bottom of the ocean. On the other hand, sulfur may be returned to the atmosphere, released by spray from saltwater. In addition, sulfur can pass into the atmosphere as the result of volcanic activity or through the action of bacteria, which release it in the form of hydrogen sulfide, the foul-smelling gas discussed earlier.

As with all biogeochemical cycles, humans play a part in the sulfur cycle, and the role of modern industrial society is generally less than favorable, as is true of most such cycles. A particularly potent example is the production of acid rain. Among the impurities in coal is sulfur, and when coal is burned (as it still is, for instance, in electric power plants), it results in the production of sulfur dioxide and sulfur trioxide—SO_2

and SO_3, respectively. Sulfur trioxide reacts with water in the air to produce sulfuric acid, or H_2SO_4. This mixes with moisture in the atmosphere to create acid rain, which is hazardous to both plant and animal life.

WHERE TO LEARN MORE

Beatty, Richard. *Sulfur.* New York: Benchmark Books, 2000.

———. *Phosphorus.* New York: Benchmark Books, 2001.

Biogeochemical Cycles (Web site). <http://www.bsi.vt.edu/chagedor/biol_4684/Cycles/cycles.html>.

General Chemistry Online (Web site). <http://antoine.fsu.umd.edu/chem/senese/101/index.shtml>.

Geochemistry on the World Wide Web (Web site). <http://www.geo.cornell.edu/geology/classes/Geochemweblinks.HTML>.

Hancock, Paul L., Brian J. Skinner. *The Oxford Companion to the Earth.* New York: Oxford University Press, 2000.

Kump, Lee R., James F. Kasting, and Robert G. Crane. *The Earth System.* Upper Saddle River, NJ: Prentice Hall, 2000.

Life and Biogeochemical Cycles (Web site). <http://essp.csumb.edu/esse/climate/climatebiogeo.html>.

Skinner, Brian J., Stephen C. Porter, and Daniel B. Botkin. *The Blue Planet: An Introduction to Earth System Science.* 2d ed. New York: John Wiley and Sons, 1999.

WebElements Periodic Table of the Elements (Web site). <http://www.webelements.com/webelements/scholar/index.html>.

THE CARBON CYCLE

CONCEPT

If a person were asked to name the element most important to sustaining life, chances are he or she would say oxygen. It is true that many living things depend on oxygen to survive, but, in fact, carbon is even more fundamental to the sustenance of life. Indeed, in a very real sense, carbon *is* life, since every living thing contains carbon and the term *organic* refers to certain varieties present in all life-forms. Yet carbon, in the form of such oxides as carbon dioxide as well as carbonates like calcium carbonate, is a vital part of the inorganic realm as well. Hence, the carbon cycle, by which the element is circulated through the biosphere, geosphere, atmosphere, and hydrosphere, is among the most complex of biogeochemical cycles.

HOW IT WORKS

GEOCHEMISTRY

Chemistry is concerned with the composition, structure, properties, and changes of substances, including elements, compounds, and mixtures. Central to the discipline is the atomic model, or the idea that all matter is composed of atoms, each of which represents one and only one chemical element. An element thus is defined as a substance made up of only one kind of atom, which cannot be broken chemically into other substances. A chemical reaction involves either the bonding of one atom with another or the breaking of chemical bonds between atoms.

Geochemistry brings together geology and chemistry, though as the subdiscipline has matured in the period since the 1940s, its scope has widened to take in aspects of other disciplines and subdisciplines. With its focus on such issues as the recycling of elements between the various sectors of the earth system, especially between living and nonliving things, geochemistry naturally encompasses biology, botany, and a host of earth science subdisciplines, such as hydrology.

BIOGEOCHEMICAL CYCLES. Among the most significant areas of study within the realm of geochemistry are biogeochemical cycles. These are the changes that a particular element undergoes as it passes back and forth through the various earth systems—particularly between living and nonliving matter. As we shall see, this transition between the worlds of the living and the nonliving is particularly interesting where carbon is concerned.

Along with carbon, five other elements—hydrogen, nitrogen, oxygen, phosphorus, and sulfur—are involved in biogeochemical cycles. With the exception of phosphorus, which plays little part in the atmosphere, these elements move through all four earth systems, including the atmosphere, the biosphere (the sum of all living things as well as formerly living things that have not yet decomposed), the hydrosphere (Earth's water, except for water vapor in the atmosphere), and the geosphere, or the upper part of Earth's continental crust.

Earth systems and biogeochemical cycles are discussed in greater depth within essays devoted to those topics (see Earth Systems and Biogeochemical Cycles). Likewise, the nitrogen cycle is treated separately (see Nitrogen Cycle). The role of hydrogen and oxygen, which chemically bond

THE NAME CARBON COMES FROM THE LATIN WORD FOR CHARCOAL, CARBO. COAL HAS A WIDE VARIETY OF USES, FROM MANUFACTURING STEEL TO GENERATING ELECTRICITY. *(© Andrew J. Martinez/Photo Researchers. Reproduced by permission.)*

to form water, is discussed in the context of the hydrosphere (see Hydrologic Cycle).

ELEMENTS AND COMPOUNDS

We have referred to elements and compounds, which are essential to the study of chemistry; now let us examine them briefly before going on to the subject of a specific and very important element, carbon. An element is defined not by outward characteristics, though elements do have definable features by which they are known; rather, the true meaning of the term *element* is discernible only at the atomic level.

Every atom has a nucleus, which contains protons, or subatomic particles of positive electric charge. The identity of an element is defined by the number of protons in the nucleus: for instance, if an atom has only a single proton, by definition it must be hydrogen. An atom with six protons in the nucleus, on the other hand, is always an atom of carbon. Thus, the elements are listed on the periodic table of elements by atomic number, or the number of protons in the atomic nucleus.

ELECTRONS AND CHEMICAL REACTIONS. While protons are essential to the definition of an element, they play no role in the bonding between atoms, which usually produces chemical compounds. (The reason for this is qualified by the modifier *usually,* in that sometimes two atoms of the same element may bond as well.) Chemical bonding involves only the electrons, which are negatively charged subatomic particles that spin around the nucleus. In fact, only certain of these fast-moving particles take part in bonding. These are the valence electrons, which occupy the highest energy levels in the atom.

One might say that valence electrons are at the "outside edge" of the atom, though the model of atomic structure, considered only in the briefest form here, is far more complex than that phrase implies. In any case, elements have characteristic valence electron patterns that affect their reactivity, or their ability to bond. Carbon is structured in such a way that it can form multiple bonds, and this feature plays a significant part in its importance as an element.

When an element reacts with another, they join together, generally in a molecule (we will examine some exceptions), to form a compound. Though the atoms themselves remain intact, and an element can be released from a compound, a compound quite often has properties quite unlike those of the original elements. Carbon and oxygen are essential to sustaining life, but when a single atom of one bonds with a single atom of the other, they form a toxic gas, carbon monoxide. And whereas carbon in its elemental form is a black powder and hydrogen and oxygen are colorless, odorless gases, when bonded in the proper proportions and structure, the three create sugar.

CARBON

The name carbon comes from the Latin word for charcoal, *carbo.* In fact, charcoal—wood or other plant material that has been heated without enough air present to make it burn—is just one of many well-known substances that contain carbon. Others include coal, petroleum, and other fossil fuels, all of which contain hydrocarbons, or chemical compounds built around strings of carbon and hydrogen atoms. Graphite is pure carbon, and coke, a refined version of coal, is very nearly pure. Not everything made of carbon is black, however: diamonds, too, are pure carbon in another form.

Though carbon makes up only a small portion of the known elemental mass in Earth's crust, waters, and atmosphere—just 0.08%, or 1/1,250 of the whole—it is the fourteenth most abundant element on the planet. In the human body, carbon is second only to oxygen in abundance and accounts for 18% of the body's mass. Present in the inorganic rocks of the ground and in the living creatures above it, carbon is everywhere in the earth system.

CARBON BONDING. There are two elements noted for their ability to form long strings of atoms and seemingly endless varieties of molecules: one is carbon, and the other is silicon, directly below it on the periodic table. Just as carbon forms a vast array of organic compounds, silicon, found in a huge variety of minerals, is at the center of a large number of inorganic compounds. Yet carbon is capable of forming an even greater number of bonds than silicon. (For more about silicon and the silicates, see the entries Minerals and Economic Geology.)

Carbon is distinguished further by its high value of electronegativity, the relative ability of an atom to attract valence electrons. In addition, with four valence electrons, carbon is ideally suited to finding other elements (or other carbon

A DIAMOND IS AN ALLOTROPE, A CRYSTALLINE FORM, OF CARBON. ESSENTIALLY, IT IS A HUGE MOLECULE COMPOSED OF CARBON ATOMS STRUNG TOGETHER BY COVALENT BONDS. *(© V. Fleming/Photo Researchers. Reproduced by permission.*

atoms) with which to form chemical bonds. Normally, an element does not necessarily have the ability to bond with as many other elements as it has valence electrons, but carbon—with its four valence electrons—happens to be tetravalent, or capable of bonding to four other atoms at once. Additionally, carbon can form not just a single bond but also a double bond or even a triple bond with other elements.

ALLOTROPES OF CARBON. Carbon has several allotropes—different versions of the same element distinguished by their molecular structure. The first of them is graphite, a soft material that most of us regularly encounter in the form of pencil "lead." Graphite is essentially a series of one-atom-thick sheets of carbon bonded together in a hexagonal pattern, but with only very weak attractions between adjacent sheets.

Then there is that most alluring of all carbon allotropes, diamond. Neither diamonds nor graphite, strictly speaking, are formed of molecules. Their arrangement is definite, as with a molecule, but their size is not: they simply form repeating patterns that seem to stretch on forever. Whereas graphite is in the form of sheets, a diamond is basically a huge "molecule" composed of carbon atoms strung together by what are known as covalent chemical bonds.

Graphite and diamond are both crystalline—solids in which the constituent parts have a simple and definite geometric arrangement that is repeated in all directions. (All minerals are crystalline in structure. See Minerals.) A third carbon allotrope, buckminsterfullerene, discovered in 1985 and named after the American engineer and philosopher R. Buckminster Fuller (1895–1983), is also crystalline in form.

Carbon takes yet another form, distinguished from the other three allotropes in that it is amorphous in structure—lacking a definite shape—as opposed to crystalline. Though it retains some of the microscopic structures of the plant cells in the wood from which it is made, charcoal is mostly amorphous carbon. Coal and coke are particularly significant varieties of amorphous carbon. Formed by the decay of fossils, coal was the first important fossil fuel (discussed later in this essay) used to provide heat and power to human societies.

REAL-LIFE APPLICATIONS

ORGANIC CHEMISTRY

Organic chemistry is the study of carbon, its compounds (with the exception of the carbonates and oxides mentioned earlier), and their properties. At one time chemists thought that *organic* was synonymous with living, and even as recently as the early nineteenth century, they believed that organic substances contained a supernatural "life force." Then, in 1828, the German chemist Friedrich Wöhler (1800–1882) made an amazing discovery.

By heating a sample of ammonium cyanate, a material from a nonliving source, Wöhler converted it to urea, a waste product in the urine of mammals. As he later observed, "without benefit of a kidney, a bladder, or a dog," he had turned an inorganic substance into an organic one. It was almost as though he had created life. Actually, what he had discovered was the distinction between organic and inorganic material, which results from the way in which the carbon chains are arranged.

Organic chemistry encompasses the study of many things that people commonly think of as "organic"—living creatures, formerly living creatures, and the parts and products of their bodies—but it also is concerned with substances that seem quite far removed from the living world. Among these substances are rubber, vitamins, cloth, and paper, but even in these cases, it is easy to see the relationship to a formerly living organism: a rubber plant, or a tree that was cut down to make wood pulp. But it might come as a surprise to learn that plastics, which at first glance would seem completely divorced from the living world, also have an organic basis. All manner of artificial substances, such as nylon and polyester, are made from hydrocarbons.

FOSSIL FUELS. During the Mesozoic era, which began about 248.2 million years ago, dinosaurs ruled the earth; then, about 65 million years ago, a violent event brought an end to their world. The cause of this mass extinction is unknown, though it is likely that a meteorite hit the planet, sending so much dust into the atmosphere that it dramatically changed local climates, bringing about the destruction of the dinosaurs—along with a huge array of other animal and plant forms. (See Paleontology for more on this subject.)

The bodies of the dinosaurs, along with those of other organisms, were deposited in the solid earth and covered by sediment. They might well have simply rotted, and indeed many of them probably did. But many of these organisms were deposited in an anaerobic, or non-oxygen-containing, environment. Rather than simply rotting, this organic material underwent transformation into hydrocarbons and became the basis for the fossil fuels, the most important of which—from the standpoint of modern society—is petroleum. (See Economic Geology for more on this subject.)

CARBONATES

Carbonates are important forms of inorganic carbon in the geosphere. In chemical terms, a carbonate is made from a single carbon atom bonded to three oxygen atoms, but in mineralogical terms, carbonates are a class of mineral that may contain carbon, nitrogen, or boron in a characteristic molecular formation. Typically, a carbonate is transparent and light in color with a relatively high density.

CALCIUM CARBONATE, ONE OF THE MOST COMMON COMPOUNDS IN THE GEOSPHERE, IS FOUND IN SEASHELLS, EGGSHELLS, PEARLS, AND CORAL (PICTURED HERE), BRIDGING THE BOUNDARY BETWEEN THE LIVING AND THE NON-LIVING. *(© Fred McConnaughey/Photo Researchers. Reproduced by permission.)*

Among carbonate minerals, the most significant compound is calcium carbonate ($CaCO_3$). One of the most common compounds in the entire geosphere, constituting 7% of the known crustal mass, it is found in such rocks as limestone, marble, and chalk. (Just as pencil "lead" is not really lead, the "chalk" used for writing on blackboards is actually gypsum, a form of calcium sulfate.) Additionally, calcium carbonate can combine with magnesium to form dolomite, and in caves it is the material that makes up stalactites and stalagmites. Yet calcium carbonate also is found in coral, seashells, eggshells, and pearls. This is a good example of how a substance can cross the chemical boundary between the worlds of the living and nonliving.

In the oceans, calcium reacts with dissolved carbon dioxide, forming calcium carbonate and sinking to the bottom. Millions of years ago, when oceans covered much of the planet, sea creatures absorbed calcium and carbon dioxide from the water, which reacted to form calcium carbonate that went into their shells and skeletons. After they died, their bodies became sedimented in the ocean floor, forming vast deposits of limestone.

CARBON DIOXIDE AND CARBON MONOXIDE

Historically, carbon dioxide was the first gas to be distinguished from ordinary air, when in 1630 the Flemish chemist and physicist Jan Baptista van Helmont (1577?–1644) discovered that air was not a single element, as had been thought up to that time. The name perhaps most closely associated with carbon dioxide, however, is that of the English chemist Joseph Priestley (1733–1804), who created carbonated water, used today in making soft drinks. Not only does the gas add bubbles to drinks, it also acts as a preservative.

By Priestley's era, chemists had begun to glimpse a relationship between plant life and carbon dioxide. Up until that time, it had been believed that plants purify the air by day and poison it at night. Today we know that carbon dioxide is an essential component in the natural balance between plant and animal life. Animals, including humans, breathe in air, and, as a result of a chemical reaction in their bodies, the oxygen molecules (O_2) bond with carbon to produce carbon dioxide. Plants "breathe" in this carbon dioxide (which is as important to their survival as air is to animals), and a reverse reaction leads

KEY TERMS

AMORPHOUS: A term for a type of solid that lacks a definite shape. Compare with *crystalline.*

ATOMIC NUMBER: The number of protons in the nucleus of an atom. Since this number is different for each element, elements are listed on the periodic table in order of atomic number.

BIOGEOCHEMICAL CYCLES: The changes that particular elements undergo as they pass back and forth through the various earth systems and particularly between living and nonliving matter. The elements involved in biogeochemical cycles are hydrogen, oxygen, carbon, nitrogen, phosphorus, and sulfur.

CELLULAR RESPIRATION: A process that, when it takes place in the presence of oxygen, involves the intake of organic substances, which are broken down into carbon dioxide and water, with the release of considerable energy.

CHEMICAL BONDING: The joining through electromagnetic force of atoms that sometimes, but not always, represent more than one chemical element. The result is the formation of a molecule.

COMPOUND: A substance made up of atoms of more than one element, chemically bonded to one another.

CRYSTALLINE SOLID: A type of solid in which the constituent parts have a simple and definite geometric arrangement that is repeated in all directions.

DECOMPOSERS: Organisms that obtain their energy from the chemical breakdown of dead organisms as well as from animal and plant waste products. The principal forms of decomposer are bacteria and fungi.

DECOMPOSITION REACTION: A chemical reaction in which a compound is broken down into simpler compounds, or into its constituent elements. In the earth system, this often is achieved through the help of detritivores and decomposers.

ECOSYSTEM: A term referring to a community of interdependent organisms along with the inorganic components of their environment.

to the release of oxygen from the plants back into the atmosphere.

CARBON MONOXIDE. Priestley discovered another carbon-oxygen compound quite different from carbon dioxide: carbon monoxide. The latter is used today by industry for several purposes, such as the production of certain fuels, proving that this toxic gas can be quite beneficial when used in a controlled environment. Nonetheless, carbon monoxide produced in an uncontrolled environment—generated by the burning of petroleum in automobiles as well as by the combustion of wood, coal, and other carbon-containing fuels—is extremely hazardous to human health.

When humans ingest carbon monoxide, it bonds with iron in hemoglobin, the substance in red blood cells that transports oxygen throughout the body. In effect, carbon monoxide fools the body into thinking that it is receiving oxygenated hemoglobin, or oxyhemoglobin. Upon reaching the cells, carbon monoxide has much less tendency than oxygen to break down, and therefore it continues to circulate throughout the body. Low concentrations can cause nausea, vomiting, and other effects, while prolonged exposure to high concentrations can result in death.

THE GREENHOUSE EFFECT. Although we have referred to carbon monoxide

KEY TERMS CONTINUED

ELECTRON: A negatively charged particle in an atom, which spins around the nucleus.

ELEMENT: A substance made up of only one kind of atom. Unlike compounds, elements cannot be broken chemically into other substances.

FOSSIL FUELS: Fuel derived from deposits of organic material that have experienced decomposition and chemical alteration under conditions of high pressure. These nonrenewable forms of bioenergy include petroleum, coal, peat, natural gas, and their derivatives.

GEOCHEMISTRY: A branch of the earth sciences, combining aspects of geology and chemistry, that is concerned with the chemical properties and processes of Earth—in particular, the abundance and interaction of chemical elements and their isotopes.

HYDROCARBON: Any organic chemical compound whose molecules are made up of nothing but carbon and hydrogen atoms.

ORGANIC: At one time chemists used the term *organic* only in reference to living things. Now the word is applied to most compounds containing carbon and hydrogen, thus excluding carbonates (which are minerals) and oxides such as carbon dioxide.

PERIODIC TABLE OF ELEMENTS: A chart that shows the elements arranged in order of atomic number along with the chemical symbol and the average atomic mass for each particular element.

PHOTOSYNTHESIS: The biological conversion of light energy (that is, electromagnetic energy) from the Sun to chemical energy in plants.

PROTON: A positively charged particle in an atom.

REACTIVITY: A term referring to the ability of one element to bond with others. The higher the reactivity, the greater the tendency to bond.

VALENCE ELECTRONS: Electrons that occupy the highest principal energy level in an atom. These are the electrons involved in chemical bonding.

as toxic, it should be noted that carbon dioxide also would be toxic to a human or other animal—for instance, if one were trapped in a sealed compartment and forced to breathe in the carbon dioxide released from one's lungs. On a global scale, both carbon dioxide and carbon monoxide in the atmosphere, produced in excessive amounts by the burning of fossil fuels, pose a potentially serious threat.

Both gases are believed to contribute to the greenhouse effect, which, as discussed in Energy and Earth, is a mechanism by which the planet efficiently uses the heat it receives from the Sun. Human consumption of fossil fuels and use of other products, including chlorofluorocarbons in aerosol cans, however, has produced a much greater quantity of greenhouse gases than the atmosphere needs to maintain normal heat levels. As a result, some scientists believe, buildup of greenhouse gases in the atmosphere is causing global warming.

CELLULAR RESPIRATION

The burning of fossil fuels is one of three ways that carbon enters the atmosphere, the others being volcanic eruption and cellular respiration. When cellular respiration takes place in the presence of oxygen, there is an intake of organic substances, which are broken down into carbon dioxide and water, with the release of considerable energy.

When plants take in carbon dioxide from the atmosphere, they combine it with water and manufacture organic compounds, using energy they have trapped from sunlight by means of photosynthesis—the conversion of light to chemical energy through biological means. As a by-product of photosynthesis, plants release oxygen into the atmosphere, as we have noted earlier.

In the process of photosynthesis, plants produce carbohydrates, which are various compounds of carbon, hydrogen, and oxygen that are essential to life. (The other two fundamental components of a diet are fats and proteins, both of which are carbon-based as well.) Animals eat the plants or eat other animals that eat the plants and thus incorporate the fats, proteins, and sugars (a form of carbohydrate) from the plants into their bodies. In cellular respiration, these nutrients are broken down to create carbon dioxide.

DECOMPOSITION. Cellular respiration also releases carbon into the atmosphere through the action of decomposers, organisms that obtain their energy from the chemical breakdown of dead organisms as well as from animal and plant waste products. Bacteria and fungi, the principal forms of decomposer, extract energy contained in the chemical bonds of the organic matter they are decomposing and, in the process, release carbon dioxide.

Certain ecosystems, or communities of interdependent organisms, are better than others at producing carbon dioxide through decomposition. As one would expect, environments where heat and moisture are greatest—for example, a tropical rainforest—yield the fastest rates of decomposition. On the other hand, decomposition proceeds much more slowly in dry, cold climates such as that of a subarctic tundra.

WHERE TO LEARN MORE

Blashfield, Jean F. *Carbon.* Austin, TX: Raintree Steck–Vaughn, 1999.

Carbon Cycle: Exploring the Environment (Web site). <http://www.cotf.edu/ete/modules/carbon/efcarbon.html>.

Carbon Cycle Greenhouse Gases Group (CCGG) (Web site). <http://www.cmdl.noaa.gov/ccgg/>.

Chemical Carousel: A Trip Around the Carbon Cycle (Web site). <http://library.thinkquest.org/11226/>.

Frostburg State University Chemistry Helper (Web site). <http://www.chemhelper.com/>.

"Global Carbon Cycle." The Woods Hole Research Center (Web site). <http://www.whrc.org/science/carbon/carbon.htm>.

Knapp, Brian J. *Carbon Chemistry.* Illus. David Woodroffe. Danbury, CT: Grolier Educational, 1998.

Loudon, G. Marc. *Organic Chemistry.* Menlo Park, CA: Benjamin/Cummings, 1988.

Sparrow, Giles. *Carbon.* New York: Benchmark Books, 1999.

Stille, Darlene. *The Respiratory System.* New York: Children's Press, 1997.

THE NITROGEN CYCLE

CONCEPT

Contrary to popular belief, the air we breathe is not primarily oxygen; by far the greatest portion of air is composed of nitrogen. A colorless, odorless gas noted for its lack of chemical reactivity—that is, its tendency not to bond with other elements—nitrogen plays a highly significant role within the earth system. Both through the action of lightning in the sky and of bacteria in the soil, nitrogen is converted to nitrites and nitrates, compounds of nitrogen and oxygen that are then absorbed by plants to form plant proteins. The latter convert to animal proteins in the bodies of animals who eat the plants, and when an animal dies, the proteins are returned to the soil. Denitrifying bacteria break down these compounds, returning elemental nitrogen to the atmosphere.

HOW IT WORKS

CHEMISTRY AND ELEMENTS

The concepts we discuss in this essay fall under the larger heading of geochemistry. A branch of the earth sciences that combines aspects of geology and chemistry, geochemistry is concerned with the chemical properties and processes of Earth. Among particular areas of interest in geochemistry are biogeochemical cycles, or the changes that particular elements undergo as they pass back and forth through the various earth systems and particularly between living and nonliving matter. The elements involved in biogeochemical cycles are hydrogen, oxygen, carbon, nitrogen, phosphorus, and sulfur (see Biogeochemical Cycles and Carbon Cycle).

An element is a substance composed of a single type of atom, which cannot be broken down chemically into a simpler substance. Each element is distinguished by its atomic number, or the number of protons (positively charged subatomic particles) in the nucleus, or center, of the atom. On the periodic table of elements, these fundamental substances of the universe are listed in order of atomic number, from hydrogen to uranium—which has the highest atomic number (92) of any element that occurs in nature—and beyond. The elements with an atomic number higher than that of uranium, all of which have been created artificially, play virtually no role in the chemical environment of Earth and are primarily of interest only to specialists in certain fields of chemistry and physics.

Everything that exists in the universe is an element, a compound formed by the chemical bonding of elements, or a mixture of compounds. In order to bond and form a compound, elements experience chemical reactions, which are the result of attractions on the part of electrons (negatively charged subatomic particles) that occupy the highest energy levels in the atom. These electrons are known as valence electrons.

CHEMICAL CHANGES. A chemical change is a phenomenon quite different from a physical change. If liquid water boils or freezes (both of which are examples of a physical change resulting from physical processes), it is still water. Physical changes do not affect the internal composition of an item or items; a chemical change, on the other hand, occurs when the actual composition changes—that is, when one substance is transformed into another. Chemical change requires a chemical reaction, a process whereby

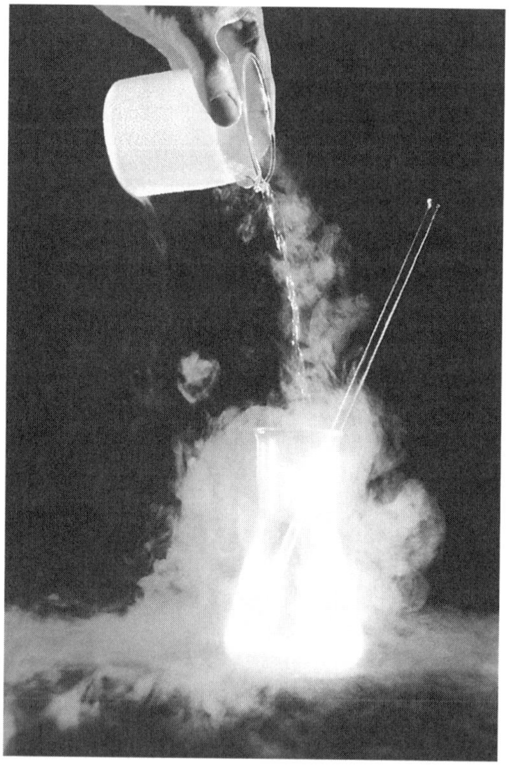

LIQUID NITROGEN. *(© David Taylor/Photo Researchers. Reproduced by permission.)*

the chemical properties of a substance are altered by a rearrangement of atoms.

There are several clues that tell us when a chemical reaction has taken place. In many chemical reactions, for instance, the substance may experience a change of state or phase—as, for instance, when liquid water is subjected to an electric current through a process known as *electrolysis,* which separates it into oxygen and hydrogen, both of which are gases. Another clue that a chemical reaction has occurred is a change of temperature. Unlike the physical change of liquid water to ice or steam, however, this temperature change involves an alteration of the chemical properties of the substances themselves. Chemical reactions also may encompass changes in color, taste, or smell.

NITROGEN'S PLACE AMONG THE ELEMENTS

With an atomic number of 7, nitrogen (chemical symbol N) is one of just 19 elements that are nonmetals. Unlike metals, nonmetals are poor conductors of heat and electricity and are not ductile—in other words, they cannot be reshaped easily. The vast majority of elements are metallic, however, the only exceptions being the nonmetals as well as six "metalloids," or elements that display characteristics of both metals and nonmetals.

Nitrogen is also one of eight "orphan" nonmetals—those nonmetals that do not belong to any family of elements, such as the halogens or noble gases. All six of the elements involved in biogeochemical cycles, in fact, are "orphan" nonmetals, with boron and selenium rounding out the list of eight orphans. Sometimes nitrogen is considered the head of a "family" of elements, all of which occupy a column or group on the periodic table.

These five elements—nitrogen, phosphorus, arsenic, antimony, and bismuth—share a common pattern of valence electrons, but otherwise they share little in terms of physical properties or chemical behavior. By contrast, chemicals that truly are related all have a common "family resemblance": all halogens are highly reactive, for instance, while all noble gases are extremely unreactive.

ABUNDANCE. The seventeenth most abundant element on Earth, nitrogen accounts for 0.03% of the planet's known elemental mass. This may seem very small, but at least nitrogen is among the 18 elements considered relatively abundant. These 18 elements account for all but 0.49% of the planet's known elemental mass, the remainder being composed of numerous other elements in small quantities. The term known elemental mass takes account of the fact that scientists do not know with certainty the elemental composition of Earth's interior, though it likely contains large proportions of iron and nickel. The known mass, therefore, is that which exists from the bottom of the crust to the top layers of the atmosphere.

Elemental proportions too small to be measured in percentage points are rendered in parts per million (ppm) or even parts per billion (ppb). Within the crust itself, nitrogen's share is certainly modest: a concentration of 19 ppm, which ties it with gallium, a metal whose name is hardly a household word, for a rank of thirty-third. On the other hand, this still makes it more abundant in the crust than many quite familiar metals, including lithium, uranium, tungsten, silver, mercury, and platinum.

In Earth's atmosphere, on the other hand, the proportion of nitrogen is much, much higher. The atmosphere is 78% nitrogen and 21% oxygen, while the noble gas argon accounts for 0.93%. The remaining 0.07% is taken up by various trace gases, including water vapor, carbon dioxide, and ozone, or O_3.

In the human body, nitrogen's share is much more modest than it is in the atmosphere but still 10 times greater than it is in relation to the planet's total mass. The element accounts for 3% of the body's mass, making it the fourth most abundant element in the human organism.

PROPERTIES AND APPLICATIONS OF NITROGEN

The Scottish chemist Daniel Rutherford (1749–1819) usually is given credit for discovering nitrogen in 1772, when he identified it as the element that remained when oxygen was removed from air. Several other scientists at about the same time made a similar discovery.

Because of its heavy presence in air, nitrogen is obtained primarily by cooling air to temperatures below the boiling points of its major components. Nitrogen boils (that is, turns into a gas) at a lower temperature than oxygen: –320.44°F (–195.8°C), as opposed to –297.4°F (–183°C). If air is cooled to –328°F (–200°C), thus solidifying it, and then allowed to warm slowly, the nitrogen boils first and therefore evaporates first. The nitrogen gas is captured, cooled, and liquefied once more.

Nitrogen also can be obtained from such compounds as potassium nitrate or saltpeter, found primarily in India, or from sodium nitrate (Chilean saltpeter), which comes from the desert regions of Chile. To isolate nitrogen chemically, various processes are undertaken in a laboratory—for instance, heating barium azide or sodium azide, both of which contain nitrogen.

REACTIONS WITH OTHER ELEMENTS. Rather than appearing as single atoms, nitrogen is diatomic, meaning that two nitrogen atoms typically bond with each other to form dinitrogen, or N_2. Nor do these atoms form single chemical bonds, as is characteristic of most elements; theirs is a *triple* bond, which effectively ties up the atoms' valence electrons, making nitrogen an unreactive element at relatively low temperatures.

Even at the temperature of combustion, a burning substance reacts with the oxygen in the air but not with the nitrogen. At very high temperatures, on the other hand, nitrogen combines with other elements, reacting with metals to form nitrides, with hydrogen to form ammonia, with O_2 (oxygen as it usually appears in nature, two atoms bonded in a molecule) to form nitrites, and with O_3 (ozone) to form nitrates. With the exception of the first-named group, all of these elements are important to our discussion of nitrogen.

NITROGEN COMBINES WITH HYDROGEN TO FORM AMMONIA. AMMONIUM NITRATE, A FERTILIZER, IS ALSO A DANGEROUS EXPLOSIVE. IT WAS USED IN APRIL OF 1995 TO BLOW UP THE ALFRED P. MURRAH FEDERAL BUILDING IN OKLAHOMA CITY, KILLING 168 PEOPLE. *(© James H. Robinson/Photo Researchers. Reproduced by permission.)*

SOME USES FOR NITROGEN. In processing iron or steel, which forms undesirable oxides if exposed to oxygen, a blanket of nitrogen is applied to prevent this reaction. The same principle is applied in making computer chips and even in processing foods, since these items, too, are affected detrimentally by oxidation. Because it is far less combustible than air (magnesium is one of the few elements that burns nitrogen in combustion), nitrogen also is used to clean tanks that have carried petroleum or other combustible materials.

As noted, nitrogen combines with hydrogen to form ammonia, used in fertilizers and cleaning materials. Ammonium nitrate, applied primarily as a fertilizer, is also a dangerous explosive, as shown with horrifying effect in the bombing of the Alfred P. Murrah Federal Building in Oklahoma City on April 19, 1995—a tragedy that took 168 lives. Nor is ammonium nitrate the only nitrogen-based explosive. Nitric acid is used in making trinitrotoluene (TNT), nitroglycerin, and dynamite as well as gunpowder and smokeless powder.

INTRODUCTION TO THE NITROGEN CYCLE

The nitrogen cycle is the process whereby nitrogen passes from the atmosphere into living things and ultimately back into the atmosphere. In the process, it is converted to nitrates and nitrites, compounds of nitrogen and oxygen that are absorbed by plants in the process of forming plant proteins. These plant proteins, in turn, are converted to animal proteins in the bodies of animals who eat the plants, and when the animal dies, the proteins are returned to the soil. Denitrifying bacteria break down these organic compounds, returning elemental nitrogen to the atmosphere.

Note what happens in the nitrogen cycle and, indeed, in all biogeochemical cycles: organic material is converted to inorganic material through various processes, and inorganic material absorbed by living organisms eventually is turned into organic material. In effect, the element passes back and forth between the realms of the living and the nonliving. This may sound a bit mystical, but it is not. To be organic, a substance must be built around carbon in certain characteristic chemical structures, and by inducing the proper chemical reaction, it is possible to break down or build up these structures, thus turning an organic substance into an inorganic one, or vice versa. (For more on this subject, see Carbon Cycle.)

STEPS IN THE CYCLE. Plants depend on biologically useful forms of nitrogen, the availability of which greatly affects their health, abundance, and productivity. This is particularly the case where plants in a saltwater ecosystem (a community of interdependent organisms) are concerned. Regardless of the specific ecosystem, however, fertilization of the soil with nitrogen has an enormous impact on the growth yield of plant life, which can be critical in the case of crops. Therefore, nitrogen is by far the most commonly applied nutrient in an agricultural setting.

There are several means by which plants receive nitrogen. They may absorb it as nitrate or ammonium, dissolved in saltwater and taken up through the roots, or as various nitrogen oxide gases. In certain situations, plants have a symbiotic, or mutually beneficial, relationship with microorganisms capable of "fixing" atmospheric dinitrogen into ammonia. In any case, plants receive nitrogen and later, when they are eaten by animals, pass these nutrients along the food chain—or rather, to use a term more favored in the earth and biological sciences, the food web.

When herbivorous or omnivorous animals consume nitrogen-containing plants, their bodies take in the nitrogen and metabolize it, breaking it down to generate biochemicals, or chemicals essential to life processes. At some point, the animal dies, and its body experiences decomposition through the activity of bacteria and other decomposers. These microorganisms, along with detritivores such as earthworms, convert nitrates and nitrites from organic sources into elemental nitrogen, which ultimately reenters the atmosphere.

REAL-LIFE APPLICATIONS

IMPORTANT FORMS OF NITROGEN

As noted earlier, dinitrogen, or N_2, is the form in which nitrogen typically appears when uncombined with other elements. This is also the form of nitrogen in the atmosphere, but it is so chemically unreactive that unlike oxygen, it plays little actual part in sustaining life. Indeed, because nitrogen in the air is essentially "filler" as far as humans are concerned, it can be substituted with helium, as is done in air tanks for divers. This prevents them from experiencing decompression sickness, or "the bends," which occurs when the diver returns too quickly to the surface, causing nitrogen in the blood to boil.

The dinitrogen in the air is a holdover from long ago in Earth's development, when volcanoes expelled elements from deep in the planet's interior to its atmosphere. Owing to its lack of reac-

Smog blankets Los Angeles in a haze. Nitric oxide reacts with oxygen in the air to form nitrogen oxide, a reddish brown gas that colors smog. *(Photograph by Walter A. Lyons. FMA Productions. Reproduced by permission.)*

tivity, dinitrogen never went anywhere. For it to play a role in the functioning of Earth cycles, it must be "fixed," as discussed later in this essay. In addition to dinitrogen, nitrogen appears in a number of other important inorganic compounds, including nitrite and nitrate; ammonia and ammonium; and nitric oxide, nitrogen dioxide, and nitrous oxide.

Nitrite and nitrate are two ionic forms of nitrogen. An ion is an atom or group of atoms that has lost or gained electrons, thus acquiring a net electric charge. Both nitrite and nitrate are anions, or negatively charged ions, designated by the use of superscript minus signs that indicate that each has a net charge of negative 1. Thus, nitrite, in which nitrogen is chemically bonded with two atoms of oxygen, is rendered as NO_2^-, while the formula for nitrate (nitrogen with three oxygen atoms), is designated as NO_3^-.

AMMONIA AND AMMONIUM. Nitrification is a process in which nitrite is produced, whereupon it undergoes a chemical reaction to form nitrate, the principal form of nitrogen nutrition for most plant species. The chemical from which the nitrite is created in the nitrification reaction is ammonium (NH_4^+), which is formed by the addition of a hydrogen cation, or a positively charged ion (H^+), to ammonia, or NH_3. The latter, which is probably familiar to most people in the form of a household cleaner, is actually an extremely abundant compound, both in natural and artificial forms.

Ammonium is soluble, or capable of being dissolved, in water and often is used as a fertilizer. It is attracted to negatively charged surfaces of clays and organic matter in soil and therefore tends to become stuck in one place rather than moving around, as nitrate does. In acidic soils, typically plants receive their nitrogen from ammonium, but most nonacidic soils can use only nitrate. As noted earlier, ammonium may be combined with nitrate to form ammonium nitrate—both a powerful fertilizer and a powerful explosive.

OXIDES. Nitrogen reenters the atmosphere in the form of the gas nitric oxide (NO), emitted primarily as the result of combustion reactions. This may occur in one of two ways. Organic nitrogen in bioenergy sources, such as biomass (organisms, their waste products, and their incompletely decomposed remains) or fossil fuels (e.g., coal or oil), may be oxidized. The latter term means that a substance undergoes a chemical reaction with oxygen: combustion itself, which requires the presence of oxygen, is an example of oxidation.

KEY TERMS

ATOM: The smallest particle of an element, consisting of protons, neutrons, and electrons. An atom can exist either alone or in combination with other atoms in a molecule.

ATOMIC NUMBER: The number of protons in the nucleus of an atom. Since this number is different for each element, elements are listed on the periodic table in order of atomic number.

BIOENERGY: Energy derived from biological sources that are used directly as fuel (as opposed to food, which becomes fuel).

BIOGEOCHEMICAL CYCLES: The changes that particular elements undergo as they pass back and forth through the various earth systems and specifically between living and nonliving matter. The elements involved in biogeochemical cycles are hydrogen, oxygen, carbon, nitrogen, phosphorus, and sulfur.

CHEMICAL BONDING: The joining, through electromagnetic force, of atoms that sometimes, but not always, represent more than one chemical element. The result is usually the formation of a molecule.

COMPOUND: A substance made up of atoms of more than one element chemically bonded to one another.

DECOMPOSERS: Organisms that obtain their energy from the chemical breakdown of dead organisms as well as from animal and plant waste products. The principal forms of decomposer are bacteria and fungi.

DECOMPOSITION REACTION: A chemical reaction in which a compound is broken down into simpler compounds or into its constituent elements. In the earth system, this often is achieved through the help of detritivores and decomposers.

DETRITIVORES: Organisms that feed on waste matter, breaking organic material down into inorganic substances that then become available to the biosphere in the form of nutrients for plants. Their function is similar to that of decomposers; however, unlike decomposers—which tend to be bacteria or fungi—detritivores are relatively complex organisms, such as earthworms or maggots.

DIATOMIC: A term describing a chemical element that typically exists as molecules composed of two atoms. Nitrogen and oxygen are both diatomic.

ECOSYSTEM: A term referring to a community of interdependent organisms along with the inorganic components of their environment.

On the other hand, nitric oxide may enter the atmosphere when atmospheric dinitrogen is combined with oxygen under conditions of high temperature and pressure, as, for instance, in an internal-combustion engine. In the atmosphere, nitric oxide reacts readily with oxygen in the air to form nitrogen dioxide (NO_2), a reddish-brown gas that adds to the tan color of smog over major cities.

Yet nitric oxide and nitrogen dioxide, usually designated together as NO_x, are also part of the life-preserving nitrogen cycle. Gaseous NO_x is taken in by plants, or oxidized to make nitrate, and circulated through the biosphere or else cycled directly to the atmosphere. In addition, denitrification, discussed later in this essay, transports nitrous oxide (N_2O) into the atmosphere from nitrate-rich soils.

KEY TERMS

ELECTRON: A negatively charged particle in an atom, which spins around the nucleus.

ELEMENT: A substance made up of only one kind of atom. Unlike compounds, elements cannot be chemically broken into other substances.

EUTROPHICATION: A state of heightened biological productivity in a body of water, which is typically detrimental to the ecosystem in which it takes place. Eutrophication can be caused by an excess of nitrogen or phosphorus in the form of nitrates and phosphates, respectively.

FOOD WEB: A term describing the interaction of plants, herbivores, carnivores, omnivores, decomposers, and detritivores, each of which consumes nutrients and passes it along to other organisms.

GEOCHEMISTRY: A branch of the earth sciences combining aspects of geology and chemistry, that is, concerned with the chemical properties and processes of Earth—in particular, the abundance and interaction of chemical elements and their isotopes.

ION: An atom or group of atoms that has lost or gained one or more electrons and thus has a net electric charge. Positively charged ions are called *cations,* and negatively charged ones are called *anions.*

LEACHING: The removal of soil materials that are in solution, or dissolved in water.

MOLECULE: A group of atoms, usually but not always representing more than one element, joined in a structure. Compounds are typically made up of molecules.

PERIODIC TABLE OF ELEMENTS: A chart that shows the elements arranged in order of atomic number along with their chemical symbols and the average atomic mass for each particular element.

PROTON: A positively charged particle in an atom.

REACTIVITY: A term referring to the ability of one element to bond with others. The higher the reactivity, the greater the tendency to bond.

SOLUBLE: Capable of being dissolved.

VALENCE ELECTRONS: Electrons that occupy the highest principal energy level in an atom. These are the electrons involved in chemical bonding.

NITROGEN PROCESSES

In order for most organisms to make use of atmospheric dinitrogen, it must be "fixed" into inorganic forms that a plant can take in through its roots and leaves. Nonbiological processes, such as a lightning strike, can bring about dinitrogen fixation. The high temperatures and pressures associated with lightning lead to the chemical bonding of atmospheric nitrogen and oxygen (both of which appear in diatomic form) to create two molecules of nitric oxide.

More often than not, however, dinitrogen fixation comes about through biological processes. Microorganisms are able to synthesize an enzyme that breaks the triple bonds in dinitrogen, resulting in the formation of two molecules of ammonia for every dinitrogen molecule thus reacted. This effect is achieved most commonly by bacteria or algae in wet or moist environments that offer nutrients other than nitrate or ammonium. In some instances, plants enjoy a symbiotic, or mutually beneficial, relationship with microorganisms capable of fixing dinitrogen.

AMMONIFICATION, NITRIFICATION, AND DENITRIFICATION. Dinitrogen fixation is just one example of a process whereby nitrogen is processed through one or more earth systems. Another is ammonification, or the process whereby nitrogen in organisms is recycled after their death. Enabled by microorganisms that act as decomposers, ammonification results in the production of either ammonia or ammonium. Thus, the soil is fertilized by the decayed matter of formerly living things.

Ammonium, as we noted earlier, also plays a part in nitrification, a process in which it first is oxidized to produce nitrite. Then the nitrite is oxidized to become nitrate, which fertilizes the soil. As previously mentioned, nitrate is useful as a fertilizer only in non-acidic soils; acidic ones, by contrast, require ammonium fertilizer.

In contrast to nitrification is denitrification, in which nitrate is reduced to the form of either nitrous oxide or dinitrogen. This takes place under anaerobic conditions—that is, in the absence of oxygen—and on the largest scale when concentrations of nitrate are highest. Flooded fields, for example, may experience high rates of denitrification.

The Role of Humans

Humans are involved in the nitrogen cycle in several ways, not all of them beneficial. One of the most significant roles people play in the nitrogen cycle is by the introduction of nitrogen-containing fertilizers to the soil. Because nitrogen has a powerful impact on plant growth, farmers are tempted to add more and more nitrate or ammonium or both to their crops, to the point that the soil becomes saturated with it and therefore unable to absorb more.

When the soil has taken in all the nitrogen it can hold, a process of leaching—the removal of soil materials dissolved in water—eventually takes place. Nitrate, in particular, leaches from agricultural sites into groundwater as well as streams and other forms of surface water. This can lead to eutrophication, a state of heightened biological productivity that is ultimately detrimental to the ecosystem surrounding a lake or other body of water. (See Biogeochemical Cycles for more about eutrophication.)

Yet another problem associated with overly nitrate-rich soils is an excessive rate of denitrification. This happens when soils that have been loaded down with nitrates become wet for long periods of time, leading to a dramatic increase in the denitrification rate. As a result, fixed nitrogen is lost, and nitrous oxide is emitted to the air. In the atmosphere nitrous oxide may contribute to the greenhouse effect, possibly helping increase the overall temperature of the planet (see Carbon Cycle and Energy and Earth).

Where to Learn More

Blashfield, Jean F. *Nitrogen.* Austin, TX: Raintree Steck–Vaughn, 1999.

Farndon, John. *Nitrogen.* New York: Benchmark Books, 1999.

Fitzgerald, Karen. *The Story of Nitrogen.* New York: Franklin Watts, 1997.

The Microbial World: The Nitrogen Cycle and Nitrogen Fixation (Web site). <http://helios.bto.ed.ac.uk/bto/microbes/nitrogen.htm>.

The Nitrogen Cycle (Web site). <http://clab.cecil.cc.md.us/faculty/biology/jason/nitrc.htm>.

The Nitrogen Cycle (Web site). <http://www.ultranet.com/~jkimball/BiologyPages/N/NitrogenCycle.html>.

The Nitrogen Cycle (Web site). <http://library.thinkquest.org/11353/nitrogen.htm>.

Nutrient Overload: Unbalancing the Global Nitrogen Cycle (Web site). <http://www.wri.org/wri/wr-98-99/nutrient.htm>.

Postgate, J. R. *The Outer Reaches of Life.* New York: Cambridge University Press, 1994.

Skinner, Brian J., Stephen C. Porter, and Daniel B. Botkin. *The Blue Planet: An Introduction to Earth System Science.* 2d ed. New York: John Wiley and Sons, 1999.

SCIENCE OF EVERYDAY THINGS
REAL-LIFE EARTH SCIENCE

THE BIOSPHERE

ECOSYSTEMS

ECOLOGY AND ECOLOGICAL STRESS

ECOSYSTEMS

CONCEPT

An ecosystem is a complete community of living organisms and the nonliving materials of their surroundings. Thus, its components include plants, animals, and microorganisms; soil, rocks, and minerals; as well as surrounding water sources and the local atmosphere. The size of ecosystems varies tremendously. An ecosystem could be an entire rain forest, covering a geographical area larger than many nations, or it could be a puddle or a backyard garden. Even the body of an animal could be considered an ecosystem, since it is home to numerous microorganisms. On a much larger scale, the history of various human societies provides an instructive illustration as to the ways that ecosystems have influenced civilizations.

HOW IT WORKS

The Biosphere

Earth itself could be considered a massive ecosystem, in which the living and nonliving worlds interact through four major subsystems: the atmosphere, hydrosphere (all the planet's waters, except for moisture in the atmosphere), geosphere (the soil and the extreme upper portion of the continental crust), and biosphere. The biosphere includes all living things: plants (from algae and lichen to shrubs and trees), mammals, birds, reptiles, amphibians, aquatic life, insects, and all manner of microscopic forms, including bacteria and viruses. In addition, the biosphere draws together all formerly living things that have not yet decomposed.

Several characteristics unite the biosphere. One is the obvious fact that everything in it is either living or recently living. Then there are the food webs that connect organisms on the basis of energy flow from one species to another. A food web is similar to the more familiar concept food chain, but in scientific terms a food chain—a series of singular organisms in which each plant or animal depends on the organism that precedes or follows it—does not exist. Instead, the feeding relationships between organisms in the real world are much more complex and are best described as a web rather than a chain.

Food Webs

Food webs are built around the flow of energy between organisms, known as energy transfer, which begins with plant life. Plants absorb energy in two ways. From the Sun, they receive electromagnetic energy in the form of visible light and invisible infrared waves, which they convert to chemical energy through a process known as photosynthesis. In addition, plants take in nutrients from the soil, which contain energy in the forms of various chemical compounds. These compounds may be organic, which typically means that they came from living things, though, in fact, the term *organic* refers strictly to characteristic carbon-based chemical structures. Plants also receive inorganic compounds from minerals in the soil. (See Minerals. For more about the role of carbon in inorganic compounds, see Carbon Cycle.)

Contained in these minerals are six chemical elements essential to the sustenance of life on planet Earth: hydrogen, oxygen, carbon, nitrogen, phosphorus, and sulfur. These are the elements involved in biogeochemical cycles, through which they continually are circulated between the living and nonliving worlds—that is, between organisms, on the one hand, and the inorganic realms of rocks, minerals, water, and air, on the other (see Biogeochemical Cycles).

FROM PLANTS TO CARNIVORES. As plants take up nutrients from the soil, they convert them into other forms, which provide usable energy to organisms who eat the plants. (An example of this conversion process is cellular respiration, discussed in Carbon Cycle.) When an herbivore, or plant-eating organism, eats the plant, it incorporates this energy.

Chances are strong that the herbivore will be eaten either by a carnivore, a meat-eating organism, or by an omnivore, an organism that consumes both herbs and herbivores—that is, both plants and animals. Few animals consume carnivores or omnivores, at least by hunting and killing them. (Detritivores and decomposers, which we discuss presently, consume the remains of all creatures, including carnivores and omnivores.) Humans are an example of omnivores, but they are far from the only omnivorous creatures. Many bird species, for instance, are omnivorous.

As nutrients pass from plant to herbivore to carnivore, the total amount of energy in them decreases. This is dictated by the second law of thermodynamics (see Energy and Earth), which shows that energy transfers cannot be perfectly efficient. Energy is not "lost"—the total amount of energy in the universe remains fixed, though it may vary with a particular system, such as an individual ecosystem—but it is dissipated, or directed into areas that do not aid in the transfer of energy between organisms. What this means for the food web is that each successive level contains less energy than the levels that precede it.

DETRITIVORES AND DECOMPOSERS. In the case of a food web, something interesting happens with regard to energy efficiency as soon as we pass beyond carnivores and omnivores to the next level. It might seem at first that there could be no level beyond carnivores or omnivores, since they appear to be "at the top of the food chain," but this only illustrates why the idea of a food web is much more useful. After carnivores and omnivores, which include some of the largest, most powerful, and most intelligent creatures, come the lowliest of all organisms: decomposers and detritivores, an integral part of the food web.

Decomposers, which include bacteria and fungi, obtain their energy from the chemical breakdown of dead organisms as well as from animal and plant waste products. Detritivores perform a similar function: by feeding on waste matter, they break organic material down into inorganic substances that then can become available to the biosphere in the form of nutrients for plants. The principal difference between detritivores and decomposers is that the former are relatively complex organisms, such as earthworms or maggots.

Both decomposers and detritivores aid in decomposition, a chemical reaction in which a compound is broken down into simpler compounds or into its constituent elements. Often an element such as nitrogen appears in forms that are not readily usable by organisms, and therefore such elements (which may appear individually or in compounds) need to be chemically processed through the body of a decomposer or detritivore. This processing involves chemical reactions in which the substance—whether an element or compound—is transformed into a more usable version.

By processing chemical compounds from the air, water, and geosphere, decomposers and detritivores deposit nutrients in the soil. These creatures feed on plant life, thus making possible the cycle we have described. Clearly this system, of which we have sketched only the most basic outlines, is an extraordinarily complex and well-organized one, in which every organism plays a specific role. In fact, earth scientists working in the realm of biosphere studies use the term niche to describe the role that a particular organism plays in its community. (For more about the interaction of species in a biological community, see Ecology and Ecological Stress.)

REAL-LIFE APPLICATIONS

THE FATE OF HUMAN CIVILIZATIONS

An interesting place to start in investigating examples of ecosystems is with a species near and dear to all of us: *Homo sapiens.* Much has been written about the negative effect industrial civilization has, or may have, on the natural environment—a topic discussed in Ecology and Ecological Stress—but here our concern is somewhat different. What do ecosystems, and specifically the availability of certain plants and animals, teach us about specific societies?

In his 1997 bestseller *Guns, Germs, and Steel: The Fate of Human Societies,* the ethnobotanist Jared M. Diamond (1937-) explained how he came to approach this question. While he was working with native peoples in New Guinea, a young man asked him why the societies of the West enjoyed an abundance of material wealth and comforts while those of New Guinea had so little. It was a simple question, but the answer was not obvious.

Diamond refused to give any of the usual pat responses offered in the past—for example, the Marxist or socialist claim that the West prospers at the expense of native peoples. Nor, of course, could he accept the standard answer that a white descendant of Europeans might have given a century earlier, that white Westerners are smarter than dark-skinned peoples. Instead, he approached it as a question of environment, and the result was his thought-provoking analysis contained in *Guns, Germs, and Steel.*

ADVANTAGES OF GEOGRAPHY. As Diamond showed, those places where agriculture was first born were precisely those blessed with favorable climate, soil, and indigenous plant and animal life. Incidentally, none of these locales was European, nor were any of the peoples inhabiting them "white." Agriculture came into existence in four places during a period from about 8000 to 6000 B.C. In roughly chronological order, they were Mesopotamia, Egypt, India, and China. All were destined to emerge as civilizations, complete with written language, cities, and organized governments, between about 3000 and 2000 B.C.

Of course, it is no accident that civilization was born first in those societies that first developed agriculture: before a civilization can evolve, a society must become settled, and in order for that to happen, it must develop agriculture. Each of these societies, it should be noted, formed along a river, and that of Mesopotamia was born at the confluence of *two* rivers, the Tigris and Euphrates. No wonder, then, that the spot where these two rivers met was identified in the Bible as the site for the Garden of Eden or that historians today refer to ancient Mesopotamia as "the Fertile Crescent." (For a very brief analysis regarding possible reasons why modern Mesopotamia—that is, Iraq—does not fit this description, see the discussion of desertification in Soil Conservation.)

In the New World, by contrast, agriculture appeared much later and in a much more circumscribed way. The same was true of Africa and the Pacific Islands. In seeking the reasons for why this happened, Diamond noted a number of factors, including geography. The agricultural areas of the Old World were stretched across a wide area at similar latitudes. This meant that the climates were not significantly different and would support agricultural exchanges, such as the spread of wheat and other crops from one region or ecosystem to another. By contrast, the land masses of the New World or Africa have a much greater north-south distance than they do east to west.

DIVERSITY OF SPECIES. Today such places as the American Midwest support abundant agriculture, and one might wonder why that was not the case in the centuries before Europeans arrived. The reason is simple but subtle, and it has nothing to do with Europeans' "superiority" over Native Americans. The fact is that the native North American ecosystems enjoyed far less biological diversity, or biodiversity, than their counterparts in the Old World. Peoples of the New World successfully domesticated corn and potatoes, because those were available to them. But they could not domesticate emmer wheat, the variety used for making bread, when they had no access to that species, which originated in Mesopotamia and spread throughout the Old World.

Similarly, the New World possessed few animals that could be domesticated either for food or labor. A number of Indian tribes domesticated some types of birds and other creatures for food, but the only animal ever adapted for labor was the llama. The llama, a cousin of the camel found in South America, is too small to carry heavy loads. Why did the Native Americans never harness the power of cows, oxen, or horses? For the simple reason that these species were not found in the Americas. After horses in the New World went extinct at some point during the last Ice Age (see Paleontology), they did not reappear in the

THE LLAMA WAS ONE OF THE FEW DOMESTICATED ANIMALS ADAPTED FOR WORK IN THE NEW WORLD, A PLACE WITH A SMALL NUMBER OF ANIMAL AND PLANT SPECIES AND LACK OF ECOLOGICAL COMPLEXITY BEFORE THE EUROPEANS ARRIVED IN ABOUT 1500 A.D. *(© Francois Gohier/Photo Researchers. Reproduced by permission.)*

Americas until Europeans brought them after A.D. 1500.

Diamond also noted the link between biodiversity and the practice, common among peoples in New Guinea and other remote parts of the world, of eating what Westerners would consider strange cuisine: caterpillars, insects—even, in some cases, human flesh. At one time, such practices served only to brand these native peoples further as "savages" in the eyes of Europeans and their descendants, but it turns out that there is a method to the apparent madness. In places such as the highlands of New Guinea, a scarcity of animal protein sources compels people to seek protein wherever they can find it.

By contrast, from ancient times the Fertile Crescent possessed an extraordinary diversity of animal life. Among the creatures present in that region (the term sometimes is used to include Egypt as well as Mesopotamia) were sheep, goats, cattle, pigs, and horses. With the help of these animals for both food and labor—people ate horses long before they discovered their greater value as a mode of transportation—the lands of the Old World were in a position to progress far beyond their counterparts in the New.

GREATER EXPOSURE TO MICROORGANISMS. Ultimately, these societies came to dominate their physical environments and excel in the development of technology; hence the "steel" and "guns" in Diamond's title. But what about "germs"? It is a fact that after Europeans began arriving in the New World, they killed vast populations without firing a shot, thanks to the microbes they carried with them. Of course, it would be centuries before scientists discovered the existence of microorganisms. But even in 1500, it was clear that the native peoples of the New World had no natural resistance to smallpox or a host of other diseases, including measles, chicken pox, influenza, typhoid fever, and bubonic plague.

Once again the Europeans' advantage over the Native Americans derived from the ecological complexity of their world compared with that of the Indians. In the Old World, close contact with farm animals exposed humans to germs and disease. So, too, did close contact with other people in crowded, filthy cities. This exposure, of course, killed off large numbers of people, but those who survived tended to be much hardier and possessed much stronger immune systems. Therefore, when Europeans came into contact with

WITH ITS SAUNA-LIKE ENVIRONMENT AND CLOSED CANOPY, A TROPICAL CLOUD FOREST PRODUCES LUSH VEGETATION AND IS ONE OF THE MOST BIODIVERSE ECOSYSTEMS ON THE EARTH. *(© G. Dimijian/Photo Researchers. Reproduced by permission.)*

native Americans, they were like walking biological warfare weapons.

EVALUATING ECOSYSTEMS

The ease with which Europeans subdued Native Americans fueled the belief that Europeans were superior, but, as Diamond showed, if anything was superior, it was the ecosystems of the Old World. This "superiority" relates in large part to the diversity of organisms an ecosystem possesses. Many millions of years ago, Earth's oceans and lands were populated with just a few varieties of single-cell organisms, but over time increasing differentiation of species led to the development of the much more complex ecosystems we know now.

Such differentiation is essential, given the many basic types of ecosystem that the world has to offer: forests and grasslands, deserts and aquatic environments, mountains and jungles. Among the many ways that these ecosystems can be evaluated, aside from such obvious parameters as relative climate, is in terms of abundance and complexity of species.

ABUNDANCE AND COMPLEXITY. The biota (a combination of all flora and fauna, or plant and animal life, respectively) in a desert or the Arctic tundra is much less complex than that of a tropical rain forest or, indeed, almost any kind of forest, because far fewer species can live in a desert or tundra environment. For this reason, it is said that a desert or tundra ecosystem is less complex than a forest one. There may be relatively large numbers of particular species in a less complex ecosystem, however, in which case the ecosystem is said to be abundant though not complex in a relative sense.

Another way to evaluate ecosystems is in terms of productivity. This concept refers to the amount of biomass—potentially burnable energy—produced by green plants as they capture sunlight and use its energy to create new organic compounds that can be consumed by local animal life. Once again, a forest, and particularly a rain forest, has a very high level of productivity, whereas a desert or tundra ecosystem does not.

Forests

Now let us look more closely at a full-fledged ecosystem—that of a forest—in action. It might seem that all forests are the same, but this could not be less the case. A forest is simply any ecosystem dominated by tree-sized woody plants. Beyond that, the characteristics of weather, climate, elevation, latitude, topography, tree species, varieties of animal species, moisture levels, and numerous other parameters create the potential for an almost endless diversity of forest types.

In fact, the United Nations Educational, Scientific, and Cultural Organization (UNESCO) defines 24 different types of forest, which are divided into two main groups. On the one hand, there are those forests with a closed canopy at least 16.5 ft. (5 m) high. The canopy is the upper portion of the trees in the forest, and closed-canopy forests are so dense with vegetation that from the ground the sky is not visible. On the other hand, the UNESCO system encompasses open woodlands with a shorter, more sparse, and unclosed canopy. The first group tends to be tropical and subtropical (located at or near the equator), while the second typically is located in temperate and subpolar forests—that is, in a region between the two tropical latitudes and the Arctic and Antarctic circles, respectively. In the next paragraphs, we examine a few varieties of forest as classified by UNESCO.

TROPICAL AND SUBTROPICAL FORESTS. Tropical rain forests are complex ecosystems with a wide array of species. The dominant tree type is an angiosperm (a type of plant that produces flowers during sexual reproduction), known colloquially as tropical hardwoods. The climate and weather are what one would expect to find in a place called a tropical rain forest, that is, rainy and warm. When the rain falls, it cools things down, but when the sun comes back out, it turns the world of the tropical rain forest into a humid, sauna-like environment.

Naturally, the creatures that have evolved in and adapted to a tropical rain forest environment are those capable of enduring high humidity, but they are tolerant of neither extremely cool conditions nor drought. Within those parameters, however, exists one of the most biodiverse ecosystems on Earth: the tropical rain forest is home to an astonishing array of animals, plants, insects, and microorganisms. Indeed, without the tropical rain forest, terrestrial (land-based) animal life on Earth would be noticeably reduced.

In the tropics, by definition the four seasons to which we are accustomed in temperate zones—winter, spring, summer, and fall—do not exist. In their place there is a rainy season and a dry season, but there is no set point in the year at which trees shed their leaves. In a tropical and subtropical evergreen forest conditions are much drier than in the rain forest, and individual trees or tree species may shed their leaves as a result of dry conditions. All trees and species do not do so at the same time, however, so the canopy remains rich in foliage year-round—hence the term evergreen. As with a rain forest, the evergreen forest possesses a vast diversity of species.

In contrast to the two tropical forest ecosystems just described, a mangrove forest is poor in species. In terms of topography and landform, these forests are found in low-lying, muddy regions near saltwater. Thus, the climate is likely to be humid, as in a rain forest, but only organisms that can tolerate flooding and high salt levels are able to survive there. Mangrove trees, a variety of angiosperm, are suited to this environment and to the soil, which is poor in oxygen.

TEMPERATE AND SUBARCTIC FORESTS. Among the temperate and subarctic forest types are temperate deciduous forests, containing trees that shed their leaves seasonally, and temperate and subarctic evergreen conifer forests, in which the trees produce cones bearing seeds. These are forest types familiar to most people in the continental United

States. The first variety is dominated by such varieties as oak, walnut, and hickory, while the second is populated by pine, spruce, or fir as well as other types, such as hemlock.

Less familiar to most Americans outside the West Coast are temperate winter-rain evergreen broadleaf forests. These forests are dominated by evergreen angiosperms and appear in regions that have both a pronounced wet season and a summer drought season. Such forests can be found in southern California, where an evergreen oak of the *Quercus* genus is predominant. Even less familiar to Americans is the temperate and subpolar evergreen rain forest, which is found in the Southern Hemisphere. Occurring in a wet, frost-free ocean environment, these forests are dominated by such evergreen angiosperms as the southern beech and southern pine.

ANGIOSPERMS VS. GYMNOSPERMS

Several times we have referred to angiosperms, a name that encompasses not just certain types of tree but also all plants that produce flowers during sexual reproduction. The name, which comes from Latin roots meaning "vessel seed," is a reference to the fact that the plant keeps its seeds in a vessel whose name emphasizes these plants' sexual-type reproduction: an ovary.

Angiosperms are a beautiful example of how a particular group of organisms can adapt to specific ecosystems and do so in a way much more efficient than did their evolutionary forebear. Flowering plants evolved only about 130 million years ago, by which time Earth had long since been dominated by another variety of seed-producing plant, the gymnosperm, of which pines and firs are an example. Yet in a relatively short period of time, from the standpoint of the earth sciences, angiosperms have gone on to become the dominant plants in the world. Today, about 80% of all living plant species are flowering plants.

ANGIOSPERM VS. GYMNOSPERM SEEDS. How did they do this? They did it by developing a means to coexist more favorably than gymnosperms with the insect and animal life in their ecosystems. Gymnosperms produce their seeds on the surface of leaflike structures, making the seeds vulnerable to physical damage and drying as the wind whips the branches back and forth. Furthermore, insects and other animals view gymnosperm seeds as a source of nutrition.

In an angiosperm, by contrast, the seeds are tucked away safely inside the ovary. Furthermore, the evolution of the flower not only has added a great deal of beauty to the world but also has provided a highly successful mechanism for sexual reproduction. This sexual reproduction makes it possible to develop new genetic variations, as genetic material from two individuals of differing ancestry come together to produce new offspring.

GYMNOSPERM POLLINATION. Gymnosperms reproduce sexually as well, but they do so by a less efficient method. In both cases, the trees have to overcome a significant challenge: the fact that sexual reproduction normally requires at least one of the individual plants to be mobile. Gymnosperms package the male reproductive component in tiny pollen grains, which are released into the wind. Eventually, the grains are blown toward the female component of another individual plant of the same species.

This method succeeds well enough to sustain large and varied populations of gymnosperms but at a terrific cost, as is evident to anyone who lives in a region with a high pollen count in the spring. A yellow dust forms on everything. So much pollen accumulates on window sills, cars, mailboxes, and roofs that only a good rain (or a car wash) can take it away, and one tends to wonder what good all this pollen is doing for the trees.

The truth is that pollination is wasteful and inefficient. Like all natural mechanisms, it benefit the overall ecosystem, in this case, by making nutrient-rich pollen grains available to the soil. Packed with energy, pollen grains contain large quantities of nitrogen, making them a major boost to the ecosystem if not to the human environment. But it costs the gymnosperm a great deal, in terms of chemical and biological energy and material, to produce pollen grains, and the benefits are much more uncertain.

Pollen *might* make it to the right female component, and, in fact, it will, given the huge amounts of pollen produced. Yet the overall system is rather like trying to solve an economic problem by throwing a pile of dollar bills into the air and hoping that some of the money lands in the right place. For this reason, it is no surprise

KEY TERMS

ABUNDANCE: A measure of the degree to which an ecosystem possesses large numbers of particular species. An abundant ecosystem may or may not have a wide array of different species. Compare with *complexity.*

ANGIOSPERM: A type of plant that produces flowers during sexual reproduction.

BIOGEOCHEMICAL CYCLES: The changes that particular elements undergo as they pass back and forth through the various earth systems and particularly between living and nonliving matter. The elements involved in biogeochemical cycles are hydrogen, oxygen, carbon, nitrogen, phosphorus, and sulfur.

BIOTA: A combination of all flora and fauna (plant and animal life, respectively) in a region.

CANOPY: The upper portion of the trees in a forest. In a closed-canopy forest the canopy (which may be several hundred feet, or well over 50 meters, high) protects the soil and lower areas from sun and torrential rainfall.

CARNIVORE: A meat-eating organism.

COMPLEXITY: A measure of the degree to which an ecosystem possesses a wide array of species. These species may or may not appear in large numbers. Compare with *abundance.*

COMPOUND: A substance made up of atoms of more than one element chemically bonded to one another.

DECOMPOSERS: Organisms that obtain their energy from the chemical breakdown of dead organisms as well as from animal and plant waste products. The principal forms of decomposer are bacteria and fungi.

DECOMPOSITION REACTION: A chemical reaction in which a compound is broken down into simpler compounds or into its constituent elements. On Earth, this often is achieved through the help of detritivores and decomposers.

DETRITIVORES: Organisms that feed on waste matter, breaking down organic material into inorganic substances that then can become available to the biosphere in the form of nutrients for plants. Their function is similar to that of decomposers, but unlike decomposers—which tend to be bacteria or fungi—detritivores are relatively complex organisms, such as earthworms or maggots.

that angiosperms gradually are overtaking gymnosperms.

ANGIOSPERM POLLINATION. The angiosperm overcomes its own lack of mobility by making use of mobile organisms. Whereas insects and animals pose a threat to gymnosperms, angiosperms actually put bees, butterflies, hummingbirds, and other flower-seeking creatures to work aiding their reproductive process. By evolving bright colors, scents, and nectar, the flowers of angiosperms attract animals, which travel from one flower to another, accidentally moving pollen as they do.

Because of this remarkably efficient system, animal-pollinated species of flowering plants do not need to produce as much pollen as gymnosperms. Instead, they can put their resources into other important functions, such as growth and greater seed production. In this way, the angiosperm solves its own problem of reproduc-

KEY TERMS CONTINUED

ECOSYSTEM: A community of interdependent organisms along with the inorganic components of their environment.

ELEMENT: A substance made up of only one kind of atom. Unlike compounds, elements cannot be broken chemically into other substances.

ENERGY TRANSFER: The flow of energy between organisms in a food web.

FOOD WEB: A term describing the interaction of plants, herbivores, carnivores, omnivores, decomposers, and detritivores in an ecosystem. Each consumes nutrients and passes it along to other organisms. Earth scientists typically prefer this name to *food chain,* an everyday term for a similar phenomenon. A food chain is a series of singular organisms in which each plant or animal depends on the organism that precedes or follows it. Food chains rarely exist in nature.

GEOSPHERE: The upper part of Earth's continental crust, or that portion of the solid earth on which human beings live and which provides them with most of their food and natural resources.

GYMNOSPERM: A type of plant that reproduces sexually through the use of seeds that are exposed, not hidden in an ovary, as with an angiosperm.

HERBIVORE: A plant-eating organism.

HYDROSPHERE: The entirety of Earth's water, excluding water vapor in the atmosphere but including all oceans, lakes, streams, groundwater, snow, and ice.

NICHE: A term referring to the role that a particular organism plays within its biological community.

OMNIVORE: An organism that eats both plants and other animals.

ORGANIC: At one time, chemists used the term *organic* only in reference to living things. Now the word is applied to most compounds containing carbon and hydrogen, thus excluding carbonates (which are minerals) and oxides, such as carbon dioxide.

PHOTOSYNTHESIS: The biological conversion of light energy (that is, electromagnetic energy) from the Sun to chemical energy in plants.

SYSTEM: Any set of interactions that can be set apart mentally from the rest of the universe for the purposes of study, observation, and measurement.

tion—and as a side benefit adds enormously to the world's beauty.

THE COMPLEXITY OF ECOSYSTEMS

The relationships between these two types of seed-producing plant and their environments illustrate, in a very basic way, the complex interactions between species in an ecosystem. Environmentalists often speak of a "delicate balance" in the natural world, and while there is some dispute as to how delicate that balance is—nature shows an amazing resilience in recovering from the worst kinds of damage—there is no question that a balance of some kind exists.

To put it another way, an ecosystem is an extraordinarily complex environment that brings together biological, geologic, hydrologic, and atmospheric components. Among these components are trees and other plants; animals, insects, and microorganisms; rocks, soil, minerals, and landforms; water in the ground and on the surface, flowing or in a reservoir; wind, sun, rain, moisture; and all the other specifics that make up weather and climate.

In the present context, we have not attempted to provide anything even approaching a comprehensive portrait of an ecosystem, drawing together all or most of the aspects described in the preceding paragraph. A full account of even the simplest ecosystem would fill an entire book. Given that level of complexity, it is safe to say that one should be very cautious before tampering with the particulars of an ecosystem. The essay on Ecology and Ecological Stress concerns what happens when such tampering occurs.

WHERE TO LEARN MORE

Beattie, Andrew J., and Paul Ehrlich. *Wild Solutions: How Biodiversity Is Money in the Bank.* New Haven, CT: Yale University Press, 2001.

Diamond, Jared M. *Guns, Germs, and Steel: The Fates of Human Societies.* New York: W. W. Norton, 1997.

The Ecosystems Center. Marine Biological Laboratory, Woods Hole, Massachusetts (Web site). <http://ecosystems.mbl.edu/>.

Ecotopia (Web site). <http://www.ecotopia.com/>.

Jordan, Richard N. *Trees and People: Forestland, Ecosystems, and Our Future.* Lanham, MD: Regnery Publishing, 1994.

Living Things: Habitats and Ecosystems (Web site). <http://www.fi.edu/tfi/units/life/habitat/habitat.html>.

Martin, Patricia A. *Woods and Forests.* Illus. Bob Italiano and Stephen Savage. New York: Franklin Watts, 2000.

Nebel, Bernard J., and Richard T. Wright. *Environmental Science: The Way the World Works.* Upper Saddle River, NJ: Prentice Hall, 2000.

Rybolt, Thomas R., and Robert C. Mebane. *Environmental Experiments About Life.* Hillside, NJ: Enslow Publishers, 1993.

The State of the Nation's Ecosystems (Web site). <http://www.us-ecosystems.org/>.

Sustainable Ecosystems Institute (Web site). <http://www.sei.org/>.

ECOLOGY AND ECOLOGICAL STRESS

CONCEPT

Ecology is the study of the relationships between organisms and their environments. As such, it is subsumed into the larger subject of ecosystems, which encompasses both living and nonliving components of the environment. As a study of the biological aspect of ecosystems, ecology is properly a part of the biological sciences rather than the earth sciences; however, in practice it is difficult to draw a line between the disciplines. This is especially the case inasmuch as the study of the environment involves such aspects as soil science, where earth sciences and ecology meet. This fact, combined with increasing concerns over ecological stresses, such as the increase of greenhouse gases in the atmosphere, warrants the consideration of ecology in an earth sciences framework.

HOW IT WORKS

ECOSYSTEMS, BIOLOGICAL COMMUNITIES, AND ECOLOGY

An ecosystem is the complete community of living organisms and the nonliving materials of their surroundings. It therefore includes components that represent the atmosphere, the hydrosphere (all of Earth's waters, except for moisture in the atmosphere), the geosphere (the soil and extreme upper portion of the continental crust), and the biosphere. The biosphere includes all living things: plants (from algae and lichen to shrubs and trees); mammals, birds, reptiles, amphibians, aquatic life, and insects as well as all manner of microscopic forms, including bacteria and viruses. In addition, the biosphere draws together all formerly living things that have not yet decomposed.

The components of the biosphere are united not only by the fact that all of them are either living or recently living but also by the food web. The food web, discussed in much detail within the context of Ecosystems, is a complex network of feeding relationships and energy transfers between organisms. At various levels and stages of the food web are plants; herbivores, or plant-eating organisms; carnivores (meat-eating organisms); omnivores (organisms that eat both meat and plants); and, finally, decomposers and detritivores, which obtain their energy from the chemical breakdown of dead organisms.

ECOLOGY. When discussing the living components of an ecosystem—that is, those components drawn from the biosphere—the term biological community is used. This also may be called *biota,* which refers to all flora and fauna, or plant and animal life, respectively, in a particular region. The relationship between these living things and their larger environment, as we have noted, is called ecology. Pioneered by the German zoologist Ernst Haeckel (1834–1919), ecology was long held in disdain by the world scientific community, in part because it seemed to defy classification as a discipline. Though its roots clearly lie in biology, its broadly based, multidisciplinary approach seems more attuned to the earth sciences.

In any case, ecology long since has gained the respect it initially failed to receive, and much of that change has to do with a growing acceptance of two key concepts. On the one hand, there is the idea that all of life is interconnected and that the living world is tied to the nonliving, or inorganic, world. This is certainly a prevailing belief in the modern-day earth sciences, with its systems approach (see Earth Systems). On the other hand, there is the gathering awareness that certain aspects of industrial civilization may have a negative impact on the environment.

Clearly, the ecosystem as a whole is held together by tight bonds of interaction, but where the biological community is concerned, those bonds are even tighter. For the biological community to survive and thrive, a balance must be maintained between consumption and production of resources. Nature provides for that balance in numerous ways, but beginning in the late twentieth century, environmentalists in the industrialized world became increasingly concerned over the possibly negative effects their own societies exert on Earth's ecosystems and ecological communities.

CLIMAX AND SUCCESSION

One of the concerns raised by environmentalists is the issue of endangered species, or varieties of animal whose existence is threatened by human activities. In fact, nature itself sometimes replaces biological communities in a process called succession. Succession involves the progressive replacement of earlier biological communities with others over time. Coupled with succession is the idea of climax, a theoretical notion intended to describe a biological community that has reached a stable point as a result of ongoing succession.

Succession typically begins with a disturbance exerted on the preexisting ecosystem, and this disturbance usually is followed by recovery. This recovery may constitute the full extent of the succession process, at which point the community is said to have reached its climax point. Whether or not this happens depends on such particulars as climate, the composition of the soil, and the local biota.

There are two varieties of succession, primary and secondary. Primary succession occurs in communities that have never experienced significant modification of biological processes. In other words, the community affected by primary succession is "virgin," and primary succession typically involves enormous stresses. On the other hand, secondary succession happens after disturbances of relatively low intensity, such that the regenerative capacity of the local biota has not been altered significantly. Secondary succession takes place in situations where the biological community has experienced alteration.

NICHE

Whereas climax and succession apply to broad biological communities, the term niche refers to the role a particular organism or species plays within the larger community. Though the concept of niche is abstract, it is unquestionable that each organism plays a vital role and that the totality of the ecosystem would suffer stress if a large enough group of organisms were removed from it. Furthermore, given the apparent interrelatedness of all components in a biological community, every species must have a niche—even human beings.

An interesting idea related to the niche is the concept of an indicator species: a plant or animal that by its presence, abundance, or chemical composition demonstrates a particular aspect of the character or quality of the environment. Indicator species can, for instance, be plants that accumulate large concentrations of metals in their tissues, thus indicating a preponderance of metals in the soil. This metal could indicate valuable deposits nearby, or it could serve as a sign that the soil is being contaminated.

In the rest of this essay, we explore a few examples of ecological stress—situations in which the relationship between organisms and environment has been placed under duress. We do not attempt to explore the ideas of succession, climax, niche, or indicator species with any consistency or depth; rather, our purpose in briefly discussing these terms is to illustrate a few of the natural mechanisms observed or hypothesized by ecologists in studying natural systems. The vocabulary of ecology, in fact, is as complex and varied as that of any natural science, and much of it is devoted to the ways in which nature responds to ecological stress.

REAL-LIFE APPLICATIONS

DEFORESTATION

In Ecosystems, we discuss a number of forest types, whose makeup is determined by climate and the dominant tree varieties. Here let us consider what happens to a forest—particularly an old-growth forest—that experiences significant

RAIN FOREST DESTRUCTION BY FIRE IN MADAGASCAR. SUCH DEFORESTATION AFFECTS THE CARBON BALANCE IN THE ATMOSPHERE AND THE DIVERSITY OF SPECIES ON EARTH. *(© Daniel Heuclin/Photo Researchers. Reproduced by permission.)*

disturbance. Actually, the term deforestation can describe any interruption in the ordinary progression of the forest's life, including clear-cut harvesting, even if the forest fully recovers.

Deforestation can take place naturally, as a result of changes in the soil and climate, but the most significant cases of deforestation over the past few thousand years have been the result of human activities. Usually, deforestation is driven by the need to clear land or to harvest trees for fuel and, in some cases, building. Though deforestation has been a problem the world over, since the 1970s it has become more of an issue in developing countries.

DEVELOPED AND DEVELOPING NATIONS. In developed nations such as the United States, environmental activism has raised public awareness concerning deforestation and has led to curtailment of large-scale cutting in forests deemed important environmental habitats. By contrast, developing nations, such as Brazil, are cutting down their forests at an alarming rate. Generally, economics is the driving factor, with the need for new agricultural land or the desire to obtain wood and other materials driving the deforestation process.

Yet the deforestation of such valuable reserves as the Amazon rain forest is an environmental disaster in the making: as noted in Soil Conservation, the soil in rain forests is typically "old" and leached of nutrients. Without the constant reintroduction of organic material from the

OLD-GROWTH FORESTS ARE HOME TO THE NORTHERN SPOTTED OWL, RECOGNIZED AS AN ENDANGERED SPECIES BECAUSE OF THE DESTRUCTION OF ITS HABITAT. *(© T. Davis/Photo Researchers. Reproduced by permission.)*

plants and animals of the rain forests, it would be too poor to grow anything. Therefore, when nations cut down their own rain forest lands, in effect, they are killing the golden goose to get the egg. Once the rain forest is gone, the land itself is worthless.

CONSEQUENCES OF DEFORESTATION. Deforestation has several extremely serious consequences. From a biological standpoint, it greatly reduces biodiversity, or the range of species in the biota. In the case of tropical rain forests as well as old-growth forests, certain species cannot survive once the environmental structure has been ruptured. From an environmental perspective, it leads to dangerous changes in the carbon content of the atmosphere, discussed later in this essay. In the case of old-growth forests or rain forests, deforestation removes an irreplaceable environmental asset that contributes to the planet's biodiversity—and to its oxygen supply.

Even from a human standpoint, deforestation takes an enormous toll. Economically, it depletes valuable forest resources. Furthermore, deforestation in many developing countries often is accompanied by the displacement of indigenous peoples. Other political and social horrors sometimes lurk in the shadows: for example, Brazil's forests are home to charcoal plants that amount to virtual slave-labor camps. Indians are lured from cities with promises of high income and benefits, only to arrive and find that the situation is quite different from what was advertised. Having paid the potential employer for transportation to the work site, however, they are unable to afford a return ticket and must labor to repay the cost.

OLD-GROWTH FORESTS

Old-growth forests represent a climax ecosystem—one that has come to the end of its stages of succession. They are dominated by trees of advanced age (hence the name *old-growth*), and the physical structure of these ecosystems is extraordinarily complex. In some places the canopy, or "rooftop," of the forest is dense and layered, while in others it has gaps. Tree sizes vary enormously, and the forest is littered with the remains of dead trees.

An old-growth forest, by definition, takes a long time to develop. Not only must it have been free from human disturbance, but it also must have been spared various natural types of disturbance that bring about succession: catastrophic storms or wildfire, for instance. For this reason, most old-growth forests are rain forests in tropical and temperate environments. Among North American old-growth forests are those of the United States Pacific Northwest as well as those in adjoining regions of southwestern Canada.

THE SPOTTED OWL. These old-growth forests are home to a bird that, in the 1980s and 1990s, became well known both to environmentalists and to their critics: the northern spotted owl, or *Strix occidentalis caurina.* A nonmigratory bird, the spotted owl has a breeding pattern such that it requires large tracts of old-growth, moist-to-wet conifer forest—that is, a forest dominated by cone-producing trees—as its habitat. Given the potential economic value of old-growth forests in the region, the situation became one of heated controversy.

On the one hand, environmentalists insisted that the spotted owl's existence would be threatened by logging, and, on the other hand, representatives of the logging industry and the local community maintained that prevention of logging in the old-growth forests would cost jobs and livelihoods. The question was not an easy one, pitting the interests of the environment against those of ordinary human beings. By the early 1990s, the federal government had stepped in on the side of the environmentalists, having recognized the spotted owl as a threatened species under the terms of the U.S. Endangered Species Act of 1973. Nonetheless, controversy over the spotted owl—and over the proper role of environmental, economic, and political concerns in such situations—continues.

THE GREENHOUSE EFFECT

Deforestation and other activities pose potential dangers to our atmosphere. In particular, such activities have led to an increasing release of greenhouse gases, which may cause the warming of the planet. As discussed in Energy and Earth, the greenhouse effect, in fact, is a natural process. Though it is typically associated, in the popular vocabulary at least, with the destructive impact of industrial civilization on the environment, it is an extremely effective mechanism whereby Earth makes use of energy from the Sun.

Rather than simply re-radiating solar radiation, Earth traps some of this heat in the atmosphere with the help of greenhouse gases, such as carbon dioxide. As in the case of most natural processes, however, if a little bit of carbon dioxide in the atmosphere is good, this does not mean that a lot is better.

As noted in the essay Carbon Cycle, all living things contain carbon in certain characteristic structures; hence, the term *organic* refers to this type of carbon content. Though carbon dioxide is not an organic compound, it is emitted by animals: they breathe in oxygen, which undergoes a chemical reaction in their carbon-based bodies, and, as a result, carbon dioxide is released. Plants, on the other hand, receive this carbon dioxide and, through a chemical process in their own cellular structures, take in the carbon while releasing the oxygen.

THE RESULT OF CUTTING MATURE FORESTS. Mature forests, such as those of the old-growth variety, contain vast amounts of carbon in the form of living and dead organic material: plants, animals, and material in the soil. Because this quantity is much greater than in a younger forest, when deforestation occurs in a mature forest ecosystem, the mature forest will be replaced by an ecosystem that contains much smaller amounts of carbon.

Ultimately, the carbon from the former ecosystem will be released to the atmosphere in the form of carbon dioxide. This will happen quickly, if the biomass of the forest is burned, or more slowly, if the timber from the forest is used for a long periods of time, for instance, in the building of houses or other structures.

Before humans began cutting down forests, Earth's combined vegetation stored some 990 billion tons (900 billion metric tons) of carbon, 90% of it appeared in forests. Today only about 616 billion tons (560 billion metric tons) of carbon are stored in Earth's vegetation, and the amount is growing smaller as time passes. At the same time, the amount of carbon dioxide in the atmosphere has increased from about 270 parts per million (ppm) in 1850 to about 360 ppm in 2000, and, again, the increase continues.

SHOULD WE BE WORRIED? Given this rise in atmospheric carbon dioxide as a result of deforestation—not to mention the more well-known cause, burning of fossil fuels—it is no wonder that atmospheric scientists and environmentalists are alarmed. Some of these scientists hypothesize that larger concentrations of carbon dioxide in the atmosphere will lead to increased intensity of the greenhouse effect. If this is true, it is possible that global warming will ensue, an eventuality that could have enormous implications for human survival. As a worst-case scenario, the polar ice cap (see Glaciology) could melt, submerging the cities of Earth.

Before succumbing to the sort of doomsday thinking and scaremongering for which many environmentalists are criticized, however, it is important to recognize that several contingencies are involved: *if* carbon dioxide in the atmosphere causes an increase in the intensity of the greenhouse effect, it *could* cause global warming. The fact is that despite a few mild winters at the end of the twentieth century, it is far from clear that the planet is warming. The winter of 1993, for instance, produced one of the worst blizzards that the eastern United States has ever seen.

As recently as the mid-1970s some environmentalists claimed that Earth actually is *cooling*—a response to a spate of cold winters in that period. The fact of the matter is that climate cycles are difficult to determine and require the

KEY TERMS

BIOACCUMULATION: The buildup of toxic chemical pollutants in the tissues of individual organisms.

BIOLOGICAL COMMUNITY: The living components of an ecosystem.

BIOMAGNIFICATION: The increase in bioaccumulated contamination at higher levels of the food web. Biomagnification results from the fact that larger organisms consume larger quantities of food—and, hence, in the case of polluted materials, more toxins.

BIOSPHERE: A combination of all living things on Earth—plants, mammals, birds, reptiles, amphibians, aquatic life, insects, viruses, single-cell organisms, and so on—as well as all formerly living things that have not yet decomposed.

BIOTA: A combination of all flora and fauna (plant and animal life, respectively) in a region.

CANOPY: The upper portion of the trees in a forest. In a closed-canopy forest, the canopy (which may be several hundred feet, or well over 50 meters, high) protects the soil and lower areas from sun and torrential rainfall.

CARNIVORE: A meat-eating organism.

CLIMAX: A theoretical notion intended to describe a biological community that has reached a stable point as a result of ongoing succession.

DECOMPOSERS: Organisms that obtain their energy from the chemical breakdown of dead organisms as well as from animal and plant waste products. The principal forms of decomposer are bacteria and fungi.

DECOMPOSITION REACTION: A chemical reaction in which a compound is broken down into simpler compounds or into its constituent elements. On Earth, this often is achieved through the help of detritivores and decomposers.

DETRITIVORES: Organisms that feed on waste matter, breaking organic material down into inorganic substances that then can become available to the biosphere in the form of nutrients for plants. Their function is similar to that of decomposers; however, unlike decomposers—which tend to be bacteria or fungi—detritivores are relatively complex organisms, such as earthworms or maggots.

ECOLOGY: The study of the relationships between organisms and their environments.

perspective of several centuries' worth of data (at least), rather than just a few years' worth. (See Glaciology for a discussion of the Little Ice Age, which took place just a few centuries ago.)

Nonetheless, it is important to be aware of the legitimate environmental concerns raised by the increased presence of carbon dioxide in the atmosphere due to human activities. Atmospheric scientists continue to monitor levels of greenhouse gases and to form hypotheses regarding the ultimate effect of such activities as deforestation and the burning of fossil fuels.

BIOACCUMULATION AND BIOMAGNIFICATION

As we have seen in a number of ways, one of the key concepts of ecological studies is also a core principle in the modern approach to the earth sciences. In both cases, there is the idea that a disturbance in one area can lead to serious conse-

KEY TERMS CONTINUED

ECOSYSTEM: A community of interdependent organisms along with the inorganic components of their environment.

ENERGY TRANSFER: The flow of energy between organisms in a food web.

FOOD WEB: A term describing the interaction of plants, herbivores, carnivores, omnivores, decomposers, and detritivores in an ecosystem. Each of these organisms consumes nutrients and passes them along to other organisms. Earth scientists typically prefer this name to food chain, an everyday term for a similar phenomenon. A food chain is a series of singular organisms in which each plant or animal depends on the organism that precedes or follows it. Food chains rarely exist in nature.

GEOSPHERE: The upper part of Earth's continental crust, or that portion of the solid earth on which human beings live and which provides them with most of their food and natural resources.

GREENHOUSE EFFECT: Warming of the lower atmosphere and surface of Earth. This occurs because of the absorption of long-wavelength radiation from the planet's surface by certain radiatively active gases, such as carbon dioxide and water vapor, in the atmosphere. These gases are heated and ultimately re-radiate energy to space at an even longer wavelength.

HERBIVORE: A plant-eating organism.

HYDROSPHERE: The entirety of Earth's water, excluding water vapor in the atmosphere, but including all oceans, lakes, streams, groundwater, snow, and ice.

NICHE: A term referring to the role that a particular organism plays within its biological community.

OMNIVORE: An organism that eats both plants and other animals.

ORGANIC: At one time chemists used the term organic only in reference to living things. Now the word is applied to most compounds containing carbon, with the exception of carbonates (which are minerals) and oxides, such as carbon dioxide.

SUCCESSION: The progressive replacement of earlier biological communities with others over time.

SYSTEM: Any set of interactions that can be set apart from the rest of the universe for the purposes of study, observation, and measurement.

quences elsewhere. The interconnectedness of components in the environment thus makes it impossible for any event or phenomenon to be truly isolated.

A good example of this is biomagnification. Biomagnification is the result of bioaccumulation, or the buildup of toxic chemical pollutants in the tissues of individual organisms. Part of what makes these toxins dangerous is the fact that the organism cannot process them easily either by metabolizing them (i.e., incorporating them into the metabolic system, as one does food or water) or by excreting them. Yet the organism ultimately does release some toxins—by passing them on to other members of the food web. This increase in contamination at higher levels of the food web is known as biomagnification.

THE PROCESS OF BIOMAGNIFICATION. Among the most prominent examples of chemical pollutants that are bioaccumulated are such pesticides as DDT (dichlorodiphenyl-trichloroethane). DDT is a chlorinated hydrocarbon (see Economic Geology) used as an insecticide. Because of its hydrocarbon base, DDT is highly soluble in oils—and in the fat of organisms. Once pesticides such as DDT have been sprayed, rain can wash them into creeks and, finally, lakes and other bodies of water, where they are absorbed by creatures that drink or swim in the water.

Atmospheric deposition, for instance, from industrial smokestacks or automobile emissions, is another source of toxins. Sludge from a sewage treatment plant can make its way into water sources, spreading all sorts of pollutants to the food web. Whatever the case, these toxins usually enter the food web by attaching to the smallest components. Particles of pollutant may stick to algae, which are so small that the toxin does little damage at this level of the food web. But even a small herbivore, such as a zooplankton, when it consumes the algae, takes in larger quantities of the pollutant, and thus begins the cycle of biomagnification.

By the time the toxin has passed from a zooplankton to a small fish, the amount of pollutant in a single organism might be 100 times what it was at the level of the algae. The reason, again, is that the fish can consume 10 zooplankton that each has consumed 10 algae. (These particular numbers, of course, are used simply for the sake of convenience.) By the time the toxins have passed on to a few more levels in the food web, they might be appearing in concentrations as great as 10,000 times their original amount.

DDT BIOMAGNIFICATION. For a period of about two decades before 1972, DDT was used widely in the United States to help control the populations of mosquitoes and other insects. Eventually, however, it found its way into water sources and fish species through the process we have described. Predatory birds, such as osprey, peregrine falcons, and brown pelicans, consumed these fish. So, too, did the bald eagle, which has long been a protected species owing to its role as America's national symbol.

DDT levels became so high that the birds' eggshells became abnormally thin, and adult birds sitting on nests accidentially would break the shells of unhatched eggs. As a result, baby birds died, and populations of these species also died. Public awareness of this phenomenon, raised by environmentalists in the late 1960s and early 1970s, led to the banning of DDT spraying in 1972. Since that time, populations of many predatory birds have increased dramatically.

ADDRESSING ECOLOGICAL CONCERNS. In the case of DDT biomagnification, humans were not directly involved, because the species of birds affected were not ones that people consume for food. Yet bioaccumulation and biomagnification have threatened humans. For example, in the 1950s, cows fed on grass that had been exposed to nuclear radiation and this radioactive material found its way into milk. Another example occurred during the 1970s and 1980s, when fish, such as tuna, were found to contain abnormally high levels of mercury.

This led the federal government and some states to issue warnings against the consumption of certain types of fish, owing to bioaccumulated levels of toxic pollutants. Obviously, such measures, however well intentioned, are just cosmetic fixes for larger problems. In the long run, what is needed is a systemic ecological approach that attempts to address problems such as biomagnification and the accumulation of greenhouse gases by approaching the root causes.

WHERE TO LEARN MORE

Ashworth, William, and Charles E. Little. *Encyclopedia of Environmental Studies.* New York: Facts on File, 2001.

The Ecological Society of America: Issues in Ecology (Web site). <http://esa.sdsc.edu/issues.htm>.

Ecology.com—An Ecological Source of Information (Web site). <http://www.ecology.com/>.

Ecology WWW Page (Web site). <http://pbil.univ-lyon1.fr/Ecology/Ecology-WWW.html>.

Endangered Species on EE-Link (Web site). <http://216.144.194.35/EndSpp/endangeredspecies-mainpage.html>.

Markley, O. W., and Walter R. McCuan. *21st Century Earth: Opposing Viewpoints.* San Diego, CA: Greenhaven Press, 1996.

The Marshall Cavendish Illustrated Encyclopedia of Plants and Earth Sciences. New York: Marshall Cavendish, 1988.

The Need to Know Library—Ecology and Environment Page (Web site). <http://www.peak.org/~mageet/tkm/ecolenv.htm>.

Philander, S. George. *Is the Temperature Rising?: The Uncertain Science of Global Warming.* Princeton, NJ: Princeton University Press, 1998.

Schultz, Warren. *The Organic Suburbanite: An Environmentally Friendly Way to Live the American Dream.* Emmaus, PA: Rodale, 2001.

WATER AND THE EARTH

HYDROLOGY

THE HYDROLOGIC CYCLE

GLACIOLOGY

HYDROLOGY

CONCEPT

Hydrology is among the principal disciplines within the larger framework of hydrologic sciences, itself a subcategory of earth sciences study. Of particular importance to hydrology is the hydrologic cycle by which water is circulated through various earth systems above and below ground. But the hydrologic cycle is only one example of the role water plays in the operation of earth systems. Outside its effect on living things, a central aspect of the hydrologic cycle, water is important for its physical and chemical properties. Its physical influence, exerted through such phenomena as currents and floods, can be astounding but no less amazing than its chemical properties, demonstrated in the fascinating realm of karst topography.

HOW IT WORKS

THE SYSTEMS APPROACH

The modern study of earth sciences looks at the planet as a large, complex network of physical, chemical, and biological interactions. This is known as a systems approach to the study of Earth. The systems approach treats Earth as a combination of several subsystems, each of which can be viewed individually or in concert with the others. These subsystems are the geosphere, atmosphere, hydrosphere, and biosphere.

The geosphere is that part of the solid earth on which people live and from which are extracted the materials that make up our world: minerals and rocks as well as the organic products of the soil. In the latter area, the geosphere overlaps with the biosphere, the province of all living and recently living things; in fact, once a formerly living organism has decomposed and become part of the soil, it is no longer part of the biosphere and has become a component of the geosphere.

Overlap occurs between all spheres in one way or another. Thus, the hydrosphere includes all of the planet's waters, except for water that has entered the atmosphere in the form of evaporation. From the time moisture is introduced to the blanket of gases that surrounds the planet until it returns to the solid earth in the form of precipitation, water is a part of the atmosphere. This aspect of the planet's water is treated in the essay on Evapotranspiration and Precipitation.

THE HYDROSPHERE AND THE HYDROLOGIC CYCLE. All aspects of water on Earth, other than evaporation and precipitation, fall within the hydrosphere. This includes saltwater and freshwater, water on Earth's surface and below it, and all imaginable bodies of water, from mountain streams to underground waterways and from creeks to oceans. One of the fascinating things about water is that because it moves within the closed system of Earth, all the planet's water circulates endlessly. Thus, there is a chance that the water in which you take your next bath or shower also bathed Cleopatra or provided a drink to Charlemagne's horse.

On a less charming note, there is also a good chance that the water with which you brush your teeth once passed through a sewer system. Lest anyone panic, however, this has always been the case and always will be; as we have noted, water circulates endlessly, and one particular molecule may serve a million different functions.

Furthermore, as long as water continues to circulate through the various earth systems—that is, as long as it is not left to stagnate in a pond—it undergoes a natural cleansing process. Modern municipal and private water systems provide further treatment to ensure that the water that people use for washing is at least reasonably clean. In any case, it is clear that the movement of water through the hydrologic cycle is a subject complex enough to warrant study on its own (see Hydrologic Cycle).

THE HYDROLOGIC SCIENCES

As noted in Studying Earth, the earth sciences can be divided into three broad areas: the geologic, hydrologic, and atmospheric sciences. Each of these areas corresponds to one of the "spheres," or subsystems within the larger earth system, that we have discussed briefly: geosphere, hydrosphere, and atmosphere.

The hydrologic sciences are concerned with the hydrosphere and its principal component, water. These disciplines include glaciology—the study of ice in general and glaciers in particular—and oceanography. Glaciology is discussed in a separate essay, and oceanography is examined briefly in the present context. Aside from these two areas of study, the central component of the hydrologic sciences is hydrology—its most basic discipline, as geology is to the geologic sciences.

OCEANOGRAPHY. Oceanography is the study of the world's saltwater bodies—that is, its oceans and seas—from the standpoint of their physical, chemical, biological, and geologic properties. These four aspects of oceanographic study are reflected in the four basic subdisciplines into which oceanography is divided: physical oceanography, chemical oceanography, marine geology, and marine ecology. Each represents the application of a particular science to the study of the oceans.

Physical oceanography, as its name implies, involves the study of physics as applied to the world's saltwater bodies. In general, it concerns the physical properties of the oceans and seas, including currents and tides, waves, and the physical specifics of seawater itself—that is, its temperature, pressure at particular depths, density in specific areas, and so on.

Just as physical oceanography weds physics to the study of seawater, chemical oceanography is concerned with the properties of the ocean as viewed from the standpoint of chemistry. These properties include such specifics as the chemical composition of seawater as well as the role the ocean plays in the biogeochemical cycles whereby certain chemical elements circulate between the organic and inorganic realms (see Biogeochemical Cycles, Carbon Cycle, and Nitrogen Cycle).

The biogeochemical focus of chemical oceanography implies an overlap with geochemistry. Likewise, marine geology exists at the nexus of oceanography and geology, involving, as it does, such subjects as seafloor spreading (see Plate Tectonics), ocean topography, and the formation of ocean basins. Finally, there is the realm where oceanography overlaps with biology, a realm known as marine ecology or biological oceanography. This subdiscipline is concerned with the wide array of life-forms, both plant and animal, that live in the oceans as well as the food webs whereby they interact with one another.

INTRODUCTION TO HYDROLOGY

As noted earlier, hydrology is the central field of the hydrologic sciences, dealing with the most basic aspects of Earth's waters. Among the areas of focus in hydrology are the distribution of water on the planet, its circulation through the hydrologic cycle, the physical and chemical properties of water, and the interaction between the hydrosphere and other earth systems.

Among the subdisciplines of hydrology are these:

- Groundwater hydrology: The study of water resources below ground.
- Hydrography: The study and mapping of large surface bodies of water, including oceans and lakes.
- Hydrometeorology: The study of water in the lower atmosphere, an area of overlap between the hydrologic and atmospheric sciences
- Hydrometry: The study of surface water—in particular, the measurement of its flow and volume.

THE WORK OF HYDROLOGISTS. Bringing together aspects of geology, chemistry, and soil science, hydrology is of enormous practical importance. Local governments, for instance, require hydrologic studies before the commencement of any significant building project, and hydrology is applied to such areas as the designation and management of flood plains. Hydrologists also are employed in the management of water resources, wastewater systems, and irrigation projects. The public use of water for

WHIRLPOOLS ARE CREATED WHERE TWO CURRENTS MEET. WATER TENDS TO ROTATE IN CIRCLES, CLOCKWISE IN THE NORTHERN HEMISPHERE AND COUNTERCLOCKWISE IN THE SOUTHERN HEMISPHERE. *(© B. Tharp/Photo Researchers. Reproduced by permission.)*

recreation and power generation also calls upon the work of hydrologists, who assist governments and private companies in controlling and managing water supplies.

Hydrologists in the field use a variety of techniques, some of them simple and time-honored and others involving the most cutting-edge modern technology. They may make use of highly sophisticated computer models and satellite remote-sensing technology, or they may apply relatively uncomplicated methods for the measurement of snow depth or the flow of rivers and streams. Local hydrologists searching for water may even avail themselves of the services of quasi-mystics who employ a nonscientific practice called *dowsing.* The latter method, which involves the sensing of underground water with a "magic" divining rod, sometimes is used, with varying degrees of success, to find water in rural areas.

REAL-LIFE APPLICATIONS

CURRENTS

Ocean waters are continually moving, not only as waves hitting the shore (a function of the Moon's gravitational pull—see Sun, Moon, and Earth) but also in the form of currents. These are patterns of oceanic flow, many of them regular and unchanging and others susceptible to change as a result of shifts in atmospheric patterns and other parameters. Among the factors that affect the flow of currents are landmasses, wind patterns, and the Coriolis effect, or the deflection of water caused by the turning of Earth.

Landmasses on either side of the Atlantic, Pacific, and Indian Oceans act as barriers to the paths of currents. For instance, if Africa were not placed as it is, between the Atlantic and Indian Oceans, water in the equatorial region would flow uniformly from east to west, or from the Indian Ocean to the Atlantic. Likewise, the movement would be uniformly west to east at the poles, as it would be at the equator.

Such is the case just off the shores of Antarctica, where the Antarctic Circumpolar Current, without the obstruction of land barriers, consistently circles the globe in a west-to-east direction. On the other hand, at the southern extremity of Africa, far from the equator, the movement of water also would be uniform, but in this case from west to east. Because of these landmasses, however, the movement of currents is much more complex.

Wind patterns also drive currents. These patterns work in tandem with the Coriolis effect, a term that generally describes a phenomenon that occurs with all particles on a rotating sphere such as Earth. The result of the Coriolis effect is that water tends to rotate in circles, clockwise in the Northern Hemisphere and counterclockwise in the Southern Hemisphere. (At the equator and poles, by contrast, the Coriolis effect is nonexistent.) Combined with prevailing winds, the Coriolis effect creates vast elliptical (oval-shaped) circulating currents called gyres.

SURFACE CURRENTS AND THEIR EFFECT ON CLIMATE. Among the basic types of currents are surface, tidal, and deep-water currents. In addition, a fourth type of current, a turbidity current, is of interest to oceanographers, hydrologists, and underwater geologists. Surface currents such as the Gulf Stream are the most well known variety, being the major form of current by which water circulates on the ocean's surface. Caused by the friction of atmospheric patterns—another type of current—moving over the sea, these currents largely are driven by winds. (Winds, in turn, are caused by differences in temperature between packets of air at differing altitudes—see Convection.)

Running as deep as 656 ft. (200 m), surface currents can be a powerful force. As a result of the Gulf Stream, for instance, a craft that sets sail eastward from the Caribbean is likely to be pulled quickly toward England. Of even greater importance is the impact of the Gulf Stream on climate. By moving warm waters in a northeasterly direction, it causes the European climate to be much warmer than it would be normally.

For instance, Boston, Massachusetts, and Rome, Italy, are on the same latitude, but whereas Boston is known for its icy winters, the mention of Rome rightly conjures images of sunshine and warmth. Likewise, London lies north of the 50th parallel, far above any city in the continental United States—including such places as International Falls, Minnesota, and Buffalo, New York, which are noted for their cruel winters. On the other hand, London, while it is far from balmy in wintertime, is many degrees warmer, thanks to the Gulf Stream.

OTHER TYPES OF CURRENT. Among the other types of currents are tidal currents, which are horizontal movements of water associated with the changing tides of the ocean. Their effect is felt primarily in the area between the continental shelf and the shore, where tidal-current phenomena such as riptides can pose a serious danger to swimmers. Deep-water currents, while they are less noticeable to people, are responsible for 90% of the water circulation that takes place in the ocean. Caused by variations in water density, which is a function of salt content (salinity) and temperature, these are slow-moving currents that move colder, denser water toward the depths of the ocean.

Then there are turbidity currents, which result from the mixing of relatively light water with water that has been made heavy by its sediment content. Earthquakes may cause these currents, which are local, fast movements of water along the ocean floor. Another cause of turbidity currents is the piling of sediment on underwater slopes. Turbidity currents play a major role in shaping the terrain of the ocean floor.

FLOODING

Whereas currents arise in areas of Earth where water is "supposed" to be, floods, by definition, do not. They often occur in valleys or on coastlines and can be caused by various natural and man-made factors. Among natural causes are rains and the melting of snow and ice, while human-related causes can include poor engineering of irrigation or other water-management systems as well as the bursting of dams. In addition, the building of settlements too close to rivers and other bodies of water that are prone to flooding has resulted in the increase of human casualties from flooding over the centuries.

In terms of natural causes, changes in weather patterns typically are involved—but not always. For example, a low-lying coastal area may be susceptible to flooding at times when the ocean reaches high tide. (On the other hand, such weather conditions as low barometric pressure and high winds also can bring about heightened high tides.) Additionally, floods can be caused by earthquakes and other geologic phenomena that have no relation to the weather.

From ancient times people have located settlements near water. This settlement pattern resulted from the obvious benefits that accrued

THE FLOODWATERS OF THE MISSISSIPPI RISE TO 4 FT. (1.2 M), SURROUNDING THE PUMPING STATION IN HANNIBAL, MISSOURI. APART FROM NATURAL CAUSES, FLOODS CAN RESULT FROM INCONSISTENT FLOOD MANAGEMENT, POOR CIVIL ENGINEERING DESIGN, AND UNWISE AGRICULTURAL PRACTICES. *(AP/Wide World Photos. Reproduced by permission.)*

from access to water, and even though flooding was naturally a hazard, in some cases flooding itself was found to be beneficial. For the ancient Egyptians, the yearly cycles of flooding on the part of the Nile caused the deposition of rich soil, which played a major part in the fertility of the farmlands that, in turn, made possible the brilliant civilization of the pharaohs.

Along with these benefits, however, ancient peoples learned to fear the changes in weather and other circumstances that could bring about sudden flooding. This feeling is reflected, for instance, in Jesus' parable about the wise and foolish house builders. In the parable, a favorite Sunday school topic, the foolish man builds his house upon the sand, so that when the floods come, they sweep away his household. On the other hand, the wise man builds his house on rock, so that his household withstands the inevitable flood—an illustration about spiritual values that likewise reflects a reality of daily life in the ancient Near East.

HUMAN CAUSES AND EFFECTS. Humans can cause floods by such disastrous practices as clear-cutting of land and runaway grazing. Such activities remove vegetation, which holds soil in place and, in turn, keeps rivers and other bodies of water from flowing over onto the land. In addition, without vegetation to absorb rain, ground becomes saturated and thus susceptible to flooding. Not surprisingly, these unwise agricultural practices have helped bring about other disasters, such as the massive erosion of soil in the United States plains states that culminated in the dust bowl of the mid-1930s (see Soil Conservation).

Less well known than the dust bowl but still massive in its impact was the 1927 flood of the Mississippi River, which left more than a million people homeless. It, too, was in part the result of unwise practices, in this case, inconsistent flood management and civil engineering design, according to John M. Barry, the author of *Rising Tide: The Great Mississippi Flood of 1927 and How It Changed America.* As Barry indicated in his subtitle, the flood's impact went far beyond its direct effect on human lives or the landscape.

As Barry discusses in the book, the flood was a major cause behind the rise of the poor-white discontent in Louisiana that led to the governorship (and, in the opinion of some people, the dictatorship) of the notorious Huey P. Long (1893–1935). Long, who won election on promises to ease the suffering of the underclass, ultimately became the virtual ruler of his state, with a degree of power that in the opinions of some pundits rivaled that of President Franklin D. Roosevelt—if not that of his other contemporaries, Adolf Hitler and Benito Mussolini. Of even greater long-term significance, Barry maintained, the flood brought about the large-scale flight of African Americans to the north and the shift of black political allegiance from the Republican Party to the Democratic Party.

Few people alive today remember the 1927 flooding, but plenty recall the devastating floods of 1993, which killed 52 people and left over 70,000 homeless. Human mismanagement could not be blamed for the flooding itself, an outgrowth of exceptionally high soil moisture levels remaining from the fall of 1992, as well as heavy precipitation that continued in early 1993. However, once again, human attempts to control the flooding were less than successful: of some 1,300 levees or embankments that had been built (partly as a result of the 1927 flood) to keep flood waters back, all but about 200 failed. The floods, which lasted from late June to mid-August, destroyed nearly 50,000 homes and rendered over 12,000 sq mi. (31,000 sq km) of farmland useless. The overall damage estimate was in the range of $15 to $20 billion.

POWER AND PREVENTION. It is no wonder that a flood can have such a far-reaching impact, given the enormous power of water running wild in nature. Water is extremely heavy: just a bathtub full of water can weigh as much 750 lb. (340 kg). And it can travel as fast as 20 MPH (32 km/h), giving it tremendous physical force. Under certain conditions, a flood just 1 in. (2.54 cm) in depth can have as much potential energy as 60,000 tons (54,400 metric tons) of TNT. A U.S. study of persons killed in natural disasters during the 20-year period that ended in 1967 found that of 443,000 victims, nearly 40%, or about 173,000, were killed in floods. The other 60% was made up of people killed in 18 different types of other natural disaster, including hurricanes, earthquakes, and tornadoes.

Given this great destructive potential, communities—often with the help of hydrologists—have devised several means to control floods. Among such methods are the construction of dams and the diversion of floodwaters away from populated areas to flood-control reservoirs. These reservoirs then release the water at a slower rate than it would be released in the situation of a flood, thus giving the soil time to absorb the excess water. About one-third of all reservoirs in the United States are used for this purpose.

Hydrologists are particularly important in helping communities protect against flooding by methods known as hazard zoning and minimizing encroachment. By studying historical records, along with geologic maps and aerial photographs, hydrologists and other planners can make recommendations regarding the zoning laws for a particular area, so that builders will take special precautions. In addition, they can help minimize encroachment—that is, ensure that new buildings are not located in such a way that they restrict the flow of water or cause water to pool up excessively.

KARST TOPOGRAPHY

In contrast to the dramatic action of flooding or currents, karst topography is a more subtle but no less intriguing aspect of the ways in which water affects the dynamics of Earth. In this case, however, the effect is chemical rather than physical in origin. Karst topography is a particular variety of landscape created where water comes into contact with extremely soluble (easily dissolved in water) varieties of bedrock.

Karst is the German name for Kras, a region of Slovenia noted for its unusual landscape of strangely shaped white rock. In addition to the odd, funhouse forms of nightmarishly steep hills and twisting caves, much of it like something from a Dr. Seuss book, karst topography is noted for its absence of surface water, topsoil, or vegetation. The reason is that the bedrock comprises extremely soluble calcium carbonate minerals, such as limestone, gypsum, or dolomite.

Karst regions form as a result of chemical reactions between groundwater and bedrock. In the atmosphere and on the surface of the solid earth, water combines with carbon dioxide in the air, and this combination acts as a corrosive on calcium carbonate rocks. This corrosive or acidic material seeps into all crevices of the rock, developing into sinkholes and widening fissures over time. Gradually, it carves out enormous underground drainage systems and caves.

Sometimes the underground drainage structure collapses, leaving behind more odd shapes in the form of natural bridges and sink-

KEY TERMS

BIOGEOCHEMICAL CYCLES: The changes that particular elements undergo as they pass back and forth through the various earth systems and particularly between living and nonliving matter. The elements involved in biogeochemical cycles are hydrogen, oxygen, carbon, nitrogen, phosphorus, and sulfur.

BIOSPHERE: A combination of all living things on Earth—plants, mammals, birds, reptiles, amphibians, aquatic life, insects, viruses, single-cell organisms, and so on—as well as all formerly living things that have not yet decomposed.

CORIOLIS EFFECT: The deflection of water caused by the rotation of Earth. The Coriolis effect causes water currents to move in circles—clockwise in the Northern Hemisphere and counterclockwise in the Southern Hemisphere.

GEOCHEMISTRY: A branch of the earth sciences, combining aspects of geology and chemistry, that is concerned with the chemical properties and processes of Earth—in particular, the abundance and interaction of chemical elements and their isotopes.

GEOSPHERE: The upper part of Earth's continental crust, or that portion of the solid earth on which human beings live and which provides them with most of their food and natural resources.

HYDROLOGIC CYCLE: The continuous circulation of water throughout Earth and between various Earth systems.

HYDROLOGIC SCIENCES: Areas of the earth sciences concerned with the study of the hydrosphere. Among these disciplines are hydrology, glaciology, and oceanography.

HYDROLOGY: The study of the hydrosphere, including the distribution of water on Earth, its circulation through the hydrologic cycle, the physical and chemical properties of water, and the interaction between the hydrosphere and other earth systems.

HYDROSPHERE: The entirety of Earth's water, excluding water vapor in the atmosphere but including all oceans, lakes, streams, groundwater, snow, and ice.

ORGANIC: At one time, chemists used the term *organic* only in reference to living things. Now the word is applied to most compounds containing carbon, with the exception of carbonates (which are minerals) and oxides, such as carbon dioxide.

SYSTEM: Any set of interactions that can be set apart mentally from the rest of the universe for the purposes of study, observation, and measurement.

holes. This is a variety of karst topography known as doline karst. Another type is cone or tower karst, which produces tall, jagged limestone peaks, such as the sharp hills that characterize the river landscape in many parts of China. The United States is home to the world's largest karst region, which includes the Mammoth cave system in Kentucky.

WHERE TO LEARN MORE

Abbott, Patrick. *Natural Disasters.* Dubuque, IA: William C. Brown Publishers, 1996.

Arnell, Nigel. *Hydrology and Global Environmental Change.* Upper Saddle River, NJ: Prentice Hall, 2001.

Barry, John M. Rising Tide: *The Great Mississippi Flood of 1927 and How It Changed America.* New York: Simon and Schuster, 1997.

Dingman, S. Lawrence. *Physical Hydrology.* 2d ed. Upper Saddle River, NJ: Prentice Hall, 2002.

Gallant, Roy A. *Earth's Water.* New York: Benchmark Books, 2002.

Global Hydrology and Climate Center (Web site). <http://www.ghcc.msfc.nasa.gov/>.

Hydrology, Marine Science, Freshwater Science and Aquaculture Journals (Web site). <http://www.sciencekomm.at/journals/hydro.html>.

Karst Link Page (Web site). <http://hum.amu.edu.pl/~sgp/spec/linkk.html>.

Karst Waters Institute (Web site). <http://www.karstwaters.org/>.

Skinner, Brian J., Stephen C. Porter, and Daniel B. Botkin. *The Blue Planet: An Introduction to Earth System Science.* 2d ed. New York: John Wiley and Sons, 1999.

The World-Wide Web Virtual Library: Oceanography (Web site). <http://www.mth.uea.ac.uk/ocean/vl/>.

THE HYDROLOGIC CYCLE

CONCEPT

The hydrologic cycle is the continuous circulation of water throughout Earth and between Earth's systems. At various stages, water—which in most cases is synonymous with the hydrosphere—moves through the atmosphere, the biosphere, and the geosphere, in each case performing functions essential to the survival of the planet and its life-forms. Thus, over time, water evaporates from the oceans; then falls as precipitation; is absorbed by the land; and, after some period of time, makes its way back to the oceans to begin the cycle again. The total amount of water on Earth has not changed in many billions of years, though the distribution of water does. The water that we see, though vital to humans and other living things, makes up only about 0.0001% of the total volume of water on Earth; far more is underground and in other compartments of the environment.

HOW IT WORKS

WATER AND THE HYDROSPHERE

As we have noted, water and the hydrosphere are practically synonymous, but not completely so. The hydrosphere is the sum total of water on Earth, except for that portion in the atmosphere. This combines all water underground—which, as we shall see, constitutes the vast majority of water on the planet—as well as all freshwater in streams, rivers, and lakes; saltwater in seas and oceans; and frozen water in icebergs, glaciers, and other forms of ice (see Glaciology).

It is almost unnecessary to point out that water is essential to life. Human bodies, after all, are almost entirely made of water, and without water we would die much sooner than we would if we were denied food. Humans are not the only organisms dependent on water; whereas there are forms of life designated as *anaerobic*, meaning that they do not require oxygen, virtually nothing that lives can survive independent of water. Thus, the biosphere, which combines all living things and all recently deceased things, is connected intimately with the hydrosphere.

WATER ON EARTH. Throughout most of the modern era of scientific study—from the 1500s, which is to say most of the era of useful scientific study in all of human history—it has been assumed that water is unique to Earth. Presumably, if and when we found life on another planet, that planet also would contain water. But until that time, so it was assumed, we could be assured that the only planet with life was also the only planet with water.

In the latter part of the twentieth century, however, as evidence began to gather that Mars contains ice crystals on its surface, this exclusive association of water with Earth has been challenged. As it turns out, frozen water exists in several places within our solar system—as well it might, since water on Earth had to arrive from somewhere. It is believed, in fact, that water arrived on Earth at a very early stage, carried on meteors that showered the planet from space (see the entries Planetary Science and Sun, Moon, and Earth).

Since about three billion years ago, the amount of water on Earth has remained relatively constant. The majority of that water, however, is not in the biosphere, the atmosphere, or what we normally associate with the hydrosphere—

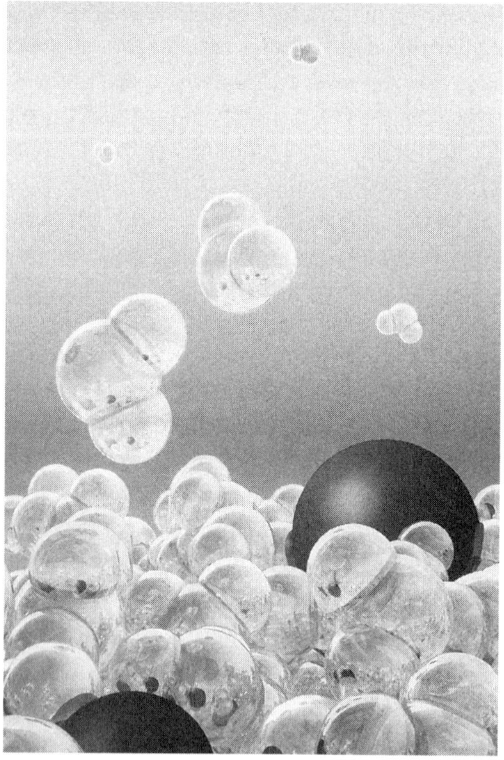

WATER MOLECULES EVAPORATING FROM A SOLUTION. *(© K. Eward/Photo Researchers. Reproduced by permission.)*

that is, the visible rivers, lakes, oceans, and ice formations on Earth's surface. Rather, the largest portion of water on Earth is hidden away in the geosphere—that is, the upper portion of Earth's crust, on which humans live and from which we obtain the minerals and grow the plants that constitute much of the world we know.

WATER COMPARTMENTS IN THE ENVIRONMENT

In the course of circulating throughout Earth, water passes from the hydrosphere to the atmosphere. It does so through the processes of evaporation and transpiration. The first of these processes, of course, is the means whereby liquid water is converted into a gaseous state and transported to the atmosphere, while the second one—a less familiar term—is the process by which plants lose water through their stomata, small openings on the undersides of leaves. Earth scientists sometimes speak of the two as a single phenomenon, evapotranspiration.

Evaporation and transpiration, as well as the process whereby such moisture is returned to the solid earth—that is, precipitation—are discussed in the essay Evapotranspiration and Precipitation. Still, the atmosphere is just one of several "compartments" in which water is stored within the larger environment. Among the other important places in which water is found are the oceans and other surface waters, ice in its many forms, and aquifers. The latter are underground rock formations in which groundwater—water resources that occupy pores in bedrock—is stored.

IMBALANCES IN THE SYSTEM. The total amount of water in all these compartments is fixed, but water moves readily between various compartments through the processes of evaporation, precipitation, and surface and subsurface flows. The hydrologic cycle is thus a system all its own, a "system" (in scientific terms) being any set of interactions that can be set apart mentally from the rest of the universe for the purposes of study, observation, and measurement. Its net input and output balance each other. There may be imbalances of input and output in particular areas, which will manifest as drought or flooding.

Flooding, as well as other aspects of the hydrosphere and its study, is discussed in the essay Hydrology. As for drought, its immediate cause is a lack of precipitation, though other causes can be responsible for the removal of water from the local environment. For instance, at present a large portion of Earth's water is tied up in glaciers and other ice formations, but at other times in the planet's history this ice has been melted, leaving much of the continental mass that we know today submerged under water (see Glaciology).

ACCOUNTING FOR EARTH'S WATER. Earth's total water supplies are so large that instead of being measured by gallons or other units of volume, they are measured in terms of tons or metric tons, designated as *tonnes.* Nonetheless, for comparison's sake, consider the following figures in light of the fact that a gallon (3.8 l) of water weighs 8.4 lb. (3.8 kg). A ton contains 238 gal., and a tonne has 1,000 l.

Just as heat from the Sun accounts for the lion's share of Earth's total energy budget (see Energy and Earth), the vast majority of water on Earth comes from the deep lithosphere, the upper layer of Earth's interior, comprising the crust and the brittle portion at the top of the mantle. In this vast region are contained 2.76×10^{19} tons (2.5×10^{19} tonnes). This figure, equal to 27.6 billion billion tons, is about 94.7% of the global total.

The next largest compartment is the oceans, which contain 1.41 $\times$ 10^{18} tons (1.38 $\times$ 10^{18} tonnes), or 5.2% of the total. Ice caps, glaciers, and icebergs contain 1.74 $\times$ 10^{16} tons (0.017 $\times$ 10^{16} tonnes), thus accounting for most of the remaining 0.1% of Earth's water. Beyond these amounts are much smaller quantities representing shallow groundwater (2.76 $\times$ 10^{14} tons, or 2.5 $\times$ 10^{14} tonnes); inland surface waters, such as lakes and rivers (2.76 $\times$ 10^{13} tons, or 2.5 $\times$ 10^{13} tonnes); and the atmosphere (1.43 $\times$ 10^{13} tons, or 1.3 $\times$ 10^{13} tonnes).

REAL-LIFE APPLICATIONS

THE LIFE OF A WATER DROPLET

Now let us follow the progress of a single water droplet as it passes through the water cycle. This particular droplet, like all others, has passed through the cycle countless times over the course of the past few billion years, and in its various incarnations it has existed as groundwater, as moisture in the atmosphere, and as ice.

For the short span of Earth's existence that humans have occupied the planet, it is conceivable that our droplet has been consumed—either directly, as liquid water, or indirectly, as part of the water content in animal or vegetable material. That would mean that it also has been excreted, after which it will have continued the cycle of circulation. In theory, it might well be part of the water in which humans bathe, brush their teeth, or wash their clothes.

WATER AND FOREIGN MATERIAL. Of course, personal hygiene as we know it today is an extremely recent development: for instance, regular toothbrushing as a practice among the whole population began in the United States only around the turn of the nineteenth century. Still, it is a bit disconcerting to think that the water in which we brush our teeth today may have floated down a sewer pipe at another time. Nonetheless, by moving water through so many locales, the hydrologic cycle has a built-in cleansing component.

This cleansing component can be illustrated by the experience of saltwater, which despite its presence in the ocean is actually a small portion of Earth's total water supply. The reason is that the salt seldom travels with the water; as soon as the water evaporates, the salt is left behind. This is why people on the proverbial desert island or in other survival situations use evaporation to make saltwater drinkable.

Likewise, saltwater as such cannot survive the transition from liquid water to ice: as the water freezes, the ice (which has a much lower freezing point) simply is precipitated and left behind. Just as salt does not travel with water as it makes its way through the various stages of the hydrologic cycle, so other varieties of foreign matter are left behind as well; as long as water is not allowed to stagnate, it usually is cleansed in the course of traveling between the ground and the atmosphere.

This is not to say that water typically exists in a pure form. Often called the *universal solvent,* water has such a capacity to absorb other substances that it is unlikely ever to appear in pure form unless it is distilled under laboratory conditions. Water in mountain streams, for instance, absorbs fragments of rock as it travels downhill, slowly eroding the surrounding rock and soil.

FROM THE WATERSHED TO POINTS BEYOND

On a particular watershed—an area of terrain from which water flows into a stream, river, or lake—our hypothetical water droplet may enter from a number of directions. In the simplest model of water flow, it comes from precipitation, including rain, snow, or even mist from clouds. The water has to go somewhere, and it may go either up or down. It may return to the atmosphere as evapotranspiration; it may enter the ground; or, if it reaches the solid earth at some elevation above sea level, it may enter a stream and flow ultimately to the ocean.

For water to enter the atmosphere generally requires an extensive surface area of vegetation, which supports high rates of transpiration. This transpiration, combined with evaporation from such inorganic surfaces as moist soil or bodies of water, puts a great deal of water into the atmosphere. Without significant evapotranspiration, however, it is necessary for the water to drain from the watershed, either as seepage to deep groundwater or as flow in the form of a stream.

THE FIVE STAGES OF THE HYDROLOGIC CYCLE. The overall process of the hydrologic cycle can be divided into five parts: condensation, precipitation, infil-

THE ARKANSAS RIVER, PART OF THE OGALLALA AQUIFER, RUNS THROUGH IRRIGATED FIELDS IN KANSAS. IRRIGATION PLACES A BIG DEMAND ON THE OGALLALA'S DIMINISHING WATER SUPPLY. *(AP/Wide World Photos. Reproduced by permission.)*

tration, runoff, and evaporation. Water vapor in the atmosphere condenses, forming clouds, which eventually become so saturated that they release the water to the solid earth in the form of precipitation. When precipitation enters the ground, it is known as infiltration.

Infiltration can be great or small, depending on the permeability of the ground. The soil of a rainforest, for instance, has so much organic matter that it is likely to be highly permeable. On the other hand, cities have large amounts of what land developers call impervious surface: roads, buildings, and other areas in which concrete and other materials prevent water from infiltrating the ground.

Assuming that water is unable to infiltrate, it becomes runoff. Runoff is simply surface water, which may take the form of streams, rivers, lakes, and oceans. If runoff occurs in an area that is not already a body of water, flood conditions may ensue. Thus, water may either infiltrate or become runoff, but as long as it remains close to the surface, it will experience evaporation.

In evaporation energy from the Sun changes liquid water into gaseous form, transporting it as a vapor into the atmosphere. Thus, the water is returned to the air, where it condenses and resumes the cycle we have described. As noted earlier, the water on or near Earth's surface is a small portion of the total. What about groundwater far below the surface? Let us now examine a particularly notable example of an aquifer, or groundwater reservoir.

THE OGALLALA AQUIFER

Located beneath the central United States, the Ogallala Aquifer provides a vast store of groundwater that supports a large portion of American agriculture. The Ogallala, also known as the High Plains Regional Aquifer, was discovered in the early years of the nineteenth century. It did not become a truly significant economic resource, however, until the second half of the twentieth century, when advanced pumping technology made possible large-scale irrigation from the aquifer's supplies. By 1980 the Ogallala supported some 170,000 wells and accounted for fully one-third of all water used for irrigation purposes in the United States.

Centered in Nebraska, the aquifer underlies parts of seven other states: South Dakota, Wyoming, Colorado, Kansas, Oklahoma, Texas, and New Mexico. It stretches 800 mi. (1,287 km) from north to south and 200 mi. (322 km) from east to west at its widest point. All told, the Ogallala covers some 175,000 sq. mi. (453,250 sq km), an area larger than Germany—all of it underground.

In Nebraska, the aquifer is between 400 ft. and 1,200 ft. (130–400 m) deep, while at the southern edges its depth extends no more than 100 ft. (30 m). Composed of porous sand, silt, and clay formations deposited by wind and water from the Rocky Mountains, the Ogallala is made up of several sections, called formations. The largest of these is the Ogallala formation, which accounts for about 77% of its total volume.

USING UP THE OGALLALA. The Ogallala is particularly important to local agriculture because the states that it serves are home to numerous dry areas. Yet high-volume pumping of the underground reserves has reduced the available groundwater, much as the pumping of oil gradually is consuming Earth's fossil-fuel reserves. Indeed, the water of the Ogallala is known as fossil water, meaning that it has been stored underground for millions of years, just as the coal, oil, and gas that runs modern industrial civilization has.

Of course, the water from the Ogallala is not simply used up in an irreversible process, as is the case with fossil fuels; nonetheless, the rapidly accelerating reduction of its water supplies is cause for some alarm. Less than 0.5% of the water removed from this aquifer is being returned to the ground, and if the current rate of pumping increases, the supplies will be 80% depleted by 2020.

The consequences of this depletion are already being felt in Kansas, where streams and rivers, dependent on groundwater to feed their flow, are running dry. In that state alone, more than 700 mi. (1,126 km) of rivers that formerly flowed year-round have been reduced to dry channels. In New Mexico and Texas, use of center-pivot irrigation, which requires a well capable of pumping 750 gal. (2,839 l) per minute, is disappearing because the local aquifer can no longer sustain such volumes.

In addition to the problem of diminishing supplies, contamination is an issue. As more and more agricultural chemicals seep into an ever shrinking reservoir, the towns of the high plains—places once known for their pure, clean groundwater—now have tap water that is considered unsafe for children and pregnant women. Overuse of the Ogallala is also exacting a financial toll, as more and more wells run dry and farms go bankrupt.

KEY TERMS

AQUIFER: An underground rock formation in which groundwater is stored.

BEDROCK: The solid rock that lies below the C horizon, the deepest layer of soil.

BIOSPHERE: A combination of all living things on Earth—plants, mammals, birds, reptiles, amphibians, aquatic life, insects, viruses, single-cell organisms, and so on—as well as all formerly living things that have not yet decomposed.

EVAPORATION: The process whereby liquid water is converted into a gaseous state and transported to the atmosphere.

EVAPOTRANSPIRATION: The loss of water to the atmosphere via the processes of evaporation and transpiration.

GEOSPHERE: The upper part of Earth's continental crust, or that portion of the solid earth on which human beings live and which provides them with most of their food and natural resources.

GROUNDWATER: Underground water resources that occupy the pores in bedrock.

HYDROLOGIC CYCLE: The continuous circulation of water throughout Earth and between various earth systems.

HYDROLOGIC SCIENCES: Areas of the earth sciences concerned with the study of the hydrosphere. Among these areas of study are hydrology, glaciology, and oceanography.

HYDROLOGY: The study of the hydrosphere, including the distribution of water on Earth, its circulation through the hydrologic cycle, the physical and chemical properties of water, and the interaction between the hydrosphere and other earth systems.

HYDROSPHERE: The entirety of Earth's water, excluding water vapor in the atmosphere but including all oceans, lakes, streams, groundwater, snow, and ice.

LITHOSPHERE: The upper layer of Earth's interior, including the crust and the brittle portion at the top of the mantle.

PRECIPITATION: When discussing the hydrologic cycle or meteorology, precipitation refers to the water, in liquid or solid form, that falls to the ground when the atmosphere has become saturated with moisture.

SYSTEM: Any set of interactions that can be set apart mentally from the rest of the universe for the purposes of study, observation, and measurement.

TRANSPIRATION: The process whereby plants lose water through their stomata, small openings on the undersides of leaves.

WATERSHED: An area of terrain from which water flows into a stream, river, lake, or other large body.

RIVERS

Despite the environmental challenges posed by such situations as the exhaustion of the Ogallala, the hydrologic cycle continues to roll on. As it does, it is sustained in large part by processes we cannot see: the movement of groundwater from the aquifer into streams or the evapotranspiration of surface waters to the atmosphere. Yet the movement of waters along the surface, because it is visible and recognizable to humans, attracts human attention in a way that many of these other components of the hydrologic cycle do not.

Rivers and other forms of surface water actually account for a relatively small portion of the planet's water supply, but they loom large in the human imagination as the result of their impact on our lives. The first human civilizations developed along rivers in Egypt, Mesopotamia, India, and China, and today many a great city lies along a river. Rivers provide us with a means of transportation and recreation, with hydroelectric power, and even—after the river water has been treated—with water for drinking and bathing.

FORMATION OF RIVERS. Rivers usually form from tributaries, such as springs. As the river flows, it is fed by more tributaries and by groundwater and continues on its way at various speeds, depending on the terrain. Finally, the river discharges into an ocean, a lake, or a desert basin.

River waters typically begin with precipitation, whether in the form of rainwater or melting snow. They also are fed by groundwater exuding from bedrock to the surface. When precipitation falls on ground that is either steeply sloped or already saturated, the runoff moves along Earth's surface, initially in an even, paper-thin sheet. As it goes along, however, it begins to form parallel rills, and its flow becomes turbulent. As the rills pass over fine soil or silt, they dig shallow channels, or runnels.

At some point in their flow, the runnels merge with one another, until there are enough of them to form a stream. Once enough streams have converged to create a continuously flowing body of water, it becomes a brook, and once the volume of water carried reaches a certain level, the brook becomes a river. As we have already noted, however, a river is really the sum of its tributaries, and thus hydrologists speak of river *systems* rather than single rivers.

RIVER SYSTEMS. A particularly impressive example of a river system is the vast Mississippi-Missouri, which drains the central United States. Most of the rivers between the Rockies and the Appalachians that do not empty directly into the Gulf of Mexico feed this system. This system includes the Ohio, itself an impressive river that divides the eastern United States. Indeed, just as the Mississippi separates east from west in America, the Ohio separates north from south.

After the Ohio and the Mississippi converge, at the spot where Illinois, Missouri, and Kentucky meet, they retain their separate identities for many miles. A strip of clear water runs along the river's eastern side, while to the west of this strip the water is a cloudy yellow—indicating a heavier amount of sediment in the Mississippi than in the Ohio. A similar phenomenon occurs where the "Blue Nile" and "White Nile," tributaries of another great river—both named for the appearance of the water—meet at Khartoum in Sudan.

WHERE TO LEARN MORE

"Groundwater Primer: Hydrologic Cycle." U.S. Environmental Protection Agency (Web site). <http://www.epa.gov/seahome/groundwater/src/cycle.htm>.

Hamilton, Kersten. *This Is the Ocean.* Honesdale, PA: Caroline House/Boyds Mills Press, 2001.

"Hydrologic Cycle." The Groundwater Foundation (Web site). <http://www.groundwater.org/GWBasics/hydro.htm>.

The Hydrologic Cycle (Web site). <http://agen521.www.ecn.purdue.edu/AGEN521/icon.html>.

Hydrologic/Water Cycle (Web site). <http://www.und.nodak.edu/instruct/eng/fkarner/pages/cycle.htm>.

Pidwirny, Michael J. "Introduction to Hydrology: The Hydrologic Cycle." *Fundamentals of Physical Geography* (Web site). <http://www.geog.ouc.bc.ca/physgeog/contents/8b.html>.

Richardson, Joy. *The Water Cycle.* Illus. Linda Costello. New York: F. Watts, 1992.

Smith, David. *The Water Cycle.* Illus. John Yates. New York: Thomson Learning, 1993.

Schaefer, Lola M. *This Is the Ocean.* Illus. Jane Wattenberg. New York: Greenwillow Books, 2001.

Walker, Sally M. *Water Up, Water Down: The Hydrologic Cycle.* Minneapolis: Carolrhoda Books, 1992.

GLACIOLOGY

CONCEPT

Glaciology is the study of ice and its effects. Since ice can appear on or in the earth as well as in its seas and other bodies of water and even its atmosphere, the purview of glaciologists is potentially very large. For the most part, however, glaciologists' attention is direction toward great moving masses of ice called glaciers, and the intervals of geologic history when glaciers and related ice masses covered relatively large areas on Earth. These intervals are known as ice ages, the most recent of which ended on the eve of human civilization's beginnings, just 11,000 years ago. The last ice age may not even be over, to judge from the presence of large ice masses on Earth, including the vast ice sheet that covers Antarctica. On the other hand, evidence gathered from the late twentieth century onward indicates the possibility of global warming brought about by human activity.

HOW IT WORKS

ICE

Ice, of course, is simply frozen water, and though it might appear to be a simple subject, it is not. Glaciologists classify differing types of ice, for instance, with regard to their levels of density, designating them with Roman numerals. The ice to which most of us are accustomed is classified as ice I. We will not be concerned with the other varieties of ice in the present context, but it should be noted that the ice in glaciers is quite different from the ice in an ice cube or even the ice on a pond in winter. These differences are a result of massive pressure, which reduces the air content of the ice in glaciers.

By definition, ice is composed of fresh water rather than saltwater. This is true even of icebergs, though they may float on the salty oceans. The reason is that water has a much higher freezing point than salt, and, therefore, when water freezes, very little of the salt remains joined to the water. Most of the salt is left behind in the form of a briny slush, and so much of Earth's fresh water supply actually is contained in great masses of ice, such as the glaciers of Antarctica.

Glaciology is defined as the study of ice, its forms, and its effects. This means that the glaciologist has a much wider scope than a geologist, meteorologist, or oceanographer, each of whom is concerned primarily with the geosphere, atmosphere, and hydrosphere, respectively. Though ice commonly is associated with the hydrosphere, where it appears on Earth's oceans, rivers, and lakes, it also is found on and even under the solid earth. There are even situations in which ice is found in the atmosphere.

GLACIOLOGY AND GLACIERS

Despite the wide distribution of ice on Earth and the many forms it takes, the work of most glaciologists is concerned primarily with ice as it appears in glaciers. A glacier is a large, typically moving mass of ice on or adjacent to a land surface. It does not flow, as water does; rather, it is moved by gravity, a consequence of its extraordinary weight.

Obviously, a glacier can form only in an extremely cold region—one so cold that the temperature never becomes warm enough for snow to melt completely. Some snow may melt as a result of contact with the ground, which is likely to be warmer than the snow itself, but when tem-

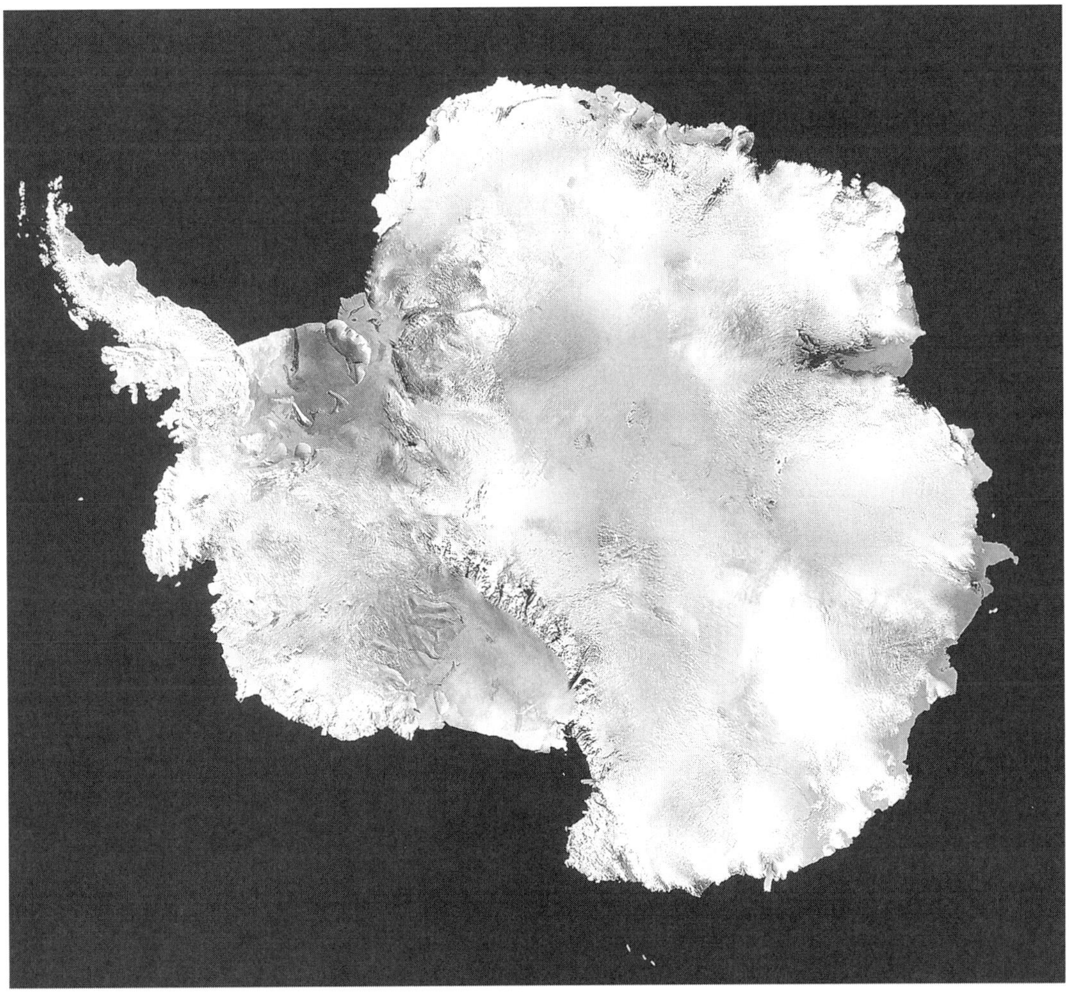

AN INFRARED SATELLITE IMAGE OF ANTARCTICA, SHOWING ICE SHELVES PROJECTING FROM THE COASTLINE. THE LARGEST ICE SHELF, THE ROSS, SITS TO THE LEFT OF THE TRANSANTARCTIC MOUNTAINS (LOWER CENTER). *(© USGS/Photo Researchers. Reproduced by permission.)*

peratures drop, it refreezes. A glacier starts with a layer of ice, on which snow gathers until refreezing gradually creates compacted layers of snow and ice.

As anyone who has ever held a snowball in his or her hand knows, snow is fluffy, or, to put it in more scientific terms, it is much less dense than ice. A sample of snow is about 80% air, but as ice accumulates over a layer of snow, the weight of the ice squeezes out most of the air. As the layers grow thicker and thicker, the weight reduces the air further, creating an extremely dense, thick layer of ice. Ultimately, the ice becomes so heavy that its weight begins to pull it downhill, at which point it becomes a glacier.

GLACIAL TEMPERATURE AND MORPHOLOGIC CHARACTERISTICS

Glaciers may be classified according to either relative temperature or morphologic characteristics (i.e., in terms of its shape). In terms of temperature, a glacier may be "warm," meaning that it is close to the pressure melting point. *Pressure melting point* is defined as the temperature at which ice begins to melt under a given amount of pressure. It is commonly known that water melts at 32°F (0°C), but only under conditions of ordinary atmospheric pressure at sea level. At higher pressures, the melting point of water is lower, which means that it can remain liquid at temperatures below its ordinary freezing point. (The melting point and freezing point of a substance are always the same.)

A "warm" glacier, such as those that appear in the Alps, is relatively mobile, because it is at the pressure melting point. This kind of glacier contrasts with a "cold," or polar, glacier, in which the temperatures are well below the pressure melting point; in other words, despite the extremely high pressure, the temperature is so low that the ice will not melt. As their name suggests, polar glaciers are found at Earth's poles, which effectively means Antarctica, since the area of the North Pole is not a land surface. A third category of glacier, in terms of temperature, is a subpolar glacier, found (not surprisingly) in regions near the poles. Examples of subpolar glaciers, or ones in which the fringes of the glacier are colder than the interior, are found in Spitsbergen, islands belonging to Norway that sit in the Arctic Ocean, well to the north of Scandinavia.

MORPHOLOGIC CLASSIFICATIONS. In the classification of geologic sciences, glaciology often is grouped with geomorphology. The latter field of study is devoted to landforms, or notable topographical features, and the forces and processes that have shaped them. Among those forces and processes are glaciers, which can be viewed in terms of their shape, the locale in which they form, and their effect on the contour of the land.

Alpine or mountain glaciers flow down a valley from a high mountainous region, typically following a path carved out by rivers or melting snow in warmer periods. They move toward valleys or the ocean, and in the process they exert considerable impact on the surrounding mountains, increasing the sharpness and steepness of these landforms. The rugged terrain in the vicinity of the Himalayas and the Andes, as well as the alpine regions of the Cascade Range and Rocky Mountains in the United States, are partly the result of weathering caused by these glaciers.

The glacial forms found in Alaska, Greenland, Iceland, and Antarctica are often piedmont glaciers, large mounds of ice that slope gently. Iceland, Greenland, and Antarctica as well as Norway are also home to cirque glaciers, which are relatively small and wide in proportion to their length. Though they experience considerable movement in place, they usually do not move out of the basinlike areas in which they are formed.

OTHER ICE FORMATIONS

There are several other significant varieties of ice formation, including ice caps, ice fields, and ice sheets. An ice cap, though much bigger than a glacier, typically has an area of less than 19,300 sq. mi. (50,000 sq km). Nonetheless, its mass is such that it exerts enormous weight on the land surface, and this exertion of force allows it to flow.

At the center of an ice cap or an ice sheet is an ice dome, and at the edges are ice shelves and outlet glaciers. Symmetrical and convex (i.e., like the outside of a bowl), an ice dome is a mass of ice often thicker than 9,800 ft. (3,000 m). An outlet glacier is a rapidly moving stream of ice that extends from an ice dome. Ice shelves, at the far outer edges, extend into the oceans, typically ending in cliffs as high as 98 ft. (30 m). Ice fields are similar to ice caps; the main difference is that the ice field is nearly level and lacks an ice dome. There are enormous variations in size for ice fields. Some may be no larger than 1.9 sq. mi. (5 sq km), while at different times in Earth's history, others have been as large as continents.

The most physically impressive of all ice formations, an ice sheet is a vast expanse of ice that gradually moves outward from its center. Ice sheets are usually at least 19,300 sq. mi. (50,000 sq km) and, like ice caps, consist of ice domes and outlet glaciers, with outlying ice shelves. Given their even greater size compared with ice caps, ice sheets exert still more force on the solid earth beneath them. They cause the rock underneath to compress, and, therefore, if an ice sheet ever melts, Earth's crust actually will rise upward in that area.

REAL-LIFE APPLICATIONS

ANTARCTICA

An example of an ice sheet is the Antarctic ice sheet, which is permanently frozen—at least for the foreseeable future. The Antarctic ice sheet covers most of Antarctica, an area of about five million sq. mi. (12.9 million sq km), the size of the United States, Mexico, and Central America combined. Within it lies 90% of the world's ice and more fresh water than in all the planet's rivers and lakes combined. By contrast, the impressive Greenland ice sheet, at 670,000 sq. mi. (1,735,000 sq km), is dwarfed, as are smaller ice sheets in Iceland, northern Canada, and Alaska.

The Antarctic ice sheet is the Sahara of ice masses, though, in fact, it is almost 50% larger than the Sahara desert and a good deal more inhospitable. Whereas the Sahara is scattered with towns and oases and has a steady population of isolated villagers, nomads, and merchants in caravans, *no one* lives on the Antarctic ice sheet except scientists on temporary missions. And whereas people have lived in the Sahara for thousands of years (it became a desert only somewhat recently, during the span of human civilization), scientific missions to Antarctica became possible only in the twentieth century. As it is, researchers spend only short periods of time on the continent and then in heavily protected environments.

ICE SHELVES AND GLACIERS. Just as the Antarctic ice sheet is the largest in the world, one of its attendant shelves also holds first place among ice shelves. The continent is shaped somewhat like a baby chick, with its head and beak pointing northward toward the Falkland Islands off the coast of South America and its two greatest ice shelves lying on either side of the "neck."

Facing the Weddell Sea, and the southern Atlantic beyond it, is the Ronne Ice Shelf, which extends about 400 mi. (640 km) over the water. The world record–holder, however, is the Ross Ice Shelf on the other side of the "neck," near Marie Byrd Land. About the same size as Texas or Spain, the Ross shelf extends some 500 mi. (800 km) into the sea and is the site of several permanent research stations.

ANTARCTIC TOPOGRAPHY. The Antarctic is also home to a vast mountain range, the Transantarctic, which stretches some 3,000 mi. (4,828 km) across the "neck" between the Ross and Weddell seas. Included in the Transantarctic Mountains is Vinson Massif, which at 16,860 ft. (5,140 m) is the highest peak on the continent. The continent as a whole is largely covered with mountain ranges, between which lie three great valleys called the Wright, Taylor, and Victoria valleys. Each is about 25 mi. (40 km) long and 3 mi. (5 km) wide.

These are the largest continuous areas of ice-free land on the continent, and they offer rare glimpses of the rocks that form the solid-earth surface deep beneath Antarctica. They are also among the strangest places on the planet, forbidding lands even by Antarctic standards. The three are known as the "dry valleys," owing to their lack of precipitation; indeed, if they lay in a more temperate zone, they would be deserts far more punishing than the Sahara. Geologists estimate that it has not rained or snowed in these three valleys for at least one million years. The reason is that ceaseless winds keep the air so dry that any falling snow evaporates before it reaches the ground. In this arid, brutally cold climate, nothing decomposes, and seal carcasses a millennium old remain fully intact.

THE THICKNESS OF THE ICE. The dry valleys are exceptional, because most of Antarctica lies under so much ice that the rocks cannot be seen. The ice in Antarctica has an *average* depth of more than a mile: the depth averages about 6,600 ft. (2,000 m), but in places on the continent it is as thick as 2 mi. (3.2 km). Thus, "ground level" on Antarctica is equivalent to a fairly high elevation in the inhabitable portion of the planet. Denver, Colorado, for instance, touts itself as the "Mile-High City," and its elevation has enough effect on a visiting flatlander that rival sports teams usually spend a few days in Denver before a game, adjusting themselves to the altitude.

The thickness of the ice has allowed glaciologists to take deep ice-core samples from Antarctica. An ice core is simply a vertical section of ice that, when studied with the proper techniques and technology, can reveal past climatic conditions in much the same way that the investigation of tree rings does. (See Paleontology for more about tree-ring research, or dendrochronology.) Ice-core samples from the Antarctic provide evidence regarding Earth's climate for the past 160,000 years and show a pattern of warming and cooling that is related directly to the presence of carbon dioxide and methane in the atmosphere. These core samples also reveal the warming effects of increases in both gases over the past two centuries.

Because of the great thickness of its ice, Antarctica has the highest average elevation of any continent on the planet. Yet beneath all that ice, the actual landmass is typically well below sea level. The reason for this is that the ice weighs it down so much; by contrast, if the ice were to melt, the land would begin to spring upward. The melting of the Antarctic ice shelf would be a disaster of unparalleled proportions. If all that ice were to melt at once, it would raise global sea levels by some 200 ft. (65 m). This would be enough to flood all the world's ports, along with vast areas of low-lying land. For instance, waters would swell over New York City and all ports on America's eastern seaboard and probably would cover an area extending westward to the Appalachian Mountains. Even if only 10% of Antarctica's ice were to melt, the world's sea level would rise by 20 ft. (6 m), enough to cause considerable damage.

WHAT GLACIERS DO TO EARTH'S SURFACE

Periodically over the past billion years, Earth's sea levels have advanced or retreated dramatically in conjunction with the beginning and end of ice ages. The latter will be discussed at the conclusion of this essay; in the present context, let us consider simply the geomorphologic effects of glaciers and ice masses. For example, as suggested in the discussion of Antarctica's ice sheet, when glaciers melt, thus redistributing their vast weight, Earth's crust rebounds. At the end of the last ice age, the crust rose upward, and in parts of North America and Europe this process of crustal rebounding is still occurring.

Glaciers move at the relatively slow speeds one would expect of massive objects made from ice: only a few feet or even a few inches per day. Friction with Earth's surface may melt the layer of ice that comes in contact with it, however, and, as a result, this layer of meltwater becomes like a lubricated surface, allowing the glacier to move much faster. The entire body of ice experiences a sudden increase of speed, called surging.

PLOWING THROUGH THE LAND. A glacier is like a huge bulldozer, plowing though rock, soil, and plants and altering every surface with which it comes into contact. It erodes the bottoms and sides of valleys, changing their V shape to a U shape. The rate at which it erodes the land is directly proportional to the depth of the glacier: the thicker the ice, the more it bears down on the land below it. As it moves, the ice pulls along rocks and soil, which are incorporated into the glacier itself. These components, in turn, make the glacier even more formidable, giving it greater weight, cutting ability, and erosive power.

The sediments left by glaciers that lack any intervening layer of melted ice are known by the general term till. In unglaciated areas, or places that have never experienced any glacial activity, sediment is formed by the weathering and decomposition of rock. On the other hand, formerly glaciated areas are distinguished by layers of till from 200 to 1,200 ft. (61–366 m) thick. Piles of till left behind by glaciers form hills called moraines, and the depressions left by these land-scouring ice masses are called kettle lakes.

North America abounds with examples of moraines and kettle lakes. Illinois, for instance, is covered with ridges, called end moraines, left behind by the melting near the conclusion of the last ice age. Visitors can take in a splendid view of moraine formations at Moraine View State Recreation Area, located astride the Bloomington moraine in central Illinois. Likewise Minnesota, Wisconsin, the Dakotas, and Wyoming are home to many a moraine. As with the Illinois recreation area, Kettle Lakes Provincial Park, near Timmins, Ontario, provides an opportunity to glimpse gorgeous natural wonders left behind by the retreat of glaciers—in this case, more than 20 deep kettle lakes. Park literature invites visitors to boat, fish, or swim in the lakes, though it would take a hardy soul indeed to brave those icy waters.

The glacier transports material from the solid earth as long as it is frozen, but wholly or partially melted glaciers leave behind sedimentary forms with their own specific names. In addition to moraines, there are piles of sediment, called eskers, left by rivers flowing under the ice. In addition, deposits of sediment may wash off the top of a glacier to form steep-sided hills called kames. If the glacier runs over moraines, eskers, or kames left by another glacier, the resulting formation is called a drumlin.

Just as rivers consist of main bodies formed by the flow of tributaries (for example, the many creeks and smaller rivers that pour into the Mississippi), so there are tributary glaciers. When a glacial tributary flows into a larger glacier, their top elevations become the same, but their bottoms do not. As a result, they carve out "hanging valleys," often the site of waterfalls. Examples include Yosemite and Bridalveil in California's Yosemite Valley.

THE KENNICOTT GLACIER IN THE WRANGELL MOUNTAINS OF ALASKA. *(© Pat and Tom Leeson/Photo Researchers. Reproduced by permission.)*

ICE AGES

The glaciers that exist today are simply the remnants of the last ice age, a time in which the size of the ice masses on Earth dwarfed even the great Antarctic ice sheet. When people speak of "the Ice Age," what they mean is the *last* ice age, which ended about 11,000 years ago. Yet it is one of only about 20 ice ages that have taken place over the past 2.5 million years, roughly coinciding with the late Pliocene and Pleistocene epochs. Actually, periods of massive glaciation (the covering of the landscape with large expanses of ice) have occurred at intervals over the past billion years. Their distribution over time has not been random; rather, they are concentrated at specific junctures in Earth's history.

Like the great mass extinctions of the past (see Paleontology), ice ages are among the markers geologists use in separating one interval of geologic time from another. In fact, there have been connections between ice ages and mass extinctions, particularly those that resulted from a recession of the seas. For example, the mass extinction that took place near the end of the Ordovician period (about 435 million years, or Ma, ago) came about as a result of a drop in the ocean level, which was caused, in turn, by an increase of glaciation that coincided with that phase in Earth's history.

The late Ordovician/early Silurian ice ages (between 460 to 430 Ma ago) are among four major phases of glaciation during the past 800

KEY TERMS

ATMOSPHERE: In general, an atmosphere is a blanket of gases surrounding a planet. Unless otherwise identified, however, the term refers to the atmosphere of Earth, which consists of nitrogen (78%), oxygen (21%), argon (0.93%), and other substances that include water vapor, carbon dioxide, ozone, and noble gases such as neon, which together comprise 0.07%.

CRUSTAL REBOUNDING: An upward movement by Earth's crust in response to the melting of a glacier, which redistributes its vast weight and causes Earth to rebound.

GEOMORPHOLOGY: An area of physical geology concerned with the study of landforms, with the forces and processes that have shaped them, and with the description and classification of various physical features on Earth.

GEOSPHERE: The upper part of Earth's continental crust, or that portion of the solid Earth on which human beings live and that provides them with most of their food and natural resources.

GLACIATION: The covering of the landscape with large expanses of ice, as during an ice age.

GLACIER: A large, typically moving mass of ice on or adjacent to a land surface.

GLACIOLOGY: An area of physical geology devoted to the study of ice, its forms, and its effects.

HYDROSPHERE: The entirety of Earth's water, excluding water vapor in the atmosphere but including all oceans, lakes, streams, groundwater, snow, and ice.

ICE AGE: A period of massive and widespread glaciation. Ice ages usually occur in series over stretches of several million years, or even several hundred million years.

ICE CAP: An ice formation bigger than a glacier but smaller than an ice sheet. An ice cap typically has an area of less than 19,300 sq. mi. (50,000 sq km) and, like an ice sheet, consists of an ice dome, with ice shelves and outlet glaciers at the edges.

ICE CORE: A vertical section of ice, usually taken from a deep ice sheet such as that in Antarctica. When studied with the proper techniques and technology, ice cores can reveal past climatic conditions in much the same way that the investigation of tree rings does.

ICE DOME: A symmetrical, convex (i.e., like the outside of a bowl) mass of ice, often thicker than 9,800 ft. (3,000 m), usually found at the center of an ice cap or an ice sheet.

million years. The first of these occurred during the late Proterozoic eon, toward the end of Precambrian time (between 800 and 600 Ma ago). Another happened during the Pennsylvanian subperiod of the Carboniferous and extended throughout the Permian period, thus lasting from about 350 to 250 million years ago. The last period of glaciation is the one in which we are living, beginning in the late Neogene period and extending into the current Quaternary.

HUMANS AND THE ICE AGES. In fact, many scientists question whether the last ice age has ended and whether we are merely living in an interglacial period. Certainly crustal rebounding is still taking place, as noted earlier.

KEY TERMS CONTINUED

ICE FIELD: A large ice formation, similar to an ice cap except that it is nearly level and lacks an ice dome. There are enormous variations in size for ice fields. Some may be no larger than 1.9 sq. mi. (5 sq km), while at different times in Earth's history, some have been as large as continents.

ICE SHEET: A vast expanse of ice, usually at least 19,300 sq. mi. (50,000 sq km), that moves outward from its center. Like the smaller ice caps, ice sheets consist of ice domes and outlet glaciers, with outlying ice shelves.

ICE SHELF: An ice formation at the edge of an ice cap or ice sheet that extends into the ocean, typically ending in cliffs as high as 98 ft. (30 m).

LANDFORM: A notable topographical feature, such as a mountain, plateau, or valley.

MA: An abbreviation used by earth scientists, meaning million years or megayears. When an event is designated as, for instance, 160 Ma, it usually means 160 million years ago.

MASS EXTINCTION: A phenomenon in which numerous species cease to exist at or around the same time, usually as the result of a natural calamity.

MORAINE: A hill-like pile of till left behind by a glacier.

MORPHOLOGY: Structure or form, or the study thereof.

OUTLET GLACIER: A rapidly moving stream of ice that extends from an ice dome.

PHYSICAL GEOLOGY: The study of the material components of Earth and of the forces that have shaped the planet. Physical geology is one of two principal branches of geology, the other being historical geology.

PRESSURE MELTING POINT: The temperature at which ice begins to melt under a given amount of pressure. The higher the pressure, the lower the temperature at which water can exist in liquid form.

RELIEF: Elevation and other inequalities on a land surface.

SEDIMENT: Material deposited at or near Earth's surface from a number of sources, most notably preexisting rock.

TILL: A general term for the sediments left by glaciers that lack any intervening layer of melted ice.

TOPOGRAPHY: The configuration of Earth's surface, including its relief as well as the position of physical features.

Inasmuch as "glacial period" refers to a time when glaciers cover significant portions of Earth and when the oceans are not at their maximum levels, we are indeed still in a glacial period.

It seems as though human existence has been bounded by ice, both in its onset and its recession. Ice ages have been a regular feature of the two million years since *Homo sapiens* came into existence, and the species had much of its formative experience in times of glaciation. The latter part of the last ice age created a land bridge that made possible the migration of Siberian peoples to the Americas, so that they are known now as *Native Americans.* (The name is well deserved: the ancestors of the Native Americans moved east from Siberia about 12,000 years ago, whereas less than half that much time has elapsed since the Indo-European ancestors of Caucasian Europeans moved west from what is now Russia. Certainly no one today questions whether Germans, Italians, British, French, and other groups are "native" Europeans.)

As an indication that ice ages have not ceased to occur, there is the Little Ice Age, which lasted from as early as 1250 to about 1850. This was a period of cooling and expansion of glaciers in the temperate latitudes on which Europe is located. Glaciers destroyed farmlands and buildings in the Alps, Norway, and Iceland, while Norse settlements in Greenland became uninhabitable. Europe as a whole suffered widespread crop failures, with a resulting loss of life. Evidence for this ice age, and indeed for all ice ages, can be discerned from the "footprints" left by glaciers from that time. Another telling sign is the transport of materials, such as rocks and fossils, from one part of Earth's surface to another.

UNDERSTANDING ICE AGES. What caused the Little Ice Age? The answer, or rather the attempt at an answer, goes to the core question of what causes ice ages in general. Earth scientists have cited both extraterrestrial and terrestrial factors. Among the leading extraterrestrial causes are an increase in sunspot activity (see Sun, Moon, and Earth for more about sunspots) as well as changes in Earth's orientation with respect to the Sun.

Contenders for a terrestrial explanation include changes in ocean circulation, as well as meteorites and volcanism. Either of these last two could have caused the atmosphere to become glutted with dust, choking out the Sun's light and cooling the planet considerably. (Such a calamity has been blamed for several instances of mass extinction, most notably, the one that wiped out the dinosaurs some 65 million years ago. See Paleontology.)

THE FUTURE. Though it appears that we are living in an interglacial period and that Earth could undergo significant cooling again thousands of years from now, there is also an even more frightening prospect of human-induced *warming*. As noted earlier, the Antarctic ice core reveals an increase of carbon dioxide and methane in the atmosphere during the past two centuries. Though these gases can be produced naturally, this excess in recent times appears to be a by-product of industrialized society. The glaciers of Europe are receding, and whether this could be the result of human activity or simply part of Earth's natural change as it comes out of the last ice age remains to be seen.

In the meantime, ice offers a great deal of potential for understanding our own planet and others. It may yet turn out that the ice sheets covering Mars contain single-cell life-forms. Furthermore, in August 1996, the National Aeronautics and Space Administration (NASA) reported that a meteorite found on the Antarctic ice may provide evidence of life on Mars. It seems that the 4.1-lb. (2 kg) meteorite contains polycyclic aromatic hydrocarbons that may have existed on that planet several billion years ago.

WHERE TO LEARN MORE

Alley, Richard B. *The Two-Mile Time Machine: Ice Cores, Abrupt Climate Change, and Our Future.* Princeton, NJ: Princeton University Press, 2000.

Bolles, Edmund Blair. *The Ice Finders: How a Poet, a Professor, and a Politician Discovered the Ice Age.* Washington, DC: Counterpoint, 1999.

Charlotte the Vermont Whale: Glaciers and the Glacial Ages (Web site). <http://www.uvm.edu/whale/GlaciersGlacialAges.html>.

Fagan, Brian M. *The Little Ice Age: How Climate Made History, 1300–1850.* New York: Basic Books, 2000.

Gallant, Roy A. *Glaciers.* New York: Franklin Watts, 1999.

"Home of Deep Glaciology." California Institute of Technology Division of Geological and Planetary Sciences (Web site). <http://skua.gps.caltech.edu/>.

Ice Ages (Web site). <http://www.museum.state.il.us/exhibits/ice_ages/>.

National Oceanic and Atmospheric Administration (NOAA) at the National Snow and Ice Data Center/World Data Center for Glaciology (Web site). <http://nsidc.org/noaa/>.

Stone, Richard. *Mammoth: The Resurrection of an Ice Age Giant.* Cambridge, MA: Perseus, 2001.

Virtually the Ice Age (Web site). <http://www.creswell-crags.org.uk/virtuallytheiceage/>.

METEOROLOGY AND THE ATMOSPHERE

EVAPOTRANSPIRATION AND PRECIPITATION

WEATHER

CLIMATE

EVAPOTRANSPIRATION AND PRECIPITATION

CONCEPT

Evaporation, along with the less well known process of transpiration, is the means by which water enters the atmosphere in the form of moisture. In evaporation liquid water from nonliving sources, such as the soil and surface waters, is converted into a gas. This conversion is driven by the power of the Sun, whose energy is also behind the process of transpiration, whereby plants lose water through their leaves. As with evaporation, transpiration places water in the atmosphere, and because the two processes work in tandem, they usually are spoken of together under the name evapotranspiration. Both make possible the formation of clouds, which, when they become saturated with moisture, produce the forms of precipitation by which water returns to the solid earth.

HOW IT WORKS

THE MOVEMENT OF WATER

The hydrosphere is the sum total of Earth's water, with the exception of water in the atmosphere. The hydrologic cycle is the continuous circulation between these two earth systems, hydrosphere and atmosphere, as well as the two other principal earth systems, biosphere and geosphere (see Earth Systems). Evapotranspiration and precipitation are principal components of this cycle, which is discussed in depth within the Hydrologic Cycle essay.

In evaporation, heat converts liquid water to a gaseous state, thus allowing it to be transferred to the atmosphere in the form of vapor. Whereas evaporation involves the loss of water to the atmosphere from nonliving sources, transpiration is the movement of water from living organisms to the atmosphere. This is achieved by the release of water through the plants' stomata, small openings on the undersides of leaves. In both cases, the Sun's electromagnetic energy, experienced as heat, is the driving mechanism, and because these phenomena are so closely related, they are normally treated together as evapotranspiration.

EARTH'S WATER BUDGET

Our present concern is primarily with the atmosphere; nonetheless, it is important to consider the other "compartments" in which water is stored. Some of them are regular way stations in the hydrologic cycle, but in the case of groundwater at least, the compartment is just that—a storage area from which it is conceivable that water might not be moved for millions of years.

The vast majority of water on Earth is indeed stored away, deep beneath the planet's surface in the form of groundwater in the lithosphere. This accounts for a staggering 94.7% of Earth's water. (For figures on the mass of this and other components in the water supply, see Hydrologic Cycles.) Much of the remainder is made up of the oceans, which account for 5.2% of the water on Earth, while glaciers and other forms of permanent and semipermanent ice make up 0.065%.

THE LAST 0.035%. We have now identified 99.965% of all water, yet we have not even approached any of the forms of water with which most of us typically come into contact. Of the remaining 0.035%, shallow groundwater, the source of most local water supplies, makes up the

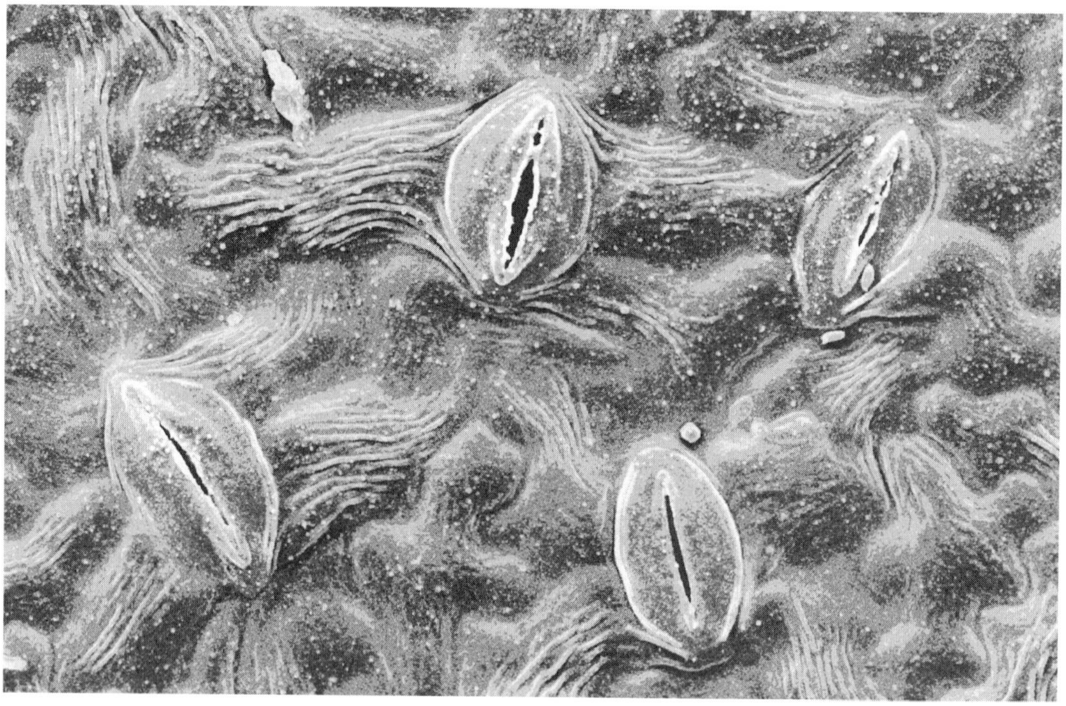

SCANNING ELECTRON MICROGRAPH OF THE STOMATA, OR PORES, ON THE UNDERSIDE OF AN APPLE TREE LEAF. LEAF PORES PULL IN CARBON DIOXIDE FOR PHOTOSYNTHESIS. *(© Scimat/Photo Researchers. Reproduced by permission.)*

bulk, 0.30% of the total. Next are the inland surface waters. All of them combined, including such vast deposits as the Great Lakes and the Caspian Sea as well as the Mississippi-Missouri, Amazon, and Nile river systems—account for just 0.03% of Earth's water.

Now we are left with only 0.02% of the total, which is the proportion occupied by moisture in the atmosphere: clouds, mist, and fog as well as rain, sleet, snow, and hail. While it may seem astounding that atmospheric moisture is such a small portion of the total, this fact says more about the vast amounts of water on Earth than it does about the small amount in the atmosphere. That "small" amount, after all, weighs 1.433×10^{13} tons (1.3×10^{13} tonnes), or 28,659,540,000,000,000 lb.

WATER TURNOVER

The smaller the compartment of water, the greater the amount of turnover—that is, the exchange of "new" water for "old"—and the shorter the turnover time. Groundwater may stay put in aquifers, underground rock formations, for millions of years; on the other hand, atmospheric water experiences an enormous amount of turnover in just a year's time.

The atmosphere receives vast inputs of evaporation from the oceans as well as evapotranspiration from terrestrial ecosystems, or land-based communities of organisms. In the course of the year, the water in the atmosphere turns over about 34 times. Thus, the inputs of evaporation and transpiration are balanced almost perfectly by outputs of precipitation, which return more than 75% of atmospheric moisture to the oceans. The remainder falls on the land, where it contributes to brooks, streams, and rivers. The land receives 67% more water as precipitation than it loses through evaporation. The difference is made up by runoff to the oceans, primarily through rivers and in much smaller portions by subterranean channels.

HOW DOES THE WATER MOVE?

Clearly, water is moving through our atmosphere, but *how?* The answer is through gradients, or differences, of energy. When you hold a stone over the side of a cliff, preparing to drop it, the stone possesses a large quantity of gravitational potential energy. This potential energy is a function of the large gradient between the position of the stone and that of the ground.

In the same way, rivers and streams are driven by the gravitational potential that exists by virtue of the gradient from their source to the place where they empty into a lake or ocean. By definition, water flows downhill, and therefore the source of the river must be at a higher elevation than its delta, or the place where it discharges.

This matter of elevation is reflected in the confusing names for the two kingdoms that united in about 3100 B.C. to form Egypt: Upper Egypt was actually to the south of Lower Egypt. The names reflect the importance of the Nile, which flowed from the higher elevations of Upper Egypt into the fertile lowlands of Lower Egypt, home of Egyptian civilization's greatest farmlands.

EVAPORATION. Evaporation likewise is driven by energy gradients; in this case, however, the energy is not gravitational but electromagnetic. Specifically, it is the energy from the Sun, which we experience in the form of light and heat. (See Sun, Moon, and Earth for more about the Sun's energy.) As surfaces on Earth absorb solar electromagnetic energy, it increases their heat content and provides a source of energy to drive evaporation.

The second law of thermodynamics, discussed in Energy and Earth, tells us that the flow of heat is always from a high-temperature reservoir or area to a low-temperature one. Thus, if you hold a snowball in your hand, you may think that the snowball is cooling your hand, but, in fact, the opposite is happening: heat is passing from your hand into the snowball and warming it. As this happens, your hand loses heat, which you perceive as cold, though no coldness has been transferred—only heat.

In the same way, the energy difference between a heated surface and the atmosphere, manifested as a difference in temperature, makes it possible for water to be vaporized and transported into the air. Though the water has risen rather than fallen, its rise is brought about by the same principle that makes it possible for objects to fall from a great height, that is, a difference in potential. Thus it can be said that in evaporating, water "falls" from an area of high heat and high energy to one of low heat and low energy.

REAL-LIFE APPLICATIONS

Transpiration

When it comes to putting moisture into the atmosphere, transpiration is at least as important as evaporation. In fact, it puts more water into the air than evaporation does: any large area of vegetation tends to evaporate much larger quantities of moisture than an equivalent nonfoliated region, such as the surface of a lake or moist soil. Physically, however, the process of transpiration is the same as that of evaporation.

The only difference is that in the case of transpiration, the source of the evaporation is a biological one—leaves, for instance, as well as skin or lungs. From an environmental standpoint, the most important kind of transpiration is that which occurs through leaves. The loss of water from foliage puts an enormous amount of moisture into the atmosphere, and for this reason, areas where foliage appears in high concentration (i.e., forests) are vital to the cycling of water through the atmosphere.

Plants lose their water through moist membranes of a tissue known as *spongy mesophyll*, found in the tiny cavities that lie beneath the microscopic leaf pores called stomata. Stomata remain open most of the time, but when they need to be closed, guard cells around their borders push them shut. Because plants depend on them to "breathe" by pulling in carbon dioxide (see Carbon Cycle), however, they keep their stomata open—just as a human's pores must remain open, or the person would die of suffocation.

Because stomata are exposed in order to receive carbon dioxide for the plant's photosynthesis, this also means that the stomata are open to allow the loss of moisture to the atmosphere. It can be said, then, that transpiration—vital as it is to the functioning of our atmosphere—is actually an unavoidable consequence of photosynthesis, an unrelated process.

TRANSPIRATION AND ANIMALS. As we have suggested, transpiration in animals (including humans) takes place for much the same reason as it does with plants, as a by-product of breathing. Animals have to keep their moist respiratory surfaces, such as the lungs, open to the atmosphere. We may not think of our own breathing as transferring moisture to the air, but with just a little consideration of the subject it becomes clear that this is the case. The presence of moisture in our lungs can be proved simply by breathing on a piece of glass and observing the misty cloud that lingers there.

On the one hand, transpiration can cause animals to become dehydrated, but it also can be important in cooling down their bodies. When human bodies become overheated, they produce perspiration, which cools the surface of the skin somewhat. If the air around us is too humid, however, it already is saturated with water, and the perspiration has no place to evaporate. Therefore, instead of continuing to cool our bodies, the perspiration simply forms a sticky film on our skin. Assuming that the air is capable of absorbing more moisture, however, the sweat will evaporate, cooling our bodies considerably.

Heat, Cold, and Evapotranspiration

The preceding discussion brings up several more points about the relationship between heat and evapotranspiration. First of all, everything that is living has some degree of heat; if it did not, it would be at a temperature known as absolute zero, or 0K (–459.67°F or –273.15°C), which is impossible to reach. Therefore, even in Greenland there is a small amount of molecular motion in plant tissues, even when they appear to be completely frozen.

Naturally, however, there is very little evapotranspiration when temperatures are extremely cold, which is the reason why it can sometimes be "too cold to snow": temperatures are too low for sufficient moisture to be moved to the atmosphere. Nonetheless, there still can be some physical evaporation as the result of the direct vaporization of solid water, a process called sublimation.

Hot, dry summers. At the height of summer, when air temperatures are warm and the trees are fully foliated (i.e., covered in leaves), a high rate of transpiration occurs. So much water is pumped into the atmosphere through foliage that the rate of evapotranspiration typically exceeds the input of water to the local environment through rainfall. The result is that soil becomes dry, some streams cease to flow, and by late summer, in extremely warm temperate areas such as the southern United States, there is a great threat of drought and related problems, such as forest fires.

Once the trees drop their leaves in the autumn, transpiration rates decline greatly. This makes it possible for the parched soil to become recharged by rainfall and for streams to flow again. Such is the case in a temperate region, which by definition is one that has the four seasons to which most people in the United States (outside Hawaii, Alaska, and extreme southern Florida and Texas) are accustomed. In a tropical region, by contrast, there is a "dry season," in which transpiration takes place, and a "rainy season," in which moisture from the atmosphere inundates the solid earth. This rainy season may be so intense that it produces floods.

Clouds

So far, we have talked mostly about the means by which water moves from the solid earth into the atmosphere. Now let us consider what happens when it gets there, at which point—assuming there is sufficient moisture—it will coalesce to form a cloud. Though most people are inclined to romanticize clouds, seeing in them shapes and colors and sometimes even reflections of their own moods, clouds are nothing more than atmospheric moisture that has condensed to form tiny droplets of water or crystals of ice.

Heated, moist air that rises from the ground is much more dense than the air that lies above it, but as it rises, it expands and becomes less dense. This expansion cools the air, so that the water vapor condenses into tiny droplets. As a result, a cloud forms, and the relative intensity of the energy gradients that brought about the formation of the cloud creates different shapes.

For example, if there is a vigorous uplift of air, resulting from sharp differences in temperature and pressure between the ground and the atmosphere, the resulting clouds will have a tall, stacked appearance. On the other hand, clouds formed by the gentle uplift of air currents have a flat, cottony appearance. Thus, the shapes of the clouds themselves, as well as the ways in which they change, assist meteorologists in predicting the weather.

Cloud Classifications

Thanks to a system developed in 1803 by the English pharmacist and amateur naturalist Luke Howard (1772–1864), it is possible to classify clouds according to three basic shapes. These shapes are known by the Latin names *cumulus* (piled heaps and puffs), *cirrus* (curly, fibrous shapes), and *stratus* (stretched and layered).

Howard combined these names with adjectival terms, such as *alto* ("high") and *nimbus* ("rain"), to describe variations on the three basic cloud types. Today, the International Cloud Classification used by meteorologists worldwide applies Howard's system to the identification of ten basic cloud types, or genera. They are divided into three high-altitude genera, two midlevel ones, three low-level genera, and two varieties of rain cloud.

HIGH-LEVEL CLOUDS. The three genera of high-level clouds appear at a range between 16,500 ft. and 45,000 ft. (5,032–13,725 m), though they usually form in a belt between 20,000 ft. and 25,000 ft. (6,000–7,500 m). All are cirrus clouds, of which the highest are pure cirrus, made of ice crystals.

Then there are cirrocumulus, the least common variety of cloud. Composed of either supercooled water droplets (i.e., water that continues to exist in liquid form even though the temperature has dropped below the freezing point) or ice crystals, these are small and white or pale gray, with a rippled appearance like oatmeal. Finally, cirrostratus clouds—which, like pure cirrus clouds, are made of ice crystals—often form a halo around the Sun or Moon in the wintertime.

MIDLEVEL CLOUDS. The two genera of midlevel clouds, which appear at 6,500-23,000 ft. (2,000–7,000 m), are named with the prefix "alto." Altostratus clouds are usually a uniform bluish or gray sheet covering large portions of the sky and either completely or nearly obscure the Sun and Moon. These complex clouds are composed of layers, with ice crystals at the top, ice and snow in the middle, and water droplets in the lower layers.

Altocumulus are oval-shaped, dense, fluffy balls that may appear either as singular units or as closely bunched groups. When sunlight or moonlight shines through these clouds, the light often is perceived from the ground in the form of rays.

LOW-LEVEL CLOUDS. There are three genera in the low level, which extends from the surface to 6,500 ft. (2,000 m). Pure stratus clouds, which are generally the lowest, blanket the sky and typically appear gray. They are formed when a large mass of air rises slowly or when cool air moves in over an area close to ground level. Stratus clouds may produce mist or drizzle, or, when they form at ground level, they may appear as fog.

WHEN MOISTURE COALESCES IN THE ATMOSPHERE, IT FORMS A CLOUD. ALTOCUMULUS CLOUDS, WHICH LOOK LIKE OVAL-SHAPED, DENSE, PUFFY BALLS, ARE MIDLEVEL CLOUDS. *(© Mark A. Schneider/Photo Researchers. Reproduced by permission.)*

Pure cumulus clouds are flat on the base and vertically thick, with a puffy appearance on top. This puffiness is the result of updrafts. Occurring primarily in warm weather, cumulus clouds are made of water droplets and look brilliant white because of the sunlight's reflection off the droplets. Last, stratocumulus clouds are large, grayish, and puffy, often looking like dark rolls.

RAIN CLOUDS. Stratocumulus clouds can transform into nimbostratus clouds, while cumulus can develop into cumulonimbus clouds. These two—nimbostratus and cumulonimbus—are the last of the ten varieties of cloud, designated not by altitude but by the fact that they give rise to precipitation. (Note that the light, airy variety known as *cirrus* never portends rain.)

Nimbostratus usually appear at midlevel altitudes. Made of water droplets that produce either rain or snow, they are thick, dark, and gray.

Micrograph of a snowflake. Snowflakes are formed by the bonding of solid ice crystals inside a cloud. *(© Kent Wood/Photo Researchers. Reproduced by permission.)*

Sometimes they are seen with virga, which are skirts of rain trailing down their sides.

Cumulonimbus, or thunderstorm, clouds arise from cumulus clouds that have reached a great height. At a certain height, the cloud flattens out, and powerful updrafts and downdrafts create a great deal of unrest inside the cloud, which contains all phases of water: gas, liquid, and solid. These clouds can cause violent storms.

Rain, Snow, Hail, and Sleet

Eventually, the amount of moisture in the cloud becomes so great that it has to fall earthward in the form of precipitation—usually rain, snow, hail, or sleet. These types of precipitation are distinguished by the form they take: liquid in the first case, lightly frozen particles in the second case, and hard, frozen nuggets in the latter two cases.

Liquid precipitation includes drizzle and raindrops, which are distinguished from each other on the basis of size. Raindrops have a radius of about 0.04 in. (1 mm), while drops of drizzle are about one-tenth that size. Note the use of the term radius, implying a sphere. Drops of liquid precipitation are spherical, and though a teardrop shape is popularly associated with raindrops, they assume that shape only when they are falling.

Snowflakes are formed by the aggregation, or physical bonding, of solid ice crystals inside a cloud, while hailstones are made of a combination of supercooled water droplets and pellets of ice. Not only are they more dense than snowflakes, but they are also more spherical. Similar to hail is sleet, pellets of pure ice that usually are much smaller than hail.

FORMING PRECIPITATION. The type of precipitation formed depends on the warmth of the cloud from which it comes. "Warm" clouds are those whose temperatures are above freezing—32°F (0°C)—and "cold" clouds are those that are at least partially below freezing temperature. These temperature values are themselves a function of altitude, since temperature in the lower atmosphere (where all precipitation occurs) decreases by about 5.32°F for every mile (6°C per kilometer) of altitude gained.

As we have suggested earlier, humidity relates to the ability of air to evaporate, and, therefore, when the air has all the water it can hold, it is said to have a relative humidity of 100%. The cooler air is, the less moisture is required for it to become saturated. Once saturation occurs, moisture molecules form around cloud condensation nuclei, such as nitrates or extremely fine particles of sea salt that have managed to evaporate.

Cold clouds form around ice crystals. Before freezing, the water in these clouds may be supercooled,. Ice nuclei are much more rare than cloud condensation nuclei and are less well understood by meteorologists.

MIXED CLOUDS. Clouds that contain both liquid water and ice are called mixed clouds. Supercooled water will freeze when it strikes an ice crystal, and if it freezes immediately, it forms what is known as opaque or rime ice. If it freezes slowly, the result is called clear ice. Eventually the ice forms a thick coating, which is the origin of hail.

Not all mixed clouds produce frozen precipitation, however. Thunderstorms, involving electric charges imparted to precipitation particles, which leads to the eventual discharge of lightning, are also the products of mixed clouds. (For more about storms and precipitation, see Weather and Climate.)

KEY TERMS

AQUIFER: An underground rock formation in which groundwater is stored.

ATMOSPHERE: In general, an atmosphere is a blanket of gases surrounding a planet. Unless otherwise identified, however, the term refers to the atmosphere of Earth, which consists of nitrogen (78%), oxygen (21%), argon (0.93%), and other substances that include water vapor, carbon dioxide, ozone, and noble gases such as neon (0.07%).

BIOSPHERE: A combination of all living things on Earth—plants, mammals, birds, amphibians, reptiles, aquatic life, insects, viruses, single-cell organisms, and so on—as well as all formerly living things that have not yet decomposed.

ECOSYSTEM: A community of interdependent organisms along with the inorganic components of their environment.

EVAPORATION: The process whereby liquid water is converted into a gaseous state and transported to the atmosphere. When discussing the atmosphere and precipitation, usually evaporation is distinguished from transpiration. In this context, evaporation refers solely to the transfer of water from nonliving sources, such as the soil or the surface of a lake.

EVAPOTRANSPIRATION: The loss of water to the atmosphere via the combined (and related) processes of evaporation and transpiration.

GEOSPHERE: The upper part of Earth's continental crust, or that portion of the solid earth on which human beings live and which provides them with most of their food and natural resources.

GROUNDWATER: Underground water resources that occupy the pores in bedrock.

HYDROLOGIC CYCLE: The continuous circulation of water throughout Earth and between various earth systems.

HYDROLOGY: The study of the hydrosphere, including the distribution of water on Earth, its circulation through the hydrologic cycle, the physical and chemical properties of water, and the interaction between the hydrosphere and other earth systems.

HYDROSPHERE: The entirety of Earth's water, excluding water vapor in the atmosphere but including all oceans, lakes, streams, groundwater, snow, and ice.

METEOROLOGY: The study of the atmosphere, weather, and weather prediction.

POTENTIAL: Position in a field, such as a gravitational force field.

POTENTIAL ENERGY: The energy that an object possesses by virtue of its position or its ability to perform work.

PRECIPITATION: When discussing the hydrologic cycle or meteorology, precipitation refers to the water, in liquid or solid form, that falls to the ground when the atmosphere has become saturated with moisture. In the context of chemistry, precipitation refers to the formation of a solid from a liquid.

SUPERCOOLED: A term for water that continues to exist in liquid form even though its temperature has dropped below the freezing point.

KEY TERMS CONTINUED

SYSTEM: Any set of interactions that can be set apart mentally from the rest of the universe for the purposes of study, observation, and measurement.

TRANSPIRATION: The process whereby moisture is transferred from living organisms to the atmosphere. A major portion of environmental moisture for precipitation comes from plants, which lose water through their stomata, small openings on the undersides of leaves.

WATERSHED: An area of terrain from which water flows into a stream, river, lake, or other large body.

WHERE TO LEARN MORE

Bad Meteorology (Web site). <http://www.ems.psu.edu/~fraser/BadMeteorology.html>.

Global Hydrology and Climate Center (Web site). <http://wwwghcc.msfc.nasa.gov/>.

Glover, David. *Flying and Floating.* New York: Kingfisher Books, 1993.

ISCCP: International Satellite Cloud Climatology Project (Web site). <http://isccp.giss.nasa.gov/isccpframe.html>.

Kahl, Jonathan D. *Wet Weather: Rain Showers and Snowfall.* Minneapolis: Lerner Publishing, 1992.

Parker, Steve. *Weather.* Illus. Kuo Kang Chen and Peter Bull. New York: Warwick Press, 1990.

Schmid, Eleonore. *The Water's Journey.* New York: North-South Books, 1990.

Tropical Rainfall Measuring Mission/TRMM (Web site). <http://www-tsdis.gsfc.nasa.gov/tsdis/tsdis.html>.

The World-Wide Web Virtual Library: Meteorology (Web site). <http://dao.gsfc.nasa.gov/DAO_people/towens/VLM/>.

Wyman, Richard L. *Global Climate Change and Life on Earth.* New York: Routledge, Chapman and Hall, 1991.

WEATHER

CONCEPT

Though people are quite accustomed to experiencing weather—sunshine and rain, wind and storms, hot and cold, fair days and foul—most have little idea how weather originates or even really what weather is. In fact, it is a condition of the atmosphere produced by one or more of six factors: air temperature, air pressure, humidity in the air, the density and variety of cloud cover, the amount and type of precipitation, and the speed and direction of the wind. Those six variables can produce the sunny, beautiful days that everyone treasures, or they can conspire to create the sort of rainy, windy, and cold day that some people despise and others (especially creative types, who do not have to be out in the elements) adore. The variables of weather also can align to manifest in the form of brutal storms, including hurricanes and tornadoes. Yet as complex as weather is, its behavior can still be forecast—with at least some accuracy—and sometimes even manipulated.

HOW IT WORKS

INTRODUCTION TO WEATHER

Weather and climate are not the same thing: weather is the condition of the atmosphere at any given time and place, while climate is the pattern of weather conditions in a particular region over an extended period. We do not speak therefore of "forecasting climate," though it is possible, on the basis of long-term climate patterns, to make some predictions concerning future climatic trends. (Climate is discussed in a separate essay.)

Nevertheless, the type of atmospheric prediction that applies much more to our daily lives is weather forecasting. The latter is the work of meteorologists, or scientists involved in studying the atmosphere and weather and in making weather predictions. In the 1970s most local television news programs still included a "weatherman" on their lineup. This was a television personality, much like the anchors and the specialists in sports or other topics, whose job it was to report the weather. During the late 1970s and early 1980s, however, most stations and networks changed their designation of weatherman to that of meteorologist.

In part this change had to do with gender politics, since the title weatherman seemed sexist but weatherwoman or weatherperson sounded a bit absurd. On the other hand, it also was made in recognition of the work that weather forecasters perform: unlike ordinary reporters, they are not so much journalists as they are scientists. (Though a nice smile, charming personality, and good looks never hurt any TV meteorologist!) And while a meteorologist, like every member of the TV news team, reports what has happened—for instance, a record snowfall—his or her job also is concerned heavily with explaining what *will* happen.

SIX FACTORS. As we have noted, weather is determined by six factors, all of which involve the atmosphere, and several of which also relate to water in the atmosphere. These factors are as follows:

- Air temperature
- Air pressure
- Wind
- Humidity

- Cloud cover
- Precipitation

The role of air temperature in weather is fairly obvious, since that is one factor of which we are used to taking note on a regular basis. Air pressure, though much less familiar to most of us, has a powerful influence on air currents, and air pressure is linked closely with wind. Likewise, there are clear linkages among humidity (moisture in the air), the amount and type of cloud cover (formed, of course, from moisture in the air), and the amount and type of precipitation, which begins to fall when the clouds are saturated with moisture.

WHAT MAKES THE WEATHER?

So much for the factors involved in weather, which go into making the weather we experience at a given time. But if we want to search for the ultimate causes behind the weather, we have to look toward larger phenomena: the Sun and Earth as well as Earth's surface and atmosphere and the interactions among all of these factors.

At the heart of weather is the Sun and the energy it transmits to Earth. As discussed in Energy and Earth, the Sun sends out vast quantities of energy, only a small portion of which comes anywhere near Earth—and a much smaller portion reaches the surface of the planet as usable energy. Yet that "tiny" amount, which is the vast majority of the energy available to Earth in any form, is enough to light and warm the planet and to drive a variety of processes in the biosphere and atmosphere.

MOVING WATER THROUGH THE ATMOSPHERE. The Sun's most direct and obvious influence on weather is its impact on temperature, but this is only one aspect of a larger and more complex picture. As discussed in the essay Evapotranspiration and Precipitation, differences in temperature—a result, primarily, of the Sun's heat—make possible the conditions for both evaporation and transpiration, or the loss of moisture from living organisms to the atmosphere.

The amount of water that evaporates into the atmosphere in a given area is its humidity, which exerts a powerful impact on the human experience of weather. As most people who have traveled widely will attest, high temperatures in a dry climate, such as that of the American Southwest, are far easier to endure than equally high temperatures (or even lower ones) in Florida or other parts of the Deep South, a much more humid region.

As water moves from Earth's surface into the atmosphere through evapotranspiration (a combination of evaporation and the related process of transpiration), the air through which it passes becomes increasingly cooler. Eventually, the vaporized water comes to an altitude at which the air around it is cold enough to cause condensation—that is, the formation of a liquid. When enough vapor has condensed into tiny droplets of water, or tiny ice crystals, the result is the formation of clouds.

Clouds, discussed in detail within the context of Evapotranspiration and Precipitation, are integral to the formation of weather patterns, which is why TV meteorologists almost always show their audiences satellite maps illustrating cloud movements. Not only do clouds reflect sunlight into space, meaning that a sufficient accumulation of cloud cover will bring about a decrease in temperature, but clouds also manufacture precipitation. Forms of precipitation include rain in all its variations, from a fine mist to a downpour, as well as snow, sleet, and hail.

AIR PRESSURE AND WIND

We have seen how the Sun affects the movement of water through the atmosphere, thus driving three of the six key weather factors we named earlier: humidity, cloud cover, and precipitation. In addition to its effect on these factors and on temperature itself, the Sun and its energy are also behind the other two important factors—air pressure and the movement of air through the atmosphere, that is, winds.

Because it does not change as frequently or as dramatically as temperature (fortunately!), air pressure is something of which we are hardly aware. But if a person accustomed to the sea-level air pressure of a place such as New York City suddenly had to travel to a high-altitude locale such as Denver, Colorado, he or she immediately would become aware of the difficulty in breathing and the other changes that attend a change in pressure.

For instance, the higher the altitude (and, hence, the lower the air pressure), the lower the temperature at which water boils. For this reason, cake mixes and similar products have special instructions for kitchens at high altitudes. Taken to an extreme, this means that with absolutely no pressure, liquid could boil—that is, turn into a vapor—at extremely low temperatures. This is one of the reasons, along with lack of oxygen and the presence of harmful rays, that an astronaut wears a protective suit; without it, the pressureless environment of space would cause a person's blood to boil!

In the English system, normal atmospheric pressure at sea level is 14.7 lb. per sq. in., or 1.013×10^5 pascals (Pa) in the metric or SI (Scientific International) system. This amount of pressure constitutes a unit in its own right, an atmosphere (atm). Atmospheric pressure, however, usually is measured in terms of the bar, an SI unit equal to 10^5 Pa. Thus, 1 atm = 1.013 bars. Meteorologists often use the millibar (mb), which, as its name implies, is equal to 0.001 bars—roughly 1/1,000 of ordinary air pressure at sea level.

WIND: FROM HIGH PRESSURE TO LOW. Heat from the Sun does not fall evenly on all places on Earth. Aside from the obvious fact that it makes a limited impact on polar regions, there are the differences in color and texture between various areas, even in the most tropical latitudes—that is, those closest to the direct path of sunlight. For example, soil, since it is almost always darker than water, tends to attract more sunlight than bodies of water do.

When an area is heated, the air above it heats up as well, and this makes that region of air less dense. As a result, the air rises, a phenomenon known as convection. Convection currents also carry other masses of air downward from the upper atmosphere toward Earth's surface. In regions where warm air moves upward, the atmospheric pressure tends to be low, whereas downward air movements are associated with higher atmospheric pressures.

These higher or lower atmospheric pressures can be measured by a barometer, an instrument that registers pressure just as a thermometer measures temperature. The fact that there are differences in pressure between areas brings about wind, which is the movement of air from a region of high pressure to one of lower pressure. (Later in this essay, we discuss a way to "create" wind and thus test this statement.)

OTHER INFLUENCES ON WEATHER

We have focused primarily on the influence that the Sun and its energy exerts on weather, but it should be noted that certain aspects of Earth itself determine weather patterns. Among these factors, as noted earlier, are color and texture, whereby certain places on the planet are more apt to receive the Sun's energy than others.

Also important is Earth's position in space. First of all, there is the tilt of its axis relative to the plane of its rotation around the Sun, which accounts for the seasons. In addition, the planet follows an elliptical, or oval-shaped, orbital pattern around the Sun, which means that there are certain times when the planet is closer to the Sun and its energy than at others.

Earth's rotation brings about complicated patterns of movement on the part of air masses heated at the equator. If the planet were not rotating, these masses of air would simply move from the equator toward the poles, where they would be cooled. Because the planet is rotating, however, the movement of global winds is much more complex and is characterized by circular patterns known as *cells.*

In addition to these factors that relate to the planet's movement in space, there is also the matter of irregularities on Earth's surface. An example of this is the effect mountains can have on cloud movements, creating a perpetually rainy climate on one side and a dry "rain shadow" on the other. Rain shadows are discussed in the essay on Mountains.

REAL-LIFE APPLICATIONS

CREATING WIND

Earlier we discussed the fact that wind results from differences in pressure. This is a statement you can test for yourself—and perhaps you already have, without knowing it. Suppose you are in a room where the heat is on too high and there is no way to adjust the thermostat. Outside the air is cold, so you open a window, hoping to cool down the room. Does it do the trick? Not likely. But if you open the door leading from the room to the hallway, a nice cool breeze will blow through. You have, in effect, created wind or at least manipulated the conditions to make wind possible.

With the door closed, the room constitutes an area of high pressure in comparison with the area outside the window. Once the window alone is opened, it is theoretically possible for air to flow into the room, but that does not mean the air actually will flow. The reason is that fluids such as air—whether in a room or in the sky—tend to move from areas of high pressure to areas of low pressure. Because the room is of relatively high pressure compared with the outside, there is no reason for the air to move into the room.

Furthermore, in line with the second law of thermodynamics (see Energy and Earth), the flow of heat always will be from an area of relatively high temperature to one of relatively low temperature. Therefore, if there is going to be any air movement in this situation, it will have to be movement of hot air *out* of the room rather than cool air *into* the room. (Even an air conditioner works by taking out the heat, not by bringing in "cold"—even though we may perceive it otherwise.)

In the case of the overheated room, there is only one way to solve the problem: by opening the door into the hallway, or whatever else lies outside the door. As soon as the door is opened, the relatively high-pressure air of the room flows into the relatively low-pressure area of the hallway, just as the laws of physics say it should. This is exactly the same principle whereby wind blows from high-pressure regions into low pressure ones. By setting up a cross-draft, in effect, we have "created" wind.

THUNDERSTORMS AND WORSE

Wind and clouds combined with rain (discussed, like clouds, in Evapotranspiration and Precipitation) and a few other factors create a thunderstorm. Those other factors, of course, are lightning and that key difference between an ordinary storm and a thunderstorm: thunder.

Inside a thunderstorm are updrafts and downdrafts of air, which bring about a buildup of static electric charges within the thunderstorm clouds. Over time, there is a large buildup of separate electric charges inside the cloud, with positive charges near the top and negative charges in the middle. This separation of charges creates huge voltage differences, which require something to equalize them: a bolt of lightning, which suddenly passes between the areas of differing charge.

Lightning produces a spark, which heats the air in an instant to more than 54,000°F (30,000°C). As a result, the molecules of air and moisture in the cloud experience an extremely sudden expansion, and this expansion is accompanied by a release of energy. The energy in this situation takes the form of sound, which we experience as thunder.

ANATOMY OF A THUNDERSTORM. Let us go back now to the point when the thunderstorm came into being. First solar energy, or some other influence, such as the presence of a mountain range, causes water to enter the atmosphere as vapor. As this air rises, it expands and cools, eventually condensing and coalescing to form a cloud. The cloud thus formed is a cumulus cloud, which may appear as the puffy, benign cloud of a clear summer's day or may turn into a cumulonimbus—a thunderstorm—cloud.

On its way to becoming a thunder cloud, a cumulus undergoes a transformation into a convective cloud, or one formed by convection. It may never become a thunder cloud at all, assuming that vertical movement stops, in which case fair weather prevails. But if the atmosphere is unstable, meaning that the air temperature drops rapidly with altitude, packets of air that begin rising and cooling inevitably will be warmer than the air around them. The rising air packet is like a balloon, weighing less per unit of volume than the surrounding area, and thus it continues to rise.

As we have noted, in order to evaporate, water needs to receive a certain infusion of energy, or heat, from the Sun. Once it condenses and forms a cloud, it releases that heat, warming the cloud and causing it to rise still higher. As long as these updrafts can support the cloud, it continues to grow and to rise, until, in the case of a thunderstorm, it reaches very great atmospheric heights—about 40,000 ft. (12 km) above the surface. In the course of rising and growing, large raindrops and hailstones can form.

TWO TYPES OF THUNDERSTORM. Over western Florida and other areas around the Gulf of Mexico, thunderstorms grow as we have described, with cumulus clouds rising and cooling. These are called air-mass storms. Once the cloud reaches a certain point of moisture saturation, rain begins to fall from the upper part of the cloud, producing precipitation and with it downdrafts. Because the downdraft in such a situation is usually stronger than the updraft, the cloud is likely to dissipate before the thunderstorm has caused much damage. Certainly there will be rain showers, thunder, and lightning, but the storm is unlikely to produce extreme wind damage or hail.

On the other hand, parts of the central and eastern United States are prone to what are called frontal thunderstorms, and these storms are much more severe. They typically form just ahead of a cold air mass, or cold front. The cold air, much more dense than the warm, humid, and unstable air of the cloud system, pushes the clouds ahead of it, which rise and form convective clouds. Eventually, these clouds produce rain, which causes downdrafts. But instead of dissipating the storm's impact (as in the case of the air-mass storm), the downdrafts only increase its intensity.

In a frontal thunderstorm, strong downdrafts hit the ground, spreading out and sending more warm humid air rising into the storm. This gives the storm clouds more latent heat, increasing the updrafts, which in turn gives more speed to the wind and improves the chances of heavy rain or even hail. The latter can appear even in summertime or in warm climates, since the cloud forms at a great height.

When ice crystals form in the cloud, the updrafts and downdrafts circulate them back and forth, and as this happens, the crystals pick up more and more moisture. Eventually, tiny ice crystals gather rainwater around them, until they have become hailstones. Depending on the conditions of the storm, hailstones can become extremely large—as big as 5.5 in. (14 cm) in diameter.

AS CLOUDS RISE AND GROW, ICE CRYSTALS FORM AND GATHER RAINWATER AROUND THEM UNTIL THEY BECOME HAILSTONES. DEPENDING ON THE CONDITIONS OF THE STORM, HAILSTONES CAN GROW EXTREMELY LARGE—MORE THAN TWICE THE SIZE OF THE GOLF BALL–SIZE ONES SHOWN HERE. *(© Gary Meszaros/Photo Researchers. Reproduced by permission.)*

TORNADOES

Some forms of weather make thunderstorms seem minor by comparison. Among these are tornadoes, a rapidly spinning column of air formed in a severe thunderstorm. The vortex, or rotating center, of the column, forms inside the storm cloud and begins to grow downward until it touches Earth's surface. The United States is particularly prone to tornadoes, owing to specific factors of its location and its larger climate patterns. (See Climate for more about the distinction between weather and climate.)

The type of severe frontal thunderstorm that can produce a tornado typically is associated with an extremely unstable atmosphere and with moving systems of low pressure, which bring masses of cold air into contact with warmer, more humid air masses. It so happens that such storms occur frequently across a wide swath of North America, from the plains states to the eastern seaboard (i.e., more or less from Kansas to North Carolina) during a period from about June to October each year.

As updrafts in a severe thunderstorm cloud become stronger, they pull more air into the base of the cloud to replace the rising air. As the air from the surface moves into a smaller area, it begins to rotate faster, owing to what physicists call the conservation of angular momentum. The latter can be illustrated by the example of an ice skater who, while spinning with her arms outstretched, pulls her arms inward. As she does so, the speed of her rotation increases. The same happens with rotating air as it moves from the large space of the ground to the smaller space of the cloud.

Thus, a funnel cloud is produced and grows, and if it becomes large enough, it may touch ground—with devastating results. Within the vortex of a tornado, which is typically about 328 ft. (100 m) in diameter, wind speeds may be greater than 220 MPH (100 m/s). The tornado itself travels at speeds of 10–30 MPH (15–45 km/h), making a sound like a freight train or a jet engine and wreaking havoc along a path as long as 200 mi. (321 km). Nothing can stand in the way of a tornado with enough force: buildings shatter, roofs and whole houses take to the air, and pieces of straw can be blown through solid wood.

TORNADO ALLEY. No country in the world is more prone to tornado activity than the United States, a large portion of which is known as "Tornado Alley." The latter area has no specific boundaries, though it is more or less contiguous with the wide Kansas-to-Carolina swath mentioned earlier. Some authorities describe Tornado Alley as including northern Texas, Oklahoma, Kansas, and southern Nebraska. However, the American Meteorological Society's *Glossary of Weather and Climate* defines Tornado Alley as "The area of the United States in which tornadoes are most frequent. It encompasses the great lowland areas of the Mississippi, the Ohio, and lower Missouri River valleys. Although no state is entirely free of tornadoes, they are most frequent in the plains area between the Rocky Mountains and the Appalachians."

It is no accident that a wide, flat region, over which heavy winds can blow from numerous directions—including winds off of the Gulf of Mexico and Atlantic Ocean—would be home to such enormous tornado activity. Tornado Alley, or parts of it, has seen numerous extraordinary weather events in which not one or two tornadoes struck, but many dozens—a phenomenon known as a tornado outbreak, or a super outbreak.

The worst super outbreak in recent memory occurred on April 3–4, 1974, when 148 rampaged through 14 states. It began on the morning of the 3rd, when a low-pressure system moved over central Kansas, spreading a cold front as far south as Texas. Meanwhile, a warm front clung to the lower Ohio River Valley, and various unstable patterns covered the South. Into this motley mix came strong winds, which soon took hold over much of the eastern United States. By afternoon, thunderstorms began raging over much of the Midwest and South, including the Ohio and Tennessee River valleys.

The result was a series of nearly 150 tornadoes, including 48 that were classified as F4 or F5, the highest levels on the Fujita-Pearson scale used to rate the intensity or tornadoes. Some other significant tornado events since that time include the Carolina outbreak, which killed 57 people, injured 1,248, and caused $200 million in damage, on March 28, 1984. A year later, on May 31, 1985, an outbreak of 41 tornadoes in Ohio, Pennsylvania, New York, and Ontario killed almost 90 people and caused over $450 million in damage.

CYCLONES

Another form of extreme weather is a cyclone, a general term for what is sometimes called a hurricane or a typhoon. These are vast circulating storm systems characterized by bands of showers, thunderstorms, and winds. They develop over warm tropical oceans, generally as isolated thunderstorms, and turn into monsters. A cyclone may have a diameter as great as 403 mi. (650 km) and be more than 7.5 mi. (12 km) in height.

Near the center of the cyclone, winds may be as high as 110 MPH (50 m/s), yet the very center is an area of complete calm, known as the eye. This is a region of descending air surrounded by the updrafts that characterize the cyclone, making the eye a little funnel of peace in the middle of a terrible storm. If it were possible to stay in the eye of the cyclone as it moves—at speeds comparable to that of a tornado—it is conceivable one could come away unscathed.

THE IMPACT OF CYCLONES. Once the cyclone blows inland and reaches population centers, it can cause massive death and destruction. The southern United States witnessed this with hurricanes such as Hugo, which devastated the Carolinas and nearby regions in 1989. The damage on Puerto Rico alone (one of the places Hugo touched down before moving north) amounted to 12 lives lost, and $2 billion in property. Thanks to evacuation measures and a bit of good fortune (the hurricane missed Charleston, South Carolina), loss of life was minimal on the continental United States, and loss of property was kept to $5 billion. By contrast, Hurricane Andrew, which struck Florida and other parts of the southern United States in August

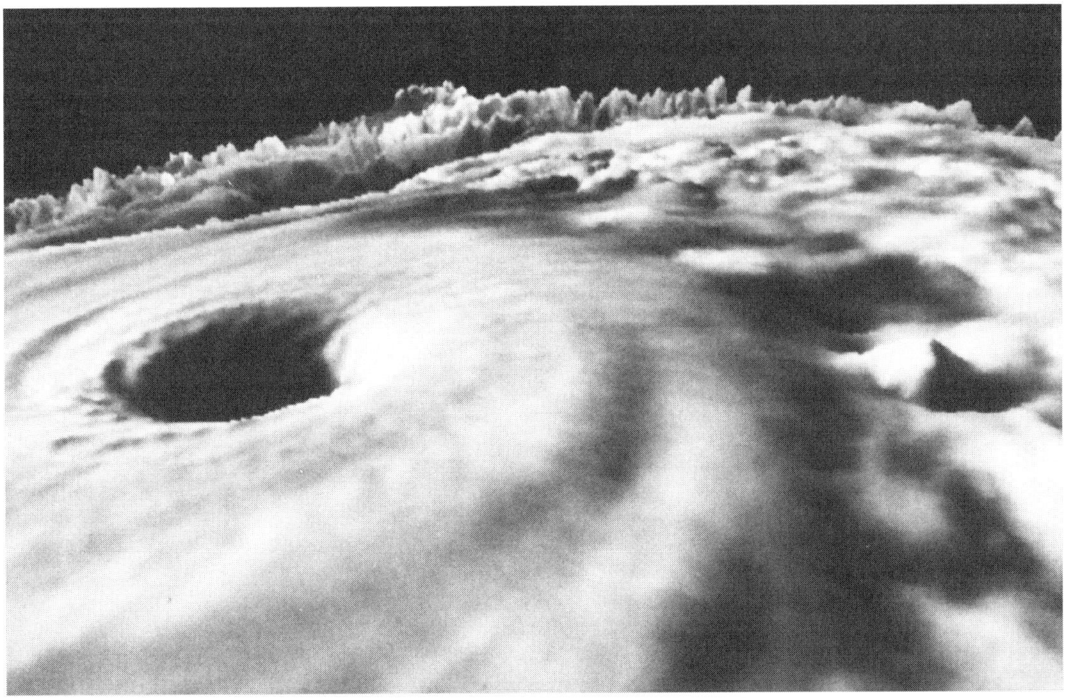

A GLOBAL NETWORK OF WEATHER SATELLITES MAKES IT POSSIBLE TO IDENTIFY AND TRACK TROPICAL CYCLONES FROM THEIR EARLIEST APPEARANCE AS DISTURBANCES OVER THE OCEAN. A CYCLONE CAN GENERATE WINDS AS HIGH AS 110 MPH (50 M/SEC), YET THE VERY CENTER IS AN AREA OF COMPLETE CALM, KNOWN AS THE EYE. *(U.S. National Aeronautics and Space Administration [NASA].)*

1992, was the costliest natural disaster in U.S. history, with over $25 billion in damage reported.

But the impact of storms in the United States is dwarfed by the destruction of hurricanes and typhoons in third world countries. Bangladesh, for instance, is a country where a population half that of the United States is crammed into an area the size of Wisconsin. (And, it should be noted, a country so poor that the income of the average full-time working man is far less than that of an American teenager working a part-time job.)

In such a situation, the impact of cyclones is bound to be devastating. Though dollar figures of property damage from the country's many cyclones are not available, the human toll is better known. Since 1963, when it was still called East Pakistan, Bangladesh, has seen seven cyclones in which 10,000 or more died. The worst, in 1970, killed half a million. Government estimates of deaths from a cyclone on April 10, 1991, were 150,000, though it is likely that many more died from disease, starvation, and exposure.

Likewise, typhoons in such countries as the Philippines cause massive power outages, flooding, landslides, and destruction of life and property. In part this is because a number of these countries are located in or near tropical zones that are especially susceptible to cyclones, but it is also a matter of preparedness.

RESPONDING TO CYCLONES. In the United States, with its greater material wealth and technological sophistication, it is possible to prepare people for extreme weather in a way that is simply beyond the reach of many less prosperous or powerful nations. It is not so much that Americans are capable of developing structures such as seawalls that can withstand the force of hurricanes, though this certainly helps. The seawall in Charleston, South Carolina, for instance—a massive bulwark of concrete more than 10 ft. (3 m) high—is intended to protect homes in that city's historic Battery when hurricanes produce powerful ocean swells.

Much of the effectiveness of the American response to hurricanes, however, is attributable to the ability to react to circumstances rather than to prevent them. With a large amount of the population possessing electronic communication, it is possible to circulate word quickly regarding the coming deluge. Furthermore, people in the United States are much more mobile

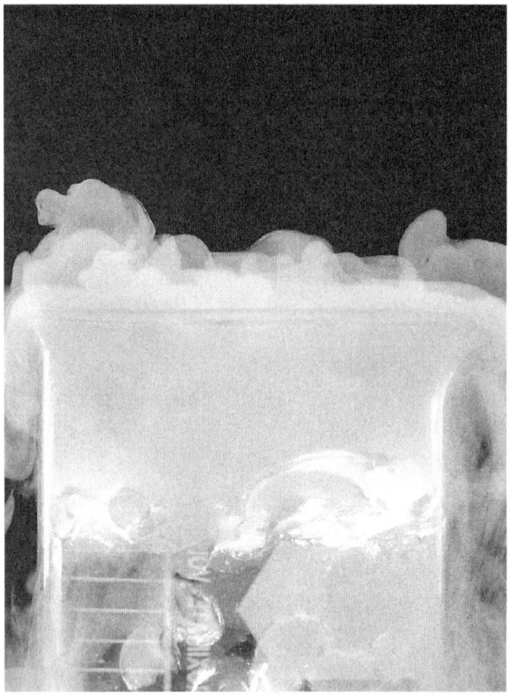

DRY ICE (SOLID CARBON DIOXIDE), SEEN HERE BEING CONVERTED INTO A VAPOR, OFTEN IS USED TO CHANGE SUPERCOOLED WATER IN CLOUDS INTO ICE CRYSTALS IN THE WEATHER MODIFICATION TECHNIQUE CALLED CLOUD SEEDING. *(© M. Meadow/Photo Researchers. Reproduced by permission.)*

than their counterparts in developing countries and can evacuate hurricane regions more easily. Additionally, a wealthier and more powerful government is able to administer relief more quickly and in greater quantities than the leadership of small, poor nations.

Even so, hurricanes and typhoons have a vast impact wherever they strike, and perhaps for this reason a certain mystery and lore have developed around them. Reflective of this fascination is the practice of personalizing hurricanes. Originally, the U.S. National Weather Service gave them women's names, but in response to protests from feminist groups, by the late 1970s it had adopted a practice of using an alphabetic list of alternating male and female first names. The Weather Service draws up new lists each year to name the cyclones of the western Pacific and the Caribbean/Gulf regions.

FORECASTING AND CONTROLLING WEATHER

Until the twentieth century, people had little warning when a cyclone was coming. Early in that century, however, the establishment of hurricane-watch services offered the hope of early warning, and by the 1930s ships and weather balloons were used to provide readings on atmospheric conditions that might portend cyclones. In the 1940s airplanes and, later, radar further increased meteorologists' ability to monitor the atmosphere. Today a global network of weather satellites makes it possible to identify and track tropical cyclones from their earliest appearance as disturbances over the remote ocean.

From the time of the Greeks, at least, people have tried to forecast the weather. They attempted to do so with varying degrees of success, using means that included folklore, superstition, old wives' tales, traditional wisdom, instinct, intuition, and even a little bit of science. Sometimes this mixed bag yielded valuable information, such as the old and essentially accurate saying "Red sky at morning, sailors take warning; red sky at night, sailor's delight." Yet without true scientific methods of forecasting, would-be meteorologists were just shooting in the dark—as the sometimes inaccurate results of today's much more scientific forecasting shows.

A turning point came in the twentieth century, with the development of such monitoring systems as the cyclone-monitoring techniques we have mentioned. It says a great deal about just how young meteorology is as a science that one of its two founders as a modern discipline died only in the 1970s. This was Jakob Bjerknes (1897–1975), a Norwegian-born American meteorologist who, in the 1920s, established a network of weather stations with his father, the Norwegian physicist Vilhelm Bjerknes (1862–1951). Vilhelm proposed the idea of air masses, a pivotal concept in meteorology, while Jakob discovered the origin of cyclones.

MODERN WEATHER FORECASTING. Weather forecasting in the United States is the responsibility of the National Weather Service (NWS), which is part of the National Oceanic and Atmospheric Administration (NOAA) within the Department of Commerce. The NWS maintains a vast network of field offices and weather stations as well as nine National Centers for Environmental Prediction, each of which is focused on specific weather-related responsibilities. The complexity of the NWS organizational system hints at the greater

KEY TERMS

ATMOSPHERE: In general, an atmosphere is a blanket of gases surrounding a planet. Unless otherwise identified, however, the term refers to the atmosphere of Earth, which consists of nitrogen (78%), oxygen (21%), argon (0.93%), and other substances that include water vapor, carbon dioxide, ozone, and noble gases such as neon (0.07%).

CLIMATE: The pattern of weather conditions in a particular region over an extended period. Compare with *weather.*

CONDENSATION: The formation of a liquid from a vapor, usually as a result of a reduction in temperature.

CONVECTION: Vertical circulation that results from differences in density ultimately brought about by differences in temperature. Convection involves the transfer of heat through the motion of hot fluid (e.g., air) from one place to another.

CONVECTION CURRENT: The flow of material heated by means of convection.

EVAPORATION: The process whereby liquid water is converted into a gaseous state and transported to the atmosphere. When discussing the atmosphere and precipitation, evaporation generally is distinguished from transpiration. In this context, evaporation refers solely to the transfer of water from nonliving sources, such as the soil or the surface of a lake.

EVAPOTRANSPIRATION: The loss of water to the atmosphere via the combined (and related) processes of evaporation and transpiration.

HUMIDITY: The amount of water vapor in the air.

METEOROLOGY: The study of the atmosphere, weather, and weather prediction.

PRECIPITATION: When discussing the hydrologic cycle or meteorology, precipitation refers to the water, in liquid or solid form, that falls to the ground when the atmosphere has become saturated with moisture.

TRANSPIRATION: The process whereby moisture is transferred from living organisms to the atmosphere. A major portion of environmental moisture for precipitation comes from plants, which lose water through their stomata, small openings on the undersides of leaves.

UNSTABLE ATMOSPHERE: A term describing a situation in which air temperature drops rapidly with altitude. As a result, a packet of air continues to move upward owing to the temperature difference between it and the surrounding air.

WEATHER: The condition of the atmosphere at a given time and place. Compare with *climate.*

complexity of weather forecasting itself, which we touch on here in only the most cursory fashion.

A number of useful techniques and forms of technology are available to the modern meteorologist. Perhaps the best known of these is Doppler radar, used to track the movement of storm systems. By detecting the direction and velocity of raindrops or hail, Doppler radar can help the meteorologist determine the direction of winds, and thus predict weather patterns that will follow in the next minutes or hours. But Doppler radar can do more than simply detect a storm in progress: Doppler technology also aids meteorologists by interpreting wind direction. Other forms of weather-forecasting technology include NEXRAD (Next Generation Radar) and GOES (Geostationary Operational Environmental Satellite).

SOME TYPES OF WEATHER FORECAST. The simplest (and usually least accurate) type of forecast is called a persistent forecast and starts with the assumption that existing patterns will continue into the future. The problem with this notion, of course, is that with the complexity of weather, patterns are always changing. More reliable is the so-called trend method, based on the relationship between the movement of air masses and the larger weather patterns. This is a type of prediction familiar to most of us from weather maps displayed by TV meteorologists.

Similar to the trend method is the analogue method, which uses analogies (hence the name) between current weather maps and similar maps from the past. If a weather map for today closely matches the map of patterns that prevailed on a particular day three years ago, it is possible to make some predictions about the weather now by referring to conditions that developed at that time.

Meteorology may employ statistical probability, and when it comes to long-range forecasting, the mathematics may become considerably more complex. This is illustrated by the fact that chaos theory, one of the most challenging branches of modern mathematics, was the brainchild not of a mathematician but of the American meteorologist Edward Lorenz (1917–). An outgrowth of Lorenz's studies in atmospheric patterns, chaos theory is applied in studying extremely complex systems whose behavior appears random. (On a pop-culture note, chaos theory was the specialization of Ian Malcolm, the mathematician played by Jeff Goldblum in the film, *Jurassic Park.*)

"MAKING" WEATHER. People at the turn of the nineteenth century would have been flabbergasted to know that it would be possible by the twenty-first century to provide a reasonably accurate 24-hour weather forecast. They would have been even more astounded at the idea of a five-day forecast, which is common (though with varying degrees of accuracy) on most local and national weather reports. And, no doubt, they would have been "blown away"—to use a popular weather metaphor—at the concept that modern technology makes it possible even to exert some control over the weather.

Weather modification includes techniques such as cloud-seeding, which originated in the 1940s. By the beginning of the twenty-first century, several other methods of weather modification existed, among them frost prevention, fog and cloud dispersal, hurricane modification, hail suppression, and lighting suppression.

Cloud seeding necessitates the conversion of supercooled water in clouds—that is, water whose temperature has dropped below the freezing point without it actually freezing—into ice crystals. Dry ice (solid carbon dioxide) and silver iodide are the substances most commonly employed for this purpose. The result is, or at least can be, snow clouds. On the other hand, cloud-seeding techniques can be used for fog dispersal, another form of weather modification useful, for instance, in the skies over airports.

WHERE TO LEARN MORE

Allaby, Michael. *DK Guide to the Weather.* New York: Dorling Kindersley, 2000.

———. *Tornadoes and Other Dramatic Weather Systems.* New York: Dorling Kindersley, 2001.

The Atmosphere (Web site). <http://liftoff.msfc.nasa.gov/academy/space/atmosphere.html>.

Folk Lore Weather Forecasting (Web site). <http://www.shoal.net.au/~seabreeze/weather.html>.

National Weather Service/National Oceanic and Atmospheric Administration (Web site). <http://www.nws.noaa.gov/>.

Reynolds, Ross, ed. *Cambridge Guide to the Weather.* New York: Cambridge University Press, 2000.

Silverstein, Alvin, Virginia B. Silverstein, and Laura Nunn. *Weather and Climate.* Brookfield, Conn.: Twenty-First Century Books, 1998.

Stein, Paul, ed. *The Macmillan Encyclopedia of Weather.* New York: Macmillan Reference USA, 2001.

The Weather Channel (Web site). <http://www.weather.com/>.

Weather Underground (Web site). <http://www.wunderground.com/>.

CLIMATE

CONCEPT

Many people think of weather and climate as concepts that are very nearly synonymous, but this is far from the truth. Whereas *weather* is a term referring to the atmospheric conditions for a particular place at a particular time, *climate* describes the overall weather pattern of a region over an extended period. A spot on the South Carolina coast, for instance, is likely to be humid and prone to hurricanes; another place, in the Rocky Mountains, might be dry and windy; while yet another locale, in Hawaii, most likely is inclined to extremely mild and equable or unvarying temperatures. In each case, what has been described is climate; for the weather in any of those particular places, one would need to check the weather report on a specific day. Despite the fact that climate is a concept that encompasses the weather in a region over an extended period, it is still possible for climate itself to vary.

HOW IT WORKS

The Atmosphere

All weather takes place in the atmosphere, the uppermost of the four earth systems. A blanket of gases created in large part by the expulsion of elements from the geosphere through volcanic eruptions, the atmosphere sustains the biosphere through its components of oxygen and carbon dioxide. But it also recirculates water from the hydrosphere, a key function in its role as a weather-generating system. (See Earth Systems for more about the interactions among geosphere, biosphere, hydrosphere, and atmosphere.)

COMPOSITION OF THE ATMOSPHERE. Though we think of the atmosphere as being composed primarily of oxygen, the vast majority of it (78%) is nitrogen. The latter is a highly unreactive gas, meaning that it does not tend to bond chemically with other elements. (Nitrogen itself is discussed in considerably greater detail within the essay Nitrogen Cycle.) Therefore, the nitrogen in our air is really just "filler," the result of volcanic eruptions billions of years ago: unlike carbon dioxide, which dissolved in the water of Earth's oceans, the nitrogen simply stuck around in the atmosphere.

Air is not a chemical compound but a mixture, of which nitrogen and oxygen together account for 99%. Another 0.93% is argon, which is a noble gas, meaning that it, too, is highly unreactive. The last 0.07% comes from trace gases (i.e., gases in very small quantities), including two that are extremely important to the operation of Earth: carbon dioxide and water vapor. The first of these is a key component in the biosphere and in biogeochemical cycles (see Carbon Cycle), while the second is of enormous importance in weather and climate.

The Troposphere and Air

It may be surprising to learn that water vapor is such a small portion of the atmosphere; even more shocking is the tiny fraction that this amount of water (equivalent to many billions of gallons) constitutes in proportion to Earth's entire water supply. Furthermore, all our weather, which plays such an integral role in our lives, is generated in a very small portion of the atmosphere.

FOR EVERY CLIMATE REGION ON EARTH, THERE ARE PARTICULAR TYPES OF PLANTS AND ANIMALS THAT HAVE ADAPTED TO THE PREVAILING CONDITIONS. ANIMALS THAT LIVE IN POLAR REGIONS, LIKE THIS BEAR, TYPICALLY HAVE VERY SMALL EARS, GIVING THEM LESS SURFACE AREA TO EXPOSE TO THE COLD; THEY ALSO TEND TO HIBERNATE AS A MEANS OF ENDURING THE WINTER. *(© John Shaw/Photo Researchers. Reproduced by permission.)*

This is the troposphere, the lowest layer, which extends to about 10 mi. (16 km) above the surface of Earth. The height of the troposphere is about one-fifth the combined height of the troposphere, stratosphere, and mesosphere, the three atmospheric layers that contain air. (Beyond these layers are the thermosphere, which eventually dissolves into the exosphere and the emptiness of space, which is characterized precisely by its lack of an atmosphere.)

A HIGH CONCENTRATION OF AIR. Small as the troposphere is within the larger atmosphere, however, it contains 80% of the atmosphere's mass. The reason is that at higher altitudes, Earth exerts less gravitational attraction on the gas molecules that make up air. Thus, if one were to travel up through the outer layers, the amount of air in the atmosphere would simply shrink to nothingness because at that height, Earth does not exert enough gravitational force to hold in place those ultralight particles.

By contrast, in the troposphere, the gravitational force—a function of the distance between two objects, in this case, a gas molecule and Earth—is extremely strong. This results in a high concentration of gas near the surface. (Note the term *gas*: remember that air is a mixture, not a compound, so there is no such thing as an "air molecule." Instead, there are only molecules of oxygen [O_2], ozone [O_3], nitrogen [N_2], carbon dioxide, and water, as well as atoms of argon and other noble gases.)

In any case, the high concentration of gas creates pressure, which, along with several other factors, causes air to move. The result is weather, and weather patterns over a long enough period of time yield what we call climate. (See Weather for more about the components that make up weather as well as their interactions.)

THE ATMOSPHERIC SCIENCES

Weather is like a daily newspaper, whereas climate is like a book of history: whereas weather reflects the atmospheric conditions for a particular place at a particular time, climate is the overall pattern of weather over an extended period. This difference is reflected in the organization of the atmospheric sciences, the area of earth sciences concerned with atmospheric phenomena.

Principal among the atmospheric sciences are meteorology, which is the study of weather, and climatology, or the study of climate. Meteorology focuses on daily or even hourly changes in conditions within the troposphere and the lower stratosphere, the region just above it. By contrast, climatology is concerned with analyzing the weather in a particular area over periods as short as a month or as long as several million years.

Whereas meteorology involves daily weather forecasts and reports, efforts in climatology are directed toward explaining differences in climate around the earth. Climatologists also investigate how these differences are related to other aspects of the natural environment. Paleoclimatology is a specialized branch of climatology devoted to the study of climatic conditions in the distant past. Such study may require investigation of fossils and other materials that provide physical or chemical clues regarding the past climate and changes it experienced.

REAL-LIFE APPLICATIONS

CLIMATES ZONES AND ORGANISMS

One of the few ancient scientific thinkers whose work is still held in high regard was the Roman geographer Pomponius Mela (*fl. ca.* A.D. 44). In *De situ orbis,* Mela introduced the idea of five climate zones on Earth: northern frigid or cold, northern temperate or mild, torrid (very hot), southern temperate, and southern frigid. In a world that knew of no lands further north than Scandinavia (a semi-mythical place the Greeks had called "Thule") or further south than the lower Nile, Mela's designation of climate zones was extraordinarily accurate.

Only in the 1910s did the Russian climatologist Wladimir Köppen (1846–1940) improve on Mela's system, developing his own five-part climate designation. By Köppen's time, it was clear that there was not necessarily any climatic difference between northern and southern seasons, except inasmuch as the seasons in the Southern Hemisphere took place at opposite times of the year from those in the Northern Hemisphere. Therefore, he dispensed with all references to latitude except insofar as they related to position relative to the equator.

LARGE CLIMATE REGIONS. The Köppen system recognizes the zones of humid tropical, dry, humid mid-latitude with mild winters, humid mid-latitude with cold winters, and polar. Each of these larger categories is then subdivided into smaller climate types: for example, among the dry-climate group, there is a distinction between deserts and steppes, which are arid but not completely barren plains of a type found in Russia and central Asia.

Two factors, the average annual temperature and amount of precipitation, serve to differentiate climate types. Humid tropical, dry, and polar zones are fairly self-explanatory; as for the two humid mid-latitude types, they are distinguished by their distance from the equator. For instance, the southern United States would be an example of a humid mid-latitude region with mild winters, while the northeastern United States—New York and New England—would be considered a humid mid-latitude region with cold winters.

For every climate region on Earth, there are particular types of plants and animals that have adapted to the prevailing conditions. Animals that live in desert regions, for example, might have large ears to add a greater surface area for perspiration. On the other hand, animals in polar regions typically have very small ears, giving them less surface area to expose to the cold. Both camels and cacti are organisms well adapted to a desert climate: the camel can store large amounts of water and food in its hump, while the cactus requires little moisture to survive. Many a desert animal is nocturnal, allowing it to survive the heat, while most polar animals tend to hibernate as a means of enduring winter.

MICROCLIMATES

Not all climates necessarily spread across a whole desert or mountain range—or, in the case of Antarctica, with its decidedly polar climate, a whole continent. A very specific area either on Earth's surface or just a few feet or meters above or below it (i.e., up in the trees or in the soil) likewise can have its own microclimate. The very existence of a microclimate, with all its complexity, serves to show just how complex the larger Earth system is.

A particular microclimate, such as that of a forest or even a particular spot within a forest, has its own specific weather conditions: temperature, humidity, wind patterns, dew, frost, and evaporation or transpiration. Soil is a major factor influencing the quality of the microclimate: if the soil

BENI ABBES DUNES, SAHARA DESERT. SOIL IS A MAJOR FACTOR INFLUENCING THE QUALITY OF A MICROCLIMATE: SOIL THAT IS SANDY IN TEXTURE AND LIGHT IN COLOR IS LIKELY TO REFLECT MORE LIGHT AND HEAT. *(© V. Engelbert/Photo Researchers. Reproduced by permission.)*

is sandy in texture, it is likely to reflect more light and heat. The same is true if it is light in color. Also important is topography, an example being the rain shadow created by mountains. (See Mountains for an explanation of rain shadows.)

MICROCLIMATES IN ACTION. A beautiful example of microclimate in action can be found when hiking on Mount Santo Tomas outside Baguio in the northern Philippines. Nestled in the mountains, with still higher regions such as Santo Tomas nearby, Baguio has long been a popular resort owing to its cool temperatures. It is one of the few places in the Philippines where one can find evergreen conifers such as pine trees, and the semi-alpine climate makes Baguio a welcome relief from the heat of Manila and other regions further south.

Even though it lies in the northern portion of Luzon, the northernmost of the major islands in the Philippine archipelago, it is not latitude that gives Baguio its cool climate; instead, it is altitude. The same is true of other places around the world: Quito, Ecuador, for instance, which has an extremely cool climate because it is high in the Andes, even though it lies on the equator. (The 2001 movie *Proof of Life,* which portrays climatic conditions ranging from mild to cold, was filmed in and around Quito.)

On Santo Tomas, however, where the altitude is even greater than that of Baguio, it is possible to experience both the heat that characterizes most of the Philippines *and* the cool conditions of the mountains in northern Luzon. Standing at a particular spot along the mountainside, one can feel the heat that falls directly on the mountain, owing to its tropical latitude—which means that the sunlight reaches the surface at a more or less perpendicular angle. At the same time, one can feel the cool breeze that blows up the mountain as the result of its altitude. It is possible, in fact, to stand in such a way that one's back, to the mountain itself, is hot and sweaty while one's face enjoys the cool, moist highland breeze.

CAN CLIMATE CHANGE?

Though we have established that climate is a long-term weather pattern, that does not mean that climate itself is fixed and unchanging. Any number of factors can affect it. For example, the "fog" in London was once such an established fact that it became associated with that city in the way that wind is with Chicago.

But just as Chicago's status as the "Windy City" is something of a myth (in fact, there are more than a dozen cities more windy, though Chicago does experience powerful winds as a result of its proximity to Lake Michigan), the London fog was not actually what it seemed to be. The fog was not really fog at all but pollution produced by the burning of coal for heat. As coal gave way to gas and other types of heat during the twentieth century, the fog in London's climate changed.

HEATING UP OR COOLING DOWN? A less advantageous condition in the global climate may be a result of other technological changes, according to some scientists. The burning of fossil fuels, such as coal, natural gas, and petroleum, during the past century or more has put large amounts of carbon dioxide into the atmosphere. That part of the situation is fairly cut and dried; as to the potential result, the scientific community is divided.

Some scientists believe that higher concentrations of carbon dioxide will lead to a greater retention of heat in the atmosphere, resulting in a higher annual average global temperature. This phenomenon, known as global warming, has attracted considerable media attention owing to its promotion by environmental activists, including celebrities who embrace environmental causes. Yet the global warming position is far from the only interpretation that the facts suggest.

Other experts contend that high levels of carbon dioxide in the atmosphere will have the opposite effect. According to this view, the concentration of carbon dioxide may heat the atmosphere in the short term, but this will result only in a larger amount of evaporation from the oceans. This evaporation, in turn, will result in the formation of much larger cloud masses, which would have the effect of reflecting sunlight back into space—thus, in fact, *lowering* Earth's temperature.

The global warming position, with its media and celebrity support, has held the lead since the 1980s. For this reason, it is easy to forget that in the mid- to late 1970s, many environmentalists held an exactly opposite position. At that time, a series of extremely cold winters led to the claim that Earth was *cooling* and that in a few more centuries, a new ice age would ensue. In fact, this is a considerably more plausible position, given the fact that Earth has experienced countless ice ages in the past.

BLOCKING OUT THE SUN. Of course, it could be maintained (as advocates of global warming do) that the present situation of massive fossil-fuel burning has never occurred in Earth's history. This is certainly true, but there is less credibility in the claim that never before has so much carbon dioxide been introduced to the atmosphere. In fact, billions of years ago, volcanoes belched vast quantities of the gas into the region above Earth's surface, but because carbon dioxide is highly water-soluble, most of it dissolved into the oceans' waters.

This points to one of the means by which a rapid climate change, in particular, a sudden cooling, is brought about: by a volcanic eruption or other phenomenon that places enormous amounts of dust and ash in the air. In so doing, it reduces the amount of solar radiation that reaches Earth's surface, causing temperatures to drop. An extreme example of this may have occurred about 65 million years ago, as the result of a meteorite hitting Earth and causing rapid climatic changes that brought about the mass extinction of the dinosaurs and other life-forms (see Paleontology).

A more benign example of atmospheric blockage occurred in 1991, after Mount Pinatubo in the Philippines erupted. The eruption appar-

NEW YORK CITY HAS A MASSIVE CONCENTRATION OF HUMANS, MACHINES, AND CONCRETE IN A VERY SMALL AREA. CITIES WITHOUT ADEQUATE GREEN SPACE, SUCH AS NEW YORK'S CENTRAL PARK (SHOWN HERE), BECOME GIANT REFLECTORS, AND THE PRESENCE OF SMOG AND HEAT FROM CARS AND OTHER MACHINES ADDS TO THE UNHEALTHY ENVIRONMENT. *(© Rafael Macia/Photo Researchers. Reproduced by permission.)*

ently released so much ash and other material into the atmosphere that it temporarily reversed a general warming trend that had prevailed during the 1980s. By the mid-1990s, however, the materials from Mount Pinatubo appeared to have settled out of the atmosphere, thus causing a return to early climate trends.

HEATING UP A MICROCLIMATE. Despite the questionable nature of some environmentalists' claims concerning the global climate, environmentalists are absolutely correct in noting the effects of human civilization, development, and technology on microclimates. An excellent example is New York City. Though it is located in the humid temperate climate zone described earlier, with cold winters, the city has less snowfall—and hotter summers—than less populated areas of New York State or even areas in Pennsylvania that lie on the same latitude.

The difference, of course, lies in the fact that New York City is one of the planet's greatest municipalities, a massive concentration of humans, machines, and concrete in a very small area. Heat-sensitive satellite imaging equipment routinely produces images of the United States that show great areas of heat around the major cities, particularly New York. Concrete, buildings, roads, and the like constitute what developers call impervious surface, meaning that rain is not supposed to leach through these surfaces to the soil; rather, it will remain on the surface as runoff. But sunlight cannot gather in puddles or run off into storm drains: it can only reflect, and thus a big city is a huge mirror to the Sun's rays.

A major city, from the standpoint of climate, is rather like a person who places a mirrored piece of metal around his or her face to "soak up the Sun's rays"—an extremely unhealthy, dangerous, and inadvisable practice. Cities without adequate green space, such as parks, become giant reflectors, and the presence of smog and heat from cars and other machines only adds to the unhealthy, artificially heated environment.

It is particularly unfortunate to see such changes in a city such as Atlanta, which until about 1980 was a fairly sleepy town noted for its abundance of trees. Atlanta has long since become a boomtown, and the trees of its metropolitan area are being massacred at an alarming rate. Not only is this denuding the city and suburbs (and in the process, some would say, stripping its last vestiges of charm), but it is raising Atlanta's average temperature—which was already high as a result of the city's latitude.

EARTH, SPACE, AND ICE AGES

If Earth were a horse on a carousel, with the Sun at the spinning center of the merry-go-round, the horse representing our planet would not be sitting upright, at a 90° angle to the carousel floor. Rather, it would be tilted off the perpendicular angle by 23.5°, a fact responsible for many features of the global climate and microclimates of Earth.

At certain times of the year, rays from the Sun strike either the Northern or Southern hemisphere more directly than at other times. Summer occurs in the Northern Hemisphere during the middle of the year, because, at that point in the planet's movement around the Sun, the rays of the Sun strike the Northern Hemisphere at an

KEY TERMS

ATMOSPHERE: In general, an atmosphere is a blanket of gases surrounding a planet. Unless otherwise identified, however, the term refers to the atmosphere of Earth, which consists of nitrogen (78%), oxygen (21%), argon (0.93%), and other substances that include water vapor, carbon dioxide, ozone, and noble gases such as neon (0.07%).

ATMOSPHERIC SCIENCES: A major division of the earth sciences, distinguished from geoscience and the hydrologic sciences by its concentration on atmospheric phenomena. Among the atmospheric sciences are meteorology and climatology.

CLIMATE: The pattern of weather conditions in a particular region over an extended period. Compare with *weather.*

CLIMATOLOGY: An area of the atmospheric sciences devoted to studying the weather in a particular region or regions over periods as short as a month or as long as several million years.

METEOROLOGY: The study of the atmosphere, weather, and weather prediction.

MICROCLIMATE: The climate of a very specific region a few feet or meters above or below Earth's surface. The size of microclimates is undefined and variable, but they are most definitely smaller than the regions (a state, a country, even a continent) over which a particular climate is said to prevail.

TROPOSPHERE: The atmospheric layer closest to the surface, extending upward approximately 10 mi. (16 km). The troposphere contains about 80% of the air in the atmosphere and is the region where weather occurs.

WEATHER: The condition of the atmosphere at a given time and place. Compare with *climate.*

angle close to the perpendicular. The same happens in the Southern Hemisphere around the end of the year, when that section of the planet receives solar radiation at a nearly perpendicular angle.

Another important factor that relates Earth's position in space to its climate and microclimates is its pattern of movement around the Sun. The carousel analogy is a flawed one, because a carousel is round; Earth's orbit, in fact, describes an ellipse, or oval. The reason is that the Sun exerts a greater gravitational pull on Earth at different parts of its orbital path, and the result is that at its closest approach to the Sun, Earth receives more solar energy than it does when it is farthest away from it. This has little do with seasons: the closest approach occurs in January, winter in the Northern Hemisphere. Conversely, in July, the hottest time of year for the Northern Hemisphere, Earth is at its furthest point from the Sun.

EXPLAINING ICE AGES. It is possible that Earth's position relative to the Sun may explain those dramatic periods of cooling known as *ice ages,* when much of Earth's water freezes, larger amounts of land are exposed, and some life-forms die off while new ones develop. There have been numerous ice ages in Earth's history, and what we call *the* Ice Age, which ended about 10,000 years ago, was just the last ice age.

That one was particularly significant, of course, not only because it was most recent, occurring as it did on the eve of civilization's beginnings, but because it was a formative juncture in human development. During those thousands of years, human hunter-gatherer society and Paleolithic technology developed to its highest point. Also, many of the significant migrations of people took place, most notably the movement of Siberian tribes eastward across the land bridge of what is now the Bering Strait.

What causes ice ages and other large-scale climate changes? In the 1930s the Serbian astrophysicist Milutin Milankovitch (1879–1958) put forward a theory maintaining that changes in the pattern of Earth's orbit around the sun, as well as in Earth's inclination to the plane of its orbit, could explain these changes in climate. The planet's axis of inclination (that is, its angle in relation to the plane of orbit) changes over a period of about 22,000 years, a cycle known as the *precession of the equinox.* As a result, first one hemisphere and then the other is be pointed toward the Sun.

Milankovitch evolved a complex theory that took into account precession of the equinoxes, changes in the shape of Earth's orbit, and changes in its orbital tilt. The theory has been improved and modified numerous times since the 1930s, but, in general, Milankovitch's ideas still provide climatologists with a basic understanding as to how large changes in Earth's climate take place.

WHERE TO LEARN MORE

The Atmosphere (Web site). <http://liftoff.msfc.nasa.gov/academy/space/atmosphere.html>.

Climate Prediction Center—National Oceanic and Atmospheric Administration/NOAA (Web site). <http://www.cpc.ncep.noaa.gov/>.

Elsner, James B., and A. Birol Kara. *Hurricanes of the North Atlantic: Climate and Society.* New York: Oxford University Press, 1999.

National Climatic Data Center—National Oceanic and Atmospheric Administration/NOAA (Web site). <http://lwf.ncdc.noaa.gov/oa/ncdc.html>.

Philander, S. George. *Is the Temperature Rising?: The Uncertain Science of Global Warming.* Princeton, NJ: Princeton University Press, 1998.

Resnick, Abraham. *Due to the Weather: Ways the Elements Affect Our Lives.* Westport, CT: Greenwood Press, 2000.

Silverstein, Alvin, Virginia B. Silverstein, and Laura Nunn. *Weather and Climate.* Brookfield, CT: Twenty-First Century Books, 1998.

Stevens, William K. *The Change in the Weather: People, Weather, and the Science of Climate.* New York: Delacorte Press, 1999.

United Nations Framework Convention on Climate Change/UNFCCC (Web site). <http://www.unfccc.de/>.

World Climate: Weather Rainfall and Temperature Data (Web site). <http://www.worldclimate.com/>.

GENERAL SUBJECT INDEX

Boldface type indicates main entry page numbers.
Italic type indicates photo and illustration page numbers.

B

C

D

G

H

I

J

K

L

M

T

INDEX OF EVERYDAY THINGS

The "everyday" items discussed in this series (volumes 1-4) have been categorized into 26 general subject areas, and are listed below. Boldface type indicates main entry volume and page numbers. Italic type indicates photo and illustration volume and page numbers.

AGRICULTURE

ANIMALS

ART

AUTOMOBILES

AVIATION

CIVIL ENGINEERING

COMMUNICATIONS

COMPUTERS

CRIME AND POLICE WORK

ELECTRONICS

ENERGY

ENVIRONMENT

FOOD AND DRINK

GEOLOGY

HOUSEHOLD PRODUCTS

MACHINES

MANUFACTURING

MEDICINE

MILITARY

MUSIC

NAVIGATION

SPORTS AND HOBBIES

TEXTILES

TRANSPORTATION

WEAPONS

WEATHER

CUMULATIVE GENERAL SUBJECT INDEX

This index contains items from volumes 1-4 of the series.
Boldface type indicates main entry volume and page numbers.
Italic type indicates photo and illustration volume and page numbers.

B

C

D

F

G

H

I

J

K

N

P

Q

R

S

T

U

V

X

Y

Z